Lecture Notes in Computer Science 8935

Commenced Publication in 1973
Founding and Former Series Editors:
Gerhard Goos, Juris Hartmanis, and Jan van Leeuwen

T0210806

Xiangjian He   Suhuai Luo   Dacheng Tao
Changsheng Xu   Jie Yang
Muhammad Abul Hasan (Eds.)

MultiMedia Modeling

21st International Conference, MMM 2015
Sydney, NSW, Australia, January 5-7, 2015
Proceedings, Part I

 Springer

Volume Editors

Xiangjian He
Dacheng Tao
Muhammad Abul Hasan
University of Technology, Sydney, NSW, Australia
E-mail:{xiangjian.he, dacheng.tao, muhammad.hasan}@uts.edu.au

Suhuai Luo
University of Newcastle, Callaghan, NSW, Australia
E-mail: suhuai.luo@newcastle.edu.au

Changsheng Xu
Chinese Academy of Sciences, Institute of Automation
National Lab of Pattern Recognition, Beijing, China
E-mail: csxu@nlpr.ia.ac.cn

Jie Yang
Shanghai Jitotong University, Shanghai, China
E-mail: jieyang@sjtu.edu.cn

ISSN 0302-9743   e-ISSN 1611-3349
ISBN 978-3-319-14444-3   e-ISBN 978-3-319-14445-0
DOI 10.1007/978-3-319-14445-0
Springer Cham Heidelberg New York Dordrecht London

Library of Congress Control Number: 2014957756

LNCS Sublibrary: SL 3 – Information Systems and Application, incl. Internet/Web
and HCI

Typesetting: Camera-ready by author, data conversion by Scientific Publishing Services, Chennai, India

Printed on acid-free paper

Springer is part of Springer Science+Business Media (www.springer.com)

Preface

These proceedings contain the papers presented at MMM 2015, the 21st International Conference on MultiMedia Modeling. The conference was organized by University of Technology, Sydney, and was held during January 5–7, 2015, at the Aerial UTS Function Centre, Sydney Australia.

We were delighted to welcome all attendees to MMM 2015. We believe that they had a wonderful stay in Australia and that their visit was both enjoyable and rewarding. We were very proud to welcome visitors from both Australia and abroad and are delighted to be able to include in the proceedings such high-quality papers for oral presentation, poster presentation, special sessions, demonstrations, and video search showcase.

MMM 2015 received 189 submissions across four categories, consisting of 136 main conference full-paper submissions, 24 special session full-paper submissions, 18 demonstration submissions, and 11 video search showcase submissions. Of these submissions, there are 27 authors (6%) from Australia, 130 (28%) from Europe, 287 (62%) from Asia, and 19 (4%) from the Americas. All main conference submissions were reviewed by at least three members of the Program Committee, to whom we owe a debt of gratitude for providing their valuable time to MMM 2015. All papers submitted to the special sessions, demonstration sessions, and video search showcase sessions were also reviewed by at least three reviewers.

Of the 136 main conference full-paper submissions, 49 were selected for oral presentation, which equates to a 36% acceptance rate. A further 24 papers were chosen for poster presentation. The accepted contributions represent the state of the art in multimedia modeling research and cover a diverse range of topics including: image and video processing, multimedia encoding and streaming, applications of multimedia modeling, and 3D and augmented reality. For the three special sessions, a total of 18 papers were accepted for MMM 2015. The three special sessions were "Personal (Big) Data Modeling for Information Access and Retrieval," "Social Geo-Media Analytics and Retrieval," and "Image or Video Processing, Semantic Analysis, and Understanding." In addition, nine demonstrations and nine video showcase papers were accepted for MMM 2015.

We would like to thank our invited keynote speakers for their stimulating contributions to the conference. Special thanks go to the Organizing Committee for their contributions and great efforts toward the success of this event, and Steering Committee for constant support and timely advice.

In addition, we wish to thank all authors who spent their time and effort to submit their work to MMM 2015, and all of the participants and student volunteers for their contributions and valuable support.

Our gratitude also goes to the MMM 2015 Program Committee members and the other invited reviewers for the large number of reviews required for MMM 2015.

We are grateful to the sponsors for providing support to the conference, including the University of Technology, Sydney, Business Events Sydney, Business Events Australia, Australia Government AusAID, and Qantas.

January 2015

Xiangjian He
Changsheng Xu
Dacheng Tao
Suhuai Luo
Jie Yang
Muhammad Abul Hasan

Organization

Steering Committee

Phoebe Chen	La Trobe University, Australia
Tat-Seng Chua	National University of Singapore, Singapore
Yang Shiqiang	Tsinghua University, China
Kiyoharu Aizawa	University of Tokyo, Japan
Noel E. O'Connor	Dublin City University, Ireland
Cess G.M. Snoek	University of Amsterdam, The Netherlands
Meng Wang	Hefei University of Technology, China
R. Manmatha	University of Massachusetts, USA
Cathal Gurrin	Dublin City University, Ireland
Klaus Schoeffmann	Klagenfurt University, Austria
Benoit Huet	Eurecom, France

Organizing Committee

Honorary Co-chairs

Massimo Piccardi	University of Technology Sydney, Australia
Phoebe Chen	La Trobe University, Australia
Tat-Seng Chua	National University of Singapore, Singapore

General Co-chairs

Xiangjian He	University of Technology Sydney, Australia
Changsheng Xu	Chinese Academy of Science, China

Program Co-chairs

Dacheng Tao	University of Technology Sydney, Australia
Suhuai Luo	University of Newcastle, Australia
Jie Yang	Shanghai Jiaotong University, China

Organizing Co-chairs

Qiang Wu	University of Technology Sydney, Australia
Tao Mei	Microsoft Research Asia, China

Local Chair

Jian Zhang	University of Technology Sydney, Australia

Special Session Co-chairs

Min Xu	University of Technology Sydney, Australia
Lexing Xie	Australian National University, Australia

Demos Co-chairs

Cathal Gurrin   Dublin City University, Ireland
Björn Þór Jónsson   Reykjavík University, Iceland

Financial Chair

Wenjing Jia   University of Technology Sydney, Australia

Tutorial Co-chairs

Richard Xu   University of Technology Sydney, Australia
Shuicheng Yan   National University of Singapore, Singapore

Publication Chair

Muhammad Abul Hasan   University of Technology Sydney, Australia

Publicity Co-chairs

Zhengjun Zha   Chinese Academy of Sciences, China
Jitao Sang   Chinese Academy of Sciences, China
Yinjie Lei   Sichuan University, China

Video Search Showcase Co-chairs

Werner Bailer   Joanneum Research, Austria
Klaus Schoeffmann   Klagenfurt University, Austria

Web Masters

Vera Chung   University of Sydney, Australia
Angus Ma   University of Technology Sydney, Australia
David Kim   University of Sydney, Australia
Feng Sha   University of Sydney, Australia
Benedict Goh   University of Sydney, Australia

Program Committee

Xiangjian He   University of Technology, Sydney, Australia
Klaus Schöffmann   University of Klagenfurt, Austria
Suhuai Luo   The University of Newcastle, Australia
Dong Liu   Columbia University, USA
Jingdong Wang   Microsoft Research Asia, China
Wenjing Jia   University of Technology Sydney, Australia
Cha Zhang   Microsoft Research, USA
Wolfgang Huerst   Utrecht University, The Netherlands

Manoranjan Mohanty   SICS, Sweden
Jiaming Li   CSIRO, Australia
Laszlo Boeszoermenyi   University of Klagenfurt, Austria
Duy-Dinh Le   National Institute of Informatics, Japan
Susanne Boll   Universität Oldenburg, Germany
Yang Yang   The University of Queensland, Australia
Matthew Cooper   FX Palo Alto Lab, Inc., USA
Shuqiang Jiang   Chinese Academy of Sciences, China
Keiji Yanai   University of Electro-Communications, Japan
Muhammad Abul Hasan   University of Technology Sydney, Australia
Wolfgang Effelsberg   University of Mannheim, Germany
Laurent Amsaleg   CNRS-IRISA, France
Vincent Oria   New Jersey Institute of Technology, USA
Ying Guo   CSIRO, Australia
Shin'Ichi Satoh   National Institute of Informatics, Japan
Xiaobai Liu   University of California Los Angeles, USA
Naoko Nitta   Osaka University, Japan
Shingo Uchihashi   Fuji Xerox Co., Ltd., Japan
Cathal Gurrin   Dublin City University, Ireland
Qiang Wu   University of Technology Sydney, Australia
Min Xu   University of Technology Sydney, Australia
Richang Hong   Hefei University of Technology, China
Tat-Seng Chua   National University of Singapore, Singapore
Yan Liu   The Hong Kong Polytechnic University,
                              Hong Kong

Zheng-Jun Zha   National University of Singapore, Singapore
Andreas Henrich   University of Bamberg, Germany
Roger Zimmermann   National University of Singapore, Singapore
Xin Jin   Pennsylvania State University, USA
Alan Smeaton   Dublin City University, Ireland
Mohan Kankanhalli   National University of Singapore, Singapore
Xiangyu Chen   Institute for Infocomm Research, Singapore
Yiannis Kompatsiaris   CERTH – ITI, Greece
Marco Paleari   Institut Eurécom, France
William Grosky   University of Michigan, USA
Liang-Tien Chia   Nanyang Technological University, Singapore
Zhu Liu   AT&T Labs, Inc., USA
Mei-Ling Shyu   University of Miami, USA
Yu-Gang Jiang   Fudan University, China
Gene Cheung   National Institute of Informatics, Japan
Michael S. Lew   Leiden University, The Netherlands
Henning Müller   University of Applied Sciences Western
                              Switzerland, Switzerland
Fernando Pereira   Instituto Superior Técnico - Instituto
                              de Telecomunicações, Portugal

Table of Contents – Part I

Image and Video Processing

Multimedia Mining and Retireval

Multimedia Encoding and Streaming

3D and Augmented Reality

Table of Contents – Part II

Applications

Video Showcase

Demonstration

Personal (Big) Data Modeling for Information Access & Retrieval

Social Geo-Media Analytics and Retrieval

Image or Video Processing, Semantic Analysis, and Understanding

An Efficient Hybrid Steganography Method Based on Edge Adaptive and Tree Based Parity Check

Hayat Al-Dmour, Noman Ali, and Ahmed Al-Ani

Faculty of Engineering and Information Technology, University of Technology,
Sydney, Ultimo NSW 2007 Australia
{HayatShahir.T.Al-Dmour,Noman.Ali}@student.uts.edu.au,
Ahmed.Al-Ani@uts.edu.au

Abstract. A major requirement for any steganography method is to minimize the changes that are introduced to the cover image by the data embedding process. Since the Human Visual System (HVS) is less sensitive to changes in sharp regions compared to smooth regions, edge adaptive has been proposed to discover edge regions and enhance the quality of the stego image as well as improve the embedding capacity. However, edge adaptive does not apply any coding scheme, and hence it embedding efficiency may not be optimal. In this paper, we propose a method that enhances edge adaptive by incorporating the Tree-Based Parity Check (TBPC) algorithm, which is a well-established coding-based steganography method. This combination enables not only the identification of potential pixels for embedding, but it also enhances the embedding efficiency through an efficient coding mechanism. More specifically, the method identifies the embedding locations according to the difference value between every two adjacent pixels, that form a block, in the cover image, and the number of embedding bits in each block is determined based on the difference between its two pixels. The incorporation of TBPC minimizes the modifications of the cover image, as it changes no more than two bits out of seven pixel bits when embedding four secret bits. Experimental results show that the proposed scheme can achieve both large embedding payload and high embedding efficiency.

Keywords: steganography, edge adaptive, human visual system (HVS), Tree-Based Parity Check (TBPC)

1 Introduction

Internet is playing an essential role in data transmission and sharing. The protection of confidential information when transmitting sensitive information over the Internet by government organizations, industry and individuals is necessary. Accordingly, intensive research has been conducted on information security [1].

X. He et al. (Eds.): MMM 2015, Part I, LNCS 8935, pp. 1–12, 2015.

Cryptography is used to add some kind of secrecy to communication channels [2]. It encrypts information into a non-readable form using substitution or permutation operations so that if obstructed, the transmitted information cannot be understood [3,4]. However, this approach attracts the attention of unauthorized intruders [5]. An alternative solution to this problem is steganography. Steganography is a method of concealing the existence of confidential data under cover media in such a way that no one has knowledge about the existence of the secret data except the authorized receiver [2,4,5,6,7]. Thus, steganography is concerned with hiding the existence of a data while cryptography hides the meaning [3,4]. Steganography algorithms aim to enhance imperceptibility, security and capacity [8].

In recent years, a large number of steganography techniques have been published. Some steganography schemes hide the secret data in the spatial domain of an image. Other steganography methods use transform domain such as Discrete Cosine Transform (DCT) and Discrete Transform Wavelet (DWT) [6]. In order to achieve a highly secured system of data hiding, a number of researches attempted to encrypt the data before embedding it in a cover media [3,4]. Most of the existing steganography algorithms are lossless because in some applications such as those related to health and military loss of confidential data is not acceptable.

This paper introduces a reversible steganography method which combines edge adaptive and Tree Based Parity Check (TBPC) to embed the secret data inside a cover image. To prevent detection, the proposed method identifies the pixel locations of sharp regions for embedding to produce minimum distortion during the embedding process. This approach takes advantage of human eyes characteristic, which are less sensitive to large changes in edge regions and more sensitive to small changes in the smooth regions. Thus, edge regions provide a good carrier for hiding data.

The reminder of this paper is divided as follows. In section 2, we briefly review some well-known Steganography methods. Section 3 describes the data embedding and extraction processes of the proposed hybrid system. Experimental results are presented in section 4. Finally, a conclusion is given in section 5.

2   Related Work

The Least Significant Bit (LSB) replacement is a well-known steganographic method. Using any digital carrier, LSB replaces the n−LSBs of each pixel by n−bits from the secret message [2,6,7,9]. However, due to its simplicity, some steganalysis methods are not only able to discover the presence of the embedded message, but can also estimate its length [10,11].

While human eyes perception is sensitive to slight modifications in smooth areas, it cannot recognize more substantial modifications in edge areas. Several Pixel Value Differencing (PVD) methods, such as [1,12,13,5] have been proposed to improve the embedding rate without introducing obvious visual artefacts. PVD is an edge adaptive technique, in which the number of hidden bits is decided by the difference values between non-overlapping blocks of two consecutive pixels. More bits of secret message can be embedded in blocks with high difference values, compared to those of small difference values. Accordingly, PVD can provide a larger embedding capacity with more than 2 bits per block.

In [9], Luo et. al. introduced edge adaptive image steganography based on LSB matching revisited to enhance the security compared with the original LSB method. EA-LSBMR divides the cover image into non-overlapping block of equal size ($bz \times bz$), and each block is rotated by random degree to discover the edge pixels in more than one direction. The difference value between two adjacent pixels is computed. If the difference is greater than the threshold value then one bit of the secret data is hidden in each pixel using LSBMR.

Crandall[14] suggested the idea of hiding data based on matrix coding to improve the embedding efficiency. It hides and retrieves the message by utilizing the parity check matrix of a linear code. Fridrich et. al. [15] introduced a steganography method based on linear codes with small dimension, which can achieve high embedding efficiency for only large embedding rates.

Li et. al. [16] proposed a data hiding method called Tree-Based Parity Check (TBPC) to improve the embedding efficiency by reducing the difference between the cover and the stego images. In order to minimize the modifications in the cover pixels, TBPC represents the LSB of the cover pixels using a complete $N-$ary tree. The method in[16] can be formulated as another specific matrix embedding, which was improved by Hou et. al. [17], where they introduced a majority-vote parity check (MPC) instead of the original matrix embedding. In [18] Liu et. al. introduced an adaptive steganography algorithm based on block complexity and matrix embedding. The embedding strategy sets are defined for seven kinds of image blocks with different complexity. The corresponding embedding strategies are determined by resolving the embedding risk minimization problem. The adaption guarantees that the message bits are mainly embedded into the regions with higher complexity values.

The next section presents our proposed method which preserves higher visual quality of the stego images. It based on an edge adaptive to detect the sharp regions and TBPC to hide the secret data into the cover image. The number of embedded bits at each block is varied based on the difference value between each two adjacent pixels of the block. It introduces minimum possible distortion during the embedding process to prevent discovering the secret data.

3   The Proposed Method

Embedding capacity is one of the major requirements of any steganography methods. However, it is important for steganography methods not to leave any noticeable changes to the human eyes after hiding the secret data. We present a hybrid image steganography method that combines edge adaptive and TBPC. The proposed method utilizes the high contrast regions of an image as embedding locations. It is well known that human eyes cannot discover modifications in the edge areas as they can do in smooth areas. Therefore, the number of hidden bits is based on the difference value between the two pixels of each block. The incorporation of

TBPC leads to a better embedding capacity. Thus, the proposed method combines the strengthes of edge adaptive and TBPC.

3.1   The Embedding Algorithm

Details of the data embedding process are described below.

Algorithm 1: The Embedding Procedure.

Inputs: Cover image (C) of size $W \times H$, secret message (M).

Output: Stego image (S) of size $W \times H$.

Step 1.   Divide the cover image into (1×2) non-overlapping blocks of two adjacent pixels $(p_i$ and $p_{i+1})$.

Step 2.   Compute the absolute difference value between the two adjacent pixels $d_i = |p_i - p_{i+1}|$.

Step 3.   Arrange the blocks into six groups as shown in the Table 1. The blocks are sorted in descending order according to the difference value between consecutive pixel pairs, as a measure for region selection. This process can help in minimizing distortion when embedding the data.

Step 4.   Construct the TBPC $2-ary$ tree. $n-$LSBs from each pixel are used in embedding to enhance the embedding rate.

Step 4.1.   A $2 - ary$ complete tree called the "master tree" is constructed to represent the $n-$LSBs from each pixel. Then the nodes in the master tree are filled up with $n-$LSB level by level, from the root to the leaf nodes.

Step 4.2.   Calculate the "master string" by performing a bitwise Exclusive-Or from the root to the leaf nodes in the master tree, as shown in Figure 1(a).

Step 4.3.   Perform a bitwise Exclusive-Or between the master string and the message bits to obtain the "toggle string", as shown in Figure 1(a).

Step 4.4.   Create a new complete $2-ary$ tree, called the "toggle tree" in a bottom$-$up order. In Figure 1(b), the leave nodes are filled up with the toggle string and the rest of the nodes are assigned a value of "0".

Step 4.5.   To reduce the number of modifications, the "1s" in the toggle tree should be minimized. Since "1" represents the number of modifications required on the master tree to embed the secret message. Level by level, from the leaf nodes to the root, each parent with its child nodes are flipped if its both children have a value of "1".

Step 4.6.   As shown in the Figure 1(c), the stego tree is constructed by performing a bitwise Exclusive-Or between the master tree and the toggle tree.

Step 5.   Check the new difference of the pixel pair after the embedding to ensure that the new difference is in the same range of the old difference. If it is not, then it can be corrected by adding or subtracting 2^{n+1}.

Table 1. Range Table

Group	G1	G2	G3	G4	G5
Range	$[0 - 7]$	$[8 - 15]$	$[16 - 31]$	$[32 - 63]$	$[64 - 255]$
Number of bits to embed (per pixels)	2	2	3	4	4

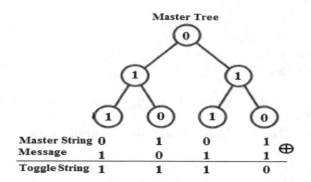

(a)

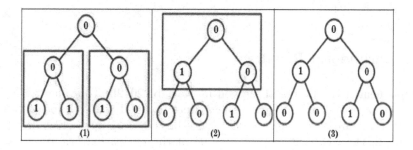

(b)

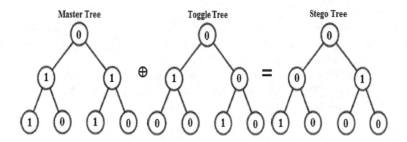

(c)

Fig. 1. (a)Master String and Toggle String of a 2-ary Master Tree with four leaves, (b.1) Toggle String, (b.2) and (b.3)Construction of a toggle tree and (c) Stego Tree

3.2   The Extraction Algorithm

Details of the data extraction algorithm are as follows:
Algorithm 2: The Extraction Procedure.
Inputs: Stego image (S) of size $W \times H$.
Output: Secret message (M).

Step 1. Divide the stego image into (1×2) non-overlapping blocks of two adjacent pixels $(p_i$ and $p_{i+1})$.

Step 2. Compute the absolute difference value between the two adjacent pixels $d_i = |p_i - p_{i+1}|$.

Step 3. Arrange the blocks into six groups as shown in the Table 1, and sort them in descending order according to the difference value between the consecutive pixel pairs.

Step 4. Construct the Stego tree from the $n-$LSBs of the stego pixels, which is filled up level by level, from top to bottom and left to right.

Step 5. Perform a bitwise Exclusive-Or from the root to the leaf nodes to retrieve the secret message.

4   Experimental Results

We implemented the proposed method in MatlabR2012b. To evaluate the proposed method, six 256×256 gray images were used ("Lena", "Baboon", "Peppers", "Camerman", "House" and "Barbara") as cover images, which are shown in Figure 2. Data capacity is used as one of the evaluation criteria, which is defined as the amount of bits that can be embedded into the cover image. The embedding capacity is computed using Eq. 1.

$$E = \frac{K}{WH} (bpp) \qquad (1)$$

where K is the number of the data message bits, while W and H are the width and height of the cover image (both cover and stego images are of the same size. For the considered images, $W = H = 256$).

The visual quality of stego images can be calculated using the Peak Signal-to-Noise Ratio (PSNR), which is calculated as shown in Eq. 2. Higher PSNR indicates better quality.

$$PSNR = 20 \log_{10} \left(\frac{255}{MSE} \right) (dB) \qquad (2)$$

where MSE is the mean square error between cover and stego images, which is defined as:

$$MSE = \frac{1}{WH} \sum_{i=1}^{W} \sum_{j=1}^{H} (c_{ij} - s_{ij})^2 \qquad (3)$$

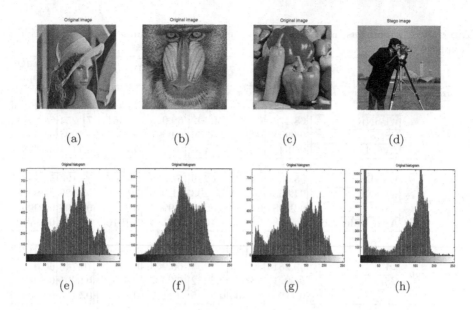

Fig. 2. The Cover Images (a) Lena,(b) Baboon (c) Pepperc, (d) Cameraman and (e) Barbara, (e-h) Corresponding histogram of the cover images

where c_{ij} and s_{ij} are the gray values of pixel (i,j) of the cover and stego images respectively. The disadvantage of the PSNR and MSE is that they simply measure how much change happened between the cover and stego image. However, they are not indicative of how the human visual framework (HVS) essentially would rate the resultant picture quality.

The weighted Peak signal-to-Noise Ratio ($wPSNR$) is an alternate measurement of imperceptibility. It utilizes an extra parameter called Noise Visibility function (NVF). $wPSNR$ is roughly equivalent to PSNR for flat areas because NVF is close to one in smooth regions. However, for regions with sharp contrasts, $wPSNR$ is higher than $PSNR$, because NVF is close to zero for complex regions. Hence, $wPSNR$ attempts to reflect how the HVS perceives images.

$$wPSNR = 10\log_{10}\left(\frac{\max(C)^2}{\|NVF(S-C)\|^2}\right)(dB) \qquad (4)$$

$$NVF = NORM\left\{\frac{1}{1+\delta^2}\right\} \qquad (5)$$

Where δ is the luminance variance for the 8×8 block and NORM is the normalization function. In this paper, we will use $wPSNR$ as a measure of imperceptibility or quality of the produced stego images.

In order to have a comprehensive comparison, we implemented two versions of PVD and the proposed algorithm. In the first version, we used 1 bit per pixel, i.e., a LSB implementation of the two algorithms. Please note that the original TBPC algorithm was also a LSB-based algorithm. In the second version, we considered

Table 2. Comparisons using 1 bpp between PVD, TBPC and the Proposed Edge Adaptive-TBPC

Image	Embedding Rate	1 bpp PVD			TBPC		Adaptive Edge TBPC (1 bpp)	
		30%	50%	80%	30%	50%	30%	50%
Lena	PSNR	54.252	52.515	50.787	57.4206	55.3404	57.287	55.354
	wPSNR	68.476	67.415	64.202	69.023	67.451	71.254	68.007
	SSIM	0.9994	0.9982	0.9969	0.9991	0.9987	0.9994	0.9988
	Avg. Difference	0.1845	0.2901	0.4456	0.1178	0.1942	0.1154	0.1818
Baboon	PSNR	55.535	52.236	50.299	57.398	55.307	57.126	55.138
	wPSNR	94.853	86.584	77.203	81.936	79.555	93.129	81.071
	SSIM	0.9998	0.9993	0.9986	0.9997	0.9995	0.9998	0.9995
	Avg. Difference	0.1818	0.2808	0.4644	0.1184	0.1916	0.1176	0.1871
Cameraman	PSNR	54.683	53.039	51.315	57.399	55.316	57.411	55.317
	wPSNR	69.2814	66.909	62.455	66.963	65.327	70.746	66.011
	SSIM	0.9991	0.9978	0.9962	0.9987	0.9983	0.9991	0.9984
	Avg. Difference	0.1795	0.2808	0.4305	0.1176	0.1912	0.1146	0.1805
Peppers	PSNR	54.244	52.495	50.792	57.427	55.39	57.333	55.388
	wPSNR	72.752	67.635	65.116	70.258	68.497	71.591	68.849
	SSIM	0.9992	0.9984	0.9969	0.9993	0.9988	0.9994	0.9988
	Avg. Difference	0.1852	0.2915	0.4468	0.1176	0.1880	0.1148	0.1814
Barbara	PSNR	54.438	52.423	50.676	57.385	55.352	57.144	55.221
	wPSNR	68.860	66.826	65.886	70.313	68.426	72.898	69.314
	SSIM	0.9994	0.9989	0.9979	0.9995	0.9991	0.9996	0.9992
	Avg. Difference	0.1833	0.2925	0.4472	0.1187	0.1896	0.1179	0.1954
House	PSNR	54.745	52.907	51.031	57.366	55.301	57.44	55.41
	wPSNR	63.541	61.918	61.231	65.574	65.05	65.88	65.134
	SSIM	0.9983	0.9972	0.9959	0.9988	0.9983	0.9989	0.9984
	Avg. Difference	0.1786	0.2819	0.4372	0.1192	0.1919	0.1134	0.1815

n bits per pixel, and hence the known PVD algorithm and our proposed Edge Adaptive-TBPC.

Table 2 presents the obtained values of $wPSNR$ and average difference for the 1 bpp implementation of the three algorithms. These results indicate that the proposed Edge Adaptive-TBPC produced higher quality stego images compared to those obtained using the other two methods. The second best algorithm is found to be the TBPC, as it has a higher embedding efficiency due to it coding capability. However, the embedding rate of the original TBPC and Edge Adaptive-TBPC cannot exceed 50% of the cover image pixels. According to Eq. 5, sharp regions is the appropraite embedding locations because weighting of the modification in high contrast regions is smaller than in smooth regions. The aim of using adaptive edge is to maintain the texture of the LSB plane. As shown in

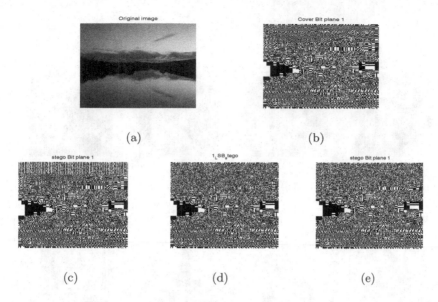

Fig. 3. (a) Cover image (b) LSB plane of the cover image (c–e) LSB planes of the stego images using using 1 bpp PVD, TBPC and Edge Adaptive-TBPC respectively with 20% embedding rate

Figure 3, any change in smooth area of the image may affect the LSB value of the pixels.

The second version of the proposed steganographic (n bits) begins with embedding the secret message in the sharp regions first according to the size of the secret data. This means the distortions will be less detectable by HVS because the modifications in edge areas. In addition, number of embedding bits on each block is based on the difference value between the adjacent pixels. However, PVD embeds the secret message in sequential order with different number of embedding bits on each block. The proposed method can reach 100% embedding rate with high visual quality.

Figures 4(a)–4(d) show the stego images when the embedding rate is 50% and the correspondence stego histograms are shown in Figure 4(e)–4(h). The produced images gave a high degree of similarity where it is quite hard to find visual differences between the cover and stego images. Figures 4(i)–4(l) show the stego images when the embedding rate is 80%. The stego histograms shown in Figure 4(m)–4(p). It is clear that the quality of the stego images decreased slightly by increasing the embedding rate.

The result of our proposed method and the original PVD are summarized in table 3. The proposed Edge Adaptive-TBPC achieved noticeably better wPSNR and average difference results compared to the PVD algorithm for all considered six images, which indicate that better imperceptibility for the same embedding capacity.

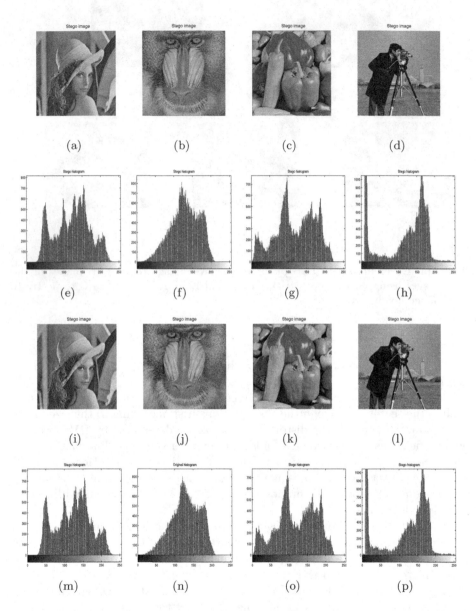

Fig. 4. (a − d) The Stego Images with embedding rate 50%, (e − h)) The histogram of the stego image with embedding rate 50%, (i − l) The Stego image with embedding rate 80% and (m − p) The histogram of the stego image with embedding rate 80%

Table 3. Comparisons of the visual quality and the embedding rate

Image	Embedding Rate	PVD			Proposed		
		30%	50%	80%	30%	50%	80%
Lena	wPSNR	63.5569	62.4805	60.8132	89.9022	73.8984	67.5874
	SSIM	0.9974	0.9961	0.9947	0.9979	0.9971	0.9967
	Avg. Difference	0.1892	0.3223	0.5315	0.1887	0.2561	0.4082
Baboon	wPSNR	89.1094	84.3872	76.0297	92.3605	88.5796	83.9867
	SSIM	0.9992	0.9987	0.9979	0.9974	0.9962	0.9953
	Avg. Difference	0.2019	0.3896	6303	0.1801	0.3217	0.5392
Cameraman	wPSNR	65.3993	62.5937	59.6606	87.4737	73.2688	66.8596
	SSIM	0.9967	0.9948	0.9929	0.9982	0.9976	0.9967
	Avg. Difference	0.2189	0.3937	0.5943	0.2158	0.3512	0.5661
Peppers	wPSNR	65.6801	63.8660	62.3860	87.5087	86.8021	72.0155
	SSIM	0.9981	0.9970	0.9959	0.9986	0.9979	0.9973
	Avg. Difference	0.2304	0.3814	0.5921	0.1924	0.3384	0.5478
Barbara	wPSNR	66.3192	64.1648	62.7694	91.3775	86.37	70.5923
	SSIM	0.9974	0.9963	0.9955	0.9986	0.9977	0.9963
	Avg. Difference	0.1776	0.3026	0.4892	0.1517	0.2822	0.4650
House	wPSNR	58.9912	57.8211	57.2199	67.5922	63.7642	62.8509
	SSIM	0.9964	0.9948	0.9929	0.9982	0.9973	0.9963
	Avg. Difference	0.1776	0.3020	0.4892	0.1517	0.2822	0.4050

5 Conclusion

This paper has introduced an image steganography method that combines the edge adaptive and TBPC algorithms to enhance the payload and imperceptibility of the stego image. Our method introduced minimum possible distortion during the embedding process to minimize the probability of discovering the secret message data from unauthorized users. We embed four bits into seven pixel bits. Due to the incorporation of an efficient coding mechanism, the probability of modifying pixel bits is 0.285. To further minimize the visual artefacts, the selection of embedding blocks is decided based on the the length of the secret message. Levels of sharp regions are categorized. The proposed method hides the secret data into the sharpest edge regions first (highest level). Then based on the message length, it moves down to the lower edge levels. Experimental results obtained by applying the proposed method to different images indicate that the proposed steganography method outperformed both of the original PVD and TBPC algorithms.

References

1. Wu, D.C., Tsai, W.-H.: A steganographic method for images by pixel-value differencing. Pattern Recognition Letters 24(9), 1613–1626 (2003)
2. Verma, N.: Review of steganography techniques. In: Proceedings of the International Conference & Workshop on Emerging Trends in Technology, pp. 990–993. ACM (2011)
3. Bailey, K., Curran, K.: An evaluation of image based steganography methods. Multimedia Tools and Applications 30(1), 55–88 (2006)
4. Sokół, B., Yarmolik, V.: Cryptography and steganography: teaching experience. In: Enhanced methods in computer security, biometric and artificial intelligence systems, pp. 83–92. Springer (2005)
5. Mukherjee, R., Ghoshal, N.: Steganography based visual cryptography (SBVC). In: Satapathy, S.C., Udgata, S.K., Biswal, B.N. (eds.) Proceedings of Int. Conf. on Front. of Intell. Comput. AISC, vol. 199, pp. 559–566. Springer, Heidelberg (2013)
6. Cheddad, A., Condell, J., Curran, K., Kevitt, P.M.: Digital image steganography: Survey and analysis of current methods. Signal Processing 90(3), 727–752 (2010)
7. Shrivastava, G., Pandey, A., Sharma, K.: Steganography and its technique: Technical overview. In: Proceedings of the Third International Conference on Trends in Information, Telecommunication and Computing, pp. 615–620. Springer (2013)
8. Johnson, N.F., Jajodia, S.: Exploring steganography: Seeing the unseen
9. Luo, W., Huang, F., Huang, J.: Edge adaptive image steganography based on lsb matching revisited. IEEE Transactions on Information Forensics and Security 5(2), 201–214 (2010)
10. Fridrich, J., Goljan, M.: On estimation of secret message length in lsb steganography in spatial domain. In: Electronic Imaging 2004, pp. 23–34. International Society for Optics and Photonics (2004)
11. Fridrich, J., Kodovský, J.: Steganalysis of LSB replacement using parity-aware features. In: Kirchner, M., Ghosal, D. (eds.) IH 2012. LNCS, vol. 7692, pp. 31–45. Springer, Heidelberg (2013)
12. Luo, W., Huang, F., Huang, J.: A more secure steganography based on adaptive pixel-value differencing scheme. Multimedia Tools and Applications 52(2-3), 407–430 (2011)
13. Agrawal, S.S., Samant, R.M.: Data hiding in gray-scale images using pixel value differencing. In: Technology Systems and Management, pp. 27–33. Springer (2011)
14. Crandall, R.: Some notes on steganography. Posted on steganography mailing list (1998)
15. Fridrich, J., Soukal, D.: Matrix embedding for large payloads. In: Electronic Imaging 2006, pp. 60721W–60721W. International Society for Optics and Photonics (2006)
16. Li, R.Y., Au, O.C., Lai, K.K., Yuk, C.K., Lam, S.Y.: Data hiding with tree based parity check. In: 2007 IEEE International Conference on Multimedia and Expo, pp. 635–638. IEEE (2007)
17. Hou, C.L., Lu, C., Tsai, S.C., Tzeng, W.G.: An optimal data hiding scheme with tree-based parity check. IEEE Transactions on Image Processing 20(3), 880–886 (2011)
18. Liu, G., Liu, W., Dai, Y., Lian, S.: Adaptive steganography based on block complexity and matrix embedding. Multimedia Systems 20(2), 227–238 (2014)

Secure Client Side Watermarking with Limited Key Size

Jia-Hao Sun[1], Yu-Hsun Lin[2,3], and Ja-Ling Wu[1,3]

[1] Dept. of CSIE, National Taiwan University
[2] Intel-NTU Connected Context Computing Center
[3] GINM, National Taiwan University

Abstract. In this paper, a novel secure client side watermarking scheme is proposed. We illustrate the reasons why a traditional watermarking scheme may not suit for a large digital-content broadcasting service and the ways how to fix the associated problems, in the proposed approach. Since the newly proposed scheme can support both visible and invisible watermarks, it is applicable to Cloud-based digital-content broadcasting services, such as Bookstores in Amazon and Google, iTunes in Apple, and Streaming Media in Netflix.

Keywords: Cryptography, Digital Right Management, Watermarking, Secure Client Side Watermarking, Digital Content Delivering.

1 Introduction

The technique of digital watermark has been used to protect the copyrights of digital contents, such as audios [15] and images [3,7,10,17]. For instance, a content seller can put a watermark into a host content before selling it to buyers. The embedded watermark can be carefully generated on the basis of some secure transaction protocols (e.g. the buyer-seller protocols [8, 12, 16]) and is tightly binded to the buyer so that he cannot disclaim his transaction activity. A malicious buyer may share his copy illegally with others. As soon as the content owner find a pirate copy which was illegally distributed by the buyer, he can claim his copyright and find out the malicious distributor according to the extracted watermark.

However, with the rapid growth of E-commerce, traditional watermarking schemes may not be applicable in reality. For example, when the amount of buyers becomes large, the heavy computational cost associated with embedding watermarks into the whole digital content for every single buyer may diminish the willingness of the service providers to adopt content protection mechanisms. On the other hands, for a service provider, direct transferring each watermarked content to each buyer is not an affordable digital-content delivering service model, because point-to-point direct connection requires huge amount of communication cost which is too heavy to be sustained. Most importantly, transferring unencrypted watermarked content via an untrusted channel (e.g the Internet) leads vulnerable contents be eavesdropped by attackers, easily. Thus, it is an urgent need to develop a new content protection scheme that has both lower computation and communication costs when large amount of buyers are involved, and of course, the valuable contents can be securely transmit over the untrusted Internet.

To satisfy all the requirements, one might consider the solution of "Secure Client Side Watermarking" as addressed in [2,4,5,9,13,14]. The basic idea of this approach

X. He et al. (Eds.): MMM 2015, Part I, LNCS 8935, pp. 13–24, 2015.

is: the seller encrypts the host content using his master key, and delivers the encrypted copy to each one of the purchased buyers via a public channel. Server then generates a specific watermarked key for each buyer and sends the generated keys to buyers through a secure channel. Since a transaction-related watermark is embedded into the watermarked key, the watermark will persist in the decrypted content when a buyer uses his specific watermarked key to decrypt the ciphertext. Under this application scenario, the communication cost will be largely reduced because the cost of transmitting watermarked key is (e.g a seed of a random number generator) relatively small. Since each buyer will get the same encrypted content, the huge cost of transmitting ciphertext can be effectively reduced by some existing broadcasting techniques, such as Content Delivering Network (CDN), Peer to Peer (P2P), and so on. The associated computational cost is also reduced because embedding the watermark into a key will be much lighter than embedding it into a whole digital content. Most of all, the security requirement is also fulfilled, in this approach, because the content is encrypted while transferring over the Internet.

Our work is based on the scheme of "Secure Client Side Watermarking" but with improvement in reducing the communication cost further. The first secure client side watermarking scheme was proposed in [2] which implemented the secure client side watermarking by a specialized stream cipher called "Chameleon". The work was then generalized in [1] and [6] . According to the original version of Chameleon, a Look-Up-Table (LUT) based method, the size of the embedded watermarked key is about hundreds of kilo-bytes (KB). Furthermore, the key size of generalized version of Chameleon [1, 6] will go up to several mega-bytes (MB). Note that one digital content may have a similar size as the above watermarked key; in other words, in Chameleon-based approaches the cost of transmitting a key is nearly the same as transmitting a content. If the purchased content is a streaming video or an electronic book of a bunch of pages, the communication cost becomes unaffordably large. From the view point of the service provider, the size of the watermarked key should be reasonably reduced. In this paper, a novel Secure Client Side Watermarking scheme with key sizes in the range of 2KB to 40KB (depend on the length of watermark), is proposed.

Since the newly proposed scheme needs a small key size, it is more appropriate to be applied to the modern digital-content broadcasting or downloading services such as E-books in Amazon and Google, musics in iTunes, and movies in Netflix.

The rest of this paper is organized as follows: Section 2 will introduce the basic idea of Secure Client Side Watermarking which solved the pre-described secure content delivering problem, briefly. Moreover, issues that made the traditional related approaches hard to be applied for reality will also be mentioned. Section 3 will review the technique of Lagrange Polynomial Interpolation that will be used in constructing our invisible scheme. In Section 4, the formulation of the proposed visible and invisible watermarking schemes will be presented in detail. The corresponding performance analyses and a short discussion about them are also provided. The experimental results will be shown in Section 5. In this section, further observations and analyses will be presented. Finally, conclusions and future work directions are given in Section 6.

2 Related Work

In this section, we will introduce the first secure client side watermarking scheme which was proposed in [2]. Let I_i be the i-th plaintexts with $0 \leq i \leq ||I|| - 1$, where $||I||$ denotes the cardinality of the set I and in this paper I will be used to denote an image or a given plaintext content. To encrypt messages, the encryptor must construct a Look-Up-Table (LUT) $\mathbf{E}[i]$ of size L where each columns in $\mathbf{E}[i]$ is a random number. Then, the encryptor calculates

$$c_i = I_i + \sum_{j=0}^{S-1} \mathbf{E}[t_{ij}] \tag{1}$$

, for all $0 \leq i \leq ||I|| - 1$, where t_{ij} is a random mapping within the range $0 \leq t_{ij} \leq L - 1$, and S is a predetermined parameter. The output c_i is the ciphertext which is going to be distributed publicly.

The process of key generation is very simple. To generate a watermarked key for the k-th customer, the key generator calculates

$$\mathbf{K'}_k[i] = -\mathbf{E}[i] + \mathbf{W}_k[i] \tag{2}$$

for every column of LUT, where k is the index of the target customer, and $\mathbf{W}_k$ is the watermark for indicating the k-th customer. The output $\mathbf{K'}_k$ is the watermarked key for the customer k which is going to be sent to the buyer(i.e., the customer).

As soon as the customer receive ciphertext c_i and watermarked key $\mathbf{K'}_k$, he then calculates

$$I'_i = c_i + \sum_{j=0}^{S-1} \mathbf{K'}_k[t_{ij}]. \tag{3}$$

It follows that the decrypted content can be represented as

$$I_i + \sum_{j=0}^{S-1} \mathbf{W}_k[t_{ij}] = I_i + w_{k,i}. \tag{4}$$

That is, the decrypted content consists of the original content I_i and the watermark $w_{k,i}$. In other words, the watermark $w_{k,i}$ has be embedded into the decrypted content automatically. Since the watermark is embedded into the content wherever the customer decrypts the encrypted content, this approach is called "Secure Client Side Watermarking" scheme.

In this work, the encryption, decryption, and key generation can be done very fast (just table look-up operations are required). Furthermore, it has been proved that the work can resist the collusion attack. However, we found that the key size is too large to be used in practice. This is because the pre-described scheme is LUT-based, the transmission of a large table cannot be avoided. According to original Secure Client Side Watermarking version given in [2], the key size will be at least 512KBs, and that of the generalized version will reach to several MBs. Thus, we will propose another Secure Client Side Watermarking approach which is not LUT-based, and therefore, we can reduce the key size significantly.

3   Preliminaries

3.1   Lagrange Polynomial Interpolation

Lagrange polynomial interpolation solves the following problem:

Given $x_0, x_1, ..., x_d, y_0, y_1, ..., y_d$, find a polynomial $q(x)$ of degree d such that $q(x_j) = y_j$, for all $j = 0, 1, ..., d$.

Lagrange Polynomial Interpolation gives a quick close-form formula to solve the above mentioned problem, in which the polynomial $q(x)$ is given by:

$$q(x) := \sum_{j=0}^{d} y_j \cdot \ell_j(x), \tag{5}$$

where

$$\ell_j(x) := \prod_{\substack{0 \le m \le d \\ m \ne j}} \frac{x - x_m}{x_j - x_m}. \tag{6}$$

It is clear that $\ell_j(x_{j^*}) = 0$ for all $j^* \ne j$ and $\ell_j(x_j) = 1$. Therefore $q(x_j) = y_j$ for all $j = 0, 1, ..., d$ can then be verified.

3.2   The Modified Version of Lagrange Polynomial Interpolation

Although Lagrange Polynomial Interpolation is a convenient method for fitting a given curve, it still has several drawbacks. Runge's phenomenon states that as the dimension d becomes larger, the interpolated polynomial will oscillate strongly at the two edges of the interpolated interval. Figure 1 illustrates that the Runge's phenomenon does occur if d is getting larger.

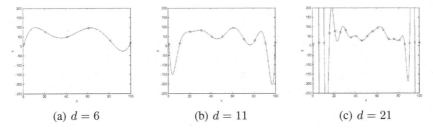

(a) $d = 6$ (b) $d = 11$ (c) $d = 21$

Fig. 1. Runge's phenomenon on different interpolated dimensions

There are several ways to reduce Runge's phenomenon. One famous solution is to choose interpolated points at Chebyshev nodes. The number of Chebyshev nodes is more dense on the two sides of the interpolated interval and is more sparse on the middle of the interpolated interval. Therefore, the two sides of the interpolated interval will have higher interpolation accuracy and lower Runge's phenomenon. However, this approach will reduced the flexibility of the interpolation scheme because the choices of $x_0, x_1, ..., x_d$ are much more restricted.

To release the above restriction a little bit, the Splined Lagrange Polynomial Interpolation is adopted in our work. The only difference from the original Lagrange polynomial interpolation is in conversing Equations (5) and (6) into:

$$q(x) := \sum_{j \in \lambda(x)} y_j \cdot \ell_j(x),$$ (7)

$$\ell_j(x) := \prod_{\substack{m \in \lambda(x) \\ m \neq j}} \frac{x - x_m}{x_j - x_m},$$ (8)

where

$$\lambda(x) = \left\{ x_m, x_{m+1}, x_{m+2} \middle| \begin{array}{l} x_{m+1} \leq x < x_{m+2} \, , if \, x_{m+2} = x_d \\ x_m \leq x < x_{m+1} \quad , otherwise. \end{array} \right\},$$ (9)

with the assumption that the given $x_0, x_1, ..., x_d$ are monotonically increased. Under these modifications, the interpolated polynomial in each interval $[x_j, x_{j+1}]$ is a spline which fits the curve in the interval with only three given points, it has been shown that this approach will reduce the Runge's phenomenon.

4   The Proposed Watermarking Schemes

4.1   The Proposed Watermarking Scheme

Note that, for simplicity of explanation, we take an image as the host content. Actually, the applicability of the proposed scheme to other media modalities can be derived similarly.

4.2   Secure Client Side Visible Watermarking

Formulation. Let $\overline{I}_{i,j}$ be the DCT coefficient of the host image I on position (i, j).

- Setup Algorithm:
  Initializing two random sequences, $A_{i,j}, B_{i,j} \in \mathbb{R}$, which are generated by two Pseudo Random Number Generators (PRNGs). Let the master key $k \triangleq \{A_{i,j}, B_{i,j}\}$. Empirically determine a watermark strength, $\beta \in \mathbb{R}$, and give a predefined threshold, T.
- Encryption Algorithm:

$$Enc(\overline{I}_{i,j}, k) \triangleq \overline{C}_{i,j} = \begin{cases} \overline{I}_{i,j} + A_{i,j} \, , \text{if } (i+j) < T \\ \overline{I}_{i,j} + B_{i,j} \, , \text{if } (i+j) \geq T \end{cases}$$ (10)

– Watermark Embedding Algorithm:
Generate a watermark image W and denote its DCT coefficients as $\overline{W}_{i,j}$.

$$Emb(k, W) = \{A'_{i,j}, B_{i,j}\} \triangleq k', \tag{11}$$

where

$$A'_{i,j} = \begin{cases} A_{i,j} & \text{, if } (i,j) = (1,1) \\ A_{i,j} - \beta\overline{W}_{i,j} & \text{, otherwise} \end{cases} \tag{12}$$

– Decryption Algorithm:

$$Dec(\overline{C}_{i,j}, k') \triangleq \overline{I}'_{i,j} = \begin{cases} \overline{C}_{i,j} - A'_{i,j}, \text{ if } (i+j) < T \\ \overline{C}_{i,j} - B_{i,j}, \text{ if } (i+j) \geq T \end{cases} \tag{13}$$

Analyses of the proposed Secure Client Side Visible Watermark Scheme. Both the encryption key k and the decryption key k' are separated into two different parts. The first part is applied to lower frequency bands, and the second part to the higher frequency ones.

For higher frequency part, the decrypted contents are the same as the original host contents. The watermark is embedded into the lower frequency part, and thus, is visible to users. The decrypted lower frequency bands can be represented as $\overline{I}_{i,j} + \beta\overline{W}_{i,j}$. One important fact is that since the second part of the decryption key is generated by a PRNG, the key can be transmitted by using the corresponding PRNG's seed instead, which largely reduces the required communication cost. However, the above trick cannot be applied to the first part of the keys since $A_{i,j} - \beta\overline{W}_{i,j}$ cannot be replaced by a PRNG's seed.

The key's separation is according to a given threshold T. The threshold T should not be chosen too small in case the lower frequency bands are revealed. Note that the more the lower frequency bands are revealed, the easier for the attacker to remove the embedded watermark. However, if T is chosen too large, the first parts of keys may become too large to limit the pre-described key-size reduction benefit.

Since the decryption key is used as nonce, the security level of the ciphertext is close to that of the Stream Cipher. On the other hand, the robustness of the watermark can be adjusted by choosing a proper T. Clearly, there is a trade off on T between the watermark robustness and the size of the decryption key.

4.3   Secure Client Side Invisible Watermarking

The proposed Secure Client Side Invisible Watermark Scheme is based on the technique of (modified) Lagrange polynomial interpolation, described in Section 3.2.

Construction. As described in Section 3, Lagrange Polynomial Interpolation gives a quick formulation for generating a polynomial $q(x)$ of degree d such that $q(x_j) = y_j$ for all $j = 0, 1, ..., d$, where x_j and y_j are the given points. It follows that if we assume $\ell_j = y_j \cdot w_j$ with $w_j \in \{\beta, \frac{1}{\beta}\}$ and interpolate with respect to the modified x_j and ℓ_j, the newly generated polynomial $q'(x)$ will be close to $q(x)$ and has only a slight difference from $q(x)$. Thus, we take advantage of this feature to produce our watermark.

The basic idea is: in the encryption side, we encrypt the plaintexts by masking a sampled value with a virtual encryption polynomial. For key generation, server makes some difference on the interpolated points. In the decryption side, decryptor generates a virtual decryption polynomial with respect to the modified interpolated points, applies the same sample method, and finally removes the mask by using the sampled value from the decryption polynomial. We refer to Theorem 1 to show that the difference between the two interpolated polynomials is limited; therefore, it is applicable for watermarking applications.

Theorem 1. *Given* $x_0, x_1, ..., x_d, y_0, y_1, ..., y_d, w_0, w_1, ..., w_d$, $q(x)$ *is a Lagrange interpolated polynomial of degree d with* $q(x_j) = y_j$, *and $q'(x)$ is another that kind of polynomial with* $q'(x_j) = y_j \cdot w_j$, *then*

$$max_x \left\{ \frac{q'(x) - q(x)}{q(x)} \right\} \leq max_j \{1 - w_j\} \tag{14}$$

Proof. W.l.o.g., let's assume that $\alpha \in [min(x_j), max(x_j)]$. According to Lagrange Polynomial Interpolation,

$$q(\alpha) = \sum_{j \in \lambda(\alpha)} y_j \cdot \prod_{\substack{m \in \lambda(\alpha) \\ m \neq j}} \frac{\alpha - x_m}{x_j - x_m} \tag{15}$$

$$q'(\alpha) = \sum_{j \in \lambda(\alpha)} y_j \cdot w_j \cdot \prod_{\substack{m \in \lambda(\alpha) \\ m \neq j}} \frac{\alpha - x_m}{x_j - x_m} \tag{16}$$

can be derived. So,

$$|q(\alpha) - q'(\alpha)| = \sum_{j \in \lambda(\alpha)} y_j \cdot (1 - w_j) \cdot \prod_{\substack{m \in \lambda(\alpha) \\ m \neq j}} \frac{\alpha - x_m}{x_j - x_m} \tag{17}$$

$$\leq max_j \{1 - w_j\} \sum_{j \in \lambda(\alpha)} y_j \cdot \prod_{\substack{m \in \lambda(\alpha) \\ m \neq j}} \frac{\alpha - x_m}{x_j - x_m} \tag{18}$$

$$\leq max_j \{1 - w_j\} \cdot q(x). \tag{19}$$

Thus, $\frac{|q(\alpha) - q'(\alpha)|}{q'(\alpha)} \leq max_j \{1 - w_j\}$ is proved now.

Formulation. Let $\overline{I}_i$ be the DCT coefficient of the host image I on the Zig-Zag ordering position i.

– Setup Algorithm:
  Randomly choose $\{x_0, x_q, ..., x_d, y_0, y_1, ..., y_d, s\} \in \mathbb{R}$ where d is the selected parameter and $x_0, x_1, ..., x_d$ are monotonically increased.
  Initialize $\{a_1, a_2, ...\}$ which are generated by a PRNG.
  Determine a watermark strength β which is near 1 (For instance, $\beta = 1.01$).
  The master key is $k = \{y_0, y_1, ..., y_d, s\}$

– Encryption Algorithm:

$$Enc(\overline{I}_i, k) \triangleq \overline{C}_i = \overline{I}_i \cdot s \cdot \left[\sum_{j \in \lambda(a_i)} y_j \cdot \prod_{\substack{m \in \lambda(a_i) \\ m \neq j}} \frac{a_i - x_m}{x_j - x_m} \right]^{\phi(i)} \tag{20}$$

– Watermark Embedding Algorithm:

Generate a binary stream watermark $W \in \{0, 1\}^{(d+1)}$ and let $w_j = \begin{cases} \beta & , \text{if } W(j)=1 \\ \frac{1}{\beta} & , \text{if } W(j)=0. \end{cases}$

$$Emb(k, W) \triangleq k' = \{\ell_j = y_j \cdot w_j \cdot s, \forall j = 0, 1, ..., d\} \tag{21}$$

– Decryption Algorithm:

$$Dec(\overline{C}_i, k') \triangleq \overline{I}_i' = \overline{C}_i \div \left[\sum_{j \in \lambda(a_i)} \ell_j \cdot \prod_{\substack{m \in \lambda(a_i) \\ m \neq j}} \frac{a_i - x_m}{x_j - x_m} \right]^{\phi(i)} \tag{22}$$

– Watermark Detection Algorithm:
Let $\overline{P}_i$ be the i-th DCT coefficient of the pirate copy P on Zig-Zag ordering position i.
Let a mapping $M : \{a_i\} \mapsto \{-1, 1\}$ be defined as:

$$M(a_i) = \begin{cases} 1, & if \frac{\overline{I}_i}{\overline{P}_i} < 1 \\ -1, & otherwise. \end{cases} \tag{23}$$

Then calculate

$$\dot{w}_j = sign \left(\sum_{a_i \in neighbor(x_j)} M(a_i) \delta_i \right), \tag{24}$$

where δ is the detection weight that can be determined by system administrator. Finally, output the cosine similarity between the two vectors: $W = (w_0, w_1, ..., w_d)$ and $\dot{W} = (\dot{w}_0, \dot{w}_1, ..., \dot{w}_d)$

$$Det : (P, I, W) = \frac{W \cdot \dot{W}}{|W| \cdot |\dot{W}|}. \tag{25}$$

Analyses of the Proposed Secure Client Side Invisible Watermark Scheme. The system relies on the difficulties of real number factorization. The common studied factorization problem is focused on prime number factorization on large integers. The large integer is usually as large as thousands of bits so that it is difficult to find its prime factorization. However, the real number factorization is more difficult than integer factorization, or even we can say that real number factorization is impossible to be solved. For instance, given a real number $R \in \mathbb{R}$ and a random number $a \in \mathbb{R}$, it is always possible to find another number $b \in \mathbb{R}$ such that $R = a \cdot b$ (just choose $b = \frac{1}{a}$). So factorization on real number field has infinity many possible solutions. Thus, attackers who get the ciphertexts generated from equation (20) are not possible to find s nor $\overline{I}_i$. Similarly, clients who get watermarked keys generated from equation ()21) are not possible to find y_j and w_j.

The function $\phi(i)$ on the power of equations (20) and (22) is an enhancing factor for boosting watermarking performance. The purpose is to produce a stronger watermark on higher frequency band, and weaker watermark on lower frequency band. Therefore, $\phi(i)$ should be designed in proportional to i. The experimentally setting is to choose $\phi(i)$ as $1 + k \cdot \frac{i}{\|I\|}$ with empirical setting of $k = 1, 2, 3, \dots$.

From equation (21), the size of the watermarked key is related to y_j and w_j, that is, d. Thus, the dimension d determines the degree of polynomial, the length of watermark, and the size of watermarked key. Besides, although the watermark stream is a binary stream, it should be converted to $\beta^{\pm 1}$ to fit our system requirement.

The proposed system contains numbers of parameters. We refer to Section 5.2 to illustrate how to find a better settings that can achieve the best watermarking performance.

5 Experimental Results

The following experiments show the results of the proposed visible and invisible watermarking schemes and provide some corresponding statistical analysis results. Figure 2 shows the test images we adopted.

(a) Lena   (b) Fruits   (c) Peppers   (d) Baboon

Fig. 2. Test images

5.1 Visible Watermark Scheme

For visible watermarking, we followed the processes described in Section 4.2 with different choices of β and T. We applied the proposed Secure Client Side Visible Watermarking to all R,G, and B channels of the test images. Figure 3 shows the results for different choices of β with fixed $T = 0.063\|I\|$. Figure 4 shows the results for different choices of T with fixed $\beta = 1$. As expected, a larger β will bring a brighter watermark, while a larger T will bring a more vivid watermark. As described in Section 4.2, the size of the decryption key can be set approximately to 1% to 10% of the original content, which largely reduces the size of information that has to be transmitted.

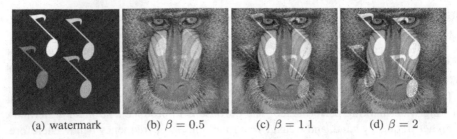

(a) watermark   (b) $\beta = 0.5$ (c) $\beta = 1.1$ (d) $\beta = 2$

Fig. 3. Results of the proposed visible scheme for the Baboon image with different β on fixed $T = 0.063\|I\|$

(a) watermark   (b) $T = 0.016\|I\|$ (c) $T = 0.098\|I\|$ (d) $T = 0.141\|I\|$

Fig. 4. Results of the proposed visible scheme for Fruits image with different T on fixed $\beta = 1$

Fig. 5. The decrypted results for the proposed invisible scheme with respect to different test images, where the best parameter settings is used

5.2   Invisible Watermark Scheme

For invisible watermarking, we followed the processes described in Section 4.3.In this section, we use the settings: { $d = 500$, Image size$= 512 * 512$, $\beta = 1.08$, $\phi(i) = 1 + 2 \cdot \frac{i}{\|I\|}$, and $\delta_i = $ Linear} to find the watermarking performance. Figure 5 shows some decrypted results for the test images (shown in Figure 2). Their average PSNR is 30.02dB. Figure 6 shows the performances with respect to JPEG attacks and Gaussian white noise addition attacks on different test images. We found that the performance order from best to worst is: Baboon, Fruits, Lena, and Peppers, respectively. The reason is that the images such as Baboon and Fruits has more complex texture so that they are easier to hide watermarks. However, images such as Lena and Peppers has more smooth texture so that they are harder to hide watermarks.

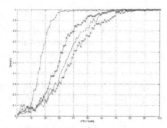 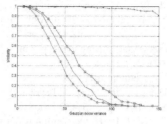

(a) JPEG attacks: results from right to (b) Gaussian white noise addition at-
left are for Baboon, Fruits, Lena, and tacks: results from right to left are for
Peppers, respectively   Peppers, Lena, Fruits, and Baboon, re-
                                         spectively

Fig. 6. Similarities on JPEG attacks and Gaussian white noise addition attacks

6   Conclusion

In this paper, a novel Secure Client Side Watermarking Scheme with Limited Key Size,
is proposed. The proposed scheme can achieve a more secure and more efficient digital-
content Broadcasting model as compared with the traditional approaches. The proposed
scheme presents some new thoughts which may solve some content delivering problems, but there are lots of different problems still remain unsolved. For instance, the
proposed scheme achieves only a light-weight secure level, how to improve its security
without sacrificing other performances is worth of further study. Besides, the robustness
of the proposed Invisible Watermarking Scheme is less than that of a standard watermarking scheme such as Improved Spread Spectrum (ISS)-based watermarking [11],
so, how to improve the robustness of the proposed scheme also needs further investigation. Moreover, the detection process of the proposed Invisible Watermarking Scheme is
an informed one, which is hard to be acceptable for many practical applications. Therefore, how to transfer the proposed informed scheme into a blind one will, of course, be
one of our future work directions.

Acknowledgment. This work was also supported by Ministry of Science and Technology, National Taiwan University and Intel Corporation under Grants MOST102-2911-
I-002-001 and NTU103R7501.

References

1. Adelsbach, A., Huber, U., Sadeghi, A.-R.: Fingercasting—joint fingerprinting and decryption of broadcast messages. In: Batten, L.M., Safavi-Naini, R. (eds.) ACISP 2006. LNCS,
   vol. 4058, pp. 136–147. Springer, Heidelberg (2006)
2. Anderson, R., Manifavas, C.: Chameleon – A new kind of stream cipher. In: Biham, E. (ed.)
   FSE 1997. LNCS, vol. 1267, pp. 107–113. Springer, Heidelberg (1997)
3. Barni, M., Bartolini, F., Cappellini, V., Piva, A.: A dct-domain system for robust image watermarking. Signal Processing 66(3), 357–372 (1998)

4. Bianchi, T., Piva, A.: Secure watermarking for multimedia content protection: A review of its benefits and open issues. IEEE Signal Processing Magazine 30(2), 87–96 (2013)
5. Celik, M., Lemma, A., Katzenbeisser, S., van der Veen, M.: Secure embedding of spread spectrum watermarks using look-up-tables. In: IEEE ICASSP 2007, vol. 2, pp. II-153–II-156 (April 2007)
6. Celik, M.U., Lemma, A.N., Katzenbeisser, S., van der Veen, M.: Lookup-table-based secure client-side embedding for spread-spectrum watermarks. IEEE Trans. on Information Forensics and Security 3(3), 475–487 (2008)
7. Hsu, C.-T., Wu, J.-L.: Hidden digital watermarks in images. IEEE Trans. on Image Processing 8(1), 58–68 (1999)
8. Lei, C.-L., Yu, P.-L., Tsai, P.-L., Chan, M.-H.: An efficient and anonymous buyer-seller watermarking protocol. IEEE Trans. on Image Processing 13(12), 1618–1626 (2004)
9. Lemma, A., Katzenbeisser, S., Celik, M.U., van der Veen, M.: Secure watermark embedding through partial encryption. In: Shi, Y.Q., Jeon, B. (eds.) IWDW 2006. LNCS, vol. 4283, pp. 433–445. Springer, Heidelberg (2006)
10. Luo, L., Chen, Z., Chen, M., Zeng, X., Xiong, Z.: Reversible image watermarking using interpolation technique. IEEE Trans. on Information Forensics and Security 5(1), 187–193 (2010)
11. Malvar, H.S., Florêncio, D.A.: Improved spread spectrum: a new modulation technique for robust watermarking. IEEE Transactions on Signal Processing 51(4), 898–905 (2003)
12. Memon, N., Wong, P.W.: A buyer-seller watermarking protocol. IEEE Trans. on Image Processing 10(4), 643–649 (2001)
13. Piva, A., Bianchi, T., De Rosa, A.: Secure client-side st-dm watermark embedding. IEEE Trans. on Information Forensics and Security 5(1), 13–26 (2010)
14. Pun, C.L.P.C.-M., Jiang, J.-J., Chen: Adaptive client-side lut-based digital watermarking. In: IEEE TrustCom 2011, pp. 795–799 (November 2011)
15. Seok, J.W., Hong, J.-W.: Audio watermarking for copyright protection of digital audio data. Electronics Letters 37(1), 60–61 (2001)
16. Zhang, J., Kou, W., Fan, K.: Secure buyer–seller watermarking protocol. IEE Proceedings-Information Security 153(1), 15–18 (2006)
17. Zhang, X.: Reversible data hiding with optimal value transfer. IEEE Trans. on Multimedia 15(2), 316–325 (2013)

Orderless and Blurred Visual Tracking
via Spatio-temporal Context

Manna Dai[1], Peijie Lin[1], Lijun Wu[1], Zhicong Chen[1], Songlin Lai[1], Jie Zhang[1],
Shuying Cheng[1,*], and Xiangjian He[2]

[1] Institute of Micro/ Nano Devices and Solar Cells,
College of Physics and Information Engineering,
Fuzhou University, Fuzhou 350108, China
[2] Faculty of Engineering and Information Technology,
University of Technology, Sydney, Australia
sycheng@fzu.edu.cn

Abstract. In this paper, a novel and robust method which exploits the spatio-temporal context for orderless and blurred visual tracking is presented. This lets the tracker adapt to both rigid and deformable objects on-line even if the image is blurred. We observe that a RGB vector of an image which is resized into a small fixed size can keep enough useful information. Based on this observation and computational reasons, we propose to resize the windows of both template and candidate target images into 2×2 and use Euclidean Distance to compute the similarity between these two RGB image vectors for the preliminary screening. We then apply spatio-temporal context based on Bayesian framework to further compute a confidence map for obtaining the best target location. Experimental results on challenging video sequences in MATLAB without code optimization show the proposed tracking method outperforms eight state-of-the-art methods.

Keywords: spatio-temporal-context, resize, Euclidean Distance, Bayesian framework.

1 Introduction

Visual object tracking is to estimate location of a target in an image sequence. It has been a long standing research topic due to its wide range of applications such as video surveillance, human computer interaction, traffic control and so on [1]. However, visual tracking is challenging due to abrupt motion, illumination change, cluttered background and occlusion. Although a significant progress has been made to overcome these challenges, developing an efficient and robust visual tracking method is still a crucial topic, particularly when a rigid or a deformable object moving disorderly occurs in a blurred image sequence.

The existing visual tracking approaches can be categorized into generative [2-4] and discriminative [5-7] methods. The generative tracking methods search for image

---

* Corresponding author.

X. He et al. (Eds.): MMM 2015, Part I, LNCS 8935, pp. 25–36, 2015.

regions that are most similar to the template, while discriminative methods aim at differentiating the target from the background. However, their main shortcomings are also remarkable as follows. Firstly, too many samples to be extracted make the computational load very heavy. Secondly, the effective searching algorithm and measured approach between template and candidate samples are difficult. Thirdly, it is hard to distinguish the target from complicated background because of the broadly varying background, the similarity between object and background, or the object which moves too fast to make the object itself and its surrounding blurred in an image.

In this paper, we propose a novel and robust tracking algorithm to exploit spatio-temporal local context information. Firstly, we use a simple and powerful work to search for object applying objectness scores. Our work is motivated by the fact that generic objects with well-defined closed boundaries [9-11] share strong correlation after resizing of their corresponding image to small fixed size. Therefore, the template and candidate model can be resized separately into 2×2 to efficiently quantify the objectness of an image. Euclidean Distance is used to compute the similarity between the template and candidate model. We choose the maximum similarity as the best result to compute the promising center of the object. Then spatio-temporal local context information is exploited to further determine the position of the object by using the previous promising center. We apply the max similarity of the template and candidate model to update the template in the next frame. The template-update in spatio-temporal context will also consider the several max confidence maps in the previous certain frames. These two update measures can bring in the current target information when the true template changes much or occlusion is occurred, and keep the original information if occlusion is removed or previous tracking results are not really exact.

The main contributions of this paper are as follows. (1) A novel and robust spatio-temporal context based orderless and blurred visual tracking method is proposed. (2) An efficient search algorithm is adopted in each tracking round by resizing an image into size 2×2 and using Euclidean Distance to compute the similarity between template and candidate image for efficiently reducing the compute load. (3) Our method makes advantage of a strong spatio-temporal relationship between the local scenes containing the object in consecutive frames. (4) The experiments show that our method is robust to appearance variations introduced by abrupt motion, occlusion, pose variations, background clutter, and illumination variation, especially in blurred and disordered scene.

2   The STC Tracker

Our approach is based on the STC tracker presented in [8]. The STC tracker formulates the spatio-temporal relationships between the object of interest and its local context based on a Bayesian framework. It models the statistical correlation between the low-level features in the target and its surrounding. Here, we provide a brief overview of this approach [8].

In STC tracker, a tracking problem is formulated by computing a confidence map that estimates the object location likelihood. In the current frame, we get the object location x^* and define the feature set as $X^c = \{c(z) = (I(z), z) | z \in \Omega_c(x^*))\}$ where $I(z)$ represents image intensity at location z and $\Omega_c(x^*)$ is the neighborhood of location x^*. We can compute the object location likelihood by

$$m(x) = P(x|o) = \sum_{c(z) \in X^c} P(x, c(z)|o) = \sum_{c(z) \in X^c} P(x|c(z), o) P(c(z)|o). \quad (1)$$

The spatial context model is a conditional probability function, which is defined as

$$P(x|c(z), o) = h^{sc}(x - z). \quad (2)$$

$h^{sc}(x - z)$ is a function regarding the relative distance and direction between object location x and its local context location z, and it encodes the spatial context relationship of the target and its spatial relation. However, $h^{sc}(x - z)$ is not a radially symmetric function.

We model the context prior probability in (2) as

$$P(c(z)|o) = I(z)w_\sigma(z - x^*), \quad (3)$$

where $I(\cdot)$ is the image intensity that represents appearance of the context.

Inspired by the biological visual system, a focus of attention function is used as a weighted function defined by

$$w_\sigma(z - x^*) = ae^{-\frac{|z - x^*|^2}{\sigma^2}}, \quad (4)$$

where a is a normalization constant which ranges from 0 to 1 to satisfy the definition of probability and σ is a scale parameter. In this weighted function, the closer the location is to the object center, the more context locations are considered.

The confidence map of an object location is in formula (1) defined as

$$m(x) = be^{-\left|\frac{x - x^*}{a}\right|^\beta}$$

$$= \sum_{z \in \Omega_c(x^*)} h^{sc}(x - z)I(z)w_\sigma(z - x^*)$$

$$= h^{sc}(x) \otimes \left(I(x)w_\sigma(x - x^*)\right). \quad (5)$$

where b is a normalization constant, α and β are a scale parameter and a shape parameter respectively. Eq. (5) can be transformed to the frequency domain for fast convolution:

$$F\left(be^{-\left|\frac{x - x^*}{\alpha}\right|^\beta}\right) = F(h^{sc}(x)) \odot F(I(x)w_\sigma(x - x^*)). \quad (6)$$

where F denotes the Fast Fourier Transform (FFT) function and $\odot$ denotes the element-wise product. The $h^{sc}(x)$ is defined as

$$h^{sc}(x) = F^{-1}\left(\frac{F(be^{-\left|\frac{x-x^*}{\alpha}\right|^\beta})}{F(I(x)w_\sigma(x-x^*))}\right). \tag{7}$$

where F^{-1} denotes the inverse FFT function. For more details, we refer to [8].

3 Orderless and Blurred Visual Tracking via Spatio-temporal Context

3.1 Framework

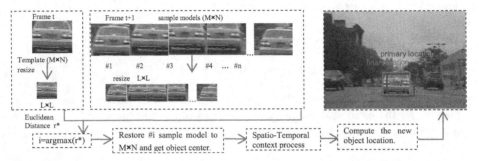

Fig. 1. Basic flow of our tracking algorithm is as shown in figure. We resize template and sample models into L×L (e.g. 2×2). Euclidean Distance is used to simply compute the similarity between template and each sample models to get the primary location. We use the best image center (center of #i image) which has the max similarity to conduct the spatio-temporal context process for obtaining the final object location.

Figure 1 shows the basic flow of our proposed tracking algorithm. The tracking process has three steps.

Step I. We resize the candidate models set $X = \{x_1, x_2, ..., x_n\}$ and template to L ×L. In the image sequences which we used, the minimum of the length and the width of the objects is 51. Therefore, the value of L is from 2 to 51 with the interval as 7. We use the different L to compute the average CLE and average FPS of our research datasets. We find that the average CLE does not highly increase and the average FPS reduces highly as the value of L increases, so we set L as 2 in order to get a good balance between CLE and FPS.

Step II. The suitable image (#i) center which is most similar to the template is selected. The similarity set $r^* = \{r_1^*, r_2^*, ..., r_n^*\}$ can be obtained by Euclidean Distance algorithm. Then, we compute the #i image center as a prior center in current frame.

Step III. We use the prior center to conduct the spatio-temporal context process to get the result image location in this frame. After separately updating the template which is used to compute by Euclidean Distance algorithm in the first step and the one which is to conduct the spatio-temporal context process in the last step, we continue to sample a new candidate models set to circulate from the first step until the end of the image sequence. The details in each step will be introduced in the following content.

3.2   Image Resizing Measure

Usually, we depend on the previous object locations to predict the promising location in next frame. However, this measure will fail if the object moves fast and disorderly. Some researchers use a sliding window fashion to search for the object [12-13] which will increase the computation burden. Therefore, we propose a local search algorithm with image resizing to help us search for the promising object location.

Objects are stand-alone things with well-defined closed boundaries and centers [9-11]. If we resize the image to a small fixed size, the little variation that closed boundaries could present in such abstracted view. In this paper, in order to process larger scale and reduce the computation burden at the same time, we decide to resize template and candidate models into 2×2 size (see in Figure 1).

3.3   Similarity by Euclidean Distance

We aim at narrowing the scope of our search for reducing the complexity in the following work. We choose the Euclidean Distance to calculate the max similarity φ of resizing template and resizing candidate models in order to provide us a promising object location and center. The Euclidean Distance is defined as follow.

$$\text{dist}(X, Y) = \sqrt{\sum_{i=1}^{n}(x_i - y_i)^2} \tag{8}$$

Here X denotes the RGB vector of template, and Y represents the RGB vector of each candidate model. The Euclidean Distance not only reflects whether two vectors are relevant, but also can calculate the level of similarity between them.

We sample the candidate models set X as

$$X = \{x : \|x - (x_t^* + d)\| < d\}, \tag{9}$$

where x is the location of the sample model, x_t^* denotes the location of the object in the t-th frame, and d represents the distance in which the center of the object moves between the $(t-1)$-th frame and $(t-2)$-th frame. Note that d can be positive and negative.

We choose RGB as the feature in this step because we will exploit the gray-scale map in the following step. We hope to introduce more information of an image into our work so that it can have a better result. Because RGB can be affected by the illumination, we use zero mean treatment to eliminate the effects of illumination and shadow.

The template in the next frame will be updated by the following function

$$M_{t+1} = (1 - \varphi^*)M_t + \varphi^* s \tag{10}$$

Here M_{t+1} is the template in the next frame, M_t is the template in current frame, φ^* is the max similarity which has been normalized in this frame, and s is the model image whose similarity is φ^* in this frame. Note that all parameters are under the condition of 2×2 image.

3.4   Further Compute with Spatio-temporal Context

We apply the spatio-temporal context in [8] to the following process. After computing the similarity between template and candidate models, we get a primary location of the object. Then, we will use this promising center to calculate the final location. Note that, we use the image in the original size from now on, not the 2×2 size.

In this process, assume that we initialize the target location in the first frame by some object detection algorithms. We learn the spatial context model $h_t^{sc}(x)$ (7) in the t-th frame, which is used to update the spatio-temporal context model $H_{t+1}^{stc}(x)$ (13) and detect the object location and center in the $(t+1)$-th frame. In [8], the object location x_{t+1}^* in the $(t+1)$-th frame is calculated by maximizing the new confidence map:

$$x_{t+1}^* = arg \max_{x \in \Omega_c(x_t^*)} m_{t+1}(x), \tag{11}$$

where $\Omega_c(x_t^*)$ is the local context region which is based on the tracked location x_t^* in the t-th frame, and we construct the corresponding context feature set $x_{t+1}^c = \{c(z) = (I_{t+1}(z), z) | z \in \Omega_c(x_t^*)\}$.

The $m_{t+1}(x)$ in (11) in [8] is defined as

$$m_{t+1}(x) = F^{-1}(F(H_{t+1}^{stc}(x)) \odot F(I_{t+1}(x) w_{\sigma_t}(x - x_t^*))), \tag{12}$$

which is deduced from (6).

We update the spatio-temporal context model in [8] by

$$H_{t+1}^{stc} = (1 - \rho)H_t^{stc} + \rho h_t^{sc}, \tag{13}$$

where ρ is considered as a learning parameter, and h_t^{sc} is the spatial context model which can be obtained from (7) in the t-th frame.

Note that we will use zero mean treatment to every frame in order to remove the effect from the illumination change. In addition, the intensity in the context region exploits a Hamming window to reduce the frequency influence from the image boundary on the FFT [14-15]. Hamming window is defined as

$$w(t) = \begin{cases} 0.54 - 0.46 \cos\left(\frac{2\pi}{\tau} t\right), & |t| \leq \frac{\tau}{2} \\ 0, & |t| > \frac{\tau}{2} \end{cases} \tag{14}$$

The tracking procedure is summarized in Algorithm 1.

Algorithm 1. The proposed tracking method

Input: Video frame f=1:F

1. For f=1:F
2. If f==1
3. Select the tracking object.
4. Compute the h^{sc}, then construct the template in spatio-temporal context as $H_1^{stc} = h^{sc}$.
5. Obtain the location x_1^* of the tracking object.
6. Resize the image of tracking object in the location x_1^* to 2×2 size, lys.
7. The template in RGB similarity is initialized as $M_1 = s$.
8. Else
9. Calculate the distance d between the (t-1)-th frame and (t-2)-th frame.
10. Sample the candidate models set $X = \{x_1, x_2, \dots, x_n\}$ by $X = \{x: \|x - (x_t^* + d)\| < d\}$.
11. Resize the template M_t and candidate models set X^* to 2×2 and calculate the similarity r^*, as the max similarity is defined as $i = \text{argmax } r^*, i = 1, 2, \dots, n$.
12. Normalize the max similarity as φ^*.
13. Restore the i-th sample model to original size and compute its center C_t.
14. Update the template M_t.
15. Use the center C_t to conduct the spatio-temporal context process to get h_t^{sc}.
16. Update the template H_{t+1}^{stc}.
17. The object location x_{t+1}^* is defined as $x_{t+1}^* = arg \max\limits_{x \in \Omega_c(x_t^*)} c_{t+1}(x)$.
18. End if
19. End for

4   Experimental Results and Analysis

4.1   Experimental Setup

In order to make our result more reasonable, we compare our method with other methods in the same experiment environment and equipment. Our approach is implemented in Matlab. The experiments are performed on an Intel(R) Core(TM) i5-2410M 2.30 GHz CPU with 2 GB RAM. In our experiments, the parameters are used in our algorithm as follows: the parameters of the map function are set to $\alpha = 1.8$ and $\beta = 1$. The learning parameter $\rho = 0.075$. Here β and ρ are set as same as inreference [8]. But in [8], $\alpha = 2.25$. In Eq. (5), the greater α is, the larger region around the center of object will be considered. On the other words, as our work focuses on the scene where image is blurred and object moves disorderly, we will pay more attention to the region around the center to avoid the blurred false image being introduced into our procedure as a noise. Therefore, we choose a smaller α.

Datasets: We use 8 color sequences namely: body, car2, car4, face, dollar, deer, shaking and david. The sequences used in our experiments pose challenging situations such as motion blur, abrupt movement, illumination changes, scale variation, occlusions, rotation, background clutter, and pose variation. Especially, we can tackle the tracking problem in image blur and fast disorder motion.

4.2   Comparison with State-of-the-Art

We compare our method with 8 different state-of-the-art trackers shown to provide excellent results in literature. The trackers used for comparison are: STC[8], WMIL[16], MIL[17], CT[5], L1[18], L1-APG[19], LOT[4] and Color Tracking[20].

In this paper, we follow the protocol used in [21] to validate our work. We will use three evaluation metrics: center location error (CLE), distance precision (DP) and overlap precision (OP). CLE is valued by the average Euclidean Distance between the estimated center location of the object and the ground-truth. DP is the relative number of frames in the sequence where CLE is smaller than a certain threshold. Here, the threshold is set as 20 pixels. OP is defined as the percentage of frames where the bounding box overlap exceeds a threshold $t \in [0,1]$. The trackers are ranked using DP scores at 20 pixels. We also present the speed of the trackers in average frames per second (FPS).

Table 1 shows CLE, where smaller CLE means more accurate tracking results. From Table 1, we can know that the quantitative results in which our tracking algorithm achieves the better performance. Figure 2 shows part of tracking results by different tracking methods. Table 2 shows a comparison with the mentioned state-of-the-art methods on 8 challenging sequences. We also present the speed in average frames per second (FPS). The best three results are shown in red, blue and green fonts respectively. Our method carries out well both in terms of speed and accuracy.

Table 1. Center location error (CLE) (in pixels). (1) indicate the best performance while (2) indicate the second best ones, and (3) indicate the third best ones.

sequence	WMIL	STC	MIL	LOT	L1	L1-APG	CT	Colortracking	ours
face	127	113	123	33.4(3)	149	183	55.8	7.5(2)	3.9(1)
shaking	12	8.2(1)	145	73.6	29.1	104	11.2(3)	13.2	10.7(2)
dollar	12.1(3)	20.4	70.5	71.6	20.5	71.5	9.3(2)	17.1	7.1(1)
deer	15.6(3)	401	202	63.7	78.1	78.9	211	5.1(1)	5.4(2)
car4	95.6	2610	140	31.3	61.6	15.3(2)	115	8.2(1)	15.6(3)
car2	163	5.41(2)	73.9	26.2(3)	49.9	156	104	86.9	5.1(1)
body	54.4	148	128	84.5	131	31.6(2)	122	36.5(3)	18.1(1)
david	34.3(3)	43.4	38.1	108	76.1	39.9	40.2	24.1(2)	10.9(1)
Average CLE	64.3	418.7	115.1	61.5(3)	74.4	84.9	83.6	24.8(2)	9.6(1)

Fig. 2. Comparison of our approach with state-of-the-art trackers in challenging situations such as motion blur, abrupt movement, illumination changes, scale variation, partial occlusions, rotation, background clutter, and pose variation. Especially we can tackle the tracking problem in image blur and fast disorder motion.

Table 2. Quantitative comparison of our trackers with 8 state-of-the-art methods on 8 challenging sequences. The results are presented in average center location error (CLE) (in pixels), average distance precision (DP) (%) and average overlap precision (OP) (%). We also provide the average frames per second (FPS). The best three results are shown with (1), (2) and (3). Note that our method is best in average CLE, average DP and average OP,and the second best in terms of speed.

	WMIL	STC	MIL	LOT	L1	L1-APG	CT	Colortracking	ours
Average CLE	64.3	418.7	115.1	61.5(3)	74.4	84.9	83.6	24.8(2)	9.6(1)
Average DP	40.6	42.5(3)	12.7	26.8	25.5	21.7	29.8	72.9(2)	88.3(1)
Average OP	44.2(3)	32.1	15.8	30.3	36.8	27.8	33.3	80.2(2)	93.0(1)
Average FPS	12.8	14.6(2)	1.1	0.3	0.2	7.8	12.1	21.1(1)	14.1(3)

Motion Blur. Figures 2 (a), (e), (f) and (g) have the motion blur. Only our method can deal with all the four sequences. In (a) and (e), our approach and Colortracking perform well in terms of CLE. L1-APG also has a good performance in (e), but it is not better than ours and Color tracking at frame #122 and #198. In (f) and (g), only our method can achieve successful tracking from beginning to end.

Abrupt Movement. Figures 2 (a), (d), (e) and (f) suffer from abrupt movement in whole sequences. In (a), (d) and (e), our approach and Colortracking can succeed to track the object. However, in (f), only our method still track the object successfully from frame #236 as we show.

Illumination Changes. In Figures 2 (b) and (h), the illumination often changes strong, some of the trackers completely fail to track. In (b) at frame #66, when the illumination changes fast and strong, only STC, Color tracking and our method still perform well as these three method all use zero mean measure to tackle the effect of illumination. In (h) at frame #83, #231, and #374, zero mean measure plays an important role when the light changes from dark to bright.

Scale Variation. In (h) at frame #231 to #496, we want to show the ability of our proposed method in disposing the scale variation. Our approach, STC and Colortracking can adapt to the scale variation of the object. Moreover, our approach has the best CLE among all 9 approaches. LOT, L1 and L1-APG completely fail to track the object while WMIL, MIL and CT suffer from sever drift.

Partial Occlusions. The object in Figure 2 (b) shaking demonstrates that the proposed method performs well in terms of position and rotation when the target undergoes partial occlusion. Our method and STC perform better than other methods at frame #105 and #253, while other methods suffer from sever drift and some of them fail to track. Thus, our method can handle occlusion and it is not sensitive to partial occlusion.

Rotation. Sequences of (g) and (h) emerge rotation of the object. In (g) from #140 to #310, our algorithm is the only one which can dispose the rotation of body of the walking man. All 8 other state-of-the-art algorithms fail. The man turns the body in (h) from #496 to #579, the trackers of STC, Colortracking and ours can deal with the rotation in this scene. In conclusion, our algorithm is the last winner in the tracking problem in the way of rotation.

Pose Variation. Figure 2 (c) at frame #70, the part of dollar is folded so that MIL, L1 and CT present their sensitivity in a certain aspect of pose variation. At frame #130, a pile of dollar is being divided into two piles, at the same time WMIL, STC and our method still exactly distinguish the right location of the object. MIL, LOT and L1-APG have total failure when the two piles of dollar are completely separated at frame #204. As showed at #290, if the two piles come close to each other, it will also have some impact on the trackers. Therefore, only our method and Colortracking can continue to develop their ability. To sum up, only our proposed method can keep accurate performance in the whole sequence.

Background Clutter. The trackers are easily confused if the object is very similar to the background. Figures 2 (b), (d) and (e) have the background clutter. Only our method can deal with this tracking problem. Other trackers drift or else completely fail to track.

4.3   Discussion

As shown in our experiments, our method can address the factors such as motion blur, abrupt movement, illumination changes, scale variation, heavy occlusions, rotation, background clutter, and pose variation. Especially we can tackle the tracking problem in image blur and fast disorder motion. The reasons are as follows. (1) Our method exploits temporal and spatial context information for tracking, which is very insensitive to multiple factors. (2) A simple but useful preliminary screening by Euclidean Distance is introduced between object template and candidate samples. (3) The measure of resizing the object template and candidate samples to a small size like 2×2 can keep enough information we need. In addition, it can abandon the redundant information to reduce the amount of calculation rapidly and can be realized easily.

5   Concluding Remarks

In this paper, a novel and robust method named orderless and blurred visual tracking is proposed. Firstly, template and candidate image are resized to small size to reduce the computation load. Then, Euclidean Distance is used to compute the similarity between these two RGB vectors from the resized template and candidate image for the preliminary screening. Finally, in order to address the shortcomings of current approaches for blurred images and orderless motion, we adopt the spatio-temporal context based on Bayesian framework to compute a confidence map for obtaining the best target location. Therefore, our method is very insensitive to appearance change. Experiments on some challenging video sequences have demonstrated the superiority of the proposed approach to 8 existing state-of-the-art ones in terms of accuracy and robustness.

References

1. Yilmaz, A., Javed, O., Shah, M.: Object tracking: A survey. ACM Computing Surveys (CSUR) 38(4) (2006)
2. Kwon, J., Lee, K.M.: Visual tracking decomposition. In: CVPR, pp. 1269–1276 (2010)
3. Kwon, J., Lee, K.M.: Tracking by sampling trackers. In: ICCV, pp. 1195–1202 (2011)

4. Oron, S., Bar-Hillel, A., Levi, D., Avidan, S.: Locally orderless tracking. In: CVPR, pp. 1940–1947 (2012)
5. Zhang, K., Zhang, L., Yang, M.-H.: Real-time compressive tracking. In: Fitzgibbon, A., Lazebnik, S., Perona, P., Sato, Y., Schmid, C. (eds.) ECCV 2012, Part III. LNCS, vol. 7574, pp. 864–877. Springer, Heidelberg (2012)
6. Hare, S., Saffari, A., Torr, P.H.: Struck: Structured output tracking with kernels. In: ICCV, pp. 263–270 (2011)
7. Zhang, K., Song, H.: Real-time visual tracking via online weighted multiple instance learning. Pattern Recognition (2012)
8. Zhang, K., Zhang, L., Liu, Q., Zhang, D., Yang, M.-H.: Fast visual tracking via dense spatio-temporal context learning. In: Fleet, D., Pajdla, T., Schiele, B., Tuytelaars, T. (eds.) ECCV 2014, Part V. LNCS, vol. 8693, pp. 127–141. Springer, Heidelberg (2014)
9. Alexe, B., Deselaers, T., Ferrari, V.: Measuring the objectness of image windows. IEEE TPAMI 34(11) (2012)
10. Heitz, G., Koller, D.: Learning spatial context: Using stuff to find things. In: Forsyth, D., Torr, P., Zisserman, A. (eds.) ECCV 2008, Part I. LNCS, vol. 5302, pp. 30–43. Springer, Heidelberg (2008)
11. Cheng, M.-M., Zhang, Z., Lin, W.-Y., Torr, P.: Bing: Binarized Normed Gradients for objectness estimation at 300fps. In: CVPR (2004)
12. Dalal, N., Triggs, B.: Histograms of oriented gradients for human detection. In: CVPR, vol. 1, pp. 886–893 (2005)
13. Felzenszwalb, P.F., Girshick, R.B., McAllester, D., Ramanan, D.: Object detection with discriminatively trained partbased models. IEEE TPAMI 32(9), 1672–1645 (2010)
14. Oppenheim, A.V., Willsky, A.S., Nawab, S.H.: Signals and systems, vol. 2. Prentice-Hall, Englewood Cliffs (1983)
15. Bolme, D.S., Draper, B.A., Beveridge, J.R.: Average of synthetic exact filters. In: CVPR, pp. 2105–2112 (2009)
16. Zhang, K., Song, H.: Real-time visual tracking via online weighted multiple instance learning. Pattern Recognition (2012)
17. Ross, D.A., Lim, J., Lin, R.-S., Yang, M.-H.: Incremental learning for robust visual tracking. International Journal of Computer Vision 77(1-3), 125–141 (2008)
18. Mei, X., Ling, H.: Robust visual tracking using 1 minimization. In: 2009 IEEE 12th International Conference on Computer Vision, pp. 1436–1443. IEEE (2009)
19. Bao, C., Wu, Y., Ling, H., Ji, H.: Real time robust L1 tracker using accelerated proximal gradient approach. In: CVPR (2012)
20. Danelljan, M., Khan, F.S., Felsberg, M., van de Weijer, J.: Adaptive color attributes for real-time visual tracking. In: CVPR, pp. 1090–1097 (2014)
21. Wu, Y., Lim, J., Yang, M.-H.: Online object tracking: A benchmark. In: CVPR (2013)

Coupled Discriminant Multi-Manifold Analysis with Application to Low-Resolution Face Recognition

Junjun Jiang[1], Ruimin Hu[1,2], Zhen Han[1,2], Liang Chen[1], and Jun Chen[1,2]

[1] National Engineering Research Center for Multimedia Software,
School of Computer, Wuhan University, Wuhan, China
[2] Research Institute of Wuhan University in Shenzhen, China
jiangjunjun@whu.edu.cn

Abstract. The problem of matching a low-resolution (LR) face image to a gallery of high-resolution (HR) face images is addressed in this letter. Previous research has focused on introducing a learning based super-resolution (LBSR) method before matching or transforming LR and HR faces into a unified feature space (UFS) for matching. To identify LR faces, we present a method called coupled discriminant multi-manifold analysis (CDMMA). In CDMMA, we first explore the neighborhood information as well as local geometric structure of the multi-manifold space spanned by the samples. And then, we explicitly learn two mappings to project LR and HR faces to a unified discriminative feature space (UDFS) through a supervised manner, where the discriminative information is maximized for classification. After that, the conventional classification method is applied in the CDMMA for final identification. Experimental results conducted on two standard face recognition databases demonstrate the superiority of the proposed CDMMA.

Keywords: Face recognition, Super-resolution, Multi-manifold, Discriminant analysis, Low-resolution.

1 Introduction

Face recognition (FR), as an important biometric technology, has been an active research area for more than two decades and many promising practical face recognition systems have been developed [1, 2]. Although most current FR systems can achieve advanced levels under controlled environments, e.g., the face region is large enough and contains sufficient information for recognition, it is well known that they confront with low-resolution (LR) problems [3]. In many surveillance scenarios, in which the individual of interest is often times far away from the cameras, a captured face image is usually LR and it lacks detailed facial features, which are of vital importance to face recognition.

In this paper, we study the problem of how to matching a LR probe image to a high-resolution (HR) gallery of enrolled faces. There are three standard approaches to this problem: (i) Downsample the entire gallery and then perform

X. He et al. (Eds.): MMM 2015, Part I, LNCS 8935, pp. 37–48, 2015.
© Springer International Publishing Switzerland 2015

matching in LR face image space; (ii) Utilize the super-resolution (SR) or inter-polation algorithm to obtain higher resolution version from the probe image and then perform matching; (iii) Reduce the gap in appearance of different resolution and then apply the traditional face recognition methods on the common space.

The first class is very intuitive and simple. Recently, many learning based super-resolution (LBSR) methods [4,5,8–14] have been developed to predict the corresponding high-resolution (HR) face from one input LR face to overcome LR problems. For example, with the assistance of training samples, Baker and Kanade [4] propose *"face hallucination"* to infer the HR face image from an input LR one under a Bayesian formulation. This is the pioneering work on hallucinating face image. Since the introduction of this work, a number of different methods and models have been introduced. With the assumption that the HR image patches and LR image patches share the same local geometric structure, Change et al. [15] utilize locally linear embedding (LLE) [16] to learn the optimal reconstruction weights of K LR base elements to estimate the target HR patch representations. Most recently, inspired by the face analysis research which states that the position information is very important for face analysis and synthesis, Ma et al. [9] propose a position-patch based face hallucination method by performing collaboratively over the whole training image patches of the same position. To overcome the unstable solution of least squares problem in [9], Jung et al. [10] and Jiang et al. [11, 12] propose to utilize sparse regularization and locality to obtain the optimal reconstruction weights for face hallucination respectively. In [13, 14], Jiang et al. further propose an iterative neighbor embedding strategy to solve face super-resolution problem.

However, although above-mentioned methods are the best and efficient in the sense of reduced mean square error (MSE) and high peak signal to noise ratio (PSNR), these methods aim at obtaining visually appealing results; moreover, these time-consuming SR methods are not suitable for real-world FR systems. Therefore, these existing LBSR based FR methods are unsuitable for the LR problems.

Inspired by some recent work of manifold learning [16], many manifold learning based common space methods have been proposed [17–19]. Huang et al. [17] apply the canonical correlation analysis (CCA) to perform a nonlinear mapping between HR and LR face images in a coherent space. Li et al. [18] propose a coupled locality preserving mapping (CLPM) method to address LR problems from the perspective of manifold learning. However, CLPM seeks to model the face space in a single manifold through an unsupervised manner (dropping the label information) and identify neighbors according to Euclidean distance, and this will easily connect samples belonging to different manifolds (Note that, following [20], we assume that faces of the same person lie on the same manifold and those of different persons are associated with different manifolds), especially those near the intersection of different manifolds. Thus, it may drop some discriminative information helpful in classification.

In order to pursue discriminability, Wilman et al. [19] propose a discriminative super-resolution (DSR) method, which incorporates the class label information

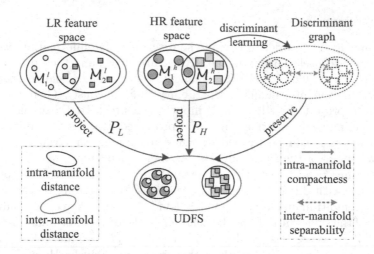

o and ● are the LR faces and corresponding HR faces of one person

▫ and ☐ are the LR faces and corresponding HR faces of another person

Fig. 1. A conceptual illustration of the proposed approach

for SR. However, DSR treats all samples in the same manner during the calculation of within-class and between-class distances and thus captures the global
geometric structure, dropping the local structure of manifold, which has recently
been shown to be effective for the classification problem [21].

To handle those drawbacks mentioned above, we develop a novel discriminative subspace learning method, namely coupled discriminant multi-manifold
analysis (CDMMA). Due to the fact that different observation sets (e.g., HR
faces and LR faces) might be located in different high-dimensional spaces, it is
difficult to match the data in their original observation spaces. However, as these
observations are from the same object, it is reasonable to assume that some common representations can be explored by in an underlying unified space. Following
CLPM, we adopt to use two explicit mappings to project the LR and HR faces
into a unified discriminative feature space (UDFS) as shown in Fig. 1. Different
from CLPM, our proposed CDMMA simultaneously learns from the neighboring information as well as the local geometric structure implied by the samples
(identifying the neighbors of each sample from the manifold space rather than
the entire Euclidean space) to minimize the intra-manifold distance and maximize the inter-manifold distance so that more discriminant information can be
exploited for FR. Experiments on standard FR databases demonstrate the effectiveness of the proposed method.

2   Coupled Discriminant Multi-Manifold Analysis

This section presents formulation of the proposed CDMMA method. It then
describes the optimization algorithm.

2.1 Preliminaries

As shown in Fig. 1, in order to exploit the discriminative information, we first learn the similarity relationship in the HR feature space, and then preserve the relationship (dotted line) for UDFS, enhancing the intra-manifold compactness and maximizing the inter-manifold separability simultaneously. Therefore, what we want is to project the HR and LR input images to a UDFS in such a way that the discriminative information in the projected space is maximized. The projection matrices can be learned by using a training set consisting of HR and LR face pairs of the same subjects. As mentioned above, we assume that faces of the same person lie on the same manifold and those of different persons are associated with different manifolds. Hence, N HR training images can be denoted by a multi-manifold space $\mathcal{M}^h = \{\mathcal{M}_c^h\}_{c=1}^C$, and the corresponding N LR images of the same subjects can be denoted by the other multi-manifold space $\mathcal{M}^l = \{\mathcal{M}_c^l\}_{c=1}^C$, where C is the class number, $\mathcal{M}_c^h = \{I_i^h\}_{i=1}^{N_c}$ and $\mathcal{M}_c^l = \{I_i^l\}_{i=1}^{N_c}$ are the HR and LR manifold of the c-th person respectively. Here, N_c is the sample number of the c-th person in the training set, $\sum_{c=1}^C N_c = N$. Let x_h and x_l denote the features of the HR and LR images respectively. Given training data, our goal is to find two projection matrices P_H and P_L to make the distances between the projected HR and LR feature vectors in the new feature space as close as possible, which means the discriminative information is maximized.

2.2 Discriminant Graph on Multi-Manifold

In CPLM [18], it directly preserves the local geometric structure of the HR feature space in an unsupervised manner, neglecting the label information and the multi-manifold structure of the training samples. Therefore, in the similarity graph of CPLM, samples on different manifolds may be connected, which will diffuse information across manifolds and be misleading.

Inspired by some recent work on spectral graph learning [22], in this letter, we try to explore the intrinsic geometric structure of the samples on multi-manifold in a supervised way to the task of detecting multiple low-dimensional manifolds embedded in high-dimensional space. The motivation of our method is to keep the label information and local geometric structure simultaneously after graph preserving (making it intrinsically different from CLPM [18] and DSR [19]). In other words, in the UDFS, we expect that the samples are still close if they are from the same manifold, and samples from different manifolds are as far from each other as possible. To this end, we define two similarity graphs on the multi-manifold, i.e., an intra-manifold similarity graph based on intra-manifold neighbors and an inter-manifold similarity graph based on inter-manifold neighbors. While a large number of graph construction approaches have been proposed in the computer vision communities recently [23], we apply the effective and efficient k-nearest-neighbor (k-NN) method to calculate the similarity matrices, intra-manifold similarity matrix W^w and inter-manifold similarity matrix W^b, as follows:

$$w_{ij}^w = \begin{cases} \frac{1}{n_w+1} & \text{if } x_i^h \in N_{n_w}(x_j^h) \text{ or } x_j^h \in N_{n_w}(x_i^h) \\ 0 & \text{otherwise} \end{cases}, \tag{1}$$

$$w_{ij}^b = \begin{cases} \frac{1}{n_b} & \text{if } x_i^h \in N_{n_b}(x_j^h) \text{ or } x_j^h \in N_{n_b}(x_i^h) \\ 0 & \text{otherwise} \end{cases}, \tag{2}$$

where $N_{n_w}(x_i^h)$ is the local set comprising n_w nearest neighbors of x_i^h belonging to the same person, while $N_{n_b}(x_i^h)$ is the set of n_b nearest neighbors of x_i^h belonging to other persons.

2.3   The Objective Function

Let $f : \mathbb{R}^D \to \mathbb{R}^d$ denote the mapping from the $\mathbb{R}^D$ input space to the $\mathbb{R}^d$ embedded space, D and d are their dimensions. We consider the mapping $f = (f_1, ..., f_d)$ as a linear combination of t basis functions of the following form

$$f_i(x; P) = \sum_{j=1}^{t} p_{ji}\phi_j(x), \tag{3}$$

where $\phi_j(x)$, $j = 1, ...t$, can be a linear or non-linear function, and $[P]_{ij} = p_{ij}$ is the weight matrix of size $t \times d$ to be determined. The mapping defined by (3) can be written in a more compact manner as follows

$$f(x; P) = P^{\mathrm{T}}\phi(x). \tag{4}$$

Here, "T" denotes the transpose of a matrix. Let P_H and P_L denote the projection matrices for the HR and LR feature vectors respectively. The basis functions for the HR and LR feature vectors are given by

$$\phi^h(x^h) = [\phi_1^h(x^h), \cdots, \phi_{t_h}^h(x^h)]^{\mathrm{T}}, \tag{5}$$

$$\phi^l(x^l) = [\phi_1^l(x^l), \cdots, \phi_{t_l}^l(x^l)]^{\mathrm{T}}, \tag{6}$$

where t_h and t_l represent the numbers of basis functions of the HR and LR feature vectors respectively, which can be same or different. Thus the mappings corresponding to the LR and HR images can be written as

$$f(x^h; P_H) = P_H^{\mathrm{T}}\phi^h(x^h), \tag{7}$$

$$f(x^l; P_L) = P_L^{\mathrm{T}}\phi^l(x^l). \tag{8}$$

The goal of CDMMA is to transform the feature vectors of the HR and LR images to make the distance between the transformed feature vectors which is called as corresponding preserving objective (J_{CP}) as close as possible. So we want to find the projection matrices P_H and P_L which minimizes the following objective function

$$J_{CP}(P_H, P_L) = \sum_{i=1}^{N} \left\| P_H^{\mathrm{T}} \phi^h(x_i^h) - P_L^{\mathrm{T}} \phi^l(x_i^l) \right\|^2. \tag{9}$$

As discussed above, in order to improve the classification performance, the objective function in (9) should be modified to include discriminative information of the training samples. Thus a discriminative analysis objective (J_{DA}) can be added to the J_{CP}, resulting in the following objective function:

$$\begin{aligned}
J(P_H, P_L) &= J_{CP}(P_H, P_L) + J_{DA}(P_H, P_L) \\
&= \sum_{i=1}^{N} \left\| P_H^{\mathrm{T}} \phi^h(x_i^h) - P_L^{\mathrm{T}} \phi^l(x_i^l) \right\|^2 \\
&+ \alpha \sum_{i=1}^{N} \sum_{j=1}^{N} \left\| P_H^{\mathrm{T}} \phi^h(x_i^h) - P_L^{\mathrm{T}} \phi^l(x_j^l) \right\|^2 w_{ij}^w \quad, \\
&- \beta \sum_{i=1}^{N} \sum_{j=1}^{N} \left\| P_H^{\mathrm{T}} \phi^h(x_i^h) - P_L^{\mathrm{T}} \phi^l(x_j^l) \right\|^2 w_{ij}^b
\end{aligned} \tag{10}$$

where α is a scalar controlling the relative contribution of intra-manifold compactness term and β is a scalar controlling the relative contribution of inter-manifold separability term.

2.4   Implementation

To solve the objective function (10), we define $\phi^h(X^h) = [\phi^h(x_1^h), \cdots, \phi^h(x_N^h)] \in \mathbb{R}^{t_h \times N}$, $\phi^l(X^l) = [\phi^l(x_1^l), \cdots, \phi^l(x_N^l)] \in \mathbb{R}^{t_l \times N}$. Here, $\phi^h(x_i^h) \in \mathbb{R}^{t_h}$ and $\phi^l(x_i^l) \in \mathbb{R}^{t_l}$ are the HR and LR feature vectors. We consider the parameterization of P_H and P_L in the forms of $P_H = V\phi^h(X^h)$ and $P_L = U\phi^l(X^l)$ respectively. With some algebraic deductions, the objective function (10) can be simplified as

$$\begin{aligned}
J(P_H, P_L) &= \mathrm{tr}\left(\begin{bmatrix} U \\ V \end{bmatrix}^T \begin{bmatrix} (\phi^l(X^l))^T & 0 \\ 0 & (\phi^h(X^h))^T \end{bmatrix} \begin{bmatrix} \phi^l(X^l) & 0 \\ 0 & \phi^h(X^h) \end{bmatrix} \right. \\
&\times \begin{bmatrix} I + \alpha D_L^w - \beta D_L^b & -I - \alpha W^w + \beta W^b \\ -I - \alpha W^w + \beta W^b & I + \alpha D_H^w - \beta D_H^b \end{bmatrix} \\
&\times \left. \begin{bmatrix} \phi^l(X^l) & 0 \\ 0 & \phi^h(X^h) \end{bmatrix} \begin{bmatrix} (\phi^l(X^l))^T & 0 \\ 0 & (\phi^h(X^h))^T \end{bmatrix} \begin{bmatrix} U \\ V \end{bmatrix} \right) \\
&= \mathrm{tr}\left(\begin{bmatrix} U \\ V \end{bmatrix}^T \begin{bmatrix} K^l & 0 \\ 0 & K^h \end{bmatrix} Q \begin{bmatrix} K^l & 0 \\ 0 & K^h \end{bmatrix}^T \begin{bmatrix} U \\ V \end{bmatrix} \right) \\
&= \mathrm{tr}(A^T Z Q Z^T A)
\end{aligned} \tag{11}$$

where $K^l = (\phi^l(X^l))^{\mathrm{T}} \phi^l(X^l)$ and $K^h = (\phi^h(X^h))^{\mathrm{T}} \phi^h(X^h)$, they can be Gaussian kernel functions, $Q = \begin{bmatrix} I + \alpha D_L^w - \beta D_L^b & -I - \alpha W^w + \beta W^b \\ -I - \alpha W^w + \beta W^b & I + \alpha D_H^w - \beta D_H^b \end{bmatrix}$, $Z = \begin{bmatrix} K^l & 0 \\ 0 & K^h \end{bmatrix}$, $A = \begin{bmatrix} U \\ V \end{bmatrix}$, I is the identity matrix, the diagonal matrices D_l^w and D_h^w are defined based on the weight matrix as $[D_L^w]_{ii} = \sum_j w_{ij}^w$ and $[D_H^w]_{jj} = \sum_i w_{ij}^w$ respectively. Similarly, $[D_L^b]_{ii} = \sum_j w_{ij}^b$ and $[D_H^b]_{jj} = \sum_i w_{ij}^b$. Therefore, U and V can be seen as the new projection matrices in the UDFS.

We introduce the scaling constraints, which remove an arbitrary scaling factor in the projection, and solve the optimization problem as

$$\min_A J(P_H, P_L) \quad \text{s.t.} \quad A^T Z Z^T A = I. \tag{12}$$

We will now switch to a Lagrangian formulation of the problem. The Lagrangian is as follows

$$L = tr(A^T Z Q Z^T A) - \lambda A^T Z Z^T A \tag{13}$$

Requiring that the gradient of L vanish gives the following eigenvector problem:

$$Z Q Z^T A = \Lambda Z Z^T A. \tag{14}$$

Therefore, given the dimension d' of the unified space, the solution of (11) with respect to A can be calculated by the d'-st smallest generalized eigenvectors of (14). Then, the obtained matrix A of size $2N \times d'$ can be divided into two projection matrices $U \in \mathbb{R}^{N \times d'}$ and $V \in \mathbb{R}^{N \times d'}$. Note that the matrix ZZ^T is usually noninvertible, and can be adjusted to $ZZ^T + \tau I$, where τ is a small positive value (say $\tau = 10^{-6}$).

The features of the LR face set $F^l = P_L \phi^l(X^l) = U^T K^l$ and the HR face set $F^h = P_H \phi^h(X^h) = V^T K^h$ are all coupled and mapped into the UDFS. In the testing phase, the kernel feature of a LR face probe x_t^l is expressed as $\phi^l(x_t^l)$. Then it is projected into the UDFS

$$f_t^l = P_L^T \phi^l(x_t^l) = \sum_{i=1}^N u_i (\phi^l(x_i^l))^T \phi^l(x_t^l) = U^T k^l(\bullet, \, x_t^l), \tag{15}$$

where $k^l(\bullet, \, x_t^l) = [(\phi^l(x_1^l))^T \phi^l(x_t^l), \cdots, (\phi^l(x_N^l))^T \phi^l(x_t^l)]$. Finally, we apply the k-NN classifier ($k=1$) to query the identity of the HR face registration list F^h.

3   Experiments

In this section, we describe the details of extensive experiments performed to evaluate the usefulness of the proposed method to face recognition.

3.1   Dataset Configuration

To investigate the performance of CDMMA, two standard FR databases are used: Extended Yale-B [24, 25] and CMU PIE face [26].

Extended Yale-B database contains 16128 face images of 38 subjects under 9 posed and 64 illumination conditions. In our experiment, we choose the frontal pose and use all the images under different illumination, thus we get 2414 image in total. All the face images are manually aligned and cropped. They are resized to 32×28 pixels, with 256 gray levels per pixel.

CMU PIE face database contains 68 individuals with 41368 face images as a whole. The face images were captured by 13 synchronized cameras and 21 flashes, under varying pose, illumination and expression. In our experiment, one near fontal poses (C27) are elected under different illuminations, lighting and expressions which leaves us 49 near frontal face images for each individual. Similarly, they are resized to 32×28 pixels, with 256 gray levels per pixel. Some samples are shown in Fig. 2.

Fig. 2. Samples of Extended Yale-B database

For each database, we randomly divide it into two equal parts. One half (containing 32 images in Extended Yale-B database and 25 images in CMU PIE face database for each person) is used as the training set (HR gallery), and the other half (LR probe is smoothed down-sampled by an averaging filter H of size 4×4 and down-sampled by a factor of 4 to 8×7 pixels) is used for testing. Some samples are shown in Fig. 3.

Fig. 3. CMU PIE face database

3.2   Comparison with Existing Methods

We compare CDMMA with CLPM [18], DSR [19] and some state-of-the-art two-step LBSR methods, which include Bicubic, Wang *et al.*'s eigentransformation based super-resolution (ESR) [5], Ma *et al.*'s position patch based super-resolution (PSR) [9] and Yang *et al.*' sparse representation based super-resolution (SSR) [8]. For CDMMA and CLPM, we directly use the face images themselves as features to be classified (we normalize each vector to unit in all the experiments). For these two-step LBSR methods, we use the PCA initial dimension reduction to obtain the feature representations. Note that we can use more discriminative features and FR engines to test our method. We simply use the PCA coefficients and 1-NN classifier in this work. For CDMMA, we empirically set $n_w \in [4,7]$, $n_b \in [50,200]$, $\alpha \in [4,40]$, $\beta \in [1,10]$. From our empirical study, good performance can be achieved when β/α takes a value between 0.2 and 0.5. This indicates that the intra-manifold compactness term is more important than the inter-manifold separability term. The kernel mapping ϕ of CDMMA is set to identity (i.e., $\phi(x) = x$) to highlight the performance improvement due to the proposed learning approach.

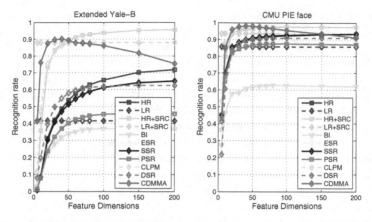

Fig. 4. Recognition results with different dimensions on Extended Yale-B database (Left) and CMU PIE face database (Right)

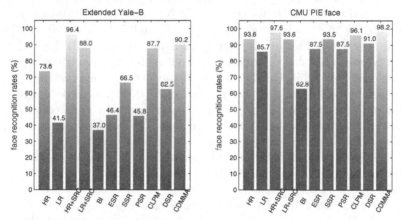

Fig. 5. Recognition results with different dimensions on Extended Yale-B database (Left) and CMU PIE face database (Right)

In addition, the performance of directly comparing HR version of probe images against HR gallery images and comparing LR probe images with down-sampled gallery images are given as baselines for comparison, which are denoted in bold **HR** and **LR** respectively. Recently, sparse representation based classification (SRC) methods [27, 28] have been widely used for FR. We also compare the performance of HR and LR gallery images with SRC FR engine, which are denoted as **HR+SRC** and **LR+SRC** respectively. Note that, in practical, the cases **HR** and **HR+SRC** do not exist given the LR observation.

Results and Analysis: In Fig. 4, the average recognition rates (all the experiments are repeated 10 times) with different feature dimensions are drawn. On the other hand, the mean recognition rates (in percents) of these 10 runs with optimal dimension are reported in Fig. 5. We can observe that:

(i) Compared with the HR image, there is a significant drop (32.1% in Extended Yale-B while 7.9% in CMU PIE face) in recognition accuracy from the LR image. This indicates that it is unwise to perform matching in the LR image space by down-sampling the gallery of HR face images.

(ii) Not all LBSR methods benefit the FR task. Existing face SR algorithms such as PSR and ESR can slightly improve the performance compared with **LR** method, but are not significant, whereas Bicubic interpolation (can be seen as a super-resolution method) drops the performance. Note that SSR gains a similar performance of the idea case **HR** and this is due to the sparse prior, which has been proven to be effective for image recognition task;

(iii) CLPM and DSR are better than existing LBSR methods except SSR and this can be attributed to the introduction of discriminative information in those methods. Through SSR is a two-step method, it can reconstruct very discriminative faces with the introduction of sparsity and surpass DSR;

(iv) The proposed CDMMA method significantly improves the performance when compared to the LBSR approaches and outperforms CLPM and DSR. When compared with CLPM and DSR, the improvement of CDMMA is around 2.5% and 27.7% in Yale-B database and 2.1% and 7.2% in CMU PIE face database respectively. This can be mainly explained by that CDMMA successfully incorporates the label information as well as local geometric structure of the training samples, while CLPM and DSR consider only one certain aspect;

(v) It can be seen that the performance of CDMMA decreased when the dimension of the feature rose. This seems to be not normal compared with other methods (e.g., the methods used to be compared) whose performance usually increases when the feature dimension increases. We attribute this to the exitance of shared UDFS of LR and HR training samples. This also proves the rationality of projecting the original LR and HR faces to the UDFS.

(vi) Last but not least, we list the performance of SRC based methods (**LR**+ SRC and the idea case **HR**+SRC) similarly in Fig. 4 and Fig. 5. They all exceed the **LR** and **HR** methods with 1-NN classifier especially in Extended Yale-B database. In addition, the performance of CDMMA is better than that of **LR**+SRC (2.2% improvement in Extended Yale-B while 4.6% improvement in CMU PIE face), and even better than that of **HR**+SRC in CMU PIE face database (0.6% improvement).

4   Conclusion

In this letter, we have proposed a novel coupled discriminant multi-manifold analysis (CDMMA) method to address the low-dimensional FR problem. Specifically, our approach employs the neighborhood information as well as local geometric structure of the manifold and derives two coupled mappings on multi-manifold space to project LR and HR feature spaces to UDFS, simultaneously maximizing a measure of discriminatory power and preserving the geometrical structure of the manifold. Experimental results on the same database have demonstrated the superiority of the proposed method over some state-of-the-art methods.

Finally, note that although all the experiments in this letter deals with low to high resolution FR problem, the proposed approach is very general and can be applied to any other heterogeneous face recognition (HFR) problem, such as photo-sketch recognition, near infrared (NIR) faces to visual (VIS) faces recognition.

Acknowledgement. The research was supported by the Nationl Key Technologies R&D Program (2013AA014602), the National Natural Science Foundation of China (61231015, 61172173, 61303114, 61170023),the technology Research Program of Ministry of Public Security (2014JSYJA016), the major Science and Technology Innovation Plan of Hubei Province (2013AAA020), the Guangdong-Hongkong Key Domain Breakthrough Project of China (2012A090200007), the China Postdoctoral Science Foundation funded project (2013M530350), and the Specialized Research Fund for the Doctoral Program of Higher Education (20130141120024), the Key Technology R&D Program of Wuhan (2013030409020109).

References

1. Zhao, W., Chellappa, R., Phillips, P.J., Rosenfeld, A.: Face recognition: A literature survey. ACM Comput. Surv. 35(4), 399–458 (2003)
2. Zhang, X., Gao, Y.: Face recognition across pose: A review. Pattern Recogn. 42(11), 2876–2896 (2009)
3. Hennings-Yeomans, P., Baker, S., Kumar, B.: Simultaneous super-resolution and feature extraction for recognition of low-resolution faces. In: CVPR 2008, pp. 1–8 (2008)
4. Baker, S., Kanade, T.: Hallucinating faces. In: Proc. IEEE Conf. on Automatic Face and Gesture (FG), pp. 83–88 (2000)
5. Wang, X., Tang, X.: Hallucinating face by eigentransformation. IEEE Trans. Systems, Man, and Cybernetics. Part C 35(3), 425–434 (2005)
6. Jia, K., Gong, S.: Multi-modal tensor face for simultaneous super-resolution and recognition. In: ICCV, pp. 1683–1690 (2005)
7. Park, J.-S., Lee, S.-W.: An example-based face hallucination method for single-frame, low-resolution facial images. IEEE Trans. Image Process. 17(10), 1806–1816 (2008)
8. Yang, J., Tang, H., Ma, Y., Huang, T.: Face hallucination via sparse coding. In: ICIP, pp. 1264–1267 (2008)
9. Ma, X., Zhang, J., Qi, C.: Hallucinating face by position-patch. Pattern Recogn. 43(6), 2224–2236 (2010)
10. Jung, C., Jiao, L., Liu, B., Gong, M.: Position-patch based face hallucination using convex optimization. IEEE Signal Proc. Let. 18(6), 367–370 (2011)
11. Jiang, J., Hu, R., Han, Z., Lu, T., Huang, K.: Position-patch based face hallucination via locality-constrained representation. In: ICME 2012, pp. 212–217 (2012)
12. Jiang, J., Hu, R., Wang, Z., Han, Z.: Noise robust face hallucination via locality-constrained representation. IEEE Trans. on Multimedia 16(5), 1268–1281 (2014)
13. Jiang, J., Hu, R., Han, Z., Wang, Z., Lu, T., Chen, J.: Locality-constraint iterative neighbor embedding for face hallucination. In: ICME, pp. 1–6 (2013)
14. Jiang, J., Hu, R., Wang, Z., Han, Z.: Graph regularized sparse coding for image representation. IEEE Trans. Image Process. 23(10), 4220–4231 (2014)

15. Chang, H., Yeung, D., Xiong, Y.: Super-resolution through neighbor embedding. In: CVPR, pp. 275–282 (2004)
16. Roweis, S.T., Saul, L.K.: Nonlinear dimensionality reduction by locally linear embedding 290(5500), 2323–2326 (2000)
17. Huang, H., He, H.: Super-resolution method for face recognition using nonlinear mappings on coherent features. IEEE Trans. Neural Netw. 22(1), 121–130 (2011)
18. Li, B., Chang, H., Shan, S., Chen, X.: Low-resolution face recognition via coupled locality preserving mappings. IEEE Signal Proc. Let. 17(1), 20–23 (2010)
19. Wilman, W.Z., Yuen, P.C.: Very low resolution face recognition problem. IEEE Trans. Image Process. 21(1), 327–340 (2012)
20. Goldberg, A.B., Zhu, X., Singh, A., Xu, Z., Nowak, R.: Multi-manifold semi-supervised learning. Journal of Machine Learning Research - Proceedings Track, 169–176 (2009)
21. Zheng, M., Bu, J., Chen, C., Wang, C., Zhang, L., Qiu, G., Cai, D.: Graph regularized sparse coding for image representation. IEEE Trans. Image Process. 20(5), 1327–1336 (2011)
22. Chung, F.: Spectral graph theory. In: Proc. Regional Conf. Series in Math.
23. Zhu, L., Zhu, S.: Face recognition based on orthogonal discriminant locality preserving projections. Neurocomputing 70(7-9), 1543–1546 (2007)
24. Georghiades, A., Belhumeur, P., Kriegman, D.: From few to many: illumination cone models for face recognition under variable lighting and pose. IEEE Trans. Pattern Anal. Mach. Intell. 23(6), 643–660 (2001)
25. Chih Lee, K., Ho, J., Kriegman, D.: Acquiring linear subspaces for face recognition under variable lighting. IEEE Trans. Pattern Anal. Mach. Intell. 27(5), 684–698 (2005)
26. Gross, R., Matthews, I., Cohn, J., Kanade, T., Baker, S.: Multi-pie, Image and Vision Computing 28(5), 807–813 (2010)
27. Wright, J., Yang, A., Ganesh, A., Sastry, S., Ma, Y.: Robust face recognition via sparse representation. IEEE Trans. Pattern Anal. Mach. Intell. 31(2), 210–227 (2009)
28. Yang, A.Y., Ganesh, A., Zhou, Z., Sastry, S., Ma, Y.: A review of fast l1-minimization algorithms for robust face recognition, CoRR abs/1007.3753

Text Detection in Natural Images Using Localized Stroke Width Transform

Wenyan Dong, Zhouhui Lian[*], Yingmin Tang, and Jianguo Xiao

Institute of Computer Science and Technology, Peking University, Beijing, China
lianzhouhui@pku.edu.cn

Abstract. How to effectively and efficiently detect texts in natural scene images is a challenging problem. This paper presents a novel text detection method using localized stroke width transform. Due to the utilization of an adaptive image binarization approach and the implementation of stroke width transform in local regions, our method markedly reduces the demand of contrast between texts and backgrounds, and becomes considerably robust against edge detection results. Experiments on the dataset of ICDAR 2013 robust reading competition demonstrate that the proposed method outperforms other state-of-the-art approaches in the application of text detection in natural scene images.

Keywords: Text detection; Image binarization; Stroke width transform.

1   Introduction

Text detection in natural scene images is of great practical significance. It can be applied to automatically extract text information and thus the computer can better interpret high-level semantics included in the images. However, quickly and accurately detecting texts in scene images is not trivial, mainly due to the fact that scene images usually contain lots of components, complex background, non-uniform illumination changes, and great variations of text font, size and orientation.

Up to now, a large amount of text detection algorithms have been proposed in the literature [1-9]. Generally, they can be classified into three categories: texture-based methods [1-3], region-based methods [4-6], and hybrid methods [7]. For instance, Kim et al. [3] introduced a texture-based method for detecting texts in images. They utilized a support vector machine (SVM) to analyze the textural information of texts which are represented by the intensities of raw pixels. One major problem of texture-based methods is that texture analyzing for scene images is typically very time-consuming. On the contrary, region-based approaches do not need to deal with all locations and scales of target images. Therefore, they are usually more efficient than the texture-based methods. Among all existing region-based approaches, the most well-known one might be the method proposed in [4] where Epshtein et al. [4] developed an image operator named Stroke Width Transform (SWT) to calculate the value of stroke width for each

---

[*] Corresponding author.

X. He et al. (Eds.): MMM 2015, Part I, LNCS 8935, pp. 49–58, 2015.

image pixel. Based on the fact that one of the most important feature which distinguishes the text from other components in a scene image is its approximately constant stroke width, the SWT operator can be applied to detect texts in natural scene images. However, when handling images with some complicated background, these kinds of region-based methods might also become computationally expensive. The other intuitive way to improve the performance of text detection is to combine different kinds of methods together. For example, Pan et al. [7] presents a hybrid method to detect and localize texts in scene images. Specifically, they first segmented candidate text components using a region-based text detector and then filtered out non-text components by a conditional random field (CRF) model. Hybrid methods perform well in the application of text detection by integrating advantages of different kinds of methods. But their performance depends largely on the manual parameter tuning or the training data used in automatic machine learning. Recently, Yao et al. [8] reported an algorithm to detect texts of arbitrary orientations in natural images and achieved the state-of-the-art performance. SWT [4] and some other features that are discriminative to separate texts from other components in images were selected in their method. However, the calculation time of SWT processing is costly when handling images with complex background.

To solve the above-mentioned problems, this paper proposes a new text detection method using localized stroke width transform (LSWT). Specifically, the proposed method implements an adaptive image binarization processing and applies SWT in local regions. During adaptive image binarization, an intersection surface is generated based on the edge map and gray map to binarize the image and thus make it easier for our method to pick up text regions. After obtaining these candidate regions, the localized stroke width transformation (LSWT) is carried out in each region to extract text features. Finally, non-text candidates are filtered out using several specially-designed text filters including a novel virtual nib filter which can effectively eliminate some kinds of non-text components that are hard to be recognized by existing methods. Experiments on the dataset of ICDAR 2013 robust reading competition (Task 2.1: text localization in real scene images) [9] demonstrate that the proposed approach obtains the state-of-the-art results and outperforms all methods evaluated in the contest in terms of Precision and F-score metrics.

2   Method Description

As shown in Fig. 1, our method consists of the following six steps: image preprocessing, adaptive image binarization, candidates generating, stroke width computing, candidates filtering and candidates to text lines. Details of each step are explicitly described in section 2.1 to section 2.6.

2.1   Image Preprocessing

First of all, images containing excessive amounts of pixels should be down-sampled. Here, upper limits of the width and height of the image are chosen as 1600 and 1200,

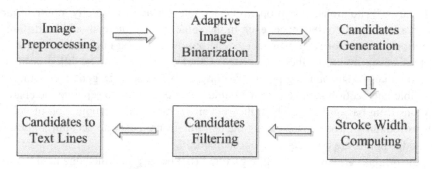

Fig. 1. The pipeline of our method

respectively. Then, the image is converted into a gray image using the algorithm proposed in [10] to reserve details with high contrast. Afterwards, Canny edge detector [11] is applied to generate an edge map for the original image (e.g., Fig. 6(a)). Moreover, the image should be smoothed again if the edge pixel rate is too high and the edge detection parameters will be automatically adjusted until the rate is lower than a given threshold. Finally, we obtain an edge map that is well-suited to be utilized in other steps of our method.

2.2 Adaptive Image Binarization

In this step, we convert the gray image to a binary image via a novel method called adaptive image binarization. With the guidance of edge map, an intersection surface is constructed to cut the gray image and classify all pixels into two categories to generate the binary image, from which it is much easier to extract character regions.

In general, there exist great differences between text areas and their backgrounds. For example, values between text and non-text pixels in a gray map often change suddenly. If we consider the gray map as a 3-dimensional surface that contains a set of "mountains" (e.g., Fig. 2(b)), typically text regions are locally conspicuous "peaks" or "valleys". Thus, character regions can be perfectly segmented from the image by cutting those "mountains" with a proper intersection surface.

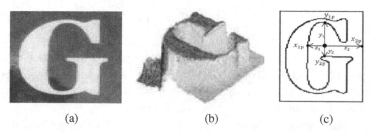

(a)   (b)   (c)

Fig. 2. The gray image patch (a), its 3D appearance (b), and its edge map (c) generated from a scene image

For various situations, the proper intersection surface is usually non-constant curved surface. As a consequence, even in the same picture there should be different intersection surfaces for each character. The basement of the intersection surface is the set of edges points in the gray image. Other positions will be filled with the weighted values of surrounding points. We originally intended to generate a smoothest possible intersection surface with the guidance of edge map to separate the character regions from backgrounds, which theoretically requires the computation of values for all surrounding edge points. Taking the computational cost into account, only four nearest edge points are considered in our method.

Let the original gray image patch (e.g., Fig. 2(b)) be Gg and the edge map be Ge. To make sure that every pixel would find its corresponding edge points in all directions, the four boundaries of edge map are marked with 1 as well. Gs which denotes the skeleton map of the intersection surface can be calculated by,

$$Gs_{(i,j)} = Gg_{(i,j)} Ge_{(i,j)} \tag{1}$$

Only the values of edge points in gray images are reserved and others are set to be 0. The intersection surface Gp is initialized to be equal to Gs, and the non-edge positions are filled with the weighted value of four nearest edge points respectively in horizontal and vertical, positive and negative directions. Specifically, Gp can be calculated by

$$Gp_{(i,j)} = \frac{x_{1p}x_2y_1y_2 + x_1x_{2p}y_1y_2 + x_1x_2y_{1p}y_2 + x_1x_2y_1y_{2p}}{x_2y_1y_2 + x_1y_1y_2 + x_1x_2y_2 + x_1x_2y_1} \tag{2}$$

where x_1, x_2, y_1 and y_2 are the distances between the current point and nearest edge points in four directions, x_{1p}, x_{2p}, y_{1p} and y_{2p} are the values of four edge points in gray map (see Fig. 2(c)), and $Gp_{(i,j)}$ denotes the value in the jth column of the ith row in the intersection surface.

Let $Gd = Gp - Gg$ be the difference map between Gp and Gg. In order to deal with two different text appearing situations, namely, bright texts on dark backgrounds and dark texts on bright backgrounds, we generate two binary images by,

$$Gb^1{}_{(i,j)} = \begin{cases} 1, & Gd_{(i,j)>pt} \\ 0, & Gd_{(i,j)\leq pt} \end{cases}, \qquad Gb^2{}_{(i,j)} = \begin{cases} 1, & Gd_{(i,j)<-nt} \\ 0, & Gd_{(i,j)\geq -nt} \end{cases} \tag{3}$$

where pt and nt denote the positive and negative thresholds which are both positive constants, and Gb^1 and Gb^2 denote the two binarization results, respectively. The following steps of our method are performed on these binary image pairs and the final output is the combination of two text detection results.

Here are some examples (shown in Fig. 3) of our adaptive image binarization results. As we can see, the method is considerably insensitive to illumination variance.

Fig. 3. Some results of our adaptive image binarization

2.3   Candidates Generating

After previous steps, the text region and the background are labeled by the value of 0 and 1, respectively. Then we associate pixels to form connected components using a simple rule that whether they are in four-connection and with value 1 in the binary image. Afterwards, a filter consisting of a set of simple rules runs on these connected pixels to eliminate some obvious non-text regions. In our experiments, a connected component will be filtered out as the non-text region if it consists of too few or too many pixels, or the rate between the width and height of its bounding box is too large or too small. Finally, remaining ones (Fig. 6(c)) are considered as candidates to be classified further.

2.4   Stroke Width Computing

SWT [4], adopted in our method due to its effectiveness and efficiency, is a local image operator which computes per pixel's width of the most likely stroke containing the pixel. The method proposed in [4] calls for perfect character edge detection: too many edge lines would markedly affect the stroke width computation, while too few edge lines would result in the leakage of stroke rays. What is more, points in crossing areas can hardly find the point with inverse gradient ray whose direction difference is less than $\pi/4$. In this paper, the restriction of the gradient direction difference is relaxed from $\pi/4$ to $\pi/2$ and a step is added to discard scattered single lines to reduce interference. Most importantly, stroke width values are calculated in local areas for each character candidates; hence it would be much more accurate than the computation in the whole image due to less mutual interference. As clustering completed, we compute stroke width values for each candidate and gather them in the SWT map to effectively eliminate neighbor edges' interference and improve the calculation accuracy (see a normalization result in Fig. 6(d)).

The Localized Stroke Width Transform (LSWT) is a new idea proposed in our paper that makes the computed stroke width results more genuine and believable. Fig. 4 shows a comparison between the original SWT [4] and our localized method. The ratio of SWT values of neighboring pixels is less than 3.0 and it's 1.8 in our method which is closer to the actual situations.

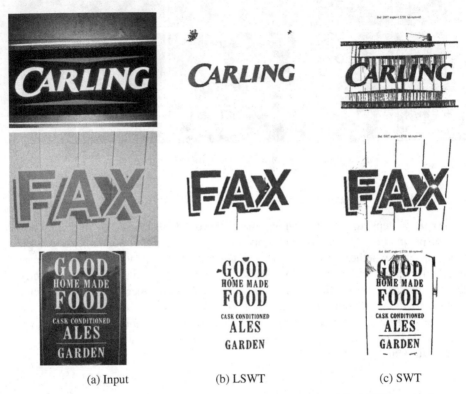

(a) Input   (b) LSWT   (c) SWT

Fig. 4. Comparison between our LSWT and the original SWT

2.5   Candidates Filtering

The purpose of this step is to identify and eliminate non-text candidates. The feature vectors of each candidate are computed, including data of the candidates' stroke width values, their bounding boxes: width and the height of bounding box; mean value, variance and median of all stroke width values, the total of character pixels. A filter consisting of a set of heuristic rules runs on a collection of statistical and geometric properties of candidates, which is very fast to compute. The candidates with one or more invalid properties will be taken as non-text regions and discarded. The statistical feature items adopted in the filter are listed as follows:

- Rate between variance and mean value
- Width rate between the bounding box and whole image
- Height rate between the bounding box and whole image

- Rate between the total of character pixels and bounding box area
- Rate between median stroke value and the diagonal of bounding box
- Rate between   the width and the height of bounding box
- Rate between median stroke value and the width of bounding box
- Reserved pixel rate after applying the virtual nib filter

The virtual nib filter can be considered as a roundness whose diameter is equal to the average stroke width of the candidate. The filter is effective due to the fact that text symbols are made of strokes with approximately constant width values. The scanning step length of the virtual nib is selected as the radius of the nib to ensure that all pixels of each candidate can be scanned. The nib moves horizontally and vertically over the candidate. In each scan, the superposition pixels are reserved if the rate of superposition and nib pixels is higher than the threshold. For some candidates in scattered distribution whose statistical data is similar to text, their finally remaining pixel rates are much lower than text candidates as shown in Fig. 5. Therefore, by limiting the pixel rate between candidates and their bounding box area, the virtual nib filter is able to eliminate a lot of non-text candidates.

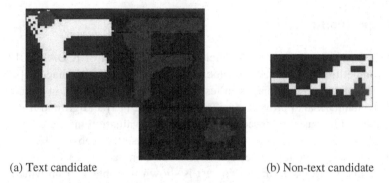

(a) Text candidate   (b) Non-text candidate

Fig. 5. Scanning procedure of the virtual nib for a text candidate (a) and a non-text candidate (b), respectively; red ball, yellow part and red part denote the virtual nib, original character and reserved part after scanning, respectively.

Through large amounts of experiments, we found that over 88% text candidates in our dataset picked up by local stroke width transform have more than one SIFT key point [12] and the majority of ones without SIFT key point are the special characters "I" , "l" or "i". Special characters possess a high rate between the total of character pixels and bounding box area, and relatively large rate between the height and width of bounding box. We train a SVM classifier with each character' pixel ratio, width and height of bounding box as input to filter out most remaining non-text candidates which have no SIFT key point and don't belong to special characters either. Candidates obtained after filtering are shown in Fig. 6(e).

2.6   Candidates to Text Lines

Candidates are aggregated into chains at this stage which also serves as a filtering step because the candidates cannot be linked into chains are treated as non-text components formed by noises or background clutters, and thus should be discarded. Here we use a similar approach as the algorithm introduced in [8].

The whole method is implemented twice for the binary image pair to handle two text appearing situations mentioned above. The final output is the combination of two text detection results (see Fig. 6(f)).

(a)   (b)   (c)   (d)   (e)   (f)

Fig. 6. Outputs obtained after each step. original input image (a), binary map (b), candidate map after clustering (c), SWT map (d), candidate map after filtering (e), the final result (f).

3   Experiments

In our experiments, we adopt the publicly available dataset of ICDAR 2013 robust reading competition (Task 2.1: text localization in real scene images) [9] which contains 462 pictures (233 images in test dataset and 229 images in training dataset). The benchmark also provides evaluating website [13] and ground truth files to evaluate the results based on three commonly-used evaluation measures (i.e., Precision, Recall and F-score). In this paper, we directly use the website [13] to evaluate the performance of our text detection approach. Comparison of our method with approaches that participated in this contest is shown in Table 1, from which we can see that our method gets the best results in both Precision and F-score among all other methods.

We also implement the original SWT approach and our method on the same platform (Matlab 2012a, Windows 7, CPU i5-2400, 4GB RAM) and compare their performance on the dataset mentioned above. As shown in Table 2, it is clear that our local-based method performs faster and more effectively.

Examples of both true and false detection results are shown in Fig.7. As we can see, our method performs well on normal pictures even with unbalanced illumination to some degree (see Fig.7(a-c)). However, there also exist various false detection results. For example, over-unbalance illumination destroys the original texts' edges and color distribution leads to false detection (see Fig.7(d)). Some handwriting characters connected as one part may lead to edge detection failure for individual characters (see Fig.7(e)). Finally, over-low contrast between the characters and their backgrounds also presents a problem for our method (see Fig.7(f)). We would like to work on these problems in the future.

Table 1. Comparison of our method with methods that took part in Task 2.1 of ICDAR 2013 robust reading competition

	Recall	*Precision*	*F-score*
Our Method	**67.82 %**	**86.56 %**	**76.05 %**
USTB_TexStar	66.45 %	88.47 %	75.89 %
CASIA_NLPR	68.24 %	78.89 %	73.18 %
Text_detector_CASIA	62.85 %	84.70 %	72.16 %
Epshtein et al.[4]	57.70 %	57.31 %	57.50 %
Baseline	34.74 %	60.76 %	44.21 %

Table 2. Comparison of our method and the original SWT method [4]

	F-score	*Time (sec.)*
Our Method	**76.05 %**	87
Epshtein et al. [4]	57.50 %	126

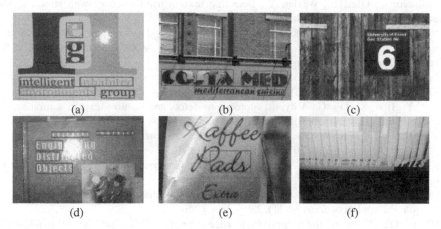

(a)   (b)   (c)

(d)   (e)   (f)

Fig. 7. Examples of text detection results of our method; true detections on the first row (a-c) and false detections on the second row (d-f)

4   Conclusion

In this paper, a text detection method is proposed using localized stroke width transform to localize texts in natural scene images, in which we apply adaptive image binarization and implement stroke width transform in local regions. The proposed text detection method is intuitive and practical, and not restricted to specific languages and fonts. We also develop a new filter called virtual nib filter to eliminate some kinds of non-text components that are hard to be recognized by existing approaches. Experiments on the dataset of ICDAR 2013 robust reading competition (Task 2.1) [9] demonstrate that the proposed method obtains state-of-the-art performance in the application of text detection in natural scene images.

Acknowledgements. This work was supported by National Natural Science Foundation of China (Grant No.: 61202230), National Hi-Tech Research and Development Program (863 Program) of China (Grant No.: 2014AA015102) and China Postdoctoral Science Foundation (Grant No.: 2013T60038).

References

1. Chen, X., Yuille, A.: Detecting and reading text in natural scenes. In: Proc. CVPR 2014 (2004)
2. Gllavata, J., Ewerth, R., Freisleben, B.: Text detection in images based on unsupervised classification of high-frequency wavelet coefficients. In: Proc. ICPR 2004, pp. 425–428 (2004)
3. Kim, K.I., Jung, K., Kim, J.H.: Texture-based approach for text detection in images using support vector machines and continuously adaptive mean shift algorithm. IEEE Trans. PAMI 25(12), 1631–1639 (2003)
4. Epshtein, B., Ofek, E., Wexler, Y.: Detecting text in natural scenes with stroke width transform. In: Proc. CVPR 2010, pp. 2963–2970 (2010)
5. Neumann, L., Matas, J.: A method for text localization and recognition in real-world images. In: Kimmel, R., Klette, R., Sugimoto, A. (eds.) ACCV 2010, Part III. LNCS, vol. 6494, pp. 770–783. Springer, Heidelberg (2011)
6. Jain, A., Yu, B.: Automatic text location in images and video frames. Pattern Recognition 31(12), 2055–2076 (1998)
7. Pan, Y., Hou, X., Liu, C.: A hybrid approach to detect and localize texts in natural scene images. IEEE Trans. IP 20(3), 800–813 (2011)
8. Yao, C., Bai, X., Liu, W., Ma, Y., Tu, Z.: Detecting texts of arbitrary orientations in natural images. In: Proc. CVPR 2012, pp. 1083–1090 (2012)
9. Karatzas, D., et al.: ICDAR 2013 Robust Reading Competition. In: Proc. ICDAR 2013, pp. 1484–1493 (2013)
10. Grundland, M., Dodgson, N.: Decolorize: fast, contrast enhancing, color to grayscale conversion. Pattern Recognition 40(11), 2891–2896 (2007)
11. Canny, J.F.: A computational approach to edge detection. IEEE Trans. PAMI 6, 679–698 (1986)
12. Lowe, D.: Object recognition from local scale-invariant features. In: Proc. ICCV 1999, pp. 1150–1157 (1999)
13. http://dag.cvc.uab.es/icdar2013competition/

Moving Object Tracking
with Structure Complexity Coefficients

Yuan Yuan[1], Yuming Fang[2], and Lin Weisi[1]

[1] School of Computer Engineering,
Nanyang Technological University, Singapore
{yyuan004,wslin}@ntu.edu.sg
[2] School of Information Technology,
Jiangxi University of Finance and Economics, Nanchang, China
leo.fangyuming@gmail.com

Abstract. Target appearance change during tracking is always a challenging problem for visual object tracking. In this paper, we present a novel visual object tracking algorithm based on Structure Complexity Coefficients (SCC) in addressing the motion related appearance change problem fundamentally. Based on our careful analysis, we found that the motion related appearance change is quite related to the SCC of target surface, where the appearance of complex structural regions is easier to change comparing with that of smooth structural regions with target motion. With the proposed SCC, a SCC-GL distance is defined in addressing both the appearance change and occlusion related problems during tracking. Moreover, an Observation Dependent Hidden Markov Model (OD-HMM) framework is designed where the observation dependency between neighboring frames is considered comparing with the standard HMM based tracking framework. The observation dependency is computed with the proposed SCC. We also present a novel outlier removing method in appearance model updating to avoid error accumulation. Experimental results on various challenging video sequences demonstrate that the proposed observation dependent tracker (**ODT**) achieves better performance than existing related tracking algorithms.

Keywords: Tracking, Structure Complexity, HMM, Outlier Removing.

1 Introduction

Visual object tracking is one of the most important techniques in smart vision systems. In most of the recent tracking studies, tracking process are treated as a recursive loop of two steps: 1) predicting target candidates in the incoming frame based on the previous target state (motion model); 2) choosing the most proper candidate to be the target based on the appearance likelihood (appearance model). Normally, since the target appearance changes gradually between neighboring frames, the target could be accurately localized in most of the frames with a relatively high appearance likelihood. However, if the target appearance changes dramatically with some sudden movements, the target could be assigned

X. He et al. (Eds.): MMM 2015, Part I, LNCS 8935, pp. 59–70, 2015.

with a low likelihood and results in drifting when the target is not distinguishable among the candidates.

With 11 normal attributes listed in [1] which may affect the tracking performance, target appearance change during tracking is mainly correlated with motion related attributes (i.e., rotation, deformation and motion blur) and illumination variation. To address illumination variation problem, existing studies have investigated illumination invariant features like adjunct pixels color ratio [2] and locality sensitive histogram [3]. In this study, we mainly focus on appearance change caused by target motion. To deal with this problem, several studies have been investigated such as finding stable local features [4,5], building appearance subspace robust to multi-view [6,7], learning part-based appearance models [8,9,10], etc. Notwithstanding the appearance change problem is somehow addressed by these methods, they hardly consider the fundamental reason of appearance change, MOTION.

When analysing video sequences, we find that the main factor of motion related appearance change is interference between moving pixels in a local region. Based on our careful analysis, we can predict that, the moving target appearance in complex structural regions is easier to be changed than that in smooth structural regions. Fig. 1(c) and 1(g) demonstrate that the pixel-wise reconstruction error of moving targets in complex structural regions is larger than that in smooth structural regions, where larger error indicates larger appearance change. Thus, in this study, we define Structure Complexity Coefficients (SCC) to predict the target appearance stability (negative related to potential appearance change) during tracking. And then, the proposed SCC is applied to compensate the pixel-wise reconstruction error by considering the reconstruction error in complex structural regions less than that in smooth structural regions. Fig. 1 demonstrates that the reconstruction error in t^{th} frame is somehow negatively correlated with the target SCC computed from $(t-1)^{th}$. This idea is also similar to contrast sensitivity of human visual system, the change in smooth structural region is more sensitive

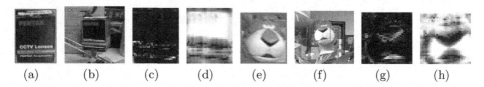

(a)	(b)	(c)	(d)	(e)	(f)	(g)	(h)

Fig. 1. Reconstruction Error Map and SCC with target movement: (a) and (e) are targets to be tracked; (b) and (f) are targets with motion blur and rotation; (c) and (g) are pixel-wise reconstruction error maps in t^{th} frame (the dark region indicates less error while the light region indicates large error); (d) and (h) are SCCs of the targets in $(t-1)^{th}$ frame (the dark region indicates complex structural region which takes high probability to change its value with motion while the light region indicates smooth structural regions which will be more stable with motion). The reconstruction error map and SCC are somehow negatively correlated, which means the SCC can be used to predict and compensate the reconstruction error caused by target motion.

than the change in in complex structural regions [11,12]. After the compensation, the target is more distinguishable among the candidates.

For simplicity, conventional HMM based object tracking studies assume that the observations between different image frames are independent and the appearance likelihood only depends on the appearance model [6,13,7,3,14]. In [15], a general training method is proposed for HMM with multiple observations where the dependency property between the observations can be characterized regarding to different applications. In object tracking, AR-HMM [16] is proposed where the dependency between neighboring frames is trained frame by frame based on the all the previous estimated target appearances. However, the error accumulation and computational complexity problems in online training can not be ignored for AR-HMM. In our study, the observation dependency is investigated with the proposed SCC by analysing the fundamental reason of motion related appearance change. Since only one frame is processed in computing SCC, the computational complexity of our proposed tracking algorithm is quite acceptable comparing to existing related algorithms.

To demonstrate the effectiveness of the proposed SCC and OD-HMM framework for visual object tracking, we design an observation dependent tracker (**ODT**) in this paper. Since Gaussian-Laplacian (GL) noise has been demonstrated to be effective in handling outliers for object tracking [13,17,8], we propose an appearance model with SCC-GL distance to measure the likelihood of candidates. Besides, a novel model updating method is presented to remove the outliers in addressing the error accumulation problem in appearance model updating. Experimental results demonstrate that the proposed algorithm can obtain much better performance in visual tracking on various challenging video sequences.

2   OD-HMM Based Tracking Framework

Let $\mathbf{x}_t$ be the state variable of the target at time t. Given a set of observed target appearance $\mathbf{y}_{1:t} = \{\mathbf{y}_1, \mathbf{y}_2, ...\mathbf{y}_t\}$ from the first to t-th frames, the aim is to estimate $\mathbf{x}_t$ by MAP estimation, $\hat{\mathbf{x}}_t = \arg\max_{\mathbf{x}_t} p(\mathbf{x}_t|\mathbf{y}_{1:t})$. Taking the Bayes rule into consideration, the posterior probability can be decomposed as:

$$p(\mathbf{x}_t|\mathbf{y}_{1:t}) \propto p(\mathbf{y}_t|\mathbf{x}_t, \mathbf{y}_{1:t-1})p(\mathbf{x}_t|\mathbf{y}_{1:t-1}). \tag{1}$$

In most HMM based tracking studies [6,13,3], the observed appearance between different frames are assumed to be independent, and the appearance model $p(\mathbf{y}_t|\mathbf{x}_t, \mathbf{y}_{1:t-1})$ is simplified to be $p(\mathbf{y}_t|\mathbf{x}_t)$.

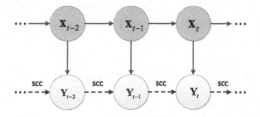

Fig. 2. Observation Dependent HMM based on Structure Complexity Coefficients

To compute the appearance likelihood more accurately, we model the tracking process as an OD-HMM by considering the observation dependency between consecutive frames. As demonstrated in Fig. 2, the observation likelihood of $\mathbf{y}_t$ not only depends on the related state $\mathbf{x}_t$, but also depends on the SCC computed from $\mathbf{y}_{t-1}$. Specifically, SCC is calculated to predict the appearance change of the tth frame with the observed appearance in the $(t-1)^{th}$ frame. Thus, within the OD-HMM framework, the posterior probability is computed as:

$$p(\mathbf{x}_t|\mathbf{y}_{1:t}) \propto p(\mathbf{y}_t|\mathbf{x}_t,\mathbf{y}_{t-1}) \int_{\mathbf{x}_{t-1}} p(\mathbf{x}_t|\mathbf{x}_{t-1})p(\mathbf{x}_{t-1}|\mathbf{y}_{1:t-1}) \qquad (2)$$

3 Appearance Stability Prediction with SCC

During object tracking, the target appearance may change dramatically due to sudden movements, and this causes reconstruction likelihood of target decrease and thus result in drifting. Since the appearance change caused by motion can be represented by interference between moving pixels in a local region, we define SCC to predict the appearance stability of different regions of the moving target based on the structure complexity of the target appearance. Here, we analyse appearance change from: rotation, deformation and motion blur.

3.1 Appearance Change Caused by Rotation

Rotation occurs frequently in object tracking. It is a global motion, which means that most of the pixels of the target shift together with the same direction and degree. By defining a certain degree and direction (i.e., motion vector defined as (m_x, m_y)) of the rotation, the pixel value of the target patch will be shifted and the pixel value variation at (x, y) can be computed as:

$$d(m_x, m_y, x, y) = \mathbf{I}(x+m_x, y+m_y) - \mathbf{I}(x, y), \qquad (3)$$

where $\mathbf{I}$ is the target patch, and (m_x, m_y) is the motion vector. By assuming a uniform distribution of the movement within a maximum velocity, the variance (i.e., expected square of the pixel value change) of the pixel value between consecutive frames can be defined as:

$$\mathrm{Var}_r(\mathbf{I}) = \sum_{m_x^2+m_y^2 \leq v_m^2} \frac{d(m_x, m_y, x, y)^2}{N_d}, \qquad (4)$$

where v_m denotes the maximum moving region between consecutive frames and N_d is the number of pixels within a near neighbour $m_x^2 + m_y^2 \leq v_m^2$. With predicted variance $\mathrm{Var}_r(\mathbf{I})$, the appearance stability $P_r(\mathbf{I})$ with rotation can be modeled as a normal distribution, $P_r(\mathbf{I}) = \exp(-\frac{\mathrm{Var}_r(\mathbf{I})}{\sigma_r^2})$, where σ_r^2 is the parameter of Gaussian kernel.

3.2 Appearance Change Caused by Deformation

Another attribute of motion related appearance change is deformation. In this case, the motion is local and varies between different regions of target appearance. For example, the facial expression variation is a kind of deformation and the movements of different parts of face are independent. Since the local movement is random, the surrounding pixels closer to center have higher possibilities to affect the center pixel. Thus, we consider a two-dimensional normal distribution within a maximum velocity to model the local movements, the variance of the pixel value between consecutive frames can be defined as:

$$\text{Var}_d(\mathbf{I}) = \sum_{m_x^2 + m_y^2 \le v_m^2} d(m_x, m_y, x, y)^2 P(m_x, m_y), \tag{5}$$

where v_m denotes the maximum region of pixels moving between consecutive frames and $P(m_x, m_y) = \mathcal{N}(0, 0, v_m^2/9, v_m^2/9, 0)$. With predicted variance $\text{Var}_d(\mathbf{I})$, the appearance stability $P_d(\mathbf{I})$ with deformation can be modeled as a normal distribution, $P_d(\mathbf{I}) = \exp(-\frac{\text{Var}_d(\mathbf{I})}{\sigma_d^2})$, where σ_d^2 is the parameter of Gaussian kernel.

3.3 Appearance Change Caused by Motion Blur

With fast motion of target or camera, pixel values of target appearance is the fusion results of radiant energy from a relatively large region, and they cannot precisely reflect the original appearance of the target. Consider with a simple movement in $x+$ direction, we can predict the blurred pixel value as $\mathbf{I}'(x, y) \propto \int_0^{\Delta T} I(x - v(t), y)dt \propto \sum_{x'=0}^{\Delta X} \mathbf{I}(x - x', y)$ [18], where $\mathbf{I}(x, y)$ is the original image, $\mathbf{I}'(x, y)$ is the predicted blurred one; $v(t)$ indicates the relative velocity; ΔT is the shutter time, and $\Delta X = v(t)\Delta T$.

Assuming an uniform distribution of the movements, for simplicity, we compute $\text{Var}_b(\mathbf{I})$ based on the average of motion blurred differences in x^+, x^-, y^+, and y^- directions.

$$\text{Var}_b(\mathbf{I}) = \frac{\text{Var}_b^{x^+}(\mathbf{I}) + \text{Var}_b^{x^-}(\mathbf{I}) + \text{Var}_b^{y^+}(\mathbf{I}) + \text{Var}_b^{y^-}(\mathbf{I})}{4}, \tag{6}$$

$$\text{Var}_b^{x^+}(\mathbf{I}) = [\frac{\sum_{n=0}^{v_m} \mathbf{I}(x - n, y)}{v_m + 1} - I(x, y)]^2, \text{Var}_b^{x^-}(\mathbf{I}) = [\frac{\sum_{n=0}^{v_m} \mathbf{I}(x + n, y)}{v_m + 1} - I(x, y)]^2,$$

$$\text{Var}_b^{y^+}(\mathbf{I}) = [\frac{\sum_{n=0}^{v_m} \mathbf{I}(x, y - n)}{v_m + 1} - I(x, y)]^2, \text{Var}_b^{y^-}(\mathbf{I}) = [\frac{\sum_{n=0}^{v_m} \mathbf{I}(x, y + n)}{v_m + 1} - \mathbf{I}(x, y)]^2,$$

where v_m denotes the assumed moving distance between consecutive frames. With predicted variance $\text{Var}_b(\mathbf{I})$, the appearance stability $P_b(\mathbf{I})$ with motion blur can be modeled as a normal distribution, $P_b(\mathbf{I}) = \exp(-\frac{\text{Var}_b(\mathbf{I})}{\sigma_b^2})$, where σ_b^2 is the parameter of Gaussian kernel.

3.4 Structure Complexity Coefficients

Looking into the fundamental reason of motion related appearance change, rotation results in pixel shift by global motion; deformation results in pixel shift by local motion; and blur results in pixel integration by fast global motion. Since these movements are independent, we formulate the appearance stability of the moving target based on the joint probability:

$$P(\mathbf{I}) = P_r(\mathbf{I}) * P_d(\mathbf{I}) * P_b(\mathbf{I}) = exp(-\frac{\text{Var}_r(\mathbf{I})}{\sigma_r^2} - \frac{\text{Var}_d(\mathbf{I})}{\sigma_d^2} - \frac{\text{Var}_b(\mathbf{I})}{\sigma_b^2}). \tag{7}$$

With the larger value in $P(\mathbf{I})$, the related pixel or region is more stable with target movement.

Motion Blur Detection: Since motion blur is detectable with Power Spectrum Slope (PSS) [19], we then decide the parameter σ_b based on the blur degree of the target. With lager motion blur, σ_b is set to be smaller to make P_b contributing more to P, and vice versa.

$$\sigma_b \propto \frac{1}{\max(q_k, \tau)}, \quad q_k = \frac{\alpha_k - \alpha_1}{\alpha_1}. \tag{8}$$

where τ is a small value to prevent dividing by zero or a negative value when $\alpha_k \le \alpha_1$ (we set τ to be 0.0001). In Eq. 8, α_k is the PSS of the target patch in frame k. Since PSS is content biased in representing blur degree, by assuming the objects are not blurred in the first frame, we derive a unique metric q_k by comparing α_k against α_1.

In this study, we set the values of σ_r and σ_d empirically in experiments.

Normalization of SCC: Finally, we normalize the probability of stability to formulate the final SCC metric, $\mathbf{C}(\mathbf{I}) = \frac{N_P \times P(\mathbf{I})}{\sum P(\mathbf{I})}$, where $\mathbf{C}(\mathbf{I})$ is SCC of the image patch I, and N_P is the number of elements in $P(\mathbf{I})$.

4 Observation Dependent Tracker with SCC

Gaussian-Laplacian noise have been demonstrated to be effective in dealing with outliers for object tracking [13,17,8], thus we model the observation vector $\mathbf{y}$ as a linear model with Gaussian-Laplacian reconstruction error:

$$\mathbf{y} = \mathbf{Ba} + \mathbf{n} + \mathbf{s}, \tag{9}$$

where $\mathbf{y} \in \mathbb{R}^{d \times 1}$ is a d-dimensional vector, $\mathbf{a} \in \mathbb{R}^{m \times 1}$ denotes the estimated m-dimensional parameter vector and $\mathbf{B} \in \mathbb{R}^{d \times m}$ represents the input data matrix (the row vector is the subspace basis in this study). Specifically, $\mathbf{n} \in \mathbb{R}^{d \times 1}$ indicates the Gaussian noise term and $\mathbf{s} \in \mathbb{R}^{d \times 1}$ indicates the Laplacian noise term. The Gaussian noise is used to model small dense noise and the Laplacian one aims to handle outliers.

4.1 SCC-Gaussian-Laplacian Distance

With Gaussian-Laplacian reconstruction error assumption, given an observation vector $\mathbf{y}$, the joint likelihood $p(\mathbf{y}, \mathbf{B})$ is defined with the optimized $\mathbf{a}$ and $\mathbf{s}$:

$$p(\mathbf{y}, \mathbf{B}) = K \exp\{-\frac{d_{GL}(\mathbf{y}, \mathbf{B})}{\sigma_N^2}\}, \quad d_{GL}(\mathbf{y}, \mathbf{B}) = \min_{\mathbf{a}, \mathbf{s}} \frac{1}{2}||\mathbf{y} - \mathbf{Ba} - \mathbf{s}||_2^2 + \lambda||\mathbf{s}||_1, \tag{10}$$

where σ_N^2 is the variance of the Gaussian error term, λ is regularization constant of Laplacian error terms and K is the normalization constant.

As shown in Eq. 10, each element in Gaussian or Laplacian noises is treated equally in L2 and L1 normalization. However, with the prediction of appearance stability by SCC, the reconstruction error of different elements should be treated

Fig. 3. Illustration with GL and SCC-GL distances, red box is a good candidate and green one is a bad candidate

differently and this means that if one pixel is predicted to be unstable with a high probability, the noise should be less considered when the final likelihood is calculated. Thus, we formulate the SCC-GL based joint reconstruction likelihood as:

$$p(\mathbf{y}, \mathbf{B}, \mathbf{C}) = K \exp\{-\frac{d_{SCC}(\mathbf{y}, \mathbf{B}, \mathbf{C})}{\sigma_N^2}\},$$

$$d_{SCC}(\mathbf{y}, \mathbf{B}, \mathbf{C}) = \min_{\mathbf{a},\mathbf{s}} \frac{1}{2}(\mathbf{y} - \mathbf{B}\mathbf{a} - \mathbf{s})^\top \mathbf{W_n}(\mathbf{y} - \mathbf{B}\mathbf{a} - \mathbf{s}) + \lambda ||\mathbf{s}\mathbf{W_s}||_1,$$

(11)

where $\mathbf{W_n}$ and $\mathbf{W_s}$ are two diagonal matrixes, which denote the error tolerance of Gaussian and Laplacian noises respectively. With the computed SCC in Section 3.4, we define $\mathbf{W_n}$ and $\mathbf{W_s}$ as $\mathbf{W_n} = \mathtt{diag}(\mathbf{C})$ and $\mathbf{W_s} = \sqrt{\mathtt{diag}(\mathbf{C})}$ respectively, where $\mathbf{C}$ is the computed SCC and $\mathtt{diag}(\mathbf{C})$ converts $\mathbf{C}$ to a diagonal matrix. We solve the optimization problem with ℓ_1-regularization inspired by [13,8].

Fig. 3 illustrates a tiger toy tracking example. In Frame 84 where a moving tiger toy exists, the SCC-GL distance of the good candidate (red bounding box) is smaller than the bad candidate (green bounding box), while the GL distance of the good candidate is larger than the bad one. It demonstrate that the algorithm using SCC-GL distance performs better in tracking moving target than that with GL distance alone.

4.2 Implementation of ODT with SCC-GL

The visual tracking task can be formulated as a Bayesian estimation problem, as demonstrated in Eq. 2.

In this work, we model the transition between consecutive frames Gaussian distribution with a diagonal covariance matrix $\sigma_\mathbf{x}$, $p(\mathbf{x}_t|\mathbf{x}_{t-1}) = \mathcal{N}(\mathbf{x}_t; \mathbf{x}_{t-1}, \sigma_\mathbf{x})$.

By assuming the target appearance is generated by a PCA subspace spanned by $\mathbf{U}$ and centered at μ, with Gaussian-Laplacian noise assumption, an observation candidate $\mathbf{y}$ is formulated as $\mathbf{y} = \mu + \mathbf{U}\mathbf{z} + \mathbf{n} + \mathbf{s}$, where $\mathbf{n}$ and $\mathbf{s}$ are Gaussian and Laplacian noise terms respectively. Given an observation candidate $\mathbf{y}_t^i$ corresponding to a predicted state variable $\mathbf{x}_t^i$, based on the discussions in Sections 4.1, the observation likelihood of the i-th candidate belonging to the target is measured by:

$$p(\mathbf{y}_t^i|\mathbf{x}_t^i, \mathbf{y}_{t-1}) = \exp(-\frac{d_{SCC}(\bar{\mathbf{y}}^i, \mathbf{U}, \mathbf{C}(\hat{\mathbf{y}}_{t-1}))}{\sigma^2}), \qquad (12)$$

where $\bar{\mathbf{y}}^i = \mathbf{y}^i - \mu$, $\hat{\mathbf{y}}_{t-1}$ is the estimated target appearance in the previous frame and σ is a constant controlling the variance of the Gaussian Kernel.

As analysed in [13,17,8], the Laplacian noise term represents the outliers. Thus, in order to eliminate the negative effect of outlier, before model updating, we reconstruct the estimated observation vector $\hat{\mathbf{y}}_t$ in each frame by subtracting its related Laplacian noise $\hat{\mathbf{s}}_t$:

$$\mathbf{y}_{recon}^t = \hat{\mathbf{y}}_t - \hat{\mathbf{s}}_t, \qquad (13)$$

where $\mathbf{y}_{recon}^t$ is the reconstructed observation vector in t-th frame and is used for updating the subspace center μ and basis matrix $\mathbf{U}$. The updating of μ and $\mathbf{U}$ is based on incremental principal component analysis (IPCA) [6].

5 Experiments

To evaluate the performance of OD-HMM and SCC in visual object tracking, we implement an observation dependent tracker (**ODT**) in MATLAB. The regularization constant λ is fixed to be 0.1. Simultaneously, the variance σ_r^2 and σ_d^2 are fixed to be 0.1, while σ_b^2 varies based on the motion blur degree. Each image observation is resized to a 32×32 image patch. The maximum moving distance between consecutive frames v_m is empirically fixed to be 4 in all the experiments. 16 eigenvectors are used to represent the PCA subspace of the target appearance. In order to obtain both efficiency and accuracy, the particle number is set as 600 and the PCA subspace is updated every 5 frames.

In this Section, we have conducted comparison experiments between the proposed **ODT** and 7 recent state-of-art algorithms, including **LSST** [13], **iVT** [6], **SCM** [14], **TLD** [20], **BLUT** [21], **Struck** [22] and **LSHT** [3]. Moreover, to verify the contribution of SCC alone, we implement **ODT-CI** by fixing the SCC matrix **C** to be an identity matrix **I**. All these algorithms are tested on thirteen published benchmark sequences with motion related challenges like rotation, deformation and motion blur. Specifically, the sequences and ground truth are collected from Visual Tracker Benchmark [1], **BLUT** [21], **PROST** [23], and **BoBoT** [24].

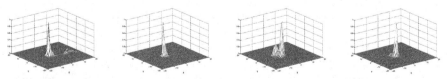

(a) GL Likelihood   (b) SCC-GL Likeli-   (c) GL Likelihood   (d) SCC-GL Likeli-
                        hood   hood

Fig. 4. The GL alone and SCC-GL likelihood maps around the target in Tiger1 sequence. (a) and (b) are from Frame 82, while (c) and (d) are from Frame 92.

Generally, the reason of losing tracking of targets is that the estimated likelihood of distracter is larger than the estimated likelihood of the target. In Fig. 4, we plot the likelihood map around the target in different frames, where X and Y axes denote the numbers of pixels shifting at the target position. It can be seen that with GL likelihood alone, a second peak occurs around the target in likelihood maps. By taking SCC into consideration, single peak likelihood maps are produced where the second peak is flattened, which makes the target more distinguishable among the candidates.

Table 1. Average per-frame overlap ratio, with the best three results shown in RED,**BLUE** and *GREEN* fonts

Sequence	ODT	ODT-CI	LSST	iVT	SCM	TLD	BLUT	Struck	LSHT
Tiger1	0.60	0.43	0.30	0.30	0.13	**0.46**	0.21	0.19	0.11
Box	**0.73**	0.65	0.68	0.70	0.25	0.13	0.13	0.82	0.34
Lemming	0.72	**0.71**	0.70	0.60	0.42	0.08	0.14	0.57	0.40
Car11	0.89	0.84	**0.84**	0.80	0.82	0.38	0.01	0.83	0.55
Body	0.63	0.71	0.68	0.08	0.71	0.52	0.67	0.18	0.68
Car4	0.89	0.74	0.83	**0.84**	**0.84**	0.72	0.77	0.26	0.71
Face	0.86	**0.85**	0.84	**0.85**	0.84	0.71	0.83	0.27	0.60
FaceOcc2	**0.78**	**0.78**	0.77	0.70	**0.78**	0.68	0.25	0.79	0.76
Jumping	**0.66**	0.40	0.64	0.28	0.67	0.63	0.02	**0.66**	0.10
CupTable	0.85	0.82	0.85	0.80	0.84	0.57	0.17	0.71	0.73
David	**0.77**	0.78	0.76	**0.77**	0.59	0.50	0.05	0.39	0.68
David2	0.82	0.81	0.52	0.74	0.74	0.61	0.02	0.88	**0.86**
Caviar2	0.75	0.61	**0.80**	0.61	0.81	0.43	0.63	0.57	0.33

5.1   Quantitative Evaluation

We use the overlap ratio criteria in performance evaluation. In Table 1, the average per-frame overlap ratio, $\tilde{r} = \frac{\sum_{t=1}^{N_q} \frac{area(R_T \cap R_G)}{area(R_T \cup R_G)}}{N_q}$, is demonstrated, where N_q is the total number of frames, and R_T and R_G are the estimated and ground truth bounding boxes respectively.

It is clear that for most sequences with motion related challenges, the proposed **ODT** algorithm ranks top three in performance for most video sequences comparing with recent studies. The processing time of the proposed **ODT** is tested on a PC with Intel E5-1650 CPU (3.2 GHz) and 16 GB memory. It runs at 5 frames per second on average, and this is comparable with existing related algorithms.

5.2   Qualitative Evaluation

In Figure 5, we show some sample frames of the comparison experiment between the proposed **ODT** and relevant existing algorithms. We give some analysis regarding to different challenges below.

Rotation: The sequences used in comparison experiments with target rotation include *David2, Box, Lemming, David, Tiger1* and *CupTable*. In most

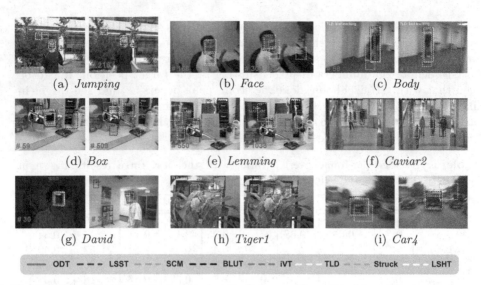

(a) *Jumping*   (b) *Face*   (c) *Body*

(d) *Box*   (e) *Lemming*   (f) *Caviar2*

(g) *David*   (h) *Tiger1*   (i) *Car4*

ODT — — — LSST — — SCM — — — BLUT — — — iVT — — — TLD — — Struck — — — LSHT

Fig. 5. Visual illustrations of some sample frames with motion blur, deformation, rotation, occlusion, etc.

rotation related sequences, the proposed **ODT** can obtain accurate tracking results by predicting the appearance change caused by rotation. Since rotation only changes the target appearance in some edge regions between consecutive frames, with local sparse representation, **Struck** and **LSHT** work well in some rotation sequences like *CupTable* and *Box*. **LSST** and **iVT** also perform well in some rotation sequences due to the robustness of the PCA in multi-view representation.

Motion Blur: In sequences of *Jumping*, *Face*, *Box*, *Lemming*, *Body* and *Car4*, the targets are blurred with different degrees by fast motion of either target or camera, which makes the target indistinguishable against its surroundings. The proposed **ODT** can obtain accurate and stable tracking results on these sequences, with blur detection and prediction. **LSST** is also able to localize the target in most frames. **SCM** performs well in some blurred sequences benefiting from its discriminative model which makes the blurred target more distinguish against the background. However, it loses tracking when there are other challenges co-existing, like occlusion and rotation in *Lemming* and *Box*. **BLUT** is specifically designed for tracking blurred objects, and it is demonstrated to perform well on *Face* and *Body*. However, without model updating, it loses tracking on other sequences with appearance variation.

Deformation: In sequences of *Body* and *Caviar2* in Fig. 5, it can be seen that the pose variation during walking causes the deformation of target pedestrians. Only the proposed **ODT**, **LSST** and **SCM** can localize the target in most frames from these two video sequences. **Struck** and **iVT** can well localize the target pedestrian in *Caviar2* since their appearance models can adjust to the appearance variation caused by deformation. However, they lose tracking on *Body* when motion blur also exists.

Occlusion: We test the tracking algorithms on sequences (*e.g.*, *Box, Lemming, FaceOcc2, Tiger1* and *Caviar2*) that the target is occluded within some time during tracking. In Fig. 5, it can be seen that the proposed **ODT** is able to obtain accurate tracking results on these sequences, since the Laplacian noise used in **ODT** can eliminate the outliers well during appearance model updating. **LSST** can also obtain good results on some sequences. However, when there are challenges like rotation and motion blur (*e.g.*, *Box* and *Tiger1*), it can not perform well stably. Since the local features are effective in dealing with partial occlusion, it can be seen that **SCM, Struck** and **LSHT** can also obtain good tracking results on some sequences with partial occlusion(*e.g.*, *Box, Caviar2* and *FaceOcc2*).

6 Conclusion

In this paper, we innovatively address the motion related appearance change problem in visual tracking by define a SCC metric to predict the appearance change. By analysing the fundamental reason of motion related appearance change, we investigate the relation between the potential appearance change and the structural complexity. Moreover, an OD-HMM based tracking framework is presented where the observation dependency is defined by the proposed SCC. An observation dependent tracker with a novel model updating mechanism is designed to test both the effectiveness and efficiency of the proposed SCC and OD-HMM. Both qualitative and quantitative comparison experiments demonstrate the better performance of the proposed **ODT** against recent state-of-art tracking algorithms on the challenging video sequences. With OD-HMM, we believe the proposed SCC can be easily integrated with other tracking algorithms as well to improve the performance in tracking moving objects.

References

1. Wu, Y., Lim, J., Yang, M.H.: Online object tracking: A benchmark. In: 2013 IEEE Conference on Computer Vision and Pattern Recognition (CVPR), pp. 2411–2418. IEEE (2013)
2. Kim, T.: k., Cho, I.M., Lee, J.H.: Illumination-invariant object tracking method and image editing system using the same, US Patent 7,171,023 (2007)
3. He, S., Yang, Q., Lau, R.W., Wang, J., Yang, M.H.: Visual tracking via locality sensitive histograms. In: 2013 IEEE Conference on Computer Vision and Pattern Recognition (CVPR), pp. 2427–2434. IEEE (2013)
4. Grabner, H., Grabner, M., Bischof, H.: Real-time tracking via on-line boosting. BMVC 1, 6 (2006)
5. Babenko, B., Yang, M.H., Belongie, S.: Visual tracking with online multiple instance learning. In: IEEE Conference on Computer Vision and Pattern Recognition, CVPR 2009, pp. 983–990. IEEE (2009)
6. Ross, D.A., Lim, J., Lin, R.S., Yang, M.H.: Incremental learning for robust visual tracking. International Journal of Computer Vision 77, 125–141 (2008)
7. Mei, X., Ling, H.: Robust visual tracking and vehicle classification via sparse representation. IEEE Transactions on Pattern Analysis and Machine Intelligence 33, 2259–2272 (2011)

8. Wang, D., Lu, H.: On-line learning parts-based representation via incremental orthogonal projective non-negative matrix factorization. Signal Processing 93, 1608–1623 (2013)
9. Kwon, J., Lee, K.M.: Highly nonrigid object tracking via patch-based dynamic appearance modeling. IEEE Transactions on Pattern Analysis and Machine Intelligence 35, 2427–2441 (2013)
10. Wu, Y., Shen, B., Ling, H.: Visual tracking via online nonnegative matrix factorization. IEEE Transactions on Circuits and Systems for Video Technology 24, 374–383 (2014)
11. Chou, C.H., Li, Y.C.: A perceptually tuned subband image coder based on the measure of just-noticeable-distortion profile. IEEE Transactions on Circuits and Systems for Video Technology 5, 467–476 (1995)
12. Ginsburg, A.: Pattern recognition techniques suggested from psychological correlates of a model of the human visual system. NAECON 73, 309–316 (1973)
13. Wang, D., Lu, H., Yang, M.H.: Least soft-threshold squares tracking. In: 2013 IEEE Conference on Computer Vision and Pattern Recognition (CVPR), pp. 2371–2378. IEEE (2013)
14. Zhong, W., Lu, H., Yang, M.H.: Robust object tracking via sparsity-based collaborative model. In: 2012 IEEE Conference on Computer Vision and Pattern Recognition (CVPR), pp. 1838–1845. IEEE (2012)
15. Li, X., Parizeau, M., Plamondon, R.: Training hidden markov models with multiple observations-a combinatorial method. IEEE Transactions on Pattern Analysis and Machine Intelligence 22, 371–377 (2000)
16. Park, D.W., Kwon, J., Lee, K.M.: Robust visual tracking using autoregressive hidden markov model. In: 2012 IEEE Conference on Computer Vision and Pattern Recognition (CVPR), pp. 1964–1971. IEEE (2012)
17. Zhang, T., Ghanem, B., Liu, S., Ahuja, N.: Robust visual tracking via multi-task sparse learning. In: 2012 IEEE Conference on Computer Vision and Pattern Recognition (CVPR), pp. 2042–2049. IEEE (2012)
18. Potmesil, M., Chakravarty, I.: Modeling motion blur in computer-generated images. ACM SIGGRAPH Computer Graphics 17, 389–399 (1983)
19. Liu, R., Li, Z., Jia, J.: Image partial blur detection and classification. In: IEEE Conference on Computer Vision and Pattern Recognition, CVPR 2008, pp. 1–8. IEEE (2008)
20. Kalal, Z., Mikolajczyk, K., Matas, J.: Tracking-learning-detection. IEEE Transactions on Pattern Analysis and Machine Intelligence 34, 1409–1422 (2012)
21. Wu, Y., Ling, H., Yu, J., Li, F., Mei, X., Cheng, E.: Blurred target tracking by blur-driven tracker. In: 2011 IEEE International Conference on Computer Vision (ICCV), pp. 1100–1107. IEEE (2011)
22. Hare, S., Saffari, A., Torr, P.H.: Struck: Structured output tracking with kernels. In: 2011 IEEE International Conference on Computer Vision (ICCV), pp. 263–270. IEEE (2011)
23. Santner, J., Leistner, C., Saffari, A., Pock, T., Bischof, H.: Prost: Parallel robust online simple tracking. In: 2010 IEEE Conference on Computer Vision and Pattern Recognition (CVPR), pp. 723–730. IEEE (2010)
24. Klein, D.A., Schulz, D., Frintrop, S., Cremers, A.B.: Adaptive real-time video-tracking for arbitrary objects. In: 2010 IEEE/RSJ International Conference on Intelligent Robots and Systems (IROS), pp. 772–777. IEEE (2010)

Real-Time People Counting across Spatially Adjacent Non-overlapping Camera Views

Ryota Akai[1], Naoko Nitta[2], and Noboru Babaguchi[2]

Graduate School of Engineering, Osaka University
akai@nanase.comm.eng.osaka-u.ac.jp,
{naoko,noboru}@comm.eng.osaka-u.ac.jp

Abstract. Counting the number of people traveling across non-overlapping camera views generally requires all persons exiting any camera view to be re-identified when they re-enter one of its spatially adjacent camera views. For their accurate re-identification, the correspondence among the exits and entries of all persons should be established so that their total correspondence confidence is maximized. In order to realize the real-time people counting, we propose to find the shortest time window to observe both the exits and entries of all persons traveling within the time window adaptively to the current people traffic flow. Further, since closely related people often travel together, the re-identification can be performed to the foreground regions to re-identify groups of people. Since the groups of people can sometimes split or merge outside the camera views, the proposed method establishes the weighted correspondence among the exits and entries of the foreground regions based on their correspondence confidence. Experimental results have shown that the adaptively determined time window was effective in terms of both the accuracy and the delay in people counting and the weighted correspondence was effective in terms of the accuracy especially when the people traffic gets congested and groups of people split/merge outside the camera views.

Keywords: people counting, non-overlapping camera views, real-time system.

1   Introduction

People counting is to measure the number and direction of people traversing a certain passage per a time period and is often performed for marketing, navigation, and crowd management purposes [1][2]. Since manual counting is very tedious and time-consuming work, computer vision techniques have been applied to surveillance cameras to automatically count the number of people who travel within a camera's field of view (FOV). A popular approach is to extract foreground regions corresponding to a group of people moving in the same directions and estimate the crowd density of the extracted regions [3][4][5][6][7]. When given a live streaming video, such approach can estimate the number of people traveling within the FOV per frame sequentially in real time. On the

X. He et al. (Eds.): MMM 2015, Part I, LNCS 8935, pp. 71–82, 2015.
© Springer International Publishing Switzerland 2015

other hand, the target of our work is to count the number of people travel-ing *outside the FOVs of multiple disjoint cameras* by estimating the number of people traversing between each pair of spatially adjacent FOVs of the cameras.

Intuitively, every person who has exited a camera view needs to be re-identified when he/she re-enters one of its spatially adjacent camera views [8]. Many ap-proaches have been proposed for tracking a person across cameras with non-overlapping FOVs [9] or for re-identifying an individual person [10][11]. However, in order to accurately re-identify all persons exiting and re-entering all spatially adjacent camera views, it would be desirable to obtain a set of exiting persons and a set of entering persons from a time window and find their correspondence so that their total correspondence confidence is maximized. Further, in order to sequen-tially provide the current people traffic flows for the live streaming video with little delay, the time window needs to be short. Thus, our proposed method firstly finds the shortest time window to observe both the exits and entries of all persons trav-eling within the time window adaptively to the current people traffic situation.

Additionally, since people often form groups with their acquaintances and keep their social distances from other groups of strangers while walking in pub-lic, we consider that each group of people exiting or entering the camera view is extracted as a foreground region from each camera view and can be re-identified as a group [12]. However, since people in a group detected in a camera view can split and merge outside the FOVs of cameras, a group of people detected as a foreground region in one camera view can be detected as several foreground regions in other camera views, which complicates their re-identification. In order to handle the split-merge problem, we propose to establish *weighted correspon-dence* among the exiting and entering persons based on the confidence levels of their correspondence and the constraints defined by the relationships among the areas of the extracted foreground regions. Finally, the number of people travers-ing between a pair of spatially adjacent FOVs of the cameras can be estimated by summing up the weighted correspondence among the persons who have exited one camera view and entered the other camera view.

The contributions of this paper can be summarized as follows;

- As far as we know, our work is the first attempt for real-time people counting for multiple spatially adjacent disjoint cameras.
- Determining the shortest time window to observe both the exits and entries of persons realizes accurate counting with little delay.
- Establishing the weighted correspondence realizes accurate counting based on simple foreground extraction.

2   People Counting across Spatially Adjacent Non-overlapping Camera Views

Assuming that an area is monitored by J cameras with *spatially adjacent non-overlapping FOVs* as shown in Fig.1, the target of our work is to count the number of people traversing across each pair of FOVs of cameras i and j, C_i and

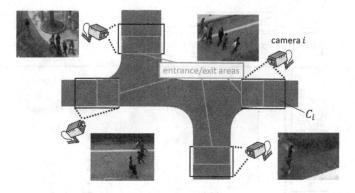

Fig. 1. Example of Multiple Cameras with Spatially Adjacent Non-Overlapping FOVs

C_j, within the previous time period of T. More concretely, we assume that the foreground regions $d_m^i (m = 1, ...)$ and $a_n^j (n = 1, ...)$ are automatically extracted from the entrance/exit areas in the camera views when persons are exiting C_i and have entered C_j and are stored as a set of exiting/entering foreground regions $D = \{d_m^i \mid i = 1, \cdots, J; \ m = 1, \cdots\}$ and $A = \{a_n^j \mid j = 1, \cdots, J; \ n = 1, \cdots\}$ respectively. The proposed approach continuously determines the shortest time window to extract the set of d_m^i and a_n^j from D and A to accurately establish their correspondence. Once the correspondence weights $L_{i,m}^{j,n}$ are estimated for all a_n^j which have entered by the time t, the number of people $V_{i,j}(t)$ traversing from C_i to C_j within the time interval from $t - T$ to t can be calculated.

Fig. 2 shows the overview of the proposed approach, which is composed of the following three steps:

1) Setting Adaptive Time Window
   The shortest time window is determined to observe both the exits and entries of all persons traveling within the time window adaptively to the current people traffic situation. The exits and entries observed within *the adaptive time window* are determined as $\hat{D} = \{d_m^i \mid i \in \{1, \cdots, J\}; \ m \in \{1, \cdots\}\}$ and $\hat{A} = \{a_n^j \mid j \in \{1, \cdots, J\}; \ n \in \{1, \cdots\}\}$, respectively.
2) Establishing Weighted Correspondence among Foreground Regions
   The correspondence among all pairs of d_m^i in $\hat{D}$ and a_n^j in $\hat{A}$ are established by estimating their correspondence weights $L_{i,m}^{j,n}$.
3) People Counting across Camera Pairs
   The number of people $V_{i,j}(t)$ traversing between the FOVs of the cameras i and j within the time interval from $t - T$ to t is calculated based on the estimated $L_{i,m}^{j,n}$.

The details of each step are described in the following subsections.

2.1   Setting Adaptive Time Window

We assume that, when persons exit from a camera view, they will enter one of its spatially adjacent camera views after a while. The travel time across the

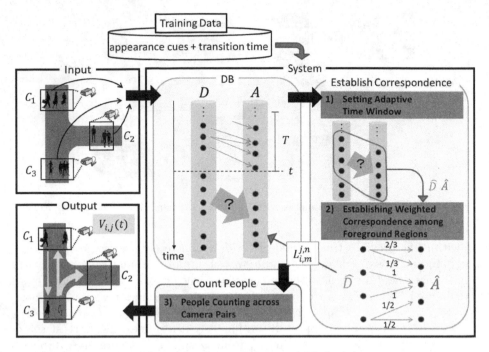

Fig. 2. Overview of Proposed Method

FOVs of the cameras depends on the distance between the FOVs and the walking speed of persons. Considering that the distance between the FOVs depend on the pair of the cameras and the walking speed of persons do not vary so much, the maximum and minimum travel time between C_i and C_j are determined as $T_{max}^{i,j}$ and $T_{min}^{i,j}$ for each pair of cameras. Then, when an exiting foreground region d_m^i is observed at the time $t_{d_m^i}$, their re-entries are expected to be observed in the view of the camera j between $T_{min}^{i,j}$ and $T_{max}^{i,j}$ after $t_{d_m^i}$. Thus, according to the d_m^i observed by the current time t_{now}, the end time of the time window t_{end} is determined so that their corresponding a_n^j are observed.

The sets of foreground regions, $\hat{D}$ and $\hat{A}$, are obtained by adaptively setting the time window as follows. Note that the start and end times of the time window, t_{start} and t_{end}, are initially set to 0.

1. t_{start} and t_{end} are set as $t_{start} = t_{end}$ and $t_{end} = \infty$. Further, $\hat{D}$ and $\hat{A}$ are initialized as the empty sets.
2. While $t_{now} < t_{end}$, whenever a_n^j is extracted, d_m^is which satisfy the condition

$$t_{d_m^i} + T_{min}^{i,j} \leq t_{a_n^j} \leq t_{d_m^i} + T_{max}^{i,j} \tag{1}$$

are determined as its candidate exits. When there are more than one candidates, a_n^j and all its candidate d_m^is are added to $\hat{A}$ and $\hat{D}$ respectively. Then, t_{end} is updated as

$$t_{end} = \max_{d_m^i \in \hat{D}, j=1,\ldots,J} (t_{d_m^i} + T_{max}^{i,j}). \tag{2}$$

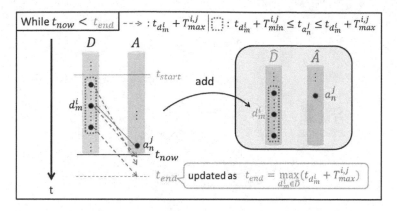

Fig. 3. How time window and foreground region sets are updated until $t_{now} = t_{end}$

Fig. 3 illustrates how the time window and the foreground region sets are updated until $t_{now} = t_{end}$.

3. When $t_{now} = t_{end}$, the weighted correspondence among the foreground regions in $\hat{D}$ and $\hat{A}$ is established as described in the next subsection. The procedure is repeated from 1. to determine the next time window.

2.2 Establishing Weighted Correspondence among Foreground Regions

Fig. 4 shows the weighted correspondence among the foreground regions in $\hat{D}$ and $\hat{A}$. Intuitively, $L_{i,m}^{j,n}$ represents the ratio of the people in d_m^i who have traveled to a_n^j. Thus, for each d_m^i, $L_{i,m}^{j,n}$ should satisfy the conditions:

$$0 \le L_{i,m}^{j,n} \le 1, \tag{3}$$

$$\sum_{a_n^j \in \hat{A}} L_{i,m}^{j,n} = 1. \tag{4}$$

In addition, when defining the areas of d_m^i and a_n^j as $S_{i,m}^d$ and $S_{j,n}^a$ respectively, the flow from d_m^i to a_n^j can be represented by $S_{i,m}^d L_{i,m}^{j,n}$. Considering that an object in the view of the camera i appears R_i^j times larger in the view of the camera j, $L_{i,m}^{j,n}$ should also satisfy the condition for each a_n^j:

$$\sum_{d_m^i \in \hat{D}} R_i^j S_{i,m}^d L_{i,m}^{j,n} = S_{j,n}^a. \tag{5}$$

Now, the person re-identification or multi-camera tracking techniques generally establish the correspondence among the individual persons based on their appearance similarity or travel time between each pair of cameras. Thus, $L_{i,m}^{j,n}$ is initially determined in a similar way. Firstly, a 2-class support vector machine (SVM) is used to predict if any pair of d_m^i and a_n^j contains the same persons or not based on the Bhattacharrya distance of their HSV color histograms as

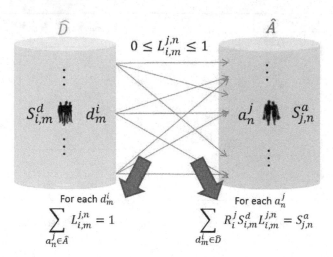

Fig. 4. Weighted Correspondence among Foreground Regions in $\hat{D}$ and $\hat{A}$

the appearance cues and the difference between their exit and entry times $t_{d_m^i}$ and $t_{a_n^j}$ as the travel time. $L_{i,m}^{j,n}$ is initially set proportionally to the prediction confidence in the range of 0 to 1 by applying the sigmoid function to the distance of the pair of d_m^i and a_n^j to the decision boundary. Then, $L_{i,m}^{j,n}$ is normalized for each d_m^i so that $L_{i,m}^{j,n}$ satisfies Eq.(4). Thus, $L_{i,m}^{j,n}$ is updated to $\hat{L}_{i,m}^{j,n}$ as follows:

$$\hat{L}_{i,m}^{j,n} = \frac{L_{i,m}^{j,n}}{\sum_{a_n^j \in \hat{A}} L_{i,m}^{j,n}}. \tag{6}$$

Now, $L_{i,m}^{j,n} = 0$ represents that d_m^i and a_n^j are highly unlikely to contain the same persons based on their appearance cues and transition time. By keeping them intact, $L_{i,m}^{j,n}$ should also satisfy Eq.(5). Thus, for each a_n^j, $\hat{L}_{i,m}^{j,n}$ is updated to $L_{i,m}^{j,n}$ as follows:

$$L_{i,m}^{j,n} = \hat{L}_{i,m}^{j,n} \times \frac{S_{j,n}^a}{\sum_{d_m^i \in \hat{D}} R_i^j S_{i,m}^d \hat{L}_{i,m}^{j,n}}. \tag{7}$$

$L_{i,m}^{j,n}$ is iteratively updated by Eqs.(6) and (7) until the total change in $L_{i,m}^{j,n}$ after an iteration converges to less than ϵ so that $L_{i,m}^{j,n}$ approximately satisfies both Eqs. (4) and (5).

2.3 People Counting across Camera Pairs

The number of people $V_{i,j}(t)$ traversing from C_i to C_j within the time interval from $t - T$ to t can be calculated as follows after $L_{i,m}^{j,n}$s are estimated for all a_n^js which have entered by the time t:

$$V_{i,j}(t) = \sum_{t-T \leq t_{a_n^j} \leq t} N_{i,m}^d L_{i,m}^{j,n}. \tag{8}$$

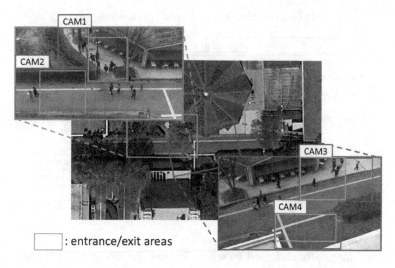

Fig. 5. Intersection Monitored by Cameras

Note that, since $S_{i,m}^d$ represents the area of persons in a camera view, it is converted to the number of persons $N_{i,m}^d$. Since d_m^i and a_n^j are extracted from the confined entrance/exit locations in the camera views, the areas $S_{i,m}^d$ can be converted by a linear function without any geometric correction within each camera view.

3   Experiments

An intersection in a university campus is monitored by two cameras as shown in Fig. 5. The top left and bottom right images are the views of the two cameras. The resolution and the frame rate of the videos are 640×360 and 30 fps, respectively. From each camera view, two rectangular regions are extracted as shown in Fig. 5 to simulate the situation where four cameras, namely CAM1, 2, 3, and 4, are monitoring the intersection. In the experiments, the numbers of people traversing between the FOVs of each pair of these four cameras are estimated.

A 120-minute video captured on February 3, 2014 and another 120-minute video captured on February 10, 2014 are used to obtain the training and test samples respectively, which are pairs of d_m^i and a_n^j. d_m^i and a_n^j were manually selected from the foreground regions which were extracted by the background subtraction. For each pair of cameras i and j, the pairs of d_m^i and a_n^j containing the same persons and the same number of pairs of d_m^i and a_n^j containing only different persons were manually prepared as the positive and negative training samples for the SVM, respectively. Moreover, by fitting the Gaussian distribution to the difference between the entry and exit times $t_{d_m^i}$ and $t_{a_n^j}$ of the positive samples, $T_{min}^{i,j}$ and $T_{max}^{i,j}$ are determined as the 1st and 99th percentile of the Gaussian distribution. Further, the number of people in the training samples

Table 1. Summary of Three Approaches

	time window	correspondence
Proposed approach: (P)	adaptive	weighted
Comparative approach 1: (C1)	fixed	weighted
Comparative approach 2: (C2)	adaptive	one-to-one

Table 2. Delay for Each Approach

	$t_{end} - t$[sec.]			computation time[sec.]			total time[sec.]		
Approach	ave.	max.	min.	ave.	max.	min.	ave.	max.	min.
(P)	23.6	159	0	0.64	3.49	0	24.2	162	0
(C1)	29.5	59	0	0.43	2.49	0.18	30.0	61.5	0.18

were specified manually and used to learn R_i^j and the linear function to convert $S_{i,m}^d$ to $N_{i,m}^d$.

Given the test samples which have exiting or are entered the camera views, the number of people traversing each pair of the four cameras is sequentially estimated by setting $T = 5$ minutes and $\epsilon = 0.01$. The number of people traversing within any $T = 5$ minutes among the four cameras vary from 37 to 108 persons with the average of 71.2 persons.

In order to separately evaluate the effectiveness of our proposed two steps, we conducted two experiments comparing the results of different approaches shown in Table 1. Firstly, in order to evaluate the effectiveness of the adaptive time window (P), we compared the results with when using the time window of a fixed length of 1 minute (C1). Secondly, in order to evaluate the effectiveness of establishing the weighted correspondence among the foreground regions (P), we compared the results with when establishing the one-to-one correspondence among the foreground regions (C2), which can be considered as a standard person re-identification approach. All three approaches were implemented on a PC with a Intel(R) Core(TM) i7-2600 (3.40 GHz) CPU and 3.49 GB memory. The experimental results are discussed in the following sections.

3.1   Evaluations of Adaptive Time Window

The time window is determined so that both the exits and entries of all persons traveling within the time window are observed. Its aim is to accurately establish the correspondence among the foreground regions which will later affect the accuracy of the people counting. On the other hand, $V_{i,j}(t)$ can only be calculated when $L_{i,m}^{j,n}$s are estimated for all a_n^js which have entered by the time t, resulting in a certain amount of delay in presenting $V_{i,j}(t)$. Thus, the effectiveness of the adaptive time window needs to be evaluated both in terms of the accuracy and the delay of the people counting.

Firstly, the delay is the sum of $t_{end} - t$ and the computation time for establishing the correspondence and for calculating $V_{i,j}(t)$. Table 2 shows how each

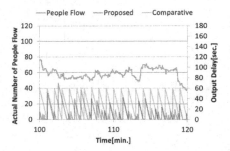 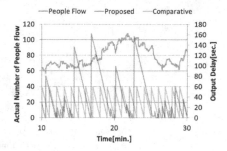

(a) From 100 minutes to 120 minutes (NOT crowded) (b) From 10 minutes to 30 minutes (crowded)

Fig. 6. Relations between Delay and People Traffic Flows

Table 3. Ratio of d_m^is in $\hat{D}$ which have the corresponding a_n^js in $\hat{A}$

Approach	$a_n^j \in \hat{A}$	$a_n^j \notin \hat{A}$
(P)	98.9% (1162/1175)	1.1%(13/1175)
(C1)	82.5% (969/1175)	17.5%(206/1175)

factor contributed to the delay for each approach. As shown in the table, most of the delay was due to $t_{end} - t$. Thus, when using the fixed time window, the delay mostly stays within the length of the time window regardless of the current people traffic situations. On the other hand, when using the adaptive time window, the delay varies depending on the traffic situations, getting shorter (11.8 seconds on average in Fig. 6(a))) when less people traffic is observed and longer (41.3 seconds on average in Fig. 6(b)) as more people traffic is observed as shown in Fig. 6. For the test video, although the maximum delay was 162 seconds, the average delay was still shorter than when using the fixed time window and stayed under 1 minute in approximately 93% of the time.

As for the accuracy, Table 3 shows the ratio of d_m^is in $\hat{D}$ which have the corresponding a_n^js in $\hat{A}$ and Fig. 7 shows the absolute errors between the actual numbers of people traversing across each camera pair and the numbers estimated by (P) and (C1). Note that the absolute errors are summed up for 12 camera pairs. As can be seen from Table 3, only either the exit or entry of 17.5% of persons was observed in the fixed time window, which will degrade the correspondence accuracy. On the other hand, both the exits and entries were observed for most persons in the adaptive time window. The errors in people counting were also lower when using the adaptive time window for most of the time regardless of the people traffic situations as shown in Fig. 7. In conclusion, the adaptive time window provided better accuracy with mostly less delay.

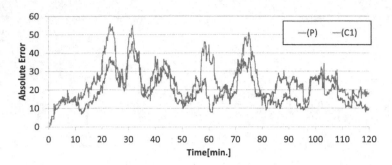

Fig. 7. Total Absolute Errors between Actual Numbers of People and Estimations by (P) and (C1) for 12 Camera Pairs

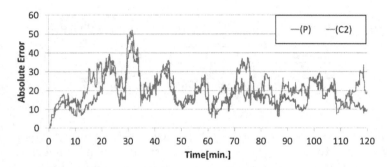

Fig. 8. Total Absolute Errors between Actual Numbers of People and Estimations by (P) and (C2) for 12 Camera Pairs

3.2   Evaluations of Weighted Correspondence

The proposed method establishes the weighted correspondence among the foreground regions for accurate people counting. Thus, its effectiveness is evaluated in terms of the accuracy by comparing the results with when establishing the one-to-one correspondence. For the one-to-one correspondence, when d_m^i has only one a_n^j which is predicted to contain the same persons by the SVM, they are considered as a match. Then, the correspondence among other d_m^is and a_n^js is established in a greedy manner in the order of the prediction confidence of the SVM.

Fig. 8 shows the absolute errors between the actual numbers of people traversing across each camera pair and the numbers estimated by (P) and (C2). The absolute errors are summed up for 12 camera pairs. Our proposed method can estimate people flows more accurately with the average errors of 1.56 persons and the standard deviation of 1.93 persons than the comparative method whose average errors and standard deviation are 1.70 and 3.22 persons. As can be seen, although the amount of errors vary according to the amount of the traffic flows, the weighted correspondence was able to decrease the maximum errors compared to the one-to-one correspondence.

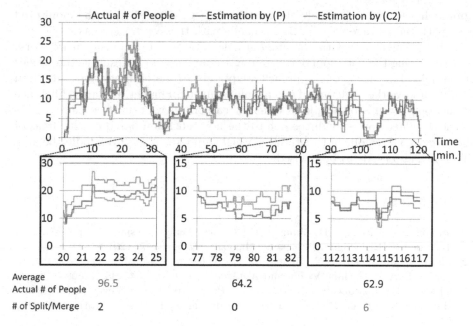

Fig. 9. Actual Numbers of People Traveling from CAM1 to CAM3 and Estimations by (P) and (C2)

In order to examine the results more closely, Fig .9 shows the actual numbers of people traveling from CAM1 to CAM3 and the estimation results. As can be seen, the weighted correspondence mostly resulted in more accurate estimation with the average errors of 1.27 persons with the standard deviation of 1.02, while the one-to-one correspondence resulted in the average errors of 1.72 persons with the standard deviation of 2.04. Especially, the weighted correspondence was able to reduce the errors when more traffic flow was observed such as from 20 to 25 minutes and when more splitting and merging were observed such as from 112 to 117 minutes in Fig. 9. By contrast, when less traffic flow was observed such as from 77 to 82 minutes in Fig. 9, the correspondence would be easily established; and therefore, the one-to-one correspondence performed better.

4   Conclusions

We proposed a method for counting the number of people traveling across multiple spatially adjacent non-overlapping camera views with little delay. The proposed method firstly sets the shortest time window to observe both the exits and entries of all groups of people traveling within the time window adaptively to the current people traffic flow. Then, in order to handle the splitting/merging of the groups outside the camera views, the weighted correspondence is established among the exits and entries within the time window. For 120-minute videos capturing an area in a university campus by four virtual cameras, the proposed approach was able to estimate the number of people traveling in the monitored area with the average delay of 24.2 seconds with the average errors of 1.56 persons per camera pair with the standard deviation of 1.93. As a future work, in establishing the correspondence among the exits and entries, selecting either the weighted or one-to-one approach adaptively to the current traffic situation would improve the accuracy of the people counting. Further, the area difference of the same groups of people in different camera views can often cause errors in the weighted correspondence due to the normalization based on the areas of the foreground regions in the iterative weight update. A way to handle the area difference also needs to be devised.

References

1. Junior, S.J., Musse, S.R., Jung, C.R.: Crowd Analysis Using Computer Vision Techniques. IEEE Signal Processing Magazine 27(5), 66–77 (2010)
2. Zhan, B., Monekosso, D.N., Remagnino, P., Velastin, S.A., Xu, L.-Q.: Crowd Analysis: A Survey. Machine Vision and Applications 19(5-6), 345–357 (2008)
3. Chan, A.B., Vasconcelos, N.: Counting People With Low-Level Features and Bayesian Regression. IEEE Transactions on Image Processing 1(4), 2160–2177 (2012)
4. Kilambi, P., Ribnick, E., Joshi, A.J., Masoud, O., Papanikolopoulos, N.: Estimating Pedestrian Counts in Groups. Computer Vision and Image Understanding 110(1), 43–59 (2008)
5. Ryan, D., Denman, S., Fookes, C., Sridharan, S.: Crowd Counting using Multiple Local Features. In: Proc. of International Conference on Digital Image Computing: Techniques and Applications, pp. 81–88 (2009)
6. Kong, D., Gray, D., Tao, H.: Counting Pedestrians in Crowds Using Viewpoint Invariant Training. In: Proc. of IEEE Conference on Computer Vision and Pattern Recognition, pp. 1–7 (2008)
7. Paragios, N., Ramesh, V.: A MRF-based Approach for Real-Time Subway Monitoring. In: Proc. of IEEE Conference on Computer Vision and Pattern Recognition, pp. I-1034–I-1040 (2001)
8. Kettnaker, V., Zabih, R.: Counting People from Multiple Cameras. In: Proc. of IEEE International Conference on Multimedia Computing and Systems, vol. 2, pp. 267–271 (1999)
9. Wang, X.: Intelligent Multi-Camera Video Surveillance: A Review. Pattern Recognition Letters 34(1), 3–19 (2013)
10. Bedagkar-Gala, A., Shah, S.K.: A Survey of Approaches and Trends in Person Re-identification. Image and Vision Computing 32(4), 270–286 (2014)
11. Vezzani, R., Baltieri, D., Rita, C.: People Reidentification in Surveillance and Forensics: A Survey. ACM Computing Surveys 46(2) article 29 (2013)
12. Nitta, N., Nakazaki, T., Akai, R., Babaguchi, N.: People Counting Across Spatially Disjoint Cameras by Flow Estimation between Foreground Regions. In: Proc. of Workshop on Activity Monitoring by Multiple Distributed Sensing, pp. 414–419 (2013)

Binary Code Learning via Iterative Distance Adjustment

Zhen-fei Ju, Xiao-jiao Mao, Ning Li, and Yu-bin Yang

State Key Laboratory for Novel Software Technology,
Nanjing University, 210093, China
yangyubin@nju.edu.cn

Abstract. Binary code learning techniques have recently been actively studied for hashing based nearest neighbor search in computer vision applications due to its merit of improving hashing performance. Currently, hashing based methods can obtain good binary codes but some data may suffer from the problem of being mapped to inappropriate Hamming codes. To address this issue, this paper proposes a novel binary code learning method via iterative distance adjustment to improve traditional hashing methods, in which we utilize very short additional binary bits to correct the spatial relationship among data points and thus enhance the similarity-preserving power of binary codes. We carry out image retrieval experiments on the well-recognized benchmark datasets to validate the proposed method. The experimental results have shown that the proposed method achieves better hashing performance than the state-of-the-art binary code learning methods.

Keywords: Hashing, nearest neighbor search, binary codes, iterative distance adjustment.

1 Introduction

Efficient similarity search in large image databases is a significant challenge in many computer vision applications. Nearest neighbor search(NNS), usually adopted to find similar objects, is one of the core technical issue involved in these applications. NNS is also a fundamental problem in data mining, machine learning and computer vision. However, the traditional way of searching nearest neighbors by scanning all the data has a linear time complexity, which is very inefficient and expensive for large databases. Fortunately in many applications, it is acceptably sufficient to return approximated nearest neighbors(ANN) instead of the exact ones.

Tree-based methods, such as KD-tree [4], metric tree [19], have been widely studied and used for NNS. These techniques attempt to decrease the complexity of nearest neighbor search, but they may degrade into linear search in the worst case [10]. Unfortunately, these technologies may not be appropriate for high dimensional data because the construction of tree structure is time-consuming and requires large memory space. Hence, hashing based techniques have been actively studied for mapping data to compact binary codes, which are then

X. He et al. (Eds.): MMM 2015, Part I, LNCS 8935, pp. 83–94, 2015.

used to establish hash tables for large databases efficiently. Since the Hamming distance between two binary codes can be computed via exclusive OR operation which is extremely fast, compact binary codes are particularly appropriate for approximating nearest neighbor search.

Hashing based methods usually preserve data similarity when mapping data points to the Hamming space appropriately. Current existing hashing based techniques can be divided into three categories based on their learning strategy: 1)unsupervised, 2)semi-supervised and 3)supervised. For unsupervised methods, Locality-Sensitive Hashing (LSH) [5] [1], which maps data to a low dimensional Hamming space via random projection, is a basic but widely used hashing based technique. After that, many other unsupervised methods have been proposed, such as Spectral Hashing SH [21], KLSH [9], ITQ [6] and AGH [12]. For semi-supervised methods, SSH [20] and Weakly-Supervised Hashing [14] are typical representatives. For supervised hashing including semantic hashing (RBM) [17], BRE [8], MLH [15] and others, they improve hashing performance by incorporating supervised information. Recently, a number of optimized hashing methods have also been proposed, such as Weighted Hamming Ranking [22], Hash Bit Selection [13] and JSD [2]. These methods further improve hashing performance based on the pre-existing methods.

In this paper, we present an iterative distance adjusting method to improve the similarity preserving power of binary codes. Motivated by the optimized methods, we propose a hashing method by correcting the Hamming spatial relationship between the base codes generated by other hashing methods. Although the traditional hashing-based method can obtain good hashing performance,there are still some data mapped to inappropriate Hamming codes. To correct these errors, we introduce a proper adjustment on the Hamming distance between the inaccurate binary code pairs. Moreover, aiming to overcome the inefficiency of building a binary code pool which consists of a number of many unnecessary codes, we resort to providing some additional bits for the base codes to enhance hashing performance, rather than selecting good bits from a binary code pool [13] [2]. It also guarantees that our method doesn't need any redundant candidate codes. An iterative distance adjustment step, which shrinks the Hamming distance of neighboring data and increase the Hamming distance of non-neighbor data, is adopted to reduce the Hamming approximation error between the Euclidean space and the Hamming space. Our method achieves good performance and outperforms several state-of-the-art methods.

The rest of this paper is organized as follows. Section 2 provides a brief introduction to the related work. Section 3 describes the details of our method. The experimental results are then provided in Section 4. Finally, Section 5 provides concluding remarks.

2   Related Work

In this section we present some existing studies related to hash-based nearest neighbor search techniques. Given a dataset $X = [x_1, x_2, \ldots, x_n], x_i \in \mathbb{R}^d$, nearest neighbor search aims to find the nearest neighbors for a query q. The objective of hash-based NNS methods is mapping data to the Hamming space while appropriately preserving the relative distances among them.

2.1 Unsupervised Hashing

LSH [5], a basic but widely used hashing technique, maps data to a low dimensional Hamming space via random projection. The main idea of LSH is mapping similar data to the same hash bucket with a high probability. A typical LSH function is denoted as:

$$h(x) = sign(\boldsymbol{w}^\top \boldsymbol{x} + b) \tag{1}$$

where $\boldsymbol{w}$ is a hyperplane and b is a random intercept. $\boldsymbol{w}$ is usually randomly sampled from a specific distribution, such as *p-stable* distribution. KLSH (Kernelized LSH) [9]incorporates LSH with kernel learning so that it can generalize similarity search from standard metric space to kernel space. Since hyperplane $\boldsymbol{w}$ is independently sampled to data, the number of hash bits may be large in order to maintain the original distances.

Motivated by spectral graph partition, Spectral Hashing improves performance over LSH, especially for compact bit lengths. But it has an assumption of underlying distribution of data. Iterative Quantization (ITQ) is motivated by the idea of data rotation to minimize quantization loss. ITQ explains that if the hyperplane coefficients $\boldsymbol{W}$ is an optimal solution, so is $\widetilde{\boldsymbol{W}} = \boldsymbol{W}\boldsymbol{R}$ for any orthogonal matrix R. The goal of ITQ is to minimize the quantization error of mapping data to the vertices of a zero-centered binary hypercube.

Anchor Graph Hashing(AGH) [12] utilizes Anchor Graphs to obtain tractable low-rank adjacency matrices. The hash functions of AGH are learned by thresholding the lower eigenfunctions of the Anchor Graph Laplacian. Meanwhile, a hierarchical hashing method are further proposed to overcome the issue that neighboring points close to the boundary are assigned to different bits due to inappropriate thresholding.

2.2 Supervised Hashing

Supervised hashing methods such as RBM [17], BRE [8] and MLH [15] have shown higher search accuracy than unsupervised ones, but they are more difficult to be optimized and slower to be trained. Semantic Hashing (RBM) leverages Restricted Boltzmann Machines [3] to construct a multi-layer autoencoder to encode a low dimensional binary code. Binary reconstructive embedding (BRE) makes a loss function that penalizes the squared error between the original distances and the reconstructed distances. Since the objective function is non-convex, a coordinate-descent algorithm is proposed for optimization. Minimal Loss Hashing (MLH) advocates a loss function similar to hinge loss used in the SVMs to learn binary hash functions.

2.3 Optimized Hashing

Some hashing methods have been proposed to improve performance of binary codes by optimizing the pre-existing ones. [22] explained that there are often lots of results sharing the same Hamming distance to a query, which makes this distance measure ambiguous. They proposed a weighted Hamming distance ranking algorithm to rank the generated binary codes so that it can differentiate the ambiguous codes. The algorithm learns both data-adaptive and query-sensitive weight for each hash bit. Hash Bit Selection [13] builds a large pool of overcomplete hash bits encoded by various hashing methods with different features. Then, good hash bits are selected through two criteria: 1)similarity preserving and 2)independence. JSD [2] seeks a set of hash functions that minimizes the total probability of Bayes decision errors. A sequential learning algorithm based on LSH [5] [1] is also provided to obtain the projection minimizing Bayes decision errors from the candidate projections.

3   Correctional Hashing

3.1   Formulation

It has been well investigated in [13] [2] [22] that some bits learned by the existing hashing methods usually carry little or redundant information. To deal with this issue, they either selected good project directions from a candidate projection pool, or set different weights to different bits [22]. To build a candidate projection pool, many candidate binary codes are needed to be learned, which is very inefficient. In this paper, we address this problem in a totally different manner by investigating whether it is possible to improve the similarity preserving power of the codes by providing some additional bits.

Learning compact binary codes has been commonly treated as a solution to similarity-preserving problem, i.e. maping similar data points to similar binary codes in hashing. But there still exists a gap between the distances among binary codes in Hamming space and the distances among features in Euclidean space. In traditional binary learning methods, most data can be mapped to proper binary codes, but some of them are usually mapped to inappropriate ones, which are either neighboring data in the original space mapping to distant codes in Hamming space, or distant data in the original space mapping to neighboring codes in Hamming space. We design an algorithm to correct such kind of error via providing very compact additional bits. This adjustment can reduce the approximating error between Euclidean space and Hamming space, with which, the learned codes may achieve higher similarity preserving power and the hashing performance can be consequently improved.

Afterwards, we propose a novel method to correct the spatial relationship between pairwise binary codes iteratively. At each iteration, additional binary bits are introduced to minimize the Hamming approximating error. The basic principle is that the Hamming distance between the points which have been assigned to inaccurate base codes can be modified and corrected through additional bits. In a word, among the inaccurate pairs of binary codes, similar additional bits are used to shrink the distance between the neighboring pairs in the original space, while opposite additional bits are adopted to increase the distance between the distant pairs in the original space. A Hamming approximating error function is defined to determine which pair of binary codes is inaccurate. The Hamming approximating error is minimized iteratively by learning the additional bits incrementally. Specifically, a few bits are learned for the base codes in each iteration and they are concatenated as the new base codes in next iteration.

3.2   Iterative Distance Adjustment

Given a dataset of n points, denoted as $X = [x_1, x_2, \ldots, x_n], x_i \in \mathbb{R}^d$, the objective of hashing method is to learn a hash function family to map X to a binary matrix $B \in \{-1, 1\}^{n \times b}$, where b denotes the length of a binary code, and kth hash function of binary bit is defined as

$$h_k(\boldsymbol{x_i}) = sgn(\boldsymbol{w}_k^\top \boldsymbol{x_i} + b_k). \tag{2}$$

Let $\boldsymbol{H} = [h_1, h_2, \ldots, h_k]$ be a sequence of k hash functions and $\boldsymbol{W} = [\boldsymbol{w}_1, \boldsymbol{w}_2, \ldots, \boldsymbol{w}_k] \in \mathcal{R}^{d \times k}$. Without loss of generality, let $\boldsymbol{X}$ be normalized to have zero mean. To make hash bit carry as much information as possible, we should balance the hash function to meet $\sum_{i=1}^n H_k(\boldsymbol{x_i}) = 0$. Like [11], we denote b_k as the median of $\{\boldsymbol{w}_k^\top \boldsymbol{x_i}\}_{i=1}^n$. By further choosing a fast alternative to the median, we calculate the mean as $b = \sum_{i=1}^n \boldsymbol{w}_k^\top \boldsymbol{x_i}/n$. We have $b = 0$ because $\boldsymbol{X}$ is zero-mean.

Given a set of base binary codes $B^{base} \in \{-1, 1\}^{n \times b}$, to adjust the Hamming distance between base codes, our objective is to learn the corresponding additional codes $B^{add} \in \{-1, 1\}^{n \times k}$. The result of final binary matrix is then denoted as $B \in \{-1, 1\}^{n \times (b+k)}$. We learn t bits at each iteration, thus the number of iterations is b/t. We use linear projection mentioned earlier as the hash function for additional codes.

3.3   Learning Additional Binary Codes

First of all, the pairs of base binary codes needed to be corrected should be selected out. Similar to K-means Hashing [7], we choose minimizing the Hamming approximation error as the objective function,which is defined as:

$$min \sum_{i=1}^n \sum_{j=1}^n (d(\boldsymbol{x_i}, \boldsymbol{x_j}) - \lambda d_h(\boldsymbol{x_i}, \boldsymbol{x_j}))^2 \tag{3}$$

This equation minimizes the difference between two n-by-n affinity matrices $d(\boldsymbol{x_i}, \boldsymbol{x_j})$ and $d_h(\boldsymbol{x_i}, \boldsymbol{x_j})$, where $d(\boldsymbol{x_i}, \boldsymbol{x_j})$ is the original Euclidean distance between two data points, and $d_h(\boldsymbol{x_i}, \boldsymbol{x_j})$ is the Hamming distance between two hashing binary codes. We introduce λ here because the Euclidean distance $d(\boldsymbol{x_i}, \boldsymbol{x_j})$ can be in arbitrary ranges, while the Hamming distance $d_h(\boldsymbol{x_i}, \boldsymbol{x_j})$ is constrained in the range of [0,b] for b bits.

Based on the objective function, we define the Hamming approximating error to distinguish inaccurate pairs which need corrections as:

$$Err_{Ha} = (d(\boldsymbol{x_i}, \boldsymbol{x_j}) - d_h(\boldsymbol{x_i}, \boldsymbol{x_j}))^2 \tag{4}$$

In our work, λ has been found hard to determine. Therefore we normalize the two matrices $d(\boldsymbol{x}_i, \boldsymbol{x}_j)$ and $d_h(\boldsymbol{x}_i, \boldsymbol{x}_j)$ as a compromise. θ is a threshold set to find the binary codes needing correction.

We adopt an intuitive way to learn additional bits, and update a small number of bits at each iteration. We split the inaccurate binary code pairs into two categories: 1) neighbor in the original space, and 2) non-neighbor in the original space. Specifically, a pair $(\boldsymbol{x}_i, \boldsymbol{x}_j) \in \mathcal{N}$ is denoted as a neighboring pair in Euclidean space, and $\mathcal{D}$ is the set of non-neighboring pairs in Euclidean space. In order to correct the spatial relationship of inaccurate binary codes in Hamming space, the Hamming distance between neighboring pairs should be reduced and the Hamming distance between non-neighboring pairs should be increased. We manage to learn a $\boldsymbol{W}$ which provides similar additional bits to the neighboring pairs and different additional bits to the non-neighbor pairs. An objective function measuring the empirical accuracy for the additional hashing functions $[h_1, h_2, \ldots, h_t]$ at each iteration can be defined as:

$$J(\boldsymbol{H}) = \sum_t \left\{ \sum_{(\boldsymbol{x}_i, \boldsymbol{x}_j) \in \mathcal{N}} h_t(\boldsymbol{x}_i) h_t(\boldsymbol{x}_j) - \sum_{(\boldsymbol{x}_i, \boldsymbol{x}_j) \in \mathcal{D}} h_t(\boldsymbol{x}_i) h_t(\boldsymbol{x}_j) \right\}, \quad (5)$$

where t is the length of bits updated at each iteration. We need k/s iterations to learn all the additional projection directions.

Then, we define a matrix $S \in \mathbb{R}^{n \times n}$ to incorporate pairwise Hamming approximation error and the original distance as

$$\boldsymbol{S}_{ij} = \begin{cases} 1 & : & Err_{Ha} > \theta, (\boldsymbol{x}_i, \boldsymbol{x}_j) \in \mathcal{N} \\ -1 & : & Err_{Ha} > \theta, (\boldsymbol{x}_i, \boldsymbol{x}_j) \in \mathcal{D} \\ 0 & : & Err_{Ha} \le \theta \end{cases} \quad (6)$$

Suppose $H(\boldsymbol{X}) \in \mathbb{R}^{t \times n}$ maps the points in $\boldsymbol{X}$ to t-bit hash codes at each iteration. Then the objective function can be presented as

$$J(\boldsymbol{H}) = \frac{1}{2} tr\{H(\boldsymbol{X}) \boldsymbol{S} H(\boldsymbol{X})^\top\} \quad (7)$$

that is

$$J(\boldsymbol{W}) = \frac{1}{2} tr\{sgn(\boldsymbol{W}^\top \boldsymbol{X}) \boldsymbol{S} sgn(\boldsymbol{W}^\top \boldsymbol{X})^\top\} \quad (8)$$

where $X = [x_1, x_2, \ldots, x_n], x_i \in \mathbb{R}^d$. We learn the optimal hyperplane matrix $\boldsymbol{W}$ by maximizing objective function $J(\boldsymbol{W})$. Since $J(\boldsymbol{W})$ is nondifferentiable, the above problem is difficult to be solved. We present an intuitive relaxation by replacing the sign of projection with its signed magnitude in Eqn.(7).

$$J(\boldsymbol{H}) = \sum_t \left\{ \sum_{(\boldsymbol{x}_i, \boldsymbol{x}_j) \in \mathcal{N}} \boldsymbol{w}_t^\top \boldsymbol{x}_i \boldsymbol{x}_j^\top \boldsymbol{w}_t - \sum_{(\boldsymbol{x}_i, \boldsymbol{x}_j) \in \lceil} \boldsymbol{w}_t^\top \boldsymbol{x}_i \boldsymbol{x}_j^\top \boldsymbol{w}_t \right\}, \quad (9)$$

The above function $J(\boldsymbol{W})$ can be represented in a matrix form:

$$J(\boldsymbol{W}) = \frac{1}{2} tr\{\boldsymbol{W}^\top \boldsymbol{X}_k \boldsymbol{S} \boldsymbol{X}^\top \boldsymbol{W}\} \quad (10)$$

Algorithm 1. Correctional Hashing

Input:

The set of training sample, $\boldsymbol{X} = \boldsymbol{x}_i \in \mathbb{R}^{d\,n}_{i=1}$; The set of base binary codes of training set, $\boldsymbol{B}^{base} \in \mathbb{R}^{n \times b}$; The length of additional correctional bits, k; The length of bits updated at each iteration, t; The threshold defining "neighbor" in Euclidean space, θ_{eu}; The threshold defining the Hamming approximating error, θ_{ha}

  for $i = 1, \ldots, k/t$ **do**

    $\boldsymbol{B}^{add} \leftarrow \boldsymbol{B}^{base}$

    for $j = 1, \ldots, n$ **do**

      for $l = 1, \ldots, n$ **do**

        $Err_{Ha} \Leftarrow (d(\boldsymbol{x}_j, \boldsymbol{x}_l) - d_h(\boldsymbol{B}^{add}_j, \boldsymbol{B}^{add}_l))^2$

        if $Err_{Ha} \leq \theta_{ha}$ **then**

          $\boldsymbol{S}_{jl} \Leftarrow 0$

        else

          if $d(\boldsymbol{x}_j, \boldsymbol{x}_k) \leq \theta_{eu}$ **then**

            $\boldsymbol{S}_{jl} \Leftarrow 1$

          else

            $\boldsymbol{S}_{jl} \Leftarrow -1$

          end if

        end if

      end for

    end for

    $\boldsymbol{M} \leftarrow \boldsymbol{X}\boldsymbol{S}\boldsymbol{X}^{\top}$;

    $\boldsymbol{W} = [\boldsymbol{w}_1, \boldsymbol{w}_2, \ldots, \boldsymbol{w}_t] \leftarrow$ the eigenvectors corresponding to top-t eigenvalues of matrix $\boldsymbol{M}$;

    $\boldsymbol{B}^{new} \leftarrow \boldsymbol{W}^{\top}\boldsymbol{X}$;

    $\boldsymbol{B}^{add} \leftarrow \boldsymbol{B}^{add}$ attach $\boldsymbol{B}^{new}$, $\boldsymbol{B}^{add} \in \mathbb{R}^{n \times (b+t)}$;

  end for

Output:

n hash codes $\boldsymbol{B}^{add} \in \mathbb{R}^{n \times (b+k)}$;

Then, the objective function is solved by using eigenvalue decomposition on matrix $\boldsymbol{M} = \boldsymbol{X}\boldsymbol{S}\boldsymbol{X}^{\top}$:

$$\max_{\boldsymbol{W}} J(\boldsymbol{W}) = \sum_{s=1}^{t} \lambda_s \tag{11}$$

where $[\lambda_1, \lambda_2, \ldots, \lambda_t]$ are the top-t eigenvalues of $\boldsymbol{M}$, and $\boldsymbol{w}_k$ are the corresponding eigenvectors.

4   Experiments

4.1   Datasets and Protocols

We evaluate our method on the well-recognized MNIST digit dataset and CIFAR-10 dataset, and make comparisons with several existing state-of-the-art hash-based method.

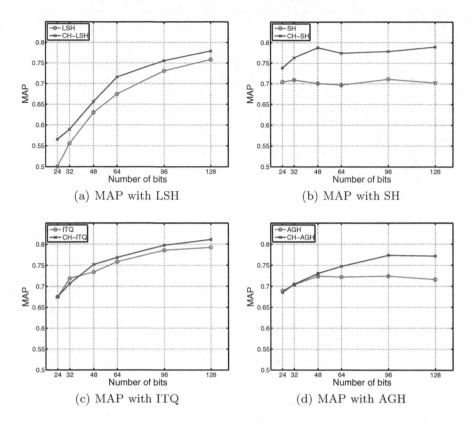

(a) MAP with LSH

(b) MAP with SH

(c) MAP with ITQ

(d) MAP with AGH

Fig. 1. Performance comparison of CH with SH, LSH, ITQ and AGH on MNIST dataset with Euclidean neighbourhood

MNIST is a well-known greyscale image dataset of handwritten digits consisting of 70,000 784-dimension digit samples from '0' to '9'. The original handwriting digit image size is 28 × 28 . CIFAR-10 dataset is a labeled subset of an 80-million tiny images collection [18]. It consists of 60,000 32 × 32 color images of 10 classes, each of which has 6,000 images. Each image in CIFAR-10 is represented by a 512-dimension GIST feature vector [16]. We randomly sample 2,000 images for training and 1000 images for testing for each dataset.

We evaluate our Correctional Hashing(CH) method by comparing the generated base binary codes with those of other four state-of-the-art methods, that is, LSH [1], SH [21], ITQ [6] and AGH [12]. For LSH, we randomly select projections from a Gaussian distribution with zero-mean to construct hash functions. To run AGH, we take the two-layer AGH(2-AGH) and fix the number of anchors as 300. The Correctional Hashing codes consists of base binary codes and additional codes as:

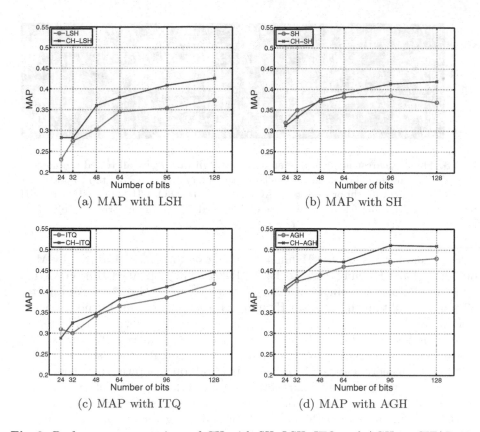

Fig. 2. Performance comparison of CH with SH, LSH, ITQ and AGH on CIFAR-10 dataset with Euclidean neighbourhood

Base Bits	16	24	32	48	64	96
Additional Bits	8	8	16	16	32	32
CH Bits	24	32	48	64	96	128

4.2 Result

We compare the codes generated by Correctional Hashing and the other four methods at the same length. The number of iterations is fixed as 2. The performance is measured by Mean Average Precision (MAP) defined as the mean precision rate of the Top n points in the ranked list of each testing query. All the points in the testing set are ranked according to the values of their Hamming distances to the query point, and top-50 points of the ranked list are taken to calculate MAP. The complexity of Hamming ranking is linear, but it is adequately fast for binary codes. We query all points in the testing set and list the mean precisions of them.

Fig. 3. The 30 nearest neighbors of an example digit output by all tested methods on the MNIST dataset. The images on the left is the query sample. From top to bottom, it shows the nearest neighbor samplesof AGH, CH-AGH, ITQ, CH-ITQ, LSH, CH-LSH, SH and CH-SH using 64-bit codes. The digit in red box is the incorrect result of the query.

Fig.1 and Fig.2 show the Hamming ranking performance measured by MAP of Correctional Hashing with LSH [1], SH [21], ITQ [6], AGH [12] on MNIST and CIFAR-10. The number of binary bits varies from 24 to 128. We take top 50 points to obtain the MAP of Hamming Ranking based on Euclidean neighborhood.

Our method adjusts the Hamming distance between inaccurate pairs so that the similarity preserving power of binary codes is enhanced. As shown in Fig.1 and Fig.2, our method outperforms other methods on generating base codes. But in some cases with short bits, our method didn't perform well, which may caused by the mediocre performance of its base binary codes. Overall, Correctional Hashing algorithm obtains higher MAP of Hamming ranking in all the tested methods.

Moreover, our method also obtains sound results when the ground-truth labels are used to evaluate semantic consistency of binary codes. We choose MNIST dataset to present the query result for the convenience of observing. The adjacent two rows are base codes learned by other hashing-based techniques and its corresponding codes learned by Correctional Hashing, respectively. As shown in Fig.3, our algorithm promotes the semantic consistency of binary code. The incorrect digits among the top-30 query results, such as digit 7 and 4 at first row, have been adjusted to latter position of the rank list. It is safe to make a conclusion that Correctional Hashing retrieves more accurate than the base methods and is capable of correcting some inaccurate binary codes.

5   Conclusion

This paper presents a novel binary code learning method via iterative distance adjustment. Considering the existence of inaccurate mapping between the data and the mapped Hamming positions, we propose to correct Hamming distance between inaccurate binary pairs to minimize Hamming approximating error. Experimental results show that the proposed method improves the hashing performance of the state-of-the-art techniques.

In the future, we will focus on learning the threshold of Hamming approximating error to achieve better hash performance.

Acknowledgment. This work is supported by the Program for New Century Excellent Talents of MOE China (Grant No.NCET-11-0213), the Natural Science Foundation of China (Grant Nos. 61273257, 61321491, 61035003), National 973 Program of China (Grant No.2010CB327903), the Program for Distinguished Talents of Jiangsu (Grant No. 2013-XXRJ-018), and the Scientific Research Foundation of Graduate School of Nanjing University (Grant No. 2014CL03).

References

1. Datar, M., Immorlica, N., Indyk, P., Mirrokni, V.S.: Locality-sensitive hashing scheme based on p-stable distributions. In: Proceedings of the Twentieth Annual Symposium on Computational Geometry, pp. 253–262. ACM (2004)
2. Fan, L.: Supervised binary hash code learning with jensen shannon divergence. In: 2013 IEEE International Conference on Computer Vision (ICCV), pp. 2616–2623. IEEE (2013)
3. Freund, Y., Haussler, D.: Unsupervised learning of distributions of binary vectors using two layer networks. Computer Research Laboratory, University of California, Santa Cruz (1994)
4. Friedman, J.H., Bentley, J.L., Finkel, R.A.: An algorithm for finding best matches in logarithmic expected time. ACM Transactions on Mathematical Software (TOMS) 3(3), 209–226 (1977)
5. Gionis, A., Indyk, P., Motwani, R., et al.: Similarity search in high dimensions via hashing. VLDB 99, 518–529 (1999)
6. Gong, Y., Lazebnik, S.: Iterative quantization: A procrustean approach to learning binary codes. In: 2011 IEEE Conference on Computer Vision and Pattern Recognition (CVPR), pp. 817–824. IEEE (2011)
7. He, K., Wen, F., Sun, J.: K-means hashing: an affinity-preserving quantization method for learning binary compact codes. In: 2013 IEEE Conference on Computer Vision and Pattern Recognition (CVPR), pp. 2938–2945. IEEE (2013)
8. Kulis, B., Darrell, T.: Learning to hash with binary reconstructive embeddings. In: Advances in Neural Information Processing Systems, pp. 1042–1050 (2009)
9. Kulis, B., Grauman, K.: Kernelized locality-sensitive hashing for scalable image search. In: 2009 IEEE 12th International Conference on Computer Vision, pp. 2130–2137. IEEE (2009)
10. Liu, T., Moore, A.W., Yang, K., Gray, A.G.: An investigation of practical approximate nearest neighbor algorithms. In: Advances in Neural Information Processing Systems, pp. 825–832 (2004)
11. Liu, W., Wang, J., Ji, R., Jiang, Y.G., Chang, S.F.: Supervised hashing with kernels. In: 2012 IEEE Conference on Computer Vision and Pattern Recognition (CVPR), pp. 2074–2081. IEEE (2012)
12. Liu, W., Wang, J., Kumar, S., Chang, S.F.: Hashing with graphs. In: Proceedings of the 28th International Conference on Machine Learning (ICML 2011), pp. 1–8 (2011)
13. Liu, X., He, J., Lang, B., Chang, S.F.: Hash bit selection: a unified solution for selection problems in hashing. In: 2013 IEEE Conference on Computer Vision and Pattern Recognition (CVPR), pp. 1570–1577. IEEE (2013)

14. Mu, Y., Shen, J., Yan, S.: Weakly-supervised hashing in kernel space. In: 2010 IEEE Conference on Computer Vision and Pattern Recognition (CVPR), pp. 3344–3351. IEEE (2010)
15. Norouzi, M., Blei, D.M.: Minimal loss hashing for compact binary codes. In: Proceedings of the 28th International Conference on Machine Learning (ICML 2011), pp. 353–360 (2011)
16. Oliva, A., Torralba, A.: Modeling the shape of the scene: A holistic representation of the spatial envelope. International Journal of Computer Vision 42(3), 145–175 (2001)
17. Salakhutdinov, R., Hinton, G.E.: Learning a nonlinear embedding by preserving class neighbourhood structure. In: International Conference on Artificial Intelligence and Statistics, pp. 412–419 (2007)
18. Torralba, A., Fergus, R., Freeman, W.T.: 80 million tiny images: A large data set for nonparametric object and scene recognition. IEEE Transactions on Pattern Analysis and Machine Intelligence 30(11), 1958–1970 (2008)
19. Uhlmann, J.K.: Satisfying general proximity/similarity queries with metric trees. Information Processing Letters 40(4), 175–179 (1991)
20. Wang, J., Kumar, S., Chang, S.F.: Semi-supervised hashing for scalable image retrieval. In: 2010 IEEE Conference on Computer Vision and Pattern Recognition (CVPR), pp. 3424–3431. IEEE (2010)
21. Weiss, Y., Torralba, A., Fergus, R.: Spectral hashing. In: Advances in neural information processing systems, pp. 1753–1760 (2008)
22. Zhang, L., Zhang, Y., Tang, J., Lu, K., Tian, Q.: Binary code ranking with weighted hamming distance. In: 2013 IEEE Conference on Computer Vision and Pattern Recognition (CVPR), pp. 1586–1593. IEEE (2013)

What Image Classifiers Really See – Visualizing Bag-of-Visual Words Models

Christian Hentschel and Harald Sack

Hasso Plattner Institute for Software Systems Engineering,
Potsdam, Germany
{christian.hentschel,harald.sack}@hpi.de

Abstract. Bag-of-Visual-Words (BoVW) features which quantize and count local gradient distributions in images similar to counting words in texts have proven to be powerful image representations. In combination with supervised machine learning approaches, models for nearly every visual concept can be learned. BoVW feature extraction, however, is performed by cascading multiple stages of local feature detection and extraction, vector quantization and nearest neighbor assignment that makes interpretation of the obtained image features and thus the overall classification results very difficult. In this work, we present an approach for providing an intuitive heat map-like visualization of the influence each image pixel has on the overall classification result. We compare three different classifiers (AdaBoost, Random Forest and linear SVM) that were trained on the Caltech-101 benchmark dataset based on their individual classification performance and the generated model visualizations. The obtained visualizations not only allow for intuitive interpretation of the classification results but also help to identify sources of misclassification due to badly chosen training examples.

1 Introduction

Given a set of images and a set of concepts, the task of visual concept detection is to automatically assign one or more concepts to each of the images solely based on the visual content. An approach commonly used to solve this task is the Bag-of-Visual-Words (BoVW) model [11], where images are represented as a frequency distribution of a set of visual words (i.e. a visual vocabulary). This extends an idea from text classification where a document is described by a vector of individual word frequencies (Bag-of-Words). Image classification is then considered a learning problem of separating histograms corresponding to images of one class from those of another.

A visual word is usually described by means of local histograms of gradients (e.g. SIFT, Scale Invariant Feature Transform [9]) extracted either at specific regions of interest (keypoints) or at dense grid points. The visual vocabulary is generated by processing all local features of the training data using vector quantization approaches such as k-means and Gaussian Mixtures [12]. By assigning the local SIFT features of each image to the most similar vocabulary

X. He et al. (Eds.): MMM 2015, Part I, LNCS 8935, pp. 95–104, 2015.

vector (e.g. through nearest neighbor search), a histogram of visual word vector frequencies is generated per image. This frequency distribution is referred to as *Bag-of-Visual-Words* and provides a global image descriptor. Learning a visual concept model usually optimizes a weight vector that emphasizes different visual words depending on the classification task – very similar to learning the importance of individual words for a specific text document class. Although this approach often provides highly accurate classification results, analysis of the derived models tends to be difficult.

While in text classification each Bag-of-Words dimension corresponds to a linguistic term that carries an explicit meaning, the Bag-of-*Visual*-Words approach usually is considered a black box. This is due to the fact that visual words are much harder to interpret for a human. Each vocabulary word is a prototype for a number of local SIFT features and each SIFT feature represents local gradient distributions at a specific image region. For many image classification scenarios, e.g. computer-aided diagnosis in medical imaging, it is however crucial to provide information about how a decision is made rather than being confined to giving positive or negative classification results only.

In this work we present an approach to visualize the impact of image regions on the classification result by superposing the images with a heat map-like graphical representation of the learned visual word weights. The obtained visualization provides an intuitive way to interpret trained visual concept models by simply analyzing the image regions that contribute most (and least) to the overall result. Sources of misclassification are made explicit such as ill-chosen training examples that exhibit specific characteristics not representative for the actual concept to be classified. We present heat maps for three different classifiers, namely linear SVMs, AdaBoost and Random Forests each trained and tested on the Caltech-101 benchmark dataset [4]. Furthermore, we compare the performance of the classifiers and show that the average precision of all classifiers is comparable. This allows us to make general propositions based on the visualization.

This paper is structured as follows: Section 2 briefly reviews the related work. In Section 3 we compare the aforementioned classifiers and present the proposed approach for model visualization. Section 4 discusses the obtained results and gives an outlook to future work.

2   Related Work

While the BoVW approach for image classification has been extensively studied, considerably few works actually addresses the visualization of the trained classification models. A large body of literature focuses on the visualization of distinct regions used for local feature extraction (e.g. [2,7]). Typically, these regions are highlighted within the image by small dots or ellipses. Although this approach helps to understand which image parts were used for feature extraction it does not help to estimate to what extent each region actually influences the classification result.

Other approaches try to visualize the learned models by showing examples for visual words. In [8] the image regions assigned to the 100 vocabulary words with maximal inter-class discrimination are presented. While this gives a notion of which parts of an image contribute to distinguish one class from another, it does not show the relative importance of each image region. The authors in [14] likewise provide binary decision scores only by using bicolored dots to visualize the decision of "important" and "not important" keypoints according to the trained model. Nevertheless, both approaches show that matching local image regions and classification importances helps to gain understanding of whether a learning algorithm identified a reasonable association between features and category.

A method that tries to identify the regions in an image that, if removed, would cause the image to be wrongly classified is presented in [14]. Although this approach provides an intuitive idea of which pixels in an image are actually important for a *true positive* classification result, it does not help to identify regions that lead to a *false negative* result, which sometimes can be even more important.

When aiming at visualizing the significance of specific image regions, heat maps have been successfully applied in the past. Being intuitive while at the same time allowing for an immediate estimation to what extent each pixel in the underlying image contributes to the conveyed information they have been proven to be powerful representations. Heat maps usually superpose the original image and use color temperature to mark important (red) and less important (blue) regions. Heat maps are often used to visualize recordings of eye trackers measuring human visual attention [3]. Quite similarly, our goal is to visualize the importance of specific image pixels, however, w.r.t. to "attention" payed by a trained visual model rather than a human. In the field of image classification, heat maps have already been used to highlight patches in images selected as important by a random forest classifier [15]. However, since the patch size used is rather large, the approach offers only a very coarse estimation of region-to-classification-result importance. In this paper, the size of each region to be assigned an importance score is only limited by the size of the support region used to extract local features.

The authors in [13] propose a semantic point detector, which is based on individual feature weights of BoVW vectors learned by a linear SVM. Although not focusing on the visualization, the authors present an approach for highlighting semantically important regions using heat maps. Similarly, we pursue heat map-like visualizations for linear SVM classifiers. Additionally, we compare these to visualizations obtained from AdaBoost and Random Forest classifiers and thus enable comparison of these different models.

3   Visualization of BoVW Models

We have computed BoVW concept models for the 101 classes of the Caltech-101 benchmark dataset [4]. SIFT features are extracted at a dense grid of $s = 6$ pixels and at a fixed scale of $\sigma = 1.0$ and k-means clustering is used to quantize the SIFT features to $k = 100$ vocabulary vectors. Thus, each image is described by a 100-dimensional histogram of visual words (see [6] for implementation details).

3.1   Model Training

Kernel-based Support Vector Machines (SVM) are widely used in BoVW classification scenarios and χ^2-kernels have shown to provide good results for histogram comparisons [16]. However, due to the kernel trick, creating a direct mapping between the individual BoVW dimensions (i.e. the visual words) and the learned model weights is infeasible. Thus, visualizing the impact of individual dimensions on the overall classification result is likewise not possible. Therefore, kernel SVM results are reported only as baseline reference and in comparison to the performance of classifiers that allow for direct inference of individual feature importances.

Linear SVMs compute a linear hyperplane to best separate positive from negative examples in the original feature space. The trained model consists of a bias and a weight vector – an unknown sample is classified by computing the dot product between the weight vector and the sample's feature vector (plus the bias). While the classification results usually tend to be inferior to the results of non-linear SVMs, the linear model allows for an immediate interpretation of the weight vector dimensions as feature importance scores.

Random Forests [1] and AdaBoost [5] are usually based on decision trees where each node correlates to a specific feature in the training set. These methods typically select features based on their capability of solving the classification problem beginning with the most perfect split, e.g. by computing the decrease in entropy of the obtained class separation. We use the *mean decrease in impurity* over all decision trees in an ensemble as direct indicator for feature importance.

Our implementation uses the scikit-learn library that provides SVM Solvers and Ensemble methods [10]. We train binary classifiers for each of the 101 concept classes in the Caltech-101 dataset using the 'background' data class as negative samples. Training is performed on 50% of the dataset images while the remainder is used for validation purposes of the obtained models. Different model parameters for each classifier are optimized using a nested cross-validation[1].

As a first step, we have computed the mean average precision (MAP) each classifier achieves on all 101 concept classes. Table 1 compares the different classification models. As expected, the performance achieved by the linear SVM model is the worst (MAP is 4–13% lower than for other models). The χ^2-kernel model on the other hand outperforms all other classifiers and underlines the superiority of this approach. However, AdaBoost and Random Forests show competitive results falling only 6–9% behind. Both ensemble methods show almost identical performance which seems logical considering they both follow a very similar approach and differ only in the way predictions of individual weak learners are aggregated. As a next step we want to visualize and compare the individual visual word importance for each of the trained classification models.

---

[1] The model parameters optimized are: the number of decision trees (AdaBoost, Random Forests), the maximum depth of each tree (AdaBoost), the regularization parameter (both SVM models). The kernel-SVM width parameter is set to the average χ^2-distance of all training examples[16].

Table 1. Classifier performance on Caltech-101 dataset. Mean average precision (MAP) scores are reported.

classifier	MAP
AdaBoost	0.615
Random Forest	0.593
linear SVM	0.549
χ^2-kernel SVM	0.678

3.2 Model Visualization

As discussed in Section 2 we have decided for a heat map-like representation to visualize each pixel's importance score. This requires to map the learned importance scores of visual words to pixels in the image plane. Since each feature of a BoVW-vector corresponds to a visual word in the vocabulary and the value of each feature is generated by binning local SIFT descriptors to the most similar visual words we can extend the learned importance scores to the respective SIFT descriptors. As the support regions of neighboring descriptors overlap (by default SIFT uses a support region of 16×16 pixels and our dense sampling step size is set to $s = 6$ pixels), the importance score of each pixel is set to be the maximum of all corresponding importances since our intention was to visualize the *most important* visual words.

Figure 1 shows the obtained visualization[2] for an example of the category "airplanes", correctly classified by our AdaBoost model with a confidence score of $c = 0.995$. For reasons of clarity we limit the visualized pixel contributions to the most important visual words, i.e. only the upper quartile of the importance scores obtained per visual word are shown. Darker areas mark more important regions.

Similarly, we have computed the visualizations for Random Forests and linear SVM (see Fig. 2 for a comparison of all three models). Again, we restrict the highlighted regions to the upper quartile of the most important visual words. The visualization of the SVM model differs in that since SVMs produce negative as well as positive weights, we visualize them using different colors (blue for negative weights, red for positive) and select the upper quartile of the most important positive weights as well as the most important negative weights.

Comparison of all three visualizations confirms the closeness of AdaBoost and Random Forests. Both ensemble models produce almost identical heat maps that differ mainly in the absolute values of the respective importance scores. Surprisingly, the visualization of the trained SVM model is also very similar to those of AdaBoost and Random Forests which could explain why the classification performance of SVM and ensemble methods are comparable. Please note that

---

[2] All figures are best viewed in color and magnification.

0.03328
0.03085
0.02841
0.02598
0.02354
0.02111
0.01867
0.01624
0.01380
0.01137

Fig. 1. Visualization of feature importances of the AdaBoost classifier trained for the category "airplanes". Top left: original image. Top right: heat map of the upper quartile of the learned feature importances. Bottom: Original superposed by the semi-transparent heat map.

regions which have been assigned a high *negative* importance weight (color coded in blue) have likewise been selected by the ensemble methods as important.

When analyzing the visualizations with regard to the ability of explaining the classification results, the figures immediately convey that considerably few important regions actually coincide with pixels that belong to the airplane object. All three models assess the sky above as well as the grassy ground below the airplane as important features for classification of airplanes. While this seems reasonable ("airplanes are objects surrounded by sky"), it likewise means that other images with large sky-like areas will have a high chance of being falsely classified as "airplanes" as well (e.g. "birds").

A second aspect that also becomes immediately apparent due to the visualization is that most photos of airplanes carry a more or less dominant white frame that surrounds the actual photo (in fact only 108 out of 800 images in the Caltech-101 dataset annotated as "airplanes" are not surrounded by a more or less prominent white border). While all three classification models have correctly learned this specificity by selecting border pixels among the most important (upper quartile) features it represents most likely an unwanted characteristic and classifying photos of airplanes that do not exhibit a white border will most likely show inferior classification results. In order to validate this assumption, we have split the testing data into those images having a border ($|I_{border}| = 341$) and those that do not ($|I_{\neg border}| = 59$), applied the models (trained on both types) and computed the average precision scores separately. As expected, scores dropped significantly by up to 24% (see Table 2).

This artifactual cue of the Caltech-101 dataset is revealed in other categories as well [16]. Figure 3 (left) shows an example visualization of the AdaBoost classifier for a correctly classified image (confidence score $c = 0.85$) in the category "trilobite". All images of this category exhibit a "corner" artifact resulting from

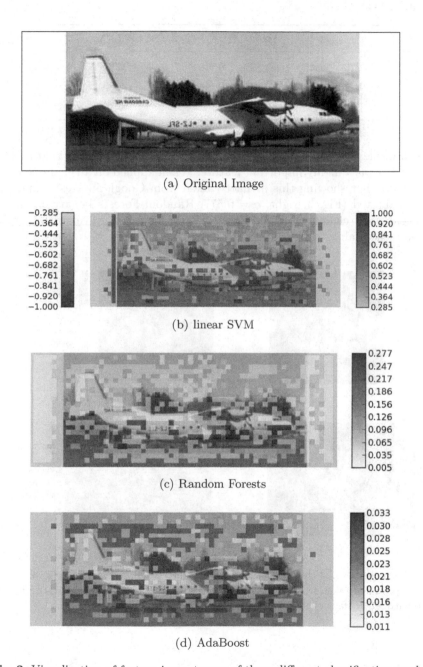

(a) Original Image

(b) linear SVM

(c) Random Forests

(d) AdaBoost

Fig. 2. Visualization of feature importances of three different classification models

Table 2. Performance of classifiers applied to images of the "airplanes" category with and without white border

classifier	AP I_{border}	AP $I_{\neg border}$
AdaBoost	0.99	0.90
Random Forest	0.98	0.83
linear SVM	0.97	0.73

artificial image rotation leaving a black background – a strong cue that all classifiers pick up as a highly important feature. When applied on an example image of a trilobite not showing this artifact (taken from Google images), AdaBoost fails as expected (Fig. 3, right, $c = 0.577$, Random Forests behave similarly). The linear SVM classifier correctly classifies the image, however with a rather low confidence value.

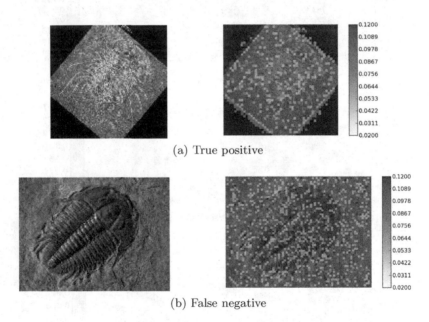

(a) True positive

(b) False negative

Fig. 3. Visualizations of the AdaBoost model trained on images of the category "trilobite". *Top:* Image taken from Caltech-101 with a prominent rotation artifact. *Bottom:* Image without artifact classified negative by the same model.

4   Summary

In this paper we have presented an approach for an intuitive visualization of different Bag-of-Visual-Words models. We have trained three different classifiers, linear SVM, Random Forests and AdaBoost, and compared the performance of these classifiers based on the Caltech-101 benchmark dataset. The visualization we propose uses a heat map-like representation of the importance scores of visual words as learned by a classifier. By providing examples from two different categories we have shown the effectiveness of our visualization. In both cases, deficits in the models' ability to generalize from the training examples as well as peculiarities within the Caltech-101 training material became immediately apparent by looking at a single testing instance enabling the user to understand *how* a decision is made.

Future work will on the one hand focus on the BoVW feature representations. Often, a multi-scale approach is taken, that extracts SIFT features at various scales in order to obtain scale-invariant descriptors. It will be interesting to see to what extent different scales will be reflected in the visualizations. Furthermore, we intend to correlate heat maps with saliency maps generated by human eye movement data. Possible similarities between image classifiers and the human perception might help to further improve image classification.

References

1. Breiman, L.: Random forests. Machine Learning (2001)
2. Csurka, G., Dance, C.R., Fan, L., Willamowski, J., Bray, C., Maupertuis, D.: Visual Categorization with Bags of Keypoints. In: Workshop on Statistical Learning in Computer Vision, ECCV (2004)
3. Cutrell, E., Guan, Z.: What are you looking for?: an eye-tracking study of information usage in web search. In: Proceedings of the SIGCHI Conference on Human Factors in Computing Systems
4. Fergus, R., Perona, P.: Learning generative visual models from few training examples: An incremental bayesian approach tested on 101 object categories. In: 2004 Conference on Computer Vision and Pattern Recognition Workshop (2004)
5. Freund, Y., Schapire, R.E.: A decision-theoretic generalization of on-line learning and an application to boosting. J. Comput. Syst. Sci. 55(1) (August 1997)
6. Hentschel, C., Gerke, S., Mbanya, E.: Classifying images at scene level: Comparing global and local descriptors. In: Detyniecki, M., García-Serrano, A., Nürnberger, A., Stober, S. (eds.) AMR 2011. LNCS, vol. 7836, pp. 72–82. Springer, Heidelberg (2013)
7. Jiang, Y.G., Yang, J., Ngo, C.W., Hauptmann, A.G.: Representations of Keypoint-Based Semantic Concept Detection: A Comprehensive Study. IEEE Transactions on Multimedia 12(1) (2010)
8. Jurie, F., Triggs, B.: Creating efficient codebooks for visual recognition. In: IEEE International Conference on Computer Vision, ICCV 2005 (2005)
9. Lowe, D.G.: Object recognition from local scale-invariant features. In: IEEE International Conference on Computer Vision, ICCV 1999 (1999)
10. Pedregosa, F., Varoquaux, G., Gramfort, et al.: Scikit-learn: Machine learning in Python. Journal of Machine Learning Research 12 (2011)
11. Sivic, J., Zisserman, A.: Video google: a text retrieval approach to object matching in videos. In: IEEE International Conference on Computer Vision, ICCV 2003 (2003)

12. Snoek, C.G.M., Worring, M.: Concept-Based Video Retrieval. Foundations and Trends in Information Retrieval 2(4) (2009)
13. Yang, K., Zhang, L., Wang, M., Zhang, H.: Semantic point detector. In: Proceedings of the 19th ACM ..., pp. 1209–1212 (2011), http://dl.acm.org/citation.cfm?id=2071976
14. Yang, L.Y.L., Jin, R.J.R., Sukthankar, R., Jurie, F.: Unifying discriminative visual codebook generation with classifier training for object category recognition. In: 2008 IEEE Conference on Computer Vision and Pattern Recognition (2008)
15. Yao, B., Khosla, A., Fei-Fei, L.: Combining randomization and discrimination for fine-grained image categorization. In: CVPR 2011 (2011)
16. Zhang, J., Marszalek, M., Lazebnik, S., Schmid, C.: Local Features and Kernels for Classification of Texture and Object Categories: A Comprehensive Study. International Journal of Computer Vision 73(2) (2006)

Coupled-View Based Ranking Optimization for Person Re-identification

Mang Ye[1], Jun Chen[1,2], Qingming Leng[3], Chao Liang[1,2],
Zheng Wang[1], and Kaimin Sun[4]

[1] National Engineering Research Center for Multimedia Software, School of
Computer, Wuhan University, Wuhan, 430072, China
[2] Research Institute of Wuhan University in Shenzhen, China
[3] School of Information Science and Technology, Jiujiang University, China
[4] State Key Laboratory of Information Engineering in Surveying, Mapping, and
Remote Sensing, Wuhan University, China
yemang@whu.edu.cn

Abstract. Person re-identification aims to match different persons observed in non-overlapping camera views. Researchers have proposed many person descriptors based on global or local descriptions, while both of them have achieved satisfying matching results, however, their ranking lists usually vary a lot for the same query person. These motivate us to investigate an approach to aggregate them to optimize the original matching results. In this paper, we proposed a coupled-view based ranking optimization method through cross KNN rank aggregation and graph-based re-ranking to revise the original ranking lists. Its core assumption is that the images of the same person should share the similar visual appearance in both global and local views. Extensive experiments on two datasets show the superiority of our proposed method with an average improvement of 20-30% over the state-of-the-art methods at CMC@1.

Keywords: Coupled-view, Ranking optimization, Person re-identification.

1 Introduction

In recent years, the person re-identification problem, namely matching people across disjoint camera views in a multi-camera system, has aroused an increasing interest in computer vision, and multimedia analysis communities [1]. The main challenges in person re-identification can be attributed to the significant visual changes in pose, illumination and viewpoint, making intra-personal variations even larger than that of inter-personal variations [7]. In addition, background clutters and occlusions cause additional difficulties [19].

To address above problems, previous works focus on constructing and selecting various distinctive and stable appearance representations for accurate identity recognition. According to the feature types of those methods, relevant work can be roughly divided into two categories: the global feature based methods [2–5] and the local feature based methods [1, 7, 8, 12, 19]. The global feature

X. He et al. (Eds.): MMM 2015, Part I, LNCS 8935, pp. 105–117, 2015.

Fig. 1. Person re-identification rank lists based on global and local methods. The red box shows the groundtruth. The global feature based method is achieved by KISSME [5], while the local feature based method is conducted by SDC [7]. Note the ranking difference between the methods while both of them achieved good results.

based methods try to integrally define a global appearance human signature with rich image features and match given reference images with the observations. For example, D. Gray et al. [2] proposed the feature ensemble to deal with viewpoint invariant recognition. To improve the performance, advanced learning techniques are employed for more reliable matching metrics, Kostinger et al. [5] projected the concatenated feature histograms into subspace by PCA to retain their global information for metric learning. The local feature based methods focus on extracting weighted patch-based features to highlight some local regions, which is decisive for constructing discriminative descriptions. For example, N. Gheissari et al. [1] adopted a decomposable triangulated graph to represent person configuration. Zhao et al. [7] extracted distinctive features by unsupervised salience learning to find the salience regions for constructing robust discriminative descriptions. More specially, the global feature based methods delineate overall feature distributions in person images, thus the retrieved candidates often appear alike at a glance but may be irrelevant, while the local feature based methods are powerful in identifying near-duplicate local regions since some local regions may be occluded or conjuncted. The complementary descriptive capability of global and local feature based methods naturally raise the question of how to aggregate them to yield better results.

As reported in existing works, both of these two category methods could achieve a relatively satisfactory results on the public datasets, e.g. VIPeR [21], CUHK01 [22] and etc. However, it's amazing to find that the ranking lists of these

Table 1. The person proportion of different overlapping ratios. The result is achieved by computing the overlapping ratios of top 20 results. Note that most overlapping ratio of the query persons are under a low value, i.e. the ranking lists vary a lot.

Overlapping ratio	0-15%	20-25%	30-45%	50-100%
Person proportion	92.08±2.21	7.59±0.91	0.32±0.15	0

two methods vary greatly. We conduct a preliminary experiment on the famous VIPeR [21] dataset to reveal the ranking difference, achieved by KISSME [5] and SDC [7] shown in Fig. 1. As shown in the figure, the global feature based yield a better result when the person possesses a single overall appearance characteristic, while the local feature based method is better when the person owns some salient local regions. Thus it can be seen that both of them can achieve relatively good results but their ranking lists vary a lot. Further more, an objective statistical data is reported in Table 1, in which the overlapping ratio is computed by the intersection size of the top 20 ranking results achieved by KISSME and SDC methods, respectively. As illustrated in the table, the overlapping ratio of most query persons is below 15%, which further verifies the ranking difference between the global and local methods. Based on these observations, their ranking lists vary a lot while both of them yield decent matching rates. These prompt us to investigate a coupled-view based ranking optimization method based on these two category methods to optimize person re-identification task.

Actually, researchers have proposed several ranking optimization methods [13–15] to revise the original ranking results, which can be categorized into two kinds: interactive relevance feedback methods and automatic re-ranking methods. For interactive relevance feedback, Ali et al [17] employed rank based constraints and convex optimization to efficiently learn the distance metric; Liu et al [14] presented a interactive error-prone post-rank visual search method at the user end. However, the interactive methods need lots of manpower which is not suitable for large scale data scenarios. For automatic re-ranking, Leng et al [13] proposed a automotive bidirectional ranking method based on the content and context similarity, but they ignore the reliability of k-nearest neighbors, and adopted the same query method for the backward re-ranking; Hirzer et al [18] proposed a similar combination method based on covariance descriptors and discriminative model, while the complementary of the two aspects is too far-fetched. In other words, existing ranking optimization methods are merely taking a single view into account. The ranking aggregation of these two category methods is seldom investigated in person re-identification.

In this paper, we proposed a novel coupled-view based rank optimization method for person re-identification based on these two complementary lines. A latent assumption is that two person images are more likely to be the same person if they share similar appearance in both global feature based and local feature based views. In particular, if a person image appeared in top-k ranking list in both global and local views, it will be more likely to be the correct match of the query image. At this point, we propose a KNN rank aggregation method by crossed-view based requery. Moreover, motivated by the relationships in social network, good friends are always trend to have more common friends [16], i.e., images captured from the same person will have more mutual k-nearest neighbors than that of different persons. Therefore, we can revise the original ranking lists by combining the nearest neighbours network similarity of the global and local feature based results. Based on this, we conduct a graph-based reranking method by constructing a two-layer complementary graph.

The main contribution of the proposed coupled-view based ranking optimization approach can be summarized as follows: (1) rank aggregation is firstly conducted in person re-identification task; (2) it presents a more reasonable combination of two category complementary methods; (3) the approach can be easily transplanted to other methods. And we have validated our approach on two public datasets, the VIPeR [21] and CUHK01 [22]. The experimental results illustrate that our method outperforms all the existing ranking optimization methods and most of the state-of-art methods.

2   Our Approach

2.1   Overview

These two category methods demonstrate distinct strengths in finding visually similar images but their ranking lists vary a lot, and the complementary descriptive capability of global and local methods makes our ranking optimization work valid. To aggregate the ranking lists achieved by different category methods, the critical issue is how to automatically measure and compare their qualities. However, the similarity scores of candidates vary largely among different queries, moreover, the similarity metrics vary a lot for different methods, all these exacerbate the directly combination difficulty of two methods. Despite these, there are many researches focus on solving this problem. More broadly, in general image retrieval, Zhang et al [20] proposed a graph fusion method, they measured the consistency among the top candidates returned by one retrieval method as the retrieval quality specific to one query, it may rely too much on the assumption that the groundtruths of the query image are always in the top k candidates which is not applied to person re-identification task, while there is only one correct match in the re-identification task [9], i.e. the top k results contain too much false matchings, it may lead to the algorithm difficult to converged and may aggravate the noise affection and increase false matching after several iterations.

The overall framework of our method is shown in Fig. 2. In order to enhance the complementarity of the two category methods, we combine them in all steps of our method. Differ to existing methods [20], we adopt conjunct neighbours in both global and local views to generate more reliable k_c nearest neighbours, instead of top k candidates directly, the sketch is shown in Fig.2(a). After getting the original ranking lists, the k_c nearest neighbours are treated as new probes to query in the original gallery set, named backward requery. More specially, to enhance the complementary of two methods, the crossed backward requery

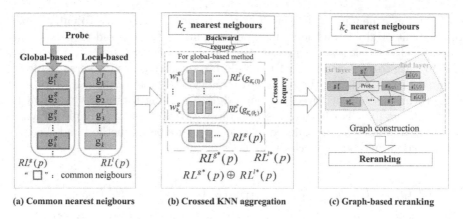

(a) Common nearest neigbours   (b) Crossed KNN aggregation   (c) Graph-based reranking

Fig. 2. The Framework of our method.(a) Common nearest neighbours; (b) Crossed KNN aggregation; (c) Graph-based reranking. Note that the superscripts denote different methods. The green and orange depict the two different methods. While the rectangle represents a person image and the rounded rectangles denote different ranking lists.

is conducted, i.e., the local-based methods is adopted for requery to revise the original global-based methods and vice versa. After that, weighted aggregation of the multi-ranking lists is introduced as shown in Fig.2(b). While the whole ranking list is revised by the crossed KNN rank aggregation, further more, the more reliable k_c neighbours are reranked by constructing a two-layer graph about the k_c common neighbours based on the two methods, the Jaccard similarity coefficient of two neighborhood images is computed, defined as graph-based reranking illustrated in Fig.2(c). The details are discussed in the following.

2.2 Common Nearest Neighbours

For the convenience of following discussion, we consider the probe person image as p and a gallery set as $G = \{g_i \mid i = 1, 2, ...n\}$, while n is the number of images in gallery set. And the global feature based method we mark it with superscript g, while local feature based method with l. We firstly get two original ranking lists achieved by two methods of query p, denote as $RL^g(p)$ and $RL^l(p)$, the k nearest neighbours of each method denote as $RL_k^g(p)$ and $RL_k^l(p)$. To get the more reliable nearest neighbours, we define their common nearest neighbours $G_{K_c}(p) = \{g_{K_c(j)} \mid j = 1, 2, ...k_c, k_c \leq k\}$ as the true neighbours of image p, k_c is the number of the common nearest neighbours, which is depicted in Fig.2(a):

$$\{G_{K_c}(p)\} = \{RL_k^g(p)\} \cap \{RL_k^g(p)\} \tag{1}$$

2.3 Crossed KNN Rank Aggregation

While we have gotten more reliable k-nearest neighbours, we choose the common nearest neighbours $G_{K_c}(p)$ for backward requery, i.e. we treat the each $g_{K_c(j)}$

in $G_{K_c}(p)$ as a new probe to search in the original gallery set. To enhance the complement of global and local methods, we adopt the crossed-view based requery, that is to say, we revise the original global feature based ranking lists $RL^g(p)$ with local feature based methods for requery and vice versa, which is depicted in Fig.2(b). For the backward requery , $(k_c + 1)$ ranking lists for each category method are achieved, i.e. one original ranking lists and k_c crossed requery ranking lists. For better presentation, we treat it as an example, of which the original ranking list is achieved by global feature based method and the local feature based method for backward requery.

$$RL^{g^*}(p) = \{RL^g(p), \quad RL^{l'}(g_{K_c(1)}), RL^{l'}(g_{K_c(2)}), \ldots, RL^{l'}(g_{K_c(k_c)})\} \qquad (2)$$

For better combination, we aggregate the ranking lists weighting by their context similarity, which represents their neighbours similarity between the query and the probe, $N_k^g(p)$ represents the k-nearset neighbours of the original ranking list and $N_k^{l'}(g_{K_c(j)})$ for the crossed backward requery neighbours:

$$w^g(g_{K_c(j)}) = \frac{\left| N_k^g(p) \cap N_k^{l'}(g_{K_c(j)}) \right|}{k} \qquad (3)$$

After computing the context similarity, the revised ranking lists of global feature based method, i.e. the eq.(2,3) can be converted to:

$$RL^{g^*}(p) = RL^g(p) + \sum_{j=1}^{k_c} w^g(g_{K_c(j)}) * RL^{l'}(g_{K_c(j)}) \qquad (4)$$

Similarly, $RL^{l^*}(p)$ can be achieved by crossed requery, and the final aggregation can be depicted as:

$$Rank^*(p) = \alpha RL^{g^*}(p) + (1 - \alpha)RL^{l^*}(p) \qquad (5)$$

where α denote the weighting parameter, we set it $\alpha = 0.8$ in this paper.

2.4   Graph-Based Re-ranking

By the crossed KNN rank aggregation, the entire ranking list is revised. While the k_c common nearest neighbours are more likely to be the correct match, further optimization of the local ranking orders is necessary, a graph-based re-ranking method is proposed. Denote the probe image as p, and $G_{K_c(p)}$ represents their common nearest neighbors for both global and local feature methods.

For each query and each common neighbour $g_{K_c(j)}$, we construct a weighted undirected graph $Graph = <G, E, w>$, the center is the query while the nodes are the neighbour images. As shown in Fig.2(c), for the green part, N_k^g denotes the forward ranking neighbours of probe image received by global-based method, while the orange part $N_k^{l'}$ expresses the backward requry neighbours of $g_{K_c(j)}$, they are linked by an edge $(p, g_{K_c(j)}) \in E$ if they are common neighbors of two layer graphs. Note that for the second layer of the graph, we adopt the other method differ to the first layer to enhance the complementarity. The revised similarity of p and $g_{K_c(j)}$ is defined as the Jaccard similarity coefficient between the neighborhoods of p and $g_{K_c(j)}$:

$$Sim(p, g_{K_c(j)}) = w(p, g_{K_c(j)}) \frac{\left| N_k^g(p) \cap N_k^{l'}(g_{K_c(j)}) \right|}{\left| N_k^g(p) \cup N_k^{l'}(g_{K_c(j)}) \right|} \tag{6}$$

where $|\cdot|$ denotes the cardinality and $w(p, g_{K_c(j)})$ is a weighting coefficient related to the original rank in $G_{K_c}(p)$, we define it as an decay factor, $rank(g_{K_c(j)}, G_{K_c}(p))$ represents the rank of $g_{K_c(j)}$ in $G_{K_c}(p)$.

$$w(p, g_{K_c(j)}) = w_0{}^{rank(g_{K_c(j)}, G_{K_c}(p))} \tag{7}$$

3   Experiments

3.1   Datasets and Evaluation Protocol

We evaluate our method on two publicly available datasets, the VIPeR dataset [21] and the CUHK01 dataset [22], as for their low image resolution, illuminations, poses and viewpoints variations, which are close to practical surveillance scene. Further more, the two datasets are the largest person re-identification datasets while much more person images are contained. VIPeR is the first dataset that mainly considering the influence of viewpoint change, and it is widely used for evaluating person re-identification methods. CUHK01 dataset get more attentions recently as for its huge number and higher resolution which is more suitable to verify the effectiveness of person re-identification approaches. All the quantitative results are exhibited in standard Cumulated Matching Characteristics (CMC) curves [10].

Following the evaluation protocol described by many predecessors, i.e., we randomly partition the dataset into two even parts, 50% for learning and 50% for testing, without overlap on person identities. All the images from Camera View A are treated as probes while those from Camera View B as gallery set. For each probe image, there is another person image matched in the gallery set, and the rank of correct match is achieved. Rank-k recognition rate is the expectation of finding the correct match within the first k ranks, and the cumulated values of recognition rate at all ranks is recorded as one-trial CMC result. For two different methods, we adopt the same configuration for experiments at each trial to get the ranking lists. Especially, we adopted KISSME [5] as the global feature based method and SDC [7] as the local feature based method. To achieve stable statistics, we looped the evaluation procedure 10 times. We set the parameter top k results as $k = 30$, and the decay factor $w_0 = 0.9$.

3.2   VIPeR Dataset

The VIPeR dataset contains 632 person image pairs captured from two different static camera views in outdoor academic environment. The dataset is challenging due to the viewpoint change of most image pairs is lager than 90 degrees, and with obvious illumination change. All the images are normalized to 128 × 48 for experiments. The experimental results obtained is shown in Fig. 3, while the Our_RA denotes the crossed KNN rank aggregation step described in Section2.3, and Our_RRA for the graph-based reranking after aggregation, while illustrative results are shown in Fig. 6. Moreover, in Table 2 we compare the performance of our method in the range of the first 25 ranks to existing rank optimization methods and the state-of-the-art methods.

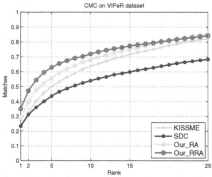

Fig. 3. Performance on the VIPeR dataset. Our approach: Our_RR and Our_RRA based on KISSME [5] and SDC [7].

As can be seen, our approach has huge promotion compared with the baseline global and local methods for both rank aggregation and graph-based reranking methods, more specially, it has nearly 55% improvement at rank 1 for graph-based reranking, and 30% at rank 5. In particular, rank 1 matching rate is around 35% for graph-based reranking and 28% for crossed KNN rank aggregation method only. Further more, compared to existing ranking optimization methods, our results are still more satisfactory, versus 24% for bidirectional ranking [13] and 19% for reranking SB [15]. The results show the efficiency of coupled-view based ranking optimization, while the single-view optimization is limited. The improvement of our approach is due to two aspects: First, we consider both the global similarity and the local similarity which can help us to get more reliable nearest neighbours for rank aggregation, optimize the results by aggregate the different weights ranking lists via the original ranking score; Second, constructing graph by two reciprocal layer with more reliable nearest neighbours give full expression to the combination of global and local feature based methods.

Moreover, comparing to existing state-of-art methods, ELF [2], SDALF [3], PCCA [6], PRDC [4] and SalMatch [19] as shown in Table 2, our results owns more outstanding performance than both metric learning methods and description-based methods in VIPeR dataset. These experimental fact shows the superiority of our methods.

Table 2. VIPeR dataset: top ranked matching rates in [%] with 316 persons

Methods	$r = 1$	$r = 2$	$r = 5$	$r = 10$	$r = 15$	$r = 25$
Our_RA	28.48	39.87	55.92	68.20	75.06	**85.41**
Our_RRA	**34.97**	**47.28**	**62.94**	**72.03**	**77.41**	85.18
SB [15]	19.32	50.70	63.37	71.28	81.18	
Bi-ranking [13]	24.57	35.28	53.38	65.18	73.32	84.27
ELF [2]	12.08	17.00	31.28	41.00	54.00	65.00
SDALF [3]	19.87	25.20	38.89	49.37	58.22	70.00
PRDC [4]	15.66	22.80	38.42	53.86	64.00	72.78
PCCA [6]	19.27	29.10	48.89	64.91	72.48	82.78
KISSME [5]	22.63	32.72	50.13	63.73	71.65	82.12
SDC [7][a]	23.32	31.27	43.73	54.05	59.87	68.45
SalMatch [19]	30.16	39.28	52.00	65.00	74.00	*

[a] It differs to [7], because they removed the complicated 98 person pairs, while we retain them as many other papers.

To further verify the availability of our method, we conduct another experiment on VIPeR shown in Fig. 4. The global feature based method we adopted is SDALF[1] [3], and the local feature based method we aopted is LDFV[2] [12]. Thus can be seen, the optimization result is satisfactory.

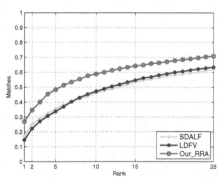

Fig. 4. Performance on the VIPeR dataset. Our approach based on SDALF [3] and LDFV [12].

---

[1] It is available at
http://www.lorisbazzani.info/code-datasets/sdalf- descriptor/
[2] It is available at http://vipl.ict.ac.cn/members/bpma

3.3   CUHK01 Dataset

The CUHK01 dataset is also obtained from two disjoint camera views in an outdoor campus environment. And it contains 971 persons much more than that of VIPeR which can be more convincible for experiments, and each person has two images in each camera. The person images in camera A are mostly captured by frontal view or back views while camera B captures the side views. All the images are normalized to 160×60 for experiments.

The experimental results based on KISSME [5] and SDC [7] are shown in Fig. 5. The results show our improvement at rank 1 is about 35%, while improved 20% at rank 10, it makes the efficiency of our method more convincible. And more comparison to existing state-of-the-art methods is shown in Table 3, since the dataset is newly reported and experiments are seldom conducted, therefore, the results are mostly from [19]. Apparently, our coupled-view ranking optimization method outperforms the existing approaches, and similar conclusions as shown in VIPeR dataset can be drawn from the comparisons.

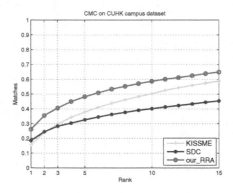

Fig. 5. Performance on the CUHK01 dataset

Table 3. CUHK01 dataset: top ranked matching rates in [%] with 485 persons

Methods	$r = 1$	$r = 2$	$r = 3$	$r = 5$	$r = 10$	$r = 15$
SDALF [3]	9.90	13.58	18.00	22.00	30.20	35.00
KISSME [5]	15.57	24.00	29.57	37.90	50.29	58.89
SDC [7]	18.52	24.52	28.29	32.62	40.14	45.42
LMNN [22]	13.45	20.15	24.50	31.00	41.50	49.00
ITML [22]	15.98	23.00	28.00	35.50	45.50	54.00
SalMatch [19]	**28.45**	**36.00**	40.00	46.00	55.50	63.00
Our_RRA	26.37	34.74	**40.66**	**48.76**	**60.58**	**68.40**

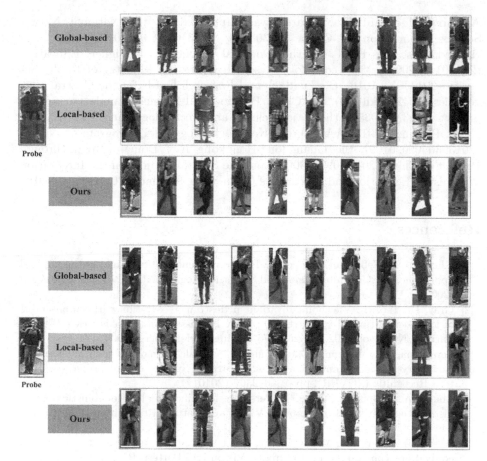

Fig. 6. Two descriptive re-identification results from VIPeR dataset. Top-10 candidates are shown for the optimization results, the red box represents the groundtruth.

4   Conclusion

In this paper, we address the ranking optimization approaches of the person re-identification problem, a novel and efficient coupled-view based ranking optimization method based on local and global feature method is conducted. The main idea is: the two images of the same person should share similar appearance in both global-based view and local-based view, the crossed KNN rank aggregation is conducted. Furthermore, motivated by the ideology in social network, if the two image share similar reciprocal neighbours, they are more likely to be the same person. Specially, constructing complementary graphs to illustrate the relations and improve the re-identification results. Extensive experiments compared to the original baseline methods and existing state-of-art methods on two publicly datasets have validated the effectiveness of our proposed method.

Acknowledgement. The research was supported by the National Nature Science Foundation of China (61303114, 61231015, 61170023), the Specialized Research Fund for the Doctoral Program of Higher Education (20130141120024), the Technology Research Project of Ministry of Public Security (2014JSYJA016), the Fundamental Research Funds for the Central Universities (2042014kf0250), the China Postdoctoral Science Foundation funded project (2013M530350), the major Science and Technology Innovation Plan of Hubei Province (2013AAA020), the Key Technology R&D Program of Wuhan (20130304090020109), the Guangdong-Hongkong Key Domain Break-through Project of China (2012A090200007), and the Special Project on the Integration of Industry, Education and Research of Guangdong Province (2011B090400601).

References

[1] Gheissari, N., Sebastian, T.B., et al.: Person re-identification using spatiotemporal appearance. In: Computer Vision and Pattern Recognition (CVPR), pp. 1528–1535 (2006)

[2] Gray, D., Tao, H.: Viewpoint invariant pedestrian recognition with an ensemble of localized features. In: Forsyth, D., Torr, P., Zisserman, A. (eds.) ECCV 2008, Part I. LNCS, vol. 5302, pp. 262–275. Springer, Heidelberg (2008)

[3] Farenzena, M., Bazzani, L., Perina, A., et al.: Person re-identification by symmetry-driven accumulation of local features. In: Computer Vision and Pattern Recognition (CVPR), pp. 2360–2367 (2010)

[4] Zheng, W.S., Gong, S., Xiang, T.: Person re-identification by probabilistic relative distance comparison. In: Computer Vision and Pattern Recognition (CVPR), pp. 649–656 (2011)

[5] Kostinger, M., Hirzer, M., Wohlhart, P., et al.: Large scale metric learning from equivalence constraints. In: Computer Vision and Pattern Recognition (CVPR), pp. 2288–2295 (2012)

[6] Mignon, A., Jurie, F.: PCCA: A new approach for distance learning from sparse pairwise constraints. In: Computer Vision and Pattern Recognition (CVPR), pp. 2666–2672 (2012)

[7] Zhao, R., Ouyang, W., Wang, X.: Unsupervised salience learning for person re-identification. In: Computer Vision and Pattern Recognition (CVPR), pp. 3586–3593 (2013)

[8] Xu, Y., Lin, L., Zheng, W.S., et al.: Human re-identification by matching compositional template with cluster sampling. In: International Conference on Computer Vision (ICCV), pp. 3152–3159 (2013)

[9] Javed, O., Shafique, K., Shah, M.: Appearance modeling for tracking in multiple non-overlapping cameras. In: Computer Vision and Pattern Recognition (CVPR), vol. 2, pp. 26–33 (2005)

[10] Wang, X., Doretto, G., Sebastian, T., et al.: Shape and appearance context modeling. In: International Conference on Computer Vision (ICCV), pp. 1–8 (2007)

[11] Hirzer, M., Roth, P.M., Köstinger, M., Bischof, H.: Relaxed pairwise learned metric for person re-identification. In: Fitzgibbon, A., Lazebnik, S., Perona, P., Sato, Y., Schmid, C. (eds.) ECCV 2012, Part VI. LNCS, vol. 7577, pp. 780–793. Springer, Heidelberg (2012)

[12] Ma, B., Su, Y., Jurie, F.: Local descriptors encoded by fisher vectors for person re-identification. In: European Conference on Computer Vision Workshops and Demonstrations (ECCV Workshop), pp. 413–422 (2012)

[13] Leng, Q., Hu, R., Liang, C., et al.: Bidirectional ranking for person re-identification. In: International Conference on Multimedia and Expo (ICME), pp. 1–6 (2013)

[14] Liu, C., Loy, C.C., Gong, S., et al.: POP: Person re-identification post-rank optimisation. In: International Conference on Computer Vision (ICCV), pp. 441–448 (2013)

[15] An, L., Chen, X., Kafai, M., et al.: Improving person re-identification by soft biometrics based reranking. In: International Conference on Distributed Smart Cameras (ICDSC), pp. 1–6 (2013)

[16] Leng, Q., Hu, R., Liang, C., et al.: Person re-identification with content and context re-ranking. In: Multimedia Tools and Applications, pp. 1–26 (2014)

[17] Ali, S., Javed, O., Haering, N., et al.: Interactive retrieval of targets for wide area surveillance. In: International Conference on Multimedia (MM), pp. 895–898 (2010)

[18] Hirzer, M., Beleznai, C., Roth, P.M., et al.: Person re-identification by descriptive and discriminative classification. In: Image Analysis (IA), pp. 91–102 (2011)

[19] Zhao, R., Ouyang, W., Wang, X.: Person re-identification by salience matching. In: International Conference on Computer Vision (ICCV), pp. 2528–2535 (2013)

[20] Zhang, S., Yang, M., Cour, T., et al.: Query specific fusion for image retrieval. In: European Conference on Computer Vision(ECCV), pp. 660–673 (2012)

[21] Gray, D., Brennan, S., Tao, H.: Evaluating appearance models for recognition, reacquisition, and tracking. In: IEEE International Workshop on Performance Evaluation of Tracking and Surveillance (2007)

[22] Li, W., Wang, X.: Locally aligned feature transforms across views. In: Computer Vision and Pattern Recognition (CVPR), pp. 3594–3601 (2013)

Wireless Video Surveillance System
Based on Incremental Learning Face Detection

Wenjuan Liao[2], Dingheng Zeng[1], Liguo Zhou[2], Shizheng Wang[3], and Huicai Zhong[1]

[1]Institute of Microelectronics of Chinese Academy of Sciences, Beijing, China 100029
[2]Chinese Academy of Sciences R&D Center for Internet of Things, Wuxi, China 214135
[3]School of Electrical & Electronic Engineering, Nanyang Technological University, 50
Nanyang Avenue, Singapore 639798
{wenjuan.liao,zhou.liguo}@outlook.com,
{zengdingheng,zhonghuicai}@ime.ac.cn, shizhengwang@ntu.edu.sg

Abstract. As an important supplement to wired video in video surveillance
applications, wireless video has taken increasing attentions and has been extensively applied into projects like "Safe City". Despite of in taxis, buses, emergen-
cy command vehicles, or temporary monitory point, there will definitely produce
massive surveillance videos. In order to retrieve and browse these videos in an
efficient way, key video browsing system and technique based on face detection
is accepted and put into promotion. Face detection is widely studied and used in
many practical applications; however, because of the distinct features of different
factors in experiments and applications, such as orientation, pose, illumination,
etc., challenges usually obstruct the practical usage. To perform successful application in wireless video browsing system, this paper proposes an incremental
learned face detection method based on auto-captured samples. Experiments
demonstrate that our proposed incremental learning algorithm has favorable face
detection performance and can work in the proposed system.

Keywords: Face detection, incremental learning, wireless video, video browsing.

1   Introduction

As one of the most novel applications in video surveillance, wireless video has attracted increasing concerns from industry and academy. In fact, wireless video is
selected as the applicable solution in most "Safe City" projects of China, for example,
Shenzhen has invested more than 280 million RMB to install wireless video surveillance system for over 13,000 buses in the city, while Dongguan has invested more
than 110 million RMB for over 10,000 buses. Such applications provide an efficient
solution of real-time storage and remote browsing for video surveillance under unfixed scene.

After the successful storage of massive surveillance video, how to retrieve and
browse these videos in an efficient and fast way has become another challenge in
addition to effective storage for surveillance video. For this reason, this paper mainly
discusses the face detection based on incremental learning in a wireless video brows-

X. He et al. (Eds.): MMM 2015, Part I, LNCS 8935, pp. 118–127, 2015.
© Springer International Publishing Switzerland 2015

ing system. This system adopts scalable coding technique [22] to encode the video with CIF resolution and 9 FPS into the base layer of surveillance video, which is transferred to remote monitoring center in a wireless manner with low-delay for real-time retrieval and browsing; meanwhile, regular video with D1 resolution and 25 FPS is encoded into complete bitstream of base layer and enhance layer, and then stored in local server of camera terminal. When there is retrieval or browsing task, videos with high resolution and frame rate will be transferred to monitoring center in a high-delayed way for browsing or checking, and the transmission mount can therefore be efficiently controlled when retrieving or browsing wireless videos. Wherein, the face detection technique oriented to wireless video is the basic supporting technique for this system.

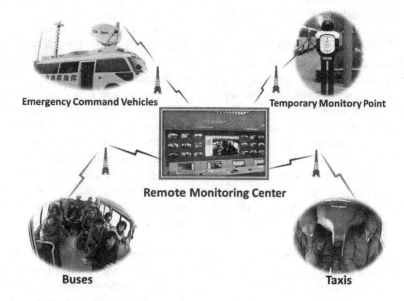

Fig. 1. The schematic diagram of wireless video system

Face detection is one of the hottest research topics in the field of computer vision; its main difficulty lies in many variations of orientation (in-plane rotation), pose (out-of-plane rotation), illumination, etc. In order to solve the problems above and reach the required speed for practical application, researchers have studied in aspects of robust feature selection, classifier improvement, detection speed promotion, and adaption enhancement. The field of face detection has significant development in the last decade. Particularly, the creative face detection algorithm based on Adaboost by Viola and Jones [1] enables the feasibility of face detection for robustness and high-speed in practical application. [2] has proposed a structured rectangle feature, [3], [4], and [5] have introduced LBP, HOG, and Edgelet feature into face detection application, respectively. [6] adds anti-spoofing technique into face detection so as to achieve reliable living face detection under the condition of low illumination as well. Since a single cascade of Haar feature can fully accomplish the frontal or near-frontal face detection task rather than straightforwardly expand to multi-view face detection. Therefore, [7] represents the parallel cascade to learn an individual classifier for each view. [2] also uses a classifiers combination with pyramid structure to solve the multi-pose problem. [8] has proposed a similar tree structure used for face detection with different orientations. In [9], a compressed-domain face detection algorithm based on neural network is proposed. [17] proposes a face image enhancement algorithm for face detection under low resolution. In addition, [10], [11], and [12] have attempted new methods to improve detection speed from the angle of feature, detector, and detection process improvement, respectively. Finally, besides different learning methods based on batch algorithm above, [13] has presented an incremental learning face detection method applicable to collision with respirator and sunglasses as well as extreme illumination change. While a threshold-based incremental learning classifier is proposed for gender recognition, face detection, and human detection [14].

However, as for wireless video, limited by view, pose, and illumination, video quality is usually low. Furthermore, no appropriate training samples can be used offline. Hence, the incremental face detection algorithm for wireless video is one of our focusing works. Our main contributions include: (1) specific to the problem of lacking offline samples with wireless video, we propose an automatic new sample selection algorithm based on region modeling, face detection, and face tracking to automatically add samples for online incremental learning; (2) merging with tree-Adaboost face detection algorithm, we propose an learning method that learns part of typical views (e.g., frontal face and 90 degree profile face) offline and then incrementally learns positive and negative samples of each view online; (3) as for the requirement of retrieval and browsing application for wireless video, we design and realize a wireless video retrieval and browsing system based on face detection, which can store detected faces for retrieving and playbacking corresponding original video according to specified face.

The rest of this paper is organized as follows. The proposed wireless video face detection method and system based on incremental learning are described in details in section 2. Some experimental results on the standard test database and scene-collected video are given to demonstrate the performance of the proposed system in section 3. Finally, section 4 concludes this paper.

2   Proposed Method and System

In this paper, oriented to the practical requirement of retrieval and browsing application for wireless video surveillance, an incremental learning based face detection and browsing system is proposed: first, an initial face detector is trained offline with standard face database; then, by region modeling, face detection, and face tracking, face tubes are extracted and some of face images are automatically selected as positive samples, while negative samples are similarly extracted from background automatically based on background modeling in selected region; next, on the basis of these new samples selected online, we perform incremental learning and online update on offline trained primary classifiers to further improve the detection performance of classifiers; finally, complete the storage of detected face images so as to retrieve and browse surveillance video according to the content and tag of face image. The functional flow diagram of face detection based on incremental learning is shown in Fig. 2.

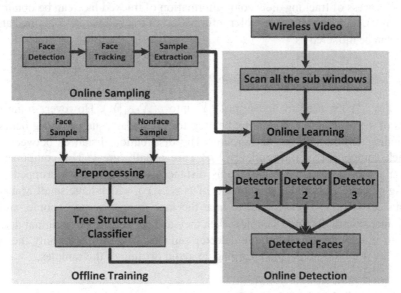

Fig. 2. The framework of incremental learning algorithm

2.1   Sample Auto-selection with Online Face Tracking

An online face tracking method facing to head rotation and pose variation is introduced in this section. As for wireless video camera, compared with fixed camera its position is unfixed, but for a certain time period (e.g., contingency command post) or a particular region (e.g., inside of buses), the background of surveillance video is relatively fixed. Therefore, Gaussian modeling method can be used for background subtraction and face tracking [15]. A gaussian mixture model (GMM) with K component densities at time t can be modeled as follow:

$$P(X_t|Y) = \sum_{k=1}^{K} w_{k,t} \cdot I(X_t, \mu_{k,t}, \sigma_{k,t}) \tag{1}$$

Where X_t denotes the appearances of each modeling pixel in the video sequence and Y denotes the background or tracked foreground tracked. $I(X_t, \mu_{k,t}, \sigma_{k,t})$ denotes the k-th Gaussian component with mean vector $\mu_{k,t}$ and covariance matrix $\sigma_{k,t}$. $w_{k,t}$ is the corresponding weight. The appearance information is the pixel value when the GMM model is applied to background modeling and face tracking.

In the algorithm of this paper, we apply background modeling, face detection and face tracking into wireless video. Through background modeling, we can narrow the range of face detection into foreground, and detect faces with offline trained face detector irrespective of false negatives rate. Then, we track the face and add the face that can be tracked but not detected into positive samples for incremental learning, while add the false alarmed face images in background region and randomly extracted images from background into negative samples for incremental learning as well.

In the process of tracking, new color information of tracked face can be obtained in each frame. Let x_{t+1} be the new color information of tracked face Y obtained at time $t + 1$. It can be modeled as:

$$P(x_{t+1}|Y) = \sum_{k=1}^{K'} w'_{k,t+1} \times I(x_{t+1}, \mu'_{k,t+1}, \sigma'_{k,t+1}) \tag{2}$$

Model $p(x_{t+1}|Y)$ is used to update $p(X_t|Y)$ into $p(X_{t+1}|Y)$. However, some components of the distribution should not be used for the update, such as those belonging to occluding objects and the background. The distribution distances between these and each component of old model $p(X_t|Y)$ are usually great. The components of $p(x_{t+1}|Y)$, which have big Mahalanobis distances to $p(X_t|Y)$, are dropped in the updating process. Only those components of $p(x_{t+1}|Y)$, which have small Mahalanobis distances to $p(X_t|Y)$, are used to treat the face from the same people, and add them to incremental learning samples when they are not detected by original detector. However, in each of the period that detector cannot detect the faces, only the first 5 samples are used for incremental learning to avoid drifting of the samples.

2.2   Online Update by Selected Sample

This paper exploits Gentle Adaboost algorithm [16] for learning and combination of each weak classifier. The goal of boosting is to combine many weak classifiers h(x) into an additive strong classifier $H(x)$:

$$H(x) = \sum_{k=1}^{K} \alpha_k h_k(x) \tag{3}$$

Where α_k are scalar weights of k classifiers. There have been many boosting algorithms proposed to learn this model; typically, this is done in a greedy algorithm where weak classifiers are trained sequentially. After each weak classifier is trained, the training examples are re-weighted so that examples misclassified previously should receive more weight.

Traditional algorithms are usually batch learning [1], training will be restarted once a new sample incoming. To this end, we propose a new sample learning method to incrementally change the existing models and classifiers. Based on the sample auto-selection method mentioned above, classifiers are capable of fully automatic online learning with new samples.

Since face detection is regarded as the problem of binary classification, in order to clearly introduce the Adaboost method combining both offline and online, we focus on the design of weak classifiers based on Gaussian distribution at first: assuming face image belongs to ω_1 class while nonface image is ω_2 class. When we adopt the same Gaussian distribution used in [14] to model each weak classifier, the distribution of features is determined by only two parameters: mean u and variance σ^2. Therefore, the classifier threshold can also be updated by Algorithm 1.

Algorithm 1: Online Learning

Input: x, one-dimensional feature.

Data: N, feature number; u_i, mean; σ_i^2, variance; g_i, minimum-error-rate classification; η, learning rate; $x(n)$, value of the n-th inputting sample on one-dimensional feature x.

Output: threshold θ.

1. Given one-dimensional feature x and its classification w_i.

   Get its mean $u_i = \dfrac{1}{N} \sum\limits_{x \in \omega_i} x$ and variance $\sigma_i^2 = \dfrac{1}{N} \sum\limits_{x \in \omega_i} (x - u_i)^2$.

2. For $g_i(x) = \ln[p(\omega_i|x)] = \ln p(x|\omega_i) + \ln P(\omega_i)$

   Then $g_i(x) = -\dfrac{1}{2}(\dfrac{x - u_i}{\sigma_i})^2 - \dfrac{1}{2}\ln 2\pi - \dfrac{1}{2}\ln \sigma^2 + \ln P(\omega_i)$

   For $g(x) = g_1(x) - g_2(x) = 0$

   (a) If $\sigma_1 = \sigma_2$

   $\theta = x = \dfrac{(u_1 + u_2)}{2}$

   (b) If $\sigma_1 \neq \sigma_2$

   $\theta = \dfrac{-(\sigma_2^2 u_1 - \sigma_1^2 u_2)^2 \pm \sqrt{(\sigma_2^2 u_1 - \sigma_1^2 u_2)^2 - (\sigma_1^2 - \sigma_2^2)(\sigma_1^2 u_2^2 - \sigma_2^2 u_1^2 + \sigma_1^2 \sigma_2^2 \ln(\sigma_2^2/\sigma_1^2))}}{(\sigma_1^2 - \sigma_2^2)}$

3. Add samples

   Update for m

   $u(n) = (1 - \eta)u(n - 1) + \eta x(n)$

   $\sigma^2(n) = (1 - \eta)\sigma^2(n - 1) + \eta(x(n) - u(t))^2$

The error bound e (Bhattacharyya bound) for the classifier is given as follows [18]:

$$e = \exp(-k(1/2)) \tag{4}$$

$$k(1/2) = \frac{1}{4}\frac{(u_+ - u_-)^2}{\sigma_+ + \sigma_-} + \frac{1}{2}\ln\frac{\sigma_+ + \sigma_-}{2\sqrt{\sigma_+ \sigma_-}} \tag{5}$$

Where (u_+, σ_+) and (u_-, σ_-) are the mean and squared variance of the positive class (ω_1) and negative classes (ω_2), respectively.

If $e_1 < e_2 < ... < e(n) < Th < ...$, we take the weak classifiers when error e is less than Th to update the threshold of corresponding classifier. Where Th is 0.8 in this paper.

3   Experimental Results

To verify the performance of method and system proposed in this paper, we adopt standard testing set MITEx face database [19], CAS-PEAL-R1 standard face database [20], and wireless video captured on buses and taxis in the procedure of project implementation to test the detection performance of incremental learning method and effectiveness of real system when retrieving and browsing wireless video, respectively.

3.1   The Effectiveness Test of Our Method

We use MITEx face database to evaluate the effectiveness and reliability of our incremental learning based face detector compared with traditional offline learned detector [1] and other incrementally learned detector [14]. The training set contains 800 face images and 1,100 nonface images of this database, in which 900 samples are randomly selected as incremental samples to gradually add into training set, while testing set contains 1,906 faces and 3,281 nonfaces with the same size of 20*20. Both positive and negative samples in this experiment are manually added.

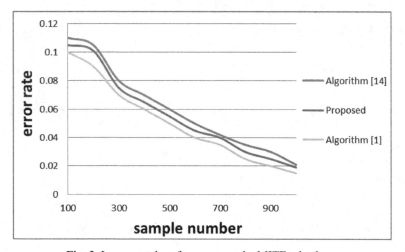

Fig. 3. Incremental performance on the MITEx database

Fig. 3 shows the detection performance improves as samples are increasing in order. Note that there are 563 features adopted in experiments. Roughly speaking, the detection performance will be always enhanced when more samples are incrementally available. Our proposed incremental algorithm performs well nearly the same as the conventional batch algorithm [1], and superior to the training method with complete online learning [14].

It can be known from experiments and data statistics that our incremental algorithm is 15 times faster than traditional algorithm [1] and the storage cost is 130 times less than it, while resources occupancy is fairly the same as other recent incremental learning algorithm [14].

3.2   The Reliability Test of Our Method

In order to verify the reliability of incremental learning algorithm, we also adopt 3500 faces and 7000 nonfaces from internet for offline learning, and employ the MITEx database [19] for testing: collecting 50 face samples from testing database at a time manually, then taking one incremental learning updating for detectors by using these 50 faces as positive samples and other 100 nonfaces as negative samples. The rest images are still reserved for testing and further incremental learning. The total time of incremental learning is 3.

In Fig. 4, the incremental learning on this testing database still can significantly improve the detection performance, where x axis is false alarm rate and y axis is detection rate.

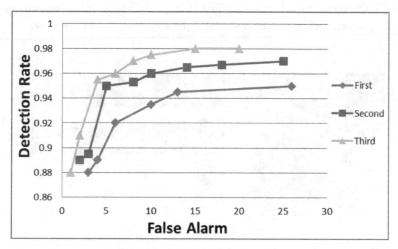

Fig. 4. Test in the MITEx databaset

3.3   The Application Test of Our Auto-incremental Learning in Real System

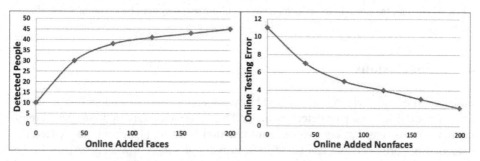

Fig. 5. Improvement of detection performance

In order to test the effectiveness of our auto-incremental learning method applied to wireless video browsing system based on face detection, we further employ the real wireless video captured in projects to perform faces detection and video browsing test. Wherein, face images in CAS-PEAL-R1 standard face database [20] are used for learning training, and then we take the 3-hours wireless video to test our incremental learning method and face detection based video browsing system (total 49 faces). Fig. 5 illustrates that, in real face detection applications with complex poses and views, incremental learning can ensure the promotion of detection performance as incremental learning time increasing by automatically adding learning other face samples, which are those faces extracted by tubes but cannot be detected at that moment. Wherein, those missed people can be detected by unremittingly adding these detected other face samples into incremental learning set.

Fig.7 describes the interface of wireless video retrieval and browsing system based on face detection. This system first detects the faces in wireless videos and then stores them. Once playback is required, just clicking the image of the certain person in database, corresponding surveillance video the one appeared can be played.

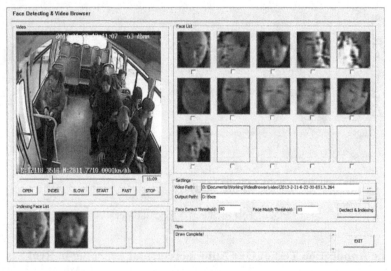

Fig. 6. The interface of wireless video retrieval and browsing system based on face detection with bus scene

4   Conclusions

Oriented to wireless video retrieval and browsing application, this paper has proposed an incremental learned face detection method based on auto-captured samples. Experiments demonstrate that our proposed incremental learning algorithm has better face detection performance and can be adopted in wireless video browsing system. In the future work, we prepare to introduce face recognition into our system to take place of manual face retrieval work, meanwhile, add corresponding feedback module to further improve the reliability of face detection and recognition rate in practical applications.

References

1. Viola, P., Jones, M.: Robust real-time face detection. Int. J. Comput. Vis. 57(2), 137–154 (2004)
2. Li, S.Z., Zhu, L., Zhang, Z., Blake, A., Zhang, H., Shum, H.-Y.: Statistical learning of multi-view face detection. In: Heyden, A., Sparr, G., Nielsen, M., Johansen, P. (eds.) ECCV 2002, Part IV. LNCS, vol. 2353, pp. 67–81. Springer, Heidelberg (2002)
3. Jin, H., Liu, Q., Lu, H., Tong, X.: Face detection using improved lbp under bayesian framework. In: Third Intl. Conf. on Image and Grahics, ICIG (2004)
4. Wang, X., Han, T.X., Yan, S.: An HOG-LBP human detector with partial occlusion handling. In: Proc. of ICCV (2009)
5. Wu, B., Nevatia, R.: Detection of multiple, partially occluded humans in a single image by bayesian combination of edgelet part detectors. In: Proc. of ICCV (2005)
6. Michelassi, P., Rocha, A.: Face Liveness Detection under Bad Illumination Conditions. In: IEEE Intl. Conference on Image Processing (2011)
7. Wu, B., Ai, H., Huang, C., Lao, S.: Fast rotation invariant multi-view face detection based on real adaboost. In: Proc. of IEEE Automatic Face and Gesture Recognition (2004)
8. Froba, B., Ernst, A.: Fast frontal-view face detection using a multi-path decision tree. In: Proc. of Audio- and Video based Biometric Person Authentication (2003)
9. Zhuang, S., Lai, S.: Face detection directly from h.264 compressed video with convolutional neural network. In: IEEE Intl. Conference on Image Processing (2009)
10. Porikli, F.: Integral histogram: A fastway to extract histograms in Cartesian spaces. In: Proc. of CVPR (2005)
11. Yan, S., Shan, S., Chen, X., Gao, W.: Locally assembled binary (LAB) feature with feature-centric cascade for fast and accurate face detection. In: Proc. of CVPR (2008)
12. Schneiderman, H.: Feature-centric evaluation for efficient cascaded object detection. In: Proc. of CVPR (2004)
13. Huang, C., Ai, H., Yamashita, T.I., Lao, S., Kawade, M.: Incremental learning of boosted face detector. In: Proceedings of the IEEE Conference on Computer Vision (2007)
14. Pang, Y., Deng, J., Yuan, Y.: Incremental threshold learning for classifier selection. Neurocomputing 89, 89–95 (2012)
15. Liu, R., Gao, X., Chu, R., Zhu, X., Li, S.Z.: Tracking and recognition of multiple faces at distances. In: Lee, S.-W., Li, S.Z. (eds.) ICB 2007. LNCS, vol. 4642, pp. 513–522. Springer, Heidelberg (2007)
16. Friedman, J., Hastie, T., Tibshirani, R.: Additive logistic regression: a statistical view of boosting. Technical report, Dept. of Statistics, Stanford University (1998)
17. Wang, Z., Hu, R., Wang, S., Jiang, J.: Face Hallucination Via Weighted Adaptive Sparse Regularization. IEEE Trans. Circuits Syst. Video Techn. 24(5), 802–813 (2014)
18. Duda, R., Hart, P.: Pattern Classification and Scene Analysis. Wiley, New York (1973)
19. http://www.ecsdn.com/search.php?keyword=MITEx-face-database
20. Zhou, D., Zhang, X., Liu, B., Gao, W.: Introduction of the JDL Large-scale Facial Database. In: The 4th Chinese Conference on Biometrics Recognition (Sinobiometrics 2003), pp. 118–121 (2003)
21. Schafer, R., Schwarz, H., Marpe, D., Schierl, T., Wiegand, T.: MCTF and scalability extension of H.264/AVC and its application to video transmission, storage and surveillance. In: Proc. VCIP, pp. 596 011-1–596 011-12 (2005)

An Automatic Rib Segmentation Method on X-Ray Radiographs

Xuechen Li[1], Suhuai Luo[1], and Qingmao Hu[2]

[1] School of Design, Communication and IT, University of Newcastle,
NSW, Australia, 2308
[2] Research Lab for Medical Imaging and Digital Surgery, Shenzhen Institutes of Advanced
Technology,
Shenzhen, P.R. China, 518000

Abstract. In this paper, an automatic rib recognition method based on image processing and data mining is presented. Firstly, multiple template matching and graph based methods are used to detect rib center line; then, the support vector machine is used to build a rib relative position model and identify the error recognition results; finally, decision trees are employed to refine the center line recognition result. The JSRT database is employed to test our method. The result of rib recognition is over 92% for sensitivity and 98% for specificity.

Keywords: rib recognition, data mining, support vector machine, decision tree.

1 Introduction

The computer aided diagnosis (CAD) system has already been widely used in the clinical practice. The purpose of CAD system is helping medical workers to make diagnosis easier and more accurate. Comparing with human, the computer has much higher computing abilities to deal with information related to the health condition of patients. Also, it has higher stability than human. Once the system has been built, it will maintain the same level and do not be affected by environment that may affect human such as mood, mental status or emotion. For these reasons, computer systems are often used to help radiologists to find potential abnormal tissues in X-ray radiographs, CT and MRI images.

In chest X-ray radiographs, all bones and organs are overlapping with each other. Different from locating tumors in CT or MIR images, the image data of X-ray images are two dimensional, having lots of noises because of the overlapping. In lung node detection, the main noises are the ribs that are overlapping with the lung field and tumors if there are. It is necessary to segment the rib and suppress them to reduce their effects. There have been many rib segmentation methods since 1970s [1-10]. Because of the poor quality of X-ray radiographs, most methods need refinement after segmentation. The main idea is to detect all potential candidates first, and using rules based on experience to identify the most probable candidates as rib border line[1-3, 6] or points[4, 5]. Also, if there are missing or extra ribs, rules based on experience are also employed to refine the result[8]. However, the rules which are designed by

X. He et al. (Eds.): MMM 2015, Part I, LNCS 8935, pp. 128–139, 2015.

experts in image processing area are imperfect and limited. The parameters in rules are mostly without optimization.

Data mining is the process of finding correlations or patterns among dozens of fields in large relational databases. It is based on artificial intelligence, machine learning, pattern recognition and statistics. It gains its popularity in recent years. The data mining technology contains classification, estimation, prediction, rule learning and clustering. With the development of data mining in last decades, many technologies are available such as support vector machine, artificial neural network, k-nearest neighbor classifier and decision tree. These classifiers can provide more accurate and robust solutions than rules. In our work, we employ support vector machine (SVM) to identify error recognition and use decision tree instead of rules to refine the recognition result.

The paper is organized as follows. Section 2 will present the rib detection method based on the feature of image by using multiple template matching and graph theory; section 3 will give a introduction of support vector machine (SVM) and decision tree; and presents the prior knowledge based result refinement by using data mining technology: SVM and decision tree; section 4 covers the experiment result and discussion, followed by section 5 which draws conclusions and points out future work.

2   Image Processing Based Rib Detection

Before rib extraction, the border of lung field should be detected first. In our previous work, the active shape model (ASM) was employed to segment lung field from X-ray image. Therefore, in this paper, we start the processing from lung field segmented X-ray images (see Fig. 1(a)).

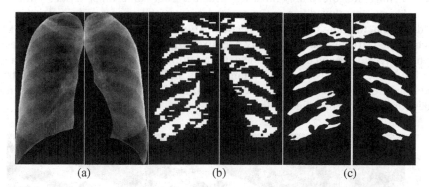

(a)   (b)   (c)

Fig. 1. Rib extraction result ((a) is the original image inside lung field; (b) is the smoothed rib image; (c) is the matching image)

2.1   Rib Extraction Based on Multiple Template Matching

If we see the ribs as signal and see the other lung tissues as noise, filters can be employed to extract the signal. In this work, we use the same methods as our previous work to extract the details. The 30×30 median filter is used to smooth the image. The detail image that contains the ribs could be obtained by estimating the difference of smoothed image and original one. A small threshold is used to binarize the image. To smooth the image, a 20×10 window is employed. The window moves pixel by pixel. If over 90% pixels in window equal 255, all pixels in the window will be set to 255, otherwise all pixels in the window will be set to 0(see Fig. 1(b)). In Fig. 1(b), there are still many noises and some ribs are connected by these noises. This condition will lead to mistaken result. Employing template to match the image under the guidance of prior knowledge can remove most noise, separate connected ribs, and still keep the accurate rib structure.

The shapes of ribs are different from each other. It is unreasonable to use one model to match all ribs even in one image. Therefore, multiple templates are used to describe the shape of ribs. The templates are straight line with different angles. The size of each model could be adjusted based on the width of lung field. The angle of each model is different so that they can fit different ribs. The correlation between image and each model are estimated, the locations with correlation over 70% maximum correlation are kept, and the template will be placed at those locations. The prior knowledge is used to guide the matching. For ribs at right side, the model slope downward should be put near the spine; model slope upward should be put far from the spine. For ribs at left side, the situation is just opposite. After matching, some non-rib tissues can be removed; also some unwanted connected rib structures can be parted (see Fig. 1(c)). In Fig. 1(c), each connected region is seen as an independent rib. To recognize the ribs, the center lines of ribs should be extracted.

2.2   Graph Based Rib Center Line Detection

Graph Building

The matching image contains most rib structures but some other tissues either. The graph theory is employed to remove none-rib tissues and extract real rib structures. To build a graph, thinning operation is needed. After thinning, the matching image keeps all connected structures but with only one pixel width (see in Fig. 2(a)). It can be seen as an undirected weighted graph. The extreme points and crossing points act as nodes (see Fig. 2(b)). The red * are the extreme points and yellow   are crossing points; the blue lines are the paths between nodes. The number of connected pixels between two nodes (the length of blue line) is the weight.

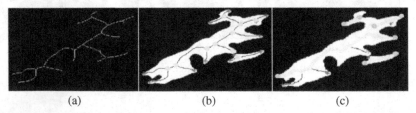

(a)   (b)   (c)

Fig. 2. Graphs of ribs((a) is the thinning image of the last rib area at right side; (b) is the graph built based on thinning image; (c) is the rib like paths after pruning operation)

The goal changes from rib extraction to selecting real extreme points of rib and finding the right path between these points.

Pruning

Since the complication of X-ray images, the graphs often contain branches. There are two main cases in the graphs. First case is the graph contains one rib only; the other is the graph contains more than one rib.

The cases that the graph only contains one rib are relatively easy to resolve. It can be dealt with by selecting the starting and ending points (extreme points and left and right, the green ● in Fig. 2(c)), and find the shortest path between these two points (the yellow line in Fig. 2(c)). There have been many algorithms to find the shortest path: Dijkstra, A*, SPFA, Bellman-Ford and Floyd-warshall, etc. the Dijkstra algorithm is the simplest method to get the shortest path. It can get the optimum solution. However, it needs traverse all nodes of graph, and consumes more time. In this case, the graphs are simple, and the number of nodes is small. The computing time is not the main problem that needs to consider. Therefore, we use Dijkstra algorithm to find the shortest path which is selected as rib path.

The cases that the graph contains more than one rib are much more difficult to handle and cannot be solved completely. In these cases, the number of ribs is unknown, and the path of rib may not be intact. What we could do is to remove non-rib path as much as possible and try to apart the graph. According to prior knowledge, for right ribs, the path that slopes downward should not appear at the middle and outer part of rib path. The slope angle of path that slopes upward should not be greater than 45° at the middle and inner part. The left ribs are opposite. The paths which do not meet the prior knowledge are removed so that the graph can be separate and become first case that one graph contains only one rib. If the graph still contains more than one rib after this operation, we leave it to later operation to refine.

The Fig. 2(c) shows the paths after pruning operation. The green ● are extreme points and yellow lines are the selected paths of each graph. They present potential rib paths.

2.3   Rib Merging

In some cases, the rib paths are not intact. One rib may be parted into two paths. Therefore, merging operation is needed. We designed rules to achieve this purpose. First, the left and right extreme points should be extracted. The distance, horizontal distance and vertical distance between the right extreme point of one rib and left extreme point of another rib will˙be calculated.

- If the distance between two extreme points is smaller than 60 pixels, these two ribs merge to one.
- If the distance is smaller than 160 pixels, vertical distance is smaller than 40 pixels and the rib at outer part is lower than inner one, these two ribs merge to one.

After merging, the average width of the connected areas will be calculated. The regions that are wider than half of average width will be kept as rib structure.

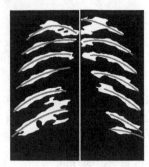

Fig. 3. Rib center line fitting result (the red lines are the fitting result)

2.4   Rib Center Line Fitting

According to prior knowledge, the shape of ribs can be expressed as parabolic curve. The equation of the curve is shown as:

$$y = a \cdot x^2 + b \cdot x + c \tag{1}$$

According to the curvature of ribs, we limit the parameter a as $a \in [0,0.003]$ and do not limit parameter b and c. We employ least square fitting to obtain the equation of rib structures (see Fig. 3). However, some rib-like areas are also kept as ribs; and on the other hand, some ribs are missing because of the imperfection of rib extraction method. Therefore, some pattern recognition based post-processing methods are needed to refine the locating result.

3   Data Mining Based Result Refinement

3.1   Support Vector Machine and Decision Tree

Support Vector Machine

Support vector machines (SVMs) are a set of related supervised learning methods used for classification and regression. Viewing input data as two sets of vectors in an n-dimensional space, an SVM will construct a separating hyperplane in that space, one which maximizes the margin between the two data sets.

Assume the training set as (x_i, y_i) $(i = 1,2, \dots, n)$, where x_i is the input with $x_i \in R_n$, y_i is the output with $y_i \in R$, here R={-1,1}, and n is the number of output samples. The liner discriminant function is $g(x) = \omega \cdot x + b$, the separating hyperplane is

$$\omega \cdot x + b = 0 \tag{2}$$

Adjusting the ω and b to normalize the discriminant function, let samples in both classes satisfied $|g(x)| \geq 1$. Among all possible hyperplanes, there is only one optimal hyperplane that maximizes the distance between the closest vectors of each class. The margin between support vectors is given as $2/\|\omega\|$. To maximize the margin, we solve the following quadratic problem with inequalities constraints:

$$\min \frac{1}{2}\|\omega\|^2 = \frac{1}{2}(\omega \cdot \omega)$$
$$s.t. \quad y_i[(\omega \cdot x_i) + b] - 1 \geq 0 \quad (i = 1, 2, \dots, n) \tag{3}$$

where ω is normal of the hyperplane, $|b|/\|\omega\|$ is the perpendicular distance from the hyperplane to the origin.

With the Lagrange multiplier method, the optimization problem of (3) is given as the saddle point of the unconstrained objective function:

$$L(\omega, b, \alpha) = \frac{1}{2}\|\omega\|^2 - \sum_{i=1}^{N} \alpha_i (y_i[(\omega \cdot x_i) + b] - 1) \tag{4}$$

where α_i denotes the Lagrange multiplier. The Lagrangian L has to be minimized with respect to ω, b and subject to the constraints $\alpha_i \geq 0$ $(i = 1, 2, \dots, n)$. The primal problem L can be transformed to its dual formulation L_D which can be optimized via easier solution. The dual problem is given as follows:

$$\min L_D(\alpha) = \frac{1}{2}\sum_{i,j=1}^{n} \alpha_i \alpha_j y_i y_j (X_i \cdot X_j) - \sum_{i=1}^{n} \alpha_i$$
$$s.t. \quad \alpha_i \geq 0 \quad (i = 1, 2, \dots, n) \quad \sum_{i=1}^{n} y_i \alpha_i = 0 \tag{5}$$

Finally, the optimal decision function is defined as:

$$f(x) = sign((\omega \cdot x) + b) = sign\left\{\sum_{i=1}^{n} \alpha_i K(x_i, x) + b\right\} \tag{6}$$

where x_i is support vector, α_i denotes Lagrange multiplier corresponding to each support vector. When generalized the optimal separating hyperplane to linearly non-separable training samples, a kernel function K is employed to map the training data to a high-dimensional space. The most common kernel function are:

- liner function: $K(x_i, x) = x_i \cdot x$;
- polynomial function: $K(x_i, x) = [(x_i \cdot x) + 1]^p$;
- radial basis function: $K(x_i, x) = e^{-\|x - xc\|^2/(2 \cdot \sigma^2)}$

Decision Tree

A decision tree is a flowchart-like structure in which internal node represents test on an attribute, each branch represents outcome of test and each leaf node represents class label (decision taken after computing all attributes). A path from root to leaf represents classification rules.

There are many specific decision-tree algorithms. Notable ones include:

- ID3 (Iterative Dichotomiser 3) [11].
- C4.5 (successor of ID3).
- CART (classification and regression tree).

Decision trees are formed by a collection of rules based on variables in the modeling data set:

- Rules based on variables' values are selected to get the best split to differentiate observations based on the dependent variable
- Once a rule is selected and splits a node into two, the same process is applied to each "child" node.
- Splitting stops when CART detects no further gain can be made, or some pre-set stopping rules are met. (Alternatively, the data are split as much as possible and then the tree is later pruned.)

Each branch of the tree ends in a terminal node. Each observation falls into one and exactly one terminal node, and each terminal node is uniquely defined by a set of rules.

3.2   SVM Based Rib Recognition Result Identification

To refine the recognition result, we first need to know which center line is error. We extract 5 features to describe the relative position of ribs and their parallelism.

- The distance between left extreme points of two neighbouring ribs (the distance of first rib is the distance between first rib and upper border of lung field, below is the same);
- The distance between middle points of two neighbouring ribs;
- The distance between right extreme points of two neighbouring ribs;
- The slope of the line that links left extreme point and middle point;
- The slope of the line that links middle point and right extreme point.

The first three features are used to measure the distance between two neighbouring ribs and last two features are used to measure the parallelism. With all these parameters together, we can build a rib relative position model. If the center line does not fit the model, it will be classified as error.

To obtain high classification accuracy, we mark all ribs and use them to train the SVM classifier. The kernel function of SVM is Radial Basis Function (RBF) and the σ in RBF equals to 0.8. The reason of choosing RBF as kernel function is that the model is highly nonlinear and RBF has better nonlinear performance than liner and polynomial function.

Since the features for ribs in different position are different, and the differences are large enough to affect the accuracy of SVM classifier, we design independent classifiers for ribs at each position. The maximum number of visible ribs is nine. Therefore, we design nine SVM classifiers. The error rate of all data is 0.0378 for left side and 0.0502 for right; the sensitivity of rib structure is 0.9619 for left side and 0.9528 for right; the sensitivity of non-rib structure is 0.9542 for left side and 0.9477 for right.

3.3   Decision Tree Based Rib Removing and Adding

After automatic marking using SVM classifiers, we can distinguish the error structures. Rules based on prior knowledge can be designed to refine the locating result by removing error recognitions and adding missed ribs[5, 6]. However, the parameter selecting of rules cannot achieve optimization by relying on experience only. Decision tree is a well-developed data mining technology. It is a nonlinear classifier and has been successful applied in many fields on pattern recognition. The goal is to create a model that predicts the value of targets based on several inputs. The structure of decision tree is similar to rules and it can be employed to achieve rib removing and adding, instead of rule based approach.

First, some parameters are defined to evaluate the relative position and distance (see Table 1).

Table 1. Parameters description

Parameters	Description
n	The number of ribs
Dism(n)	The distance of middle points between nth rib and previous neighbouring rib
MeanDism1	The average distance of middle points of first 4 ribs
MeanDism4	The average distance of middle points of 4th, 5th and 6th rib. If there are less than 6 ribs, MeanDis4 is the average distance of middle points of 4th to last rib.

Table 2. Outputs and corresponding rib removing and adding operations

Outputs	Corresponding operation
1	Remove the current rib, copy the nearest correct rib and move to the same location of middle point of the current rib.
2	Remove the current rib, copy the nearest correct rib and move to the location at 0.7 and 0.4×(Dism(n)+Dism(n+1)) from the next rib.
3	Create a new rib by calculate the average position of current rib and previous rib.
4	Remove current rib.

By using these parameters, we build 5 inputs:

- n
- Dism(n)/ MeanDism1
- Dism(n)/ MeanDism4

- Dism(n+1)/ MeanDism1
- Dism(n+1)/ MeanDism4

These 5 parameters will act as inputs of decision tree classifier, and the outputs will be numbers represent operations of adding or removing ribs. The output numbers and their corresponding operations are shown in Table 2.

The training set is built manually. It includes the input parameters and manually selected outputs. The training set is chosen randomly and it includes 435 ribs in 30 images. CART algorithm is employed to build the decision tree. The tree is shown as Fig. 4.

The ribs that are classified into errors by SVM are evaluated by the decision tree. The decision tree can make further classification and estimate which kind of error they belong to. The corresponding rib adding and removing processing will follow to refine the rib recognition result. The accuracy of decision tree is 85.33%.

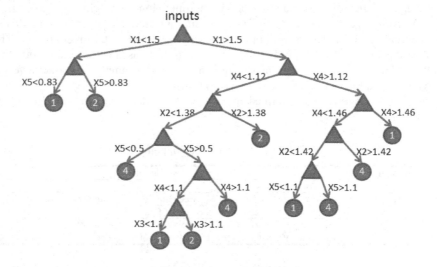

Fig. 4. The structure of decision tree (x*n* is the *n*th input defined above)

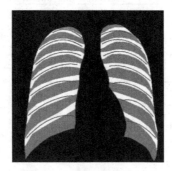

Fig. 5. Rib recognition result

The SVM and decision tree processing should be applied until no rib is classified as error. The final recognition result is shown as Fig. 5. In Fig. 5, the gray area is the lung field, the white area is rib segmented manually and the blue line is the fitted center line by using our method.

4   Result and Discussion

The method is evaluated on a database of frontal chest x-ray radiographs compiled by the Japanese Society of Radiological Technology (JSRT) [12]. This database contains 247 chest x-rays, among which, 154 x-rays are abnormal and 93 x-rays are normal. All x-ray images have a size of 2048×2048 pixels and a gray-scale color depth of 12 bits. It does not have ground truth mask for rib segmentation. We create the segmentation mask manually with the acknowledgment of radiologist.

To measure the rib recognition result, the definition of 'Recognized' has been made. If 80% of the fitted rib center line is inside the rib area, we define it as 'Recognized'. Otherwise, it is 'Unrecognized'.

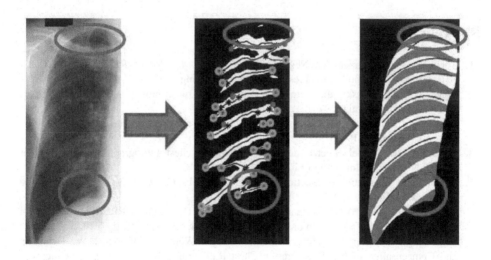

Fig. 6. The difficult example for rib locating (the red sequels are the missing and error locating rib. The first rib is missing and second rib is adjusted to error location since the first rib is missing; the visible part of last rib is too short and it is removed as noise).

We select all 93 normal data in JSRT database to test our method. They contain 1356 visible ribs. 101 of them have not been recognized and 15 center lines are error. The sensitivity and specificity of this rib recognition system are 92.27% and 98.81% respectively.

Most of the unrecognized ribs are first two ribs and last ribs (see figure 6 as an example). The reason of missing first two ribs is that first two ribs overlap with clavicle. The strong signal of clavicle becomes strong noise of rib recognition. Also, in many cases, the distance between these two ribs are very narrow. It is difficult to refine the missing rib by post-processing. The reason of missing last rib is that at lower part of lung field, pulmonary arteries are much thicker than them at other places. They overlap with last rib and they have similar shape. It will lead to missing or error recognition.

5 Conclusion and Future Work

In this paper, we develop a rib recognition method based on image processing and data mining technology; employ graph theory to extract real rib path from potential candidates; use SVM to identify error recognition results; and finally employ decision tree to refine the rib recognition result. The experiment result shows it has high sensitivity and specificity.

The future work will focus on better rib area extracting methods and more accurate and robust rib position model design.

References

1. Toriwaki, J.-I., Suenaga, Y., Negoro, T., Fukumura, T.: Pattern Recognition of Chest X-Ray Images. Computer Graphics and Image Processing 2, 252–271 (1973)
2. Wechsler, H., Sklansky, J.: Automatic detection of rib contours in chest radiographs, University of California (1975)
3. Brace, C.M., Kulick, J.H., Challis, T.W.: Automatic rib detection in chest radiographs. Queen's University (1977)
4. Ballard, D.H.: Model-directed detection of ribs in chest radiographs, Department of Computer Science, University of Rochester (1978)
5. Souza, P.D.: Automatic rib detection in chest radiographs. Computer Vision, Graphics, and Image Processing 23, 129–161 (1983)
6. Yue, Z., Goshtasby, A., Ackerman, L.V.: Automatic Detection of Rib Borders in Chest Radiographs. IEEE Transaction on Medical Imaging 14, 525–536 (1995)
7. Vogelsang, F., Weiler, F., Dahmen, J., Kilbinger, M., Wein, B., Gflnther, R.W.: Detection and Compensation of Rib Structures in Chest Radiographs for Diagnose Assistance. In: Part of the SPIE Conference on Image Processing, San Diego, California, pp. 774–785 (1998)
8. Moreira, R., Mendonça, A.M., Campilho, A.C.: Detection of Rib Borders on X-ray Chest Radiographs. In: Campilho, A.C., Kamel, M.S. (eds.) ICIAR 2004. LNCS, vol. 3212, pp. 108–115. Springer, Heidelberg (2004)
9. Loog, M., Ginneken, B.V.: Segmentation of the Posterior Ribs in Chest Radiographs Using Iterated Contextual Pixel Classificatio. IEEE Transaction on Medical Imaging 25, 602–611 (2006)
10. Lee, J.-S., Wang, J.-W., Wu, H.-H., Yuan, M.-Z.: A nonparametric-based rib suppression method for chest radiographs. Computers & Mathematics with Applications 64, 1390–1399 (2012)

11. Quinlan, J.R.: Induction of decision trees. Machine Learning 1, 81–106 (1986)
12. Shiraishi, J., Katsuragawa, S., Ikezoe, J., Matsumoto, T., Kobayashi, T., Komatsu, K.-I., et al.: Development of a Digital Image Database for Chest Radiographs With and Without a Lung Nodule: Receiver Operating Characteristic Analysis of Radiologists' Detection of Pulmonary Nodules. American Journal of Roentgenology 174, 71–74 (2000)

Content-Based Discovery of Multiple Structures from Episodes of Recurrent TV Programs Based on Grammatical Inference

Bingqing Qu[1,3], Félicien Vallet[3], Jean Carrive[3], and Guillaume Gravier[2]

[1] University of Rennes 1 – IRISA and Inria Rennes, France
[2] CNRS – IRISA and Inria Rennes, France
[3] French National Audiovisual Institute, France
{bqu,fvallet,jcarrive}@ina.fr, guillaume.gravier@irisa.fr

Abstract. TV program structuring is essential for program indexing and retrieval. Practically, various types of programs lead to a diversity of program structures. Besides, several episodes of a recurrent program might exhibit different structures. Previous work mostly relies on supervised approaches by adopting prior knowledge about program structures. In this paper, we address the problem of unsupervised program structuring with minimal domain knowledge. We propose an approach to identify multiple structures and infer structural grammars for recurrent TV programs of different types. It involves three sub-problems: i) we determine the structural elements contained in programs with minimal knowledge about which type of elements may be present; ii) we identify multiple structures for the programs if any and model the structures; iii) we generate the structural grammar for each corresponding structure. Finally, we conduct use cases on real recurrent programs of three different types to demonstrate the effectiveness of proposed approach.

1 Introduction

With the sharp increase of multimedia contents broadcasted on television, there is an emerging need for efficient audiovisual information management, including browsing, indexing and retrieval. These tasks can benefit from TV program structuring, the goal of which is to temporally segment programs into their basic constitutive elements. Practically, different programs have their own rule and methodology for edition, which leads to various structures. Most approaches for program structuring assume that the structure of the program is known as prior knowledge, for example in cases of sports [1] and news [2] which have received tremendous attention. To skirt the supervised issue, recent work exploit the properties of recurrence of some TV programs to infer the structure with minimal prior knowledge, such as [3] and [4] both focusing on recurrent TV programs.

In this paper, we follow this last path and attempt to discover the structure of programs via the analysis of a collection of episodes of the recurrent program. Recurrent programs designate programs with multiple episodes periodically broadcasted (e.g., daily, weekly). Such programs include TV news, magazines and

X. He et al. (Eds.): MMM 2015, Part I, LNCS 8935, pp. 140–154, 2015.
© Springer International Publishing Switzerland 2015

entertainments and are frequently seen on all television channels. Let's take TV news, a typical recurrent program, as an example to further understand them: TV news usually start with a brief outline of the reports (*outline*), followed by an alternation of anchorperson's announcement of the upcoming topic and news report (*news content*). Most news programs end with interview segments (*interview*), sports or program trailers (*trailer*). We designate such constitutive elements of the program as *structural elements* which refer to a video segment with a particular syntactic meaning. Most episodes of recurrent TV programs follow the same editorial structure: Structural elements repeat across episodes and appear in almost the same order with similar duration. In summary, recurrent TV programs usually have two structural properties: *element repetitiveness* and *temporal stability*. The stable temporal structure of recurrent TV programs allows generating a common model to present the overall structure of the program, as proposed in previous work [4,5] where a grammar of the program is inferred. However, across episodes, the organization of a program may partly vary according to different factors. Considering the news example above: The days when there are invited people, the right episodes usually end with the *interview*, while the days when a new film is on, the episodes usually end with the *trailer*. As a result two different structure models could be found for the same program, i.e., *multiple structures* may exist for the program, a fact that was disregarded in previous attempts to infer a grammar from a collection of episodes. With this idea in mind, we address the problem of grammatical inference to recover the structure of recurrent TV programs allowing the discovery of multiple structures. A recurrent program usually has quite a few episodes, which obviously leads to very hard workload if all episodes are processed. Since recurrent programs have stable structure across episodes, we adopt a collection of episodes from the program to discover the structures and generate the corresponding *structural grammars*, i.e., a graphical representation of the structural model representing the structure of the program, including structural elements, their temporal organizations, relative duration as well as presence probabilities. In practice, the inferred grammar can be utilized to process additional episodes from the same program, which is the application scenario of this paper, or to acess and browse particular events within the collection of episodes.

Our work is elaborated in an unsupervised way with the assumption that we have minimal prior knowledge about the program genre, the existence of multiple structures and the types of structural elements which may be present. Considering the absence of prior knowledge, we design a three-stage method to determine the structural elements and model the structure which is evidentiated by one or several grammars, depending on the number of structures found. Based on a collection of episodes, firstly we adopt a large number of audiovisual detectors to detect general events, which refer to the audiovisual segments with basic features or information of video content, such as, monochrome images, silence segments, etc. The structural elements are then discovered based on the repetitiveness property using density filtering. Secondly, in order to identify the existence of multiple structures, we adopt a hierarchical clustering method to classify the episodes into different clusters, where the number of clusters is automatically determined. For each cluster, we statistically model the structure of the episodes using a categorical distribution matrix. Finally, a graphical representation of the grammars embedded in the distribution matrix is generated by segmenting the structure of the program into coherent states.

The rest of the paper is organized as follows. Section 2 reviews the existing techniques for TV program structuring. Section 3 formally states the overall method. Section 4 describes the method to identify the structural elements. Section 5 explains the multiple structure identification and structure modeling. Section 6 introduces the structural grammar generation. Experimental evaluations are reported in Sect. 7, followed by a conclusion in Sect. 8.

2   Related Work

A number of TV program structuring techniques has been proposed in literature. Existing approaches can be classified in two categories, according to whether prior knowledge of the program structure is used or not.

In the first category, [1] and [6] target sports programs by integrating domain knowledge about the sports and editing rules to model the structure with a hidden Markov model. In the news domain, [2] and [7] automatically annotate the news content to identify typical video shots. These studies target the entire structure of a specific type of programs. Alternate works focus on detecting typical structural elements. In [8] and [9] anchorpersons are detected in newscasts. In [10] and [11] goal events are identified for soccer videos. Most of these approaches are however supervised since prior knowledge of program structure is required. In the second category, the problem of program structuring is addressed without prior knowledge of the structure and the structure of programs is discovered by generic solutions. The event repetitiveness is leveraged by considering visual recurrence in [3] to detect separators. In [12], recurrent segments exhibiting audiovisual consistency are discovered and filtered according to their relevance to the video structure, while [13] adopts a frequent pattern approach for anchorperson detection and other related purposes. These approaches focus on detecting a typical structural element by considering their inherent properties across diverse types of programs. Recent works also address the entire structure of programs in an unsupervised manner. In [4] and [5], taking advantage of the property of recurrence, the underlying structure of recurrent programs is explored by grammatical inference using multiple sequence alignment technique. The approaches of this category try to shift from supervised to unsupervised techniques for program structuring, incorporating minimal prior knowledge.

Taking the last path, our work also makes use of the property of recurrence to structure TV programs. Extending previous work on the subject, we utilize a statistical modeling method instead of multiple sequence alignment to model the structure, allowing us to further identify the existence of multiple structures within a collection of episodes. While multiple sequence alignment failed at dealing with the existence of multiple structures, in particular due to the

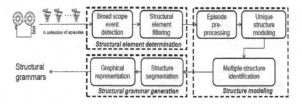

Fig. 1. General architecture for the grammatical inference of program structuring

representation adopted, the statistical modeling approach taken here allows for clustering to detect the existence of various structures which are combined in a probabilistic grammar in the end.

3   Overall Method

In this paper, we exploit grammatical inference to discover multiple structures by analyzing contents from a collection of episodes of a recurrent TV program, in the absence of prior knowledge about the program structure. Facing the challenges, a three-stage approach is proposed as depicted in Fig. 1, discovering the structural elements, modeling the structures and generating the structural grammars. Let $EP = \{ep_1, ep_2, .., ep_N\}$ be a collection of N episodes from a recurrent TV program. Firstly, structural elements, i.e., the elements relevant to the structure of the program, are discovered by leveraging a broad scope of audiovisual detectors and analyzing the temporal distribution of these general events. We denote $E = \{E_1, E_2, ..., E_d\}$ as d detected structural elements. In the second stage, the length of episodes in the collection is normalized to T, and each episode is segmented into T uniform time intervals. Relying on a hierarchical clustering method, the episodes are classified into M different clusters $C = \{C_1, C_2, ..., C_M\}$, corresponding to M different structures for the program. For each cluster, a statistical model corresponding to one of the structure of the program is obtained, yielding a collection of structural models denoted as $S = \{S_1, S_2, ..., S_M\}$. In the last stage, the structural grammar is obtained by segmenting each structure into coherent structural states and adopting symbolic representation for the purpose of concise visualization of the structural models. Hence, M corresponding grammars are generated for the program, denoted as $G = \{G_1, G_2, ..., G_M\}$. In the following sections, we detail each stage of the proposed approach for program structuring of recurrent TV programs.

4   Structural Element Determination

Discovering the elements composing the program is an essential task for program structuring. However, given the assumption of having no prior knowledge about the types of structural elements which may be present in the program, we decide to mine possible elements for each program from all existing ones, giving the priority to the basic and common elements. With this proposition, we apply a number of audiovisual detectors to the programs and filter out the valid elements by means of the property of element repetitiveness. This method for structural element determination is successfully used in [4] and [5].

4.1   Broad Scope Event Detection

In order to discover structural elements generic enough for various types of programs, a large number of event detectors should be adopted to firstly detect general purpose events which may potentially be relevant to the structure of programs. Practically, considering a trade-off between the type of programs, the complexity at run time and implementation issues, nine key detectors are applied. Among them, seven are visual detectors: shot detector, dissolve detector, monochrome image detector, text region detector, motion activity detector, person clustering, as well as shot reverse shot detector. These visual detectors are the common ones for discovering the basic visual features of TV programs. Besides, two audio detectors, i.e., speech/music/silence detector and audio recurrence detector, aim at detecting generic audio features for TV program structuring.

4.2   Structural Element Filtering

A considerable amount of general events is detected in the previous stage, but they are not all relevant to the structure of recurrent programs. For instance, a short sequence of black frames could be a separator between two successive parts of the program, however it could be also found in a night scene. In order to identify the valid structural elements, we adopted two strategies, i.e., *role recognition* and *density filtering*. On the one hand, based on person clustering, role recognition aims at identifying the persons who have an important role in the program, such as anchors, conductors, etc. The presence of these roles is always a strong cue to the structure. Role recognition is realized by comparing the time related features defined for characterizing each cluster, and identifying the person who has the most significant features as a dominant one (i.e., anchors or conductors) for each episode. On the other hand, density filtering analyzes the temporal distribution of general events, and filters out the ones which repeat with relative temporal stability across episodes. By leveraging the property of repetitiveness of structural element across episodes, the valid elements are selected based on the temporal density analysis with a Gaussian kernel function. In Fig. 2(a), we illustrate a collection of episodes, depicting the events detected (colored rectangles) as well as the structural elements resulting from density filtering (enclosed in dashed boxes).

5   Program Structure Modeling

After obtaining the structural elements composing each program along with their respective positions in each episode, we now study the task of representing and modeling the structure. As mentioned earlier, recurrent TV programs have stable

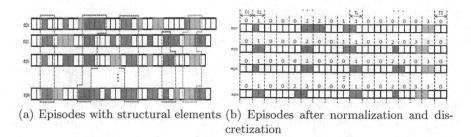

(a) Episodes with structural elements (b) Episodes after normalization and discretization

Fig. 2. An example of structural element determination and episode pre-processing

temporal structures across episodes, which allows yielding a common structure for the program. However, multiple structures may exist for one program. Taking the example given in the Sec. 1, there are two structures in the news program. Sharing the same structural elements at the beginning, one ends with *interview* when people are invited, while the other ends with *trailer* when a new film is on. This motivates us to utilize a clustering technique to identify multiple structures. The proposed structure modeling method was successfully used in [14] to model periodic behaviors for moving objects. Before identifying and modeling the program structures, pre-processing is done in order to obtain a suitable representation yielding more concise and informative structures.

5.1   Episode Pre-processing

In practice, the episodes from the same recurrent program do not have exactly the same length, or the same structural elements neither on type nor on quantity. It brings great difficulties to realize the clustering technique. In order to facilitate the task of clustering, we pre-process the episodes by normalization and discretization to translate the episodes into digital number sequences. We segment each episode into T uniform intervals after normalizing the length of the episodes to T. Let t_i ($t_1 \leq t_i \leq t_T$) denote the i-th relative time stamp in T. As illustrated in Fig. 2(b), given a set of relative time stamp $TS = \{t_1, t_2, ..., t_T\}$, we mark each time stamp t_i with a number k ($1 \leq k \leq d$), each number representing a structural element E_k from the d detected structural elements $E = \{E_1, E_2, ..., E_d\}$. We mention that the number "0" denotes the absence of any structural element. If more than one element are detected in an interval, we take the one who has longer duration in the time stamp as the valid element. In the extreme case where two elements are found in the same time stamp with the same duration, we consider the type priority to choose the valid one, because some types of elements have a stronger clue for program structures than others, such as separator or anchor. Hence each episode can be represented by a number sequence, while a collection of episodes produces a set of number sequences.

5.2   Unique Structure Modeling

While modeling the structure, we assume that a program is homogeneously structured, i.e., *unique structure*, before moving to the step of multiple structure identification. Considering a collection of episodes $EP = \{ep_1, ep_2, ..., ep_N\}$ with d discovered structural elements $E = \{E_1, E_2, ..., E_d\}$, we have a set of number sequences corresponding to the episodes. We use ep_j to denote the j-th episode, and e_i^j to denote the structural element in the i-th relative time stamp in the episode. Thus $e_i^j = k$ $(0 \leq k \leq d)$ means that the structural element is k at t_i in the j-th episode. For example, in news program, two structural elements, $E = \{E_1, E_2\}$, are determined, where E_1 represents *separators*, E_2 represents *outlines*. Therefore, $e_2^5 = 2$ means at t_2 in the 5th episode, the valid structural element is the *outline*, while $e_{10}^3 = 0$ means at t_{10} in the 3rd episode, no determined element is found.

Naturally, we may use the categorical distribution to model the element probability for TV programs. Let x_i be the categorical random variable representing the type of structural elements at time stamp t_i. We propose $\mathbf{P} = [\mathbf{P}_1, \mathbf{P}_2, ..., \mathbf{P}_T]$ to be a categorical distribution matrix, where each column $\mathbf{P}_i = [p(x_i = 0), p(x_i = 1), ..., p(x_i = d)]^T$ is an independent probability vector satisfying $\sum_{k=0}^{d} p(x_i = k) = 1$. The structure of the collection $EP = \{ep_1, ep_2, ..., ep_N\}$ modeled by some distribution matrix $\mathbf{P}$ yields

$$P(EP \mid \mathbf{P}) = \prod_{ep_j \in EP} \prod_{i=1}^{T} p(x_i = e_i^j) \tag{1}$$

considering all episodes and time intervals independent. From the estimation point of view, the best maximum likelighood model is given by

$$p(x_i = k) = \frac{\sum_{ep_j \in EP} \mathbf{1}_{e_i^j = k}}{|EP|} \tag{2}$$

where $\sum \mathbf{1}_{e_i^j = k}$ indicates the number of occurrences while $e_i^j = k$. In other words, $p(x_i = k)$ is the relative frequency of structural element E_k at t_i across all episodes from EP. As seen in Fig. 3(a), a simple example with two discovered structural elements is given. A set of number sequences issued from a collection of episodes is modeled with a probability distribution matrix. In Fig. 3(b), an illustration of the probability matrix represents the time ordered structural elements and their relative probability at each time stamp.

5.3   Multiple Structure Identification

The model representation clearly assumes that the structure is unique across episodes, which is not always the case in practice. Recurrent programs may have different structures according to the editorial rules. Given a collection of episodes from the same program, we aim at identifying multiple structures using a clustering technique. Supposing that there are M underlying structures, i.e., $S = \{S_1, S_2, ..., S_M\}$, each of which has its own structural elements and probability

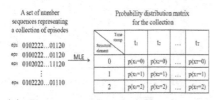 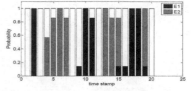

(a) Generation of probability distribu- (b) Illustration of probability distri-
tion matrix   bution matrix

Fig. 3. Structure modeling for unique structure with two structural elements

distribution matrix $\mathbf{P} = \{\mathbf{P}^1, \mathbf{P}^2, ..., \mathbf{P}^M\}$, the episodes $EP = \{ep_1, ep_2, ..., ep_N\}$ should be partitioned into M clusters, i.e., $C = \{C_1, C_2, ..., C_M\}$, so each cluster represents one structure. There are many clustering techniques to classify the episodes into M clusters, however the number of underlying structures (i.e., M) is unknown beforehand. So we propose a hierarchical agglomerative clustering method to group the episodes while at the same time determining the optimal number of structures.

In order to adopt the clustering technique, a distance measure between two clusters, i.e., two structures, needs to be defined. Since the structure is represented by a probability distribution matrix, the distance between two structures is determined by their probability distribution matrices. Several methods are available for measuring the distance between two probability distribution matrices. Here we propose to utilize the Jensen-Shannon divergence (*JS divergence*), because it measures the similarity between two probability distributions with a symmetric and bounded value, required for hierarchical clustering techniques. The JS divergence, based on the Kullback-Leibler divergence, between two probability distribution matrices $\mathbf{P}$ and $\mathbf{Q}$ is defined as:

$$D_{JS}(\mathbf{P} \parallel \mathbf{Q}) = \frac{1}{2}D_{KL}(\mathbf{P} \parallel \mathbf{M}) + \frac{1}{2}D_{KL}(\mathbf{Q} \parallel \mathbf{M}) \qquad (3)$$

where $\mathbf{M} = \frac{1}{2}(\mathbf{P} + \mathbf{Q})$, and $D_{KL}()$ denotes the Kullback-Leibler divergence. A small $D_{JS}(\mathbf{P} \parallel \mathbf{Q})$ means that two distribution matrices $\mathbf{P}$ and $\mathbf{Q}$ are similar, and conversely. We define the distance between two structures as

$$dist(S_1, S_2) = dist(C_1, C_2) = D_{JS}(\mathbf{P}^1 \parallel \mathbf{P}^2) \qquad (4)$$

During each iteration of the hierarchical clustering, two clusters with minimal distance are merged, the new cluster inheriting the episodes owned by the original clusters C_1 and C_2. The probability distribution matrix corresponding to the newly merged cluster is computed as

$$\mathbf{P}^{new} = \frac{\mid C_1 \mid}{\mid C_1 \mid + \mid C_2 \mid} \cdot \mathbf{P}^1 + \frac{\mid C_2 \mid}{\mid C_1 \mid + \mid C_2 \mid} \cdot \mathbf{P}^2 \qquad (5)$$

With such distance measure and cluster merge method, we now identify the multiple structure of programs. However without the number of underlying structures, M is unknown. We propose to determine the optimal cluster number while exhibiting hierarchical clustering by monitoring the quality of clusters using an

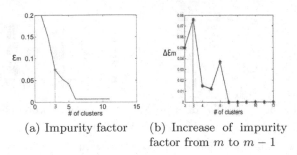

(a) Impurity factor   (b) Increase of impurity
                           factor from m to $m - 1$

Fig. 4. Determination of cluster number by impurity factor

impurity factor. At each iteration of clustering, two clusters with minimal distance are merged and the number of clusters turns from m to $m - 1$. At each iteration, an impurity factor is computed as

$$\varepsilon_m = \frac{1}{m} \sum_{i=1}^{m} IM(C_i) \tag{6}$$

where $IM(C_i)$ is the impurity [14] of each cluster, defined as

$$IM(C) = \frac{\sum\limits_{ep_j \in EP} \sum\limits_{i=1}^{T} 1_{e_i^j \neq 0} \cdot (1 - p(x_i = e_i^j))}{\sum\limits_{ep_j \in EP} \sum\limits_{i=1}^{T} 1_{e_i^j \neq 0}} \tag{7}$$

During the process of clustering, we monitor the change of ε_m. When the number of clusters turns from m to $m-1$, if the impurity factor ε_m increases significantly, it indicates that m might be the correct structure number, because that the significant increase of ε_m signifies that the newly merged cluster may contain two different structures. Therefore, the optimal structure number M is determined when the impurity factor has maximal increase while the cluster number turns from N (the total episode number of the collection) to 1. We illustrate an example of news program in Fig. 4, the impurity factor suddenly increases at $m = 3$. It indicates that there are three structures for the program.

Finally, we are able to identify multiple structures for a recurrent TV program by a hierarchical clustering method with an automatically determined structure number. These discovered structures mean to be common models to represent the structures of the recurrent program. However, verifying the clustering results obtained above, we observe the case that in certain clusters there exists just one episode. These phenomena can be explained by the fact that the isolated structure may result from an episode badly processed in earlier stages or some special made episodes, such as episode on Christmas or special issues. So we treat the episode as an isolated structure for which no grammar can be made to make sure that the discovered structure is common enough to represent the program. We can observe this case in the example in Fig. 4(a), so practically the number of structures for the news program is deemed as two instead of three.

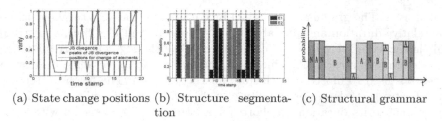

(a) State change positions (b) Structure segmenta-   (c) Structural grammar
tion

Fig. 5. Illustration of structural grammar inference

6   Structural Grammar Generation

Having discovered all structures for a recurrent program, we further generate structural grammars for the more concise visualization of structural models. Since every discovered structure for the program will be considered separately, the rest of this section will focus on one specific structure.

6.1   Structure Segmentation

In the stage of structure modeling, the episodes are segmented into uniform time stamps. Therefore the structure represented by a probability distribution matrix is based on the time stamps, as shown in Fig. 3(b). This illustration provides a global understanding of the structure, however it has very limited abstraction capabilities and is not concise enough owning to some time stamps repeating with the same structural elements and similar probability distribution. In order to better visualize structures, we propose to separate the time stamps into coherent states, each of which has the same structural elements with similar probability distribution. To achieve this goal, we verify between two successive time stamps the variations of two indicators, i.e., the JS divergence and the composition of structural elements. If one of the two indicators is changed, we consider that there is a rupture of state. More concretely, as illustrated in Fig. 5(a), the positions where the JS divergence has a peak or where the combination of structural elements are changed are deemed to be a rupture for coherent states. The peaks of JS divergence imply that in the corresponding positions the distribution vectors at two successive time stamps are far too different. Practically, while segmenting the time stamps, we ignore the elements who has small presence probability if necessary to clarify the structures. In the Fig. 5(b), the structure is divided into different coherent states according to the Fig. 5(a).

6.2   Graphical Representation

In order to generate a concise grammar with enough structural information, we leverage a graphical representation for grammars, adopting symbolic representation to present structural elements based on prior knowledge of common TV programs. For instance, a structural element corresponding to a sequence of white frames is a separator, denoted as S, while a long duration shot containing the dominant person at the beginning of a program is the conductor's opening, denoted as A. The grammar is designed by considering the structural elements, their relative duration and probabilities. As illustrated in Fig. 5(c), two structural elements, i.e., A and B, are determined, while the segments corresponding to no determined structural element are denoted as N. A rectangle filled with a symbol is used to illustrate each state for the program: The height of rectangles indicates the relative probability of each valid element while the width is proportional to its duration. In short, for a grammar, the elements with their relative duration and presence probabilities are introduced in time order.

7   Experimental Results

We conduct use cases on TV programs of different types, viz., game, news and talk show, in order to show the effectiveness of our approach for recurrent program structure discovery. Considering the applicable purpose of our work, which is to facilitate structuring the additional episodes from the same program using the inferred grammars, we firstly generate grammars for the programs and then perform two quantitative evaluations, i.e., prediction accuracy and element matching recall.

Three different programs are utilized for inference and evaluation, as given in Tab.1. We adopt 24 episodes of each program, with 12 episodes for grammar inference (*inference set*) and 12 episodes for quantitative evaluation (*test set*). *20h News* (NEWS), a daily news show of 2007, follows a standard pattern for such programs. *Que le meilleur gagne* (GAME) is a game show whose episodes were taken over two years (1991 and 1992). It has four parts divided by separators. The program, hosted by a conductor, mainly contains interview scenes and question/answer scenes with full text segments. *Le grand journal* (TALK) hosted by a conductor, including news reports, talks, weathers and musical performances. The episodes are taken from the first months of 2014.

Table 1. Description of the datasets for evaluation

Dataset	Date	Episodes	Type	Average duration
NEWS	2007	24	TV news	37.9 m
GAME	1991-1992	24	Game	31.9 m
TALK	2014	24	Talk show	71.3 m

Table 2. Quantitative evaluations

Evaluations	NEWS	GAME	TALK
PA	100%	87%	68%
MR	0.83	0.70	0.49

7.1   Use Cases

Grammar Inference. For a recurrent program, we firstly use inference set to infer the structural grammars. Figure 6 shows the inferred grammars for three different programs. Figure 6(a), 6(c) and 6(e) illustrate the structures presented by distribution matrices. NEWS has two different structures, i.e., with or without interview segment, depending on whether there are invited people or not. The game show, having unique structure, is composed of separators, anchor's opening, interview segment and full screen text scene. For TALK, two structures are

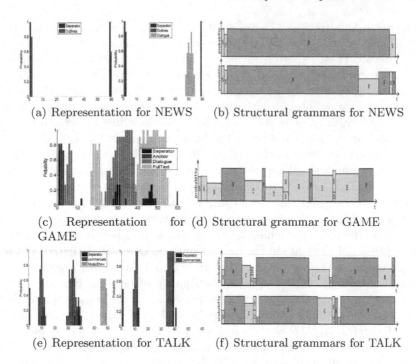

(a) Representation for NEWS   (b) Structural grammars for NEWS

(c) Representation for GAME   (d) Structural grammar for GAME

(e) Representation for TALK   (f) Structural grammars for TALK

Fig. 6. Structures presented by distribution matrix and structural grammars for real data sets

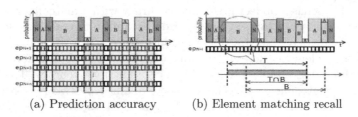

(a) Prediction accuracy   (b) Element matching recall

Fig. 7. Illustration for quantitative evaluations

identified: the structure ending with/without musical performance. The different structures for the same program have evident distinction, i.e., the presence of certain structural elements, while slight difference for the rest parts results from the episodes from which the structures are inferred. Figure 6(b), 6(d) and 6(f) are the corresponding grammars for the three programs. While inferring the grammars, we ignore the elements whose presence probability is less 0.2 to clarify the grammars. Each rectangle represents a coherent state with its structural element in symbol (*S:separator*, *T:outline*, *A:anchor*, *C:commercials*, *D: dialog*, *E:full text* and *M:music show*). The red rectangle with symbol *N* refers to the segment no determined, which will be further studied.

Quantitative Evaluation. Two quantitative evaluations are performed to further verify the effectiveness of our approach: prediction accuracy and element matching recall. These experiments aim at structuring the episodes from the *test set* using the grammars inferred by the *inference set*. Prediction accuracy computes the accuracy of structural element prediction in the test set, whose structural elements are annotated with their corresponding types. In other words, we want to verify, in the test set, how many corresponding structural elements can be found in the segments predicted by grammars. In the case of multiple grammars, the best score is chosen. As illustrated in 7(a), supposing a program has M grammars and n episodes in the test set, the prediction accuracy is $PA = max(P_1, P_2, ..., P_M)$ with $P_i = n_e/(n*N_e)$, where N_e refers to the number of rectangles in the grammar which contain the determined structural elements (segments in green), and n_e refers to the segments in the test episodes which contain the corresponding elements. So the proportion between the number of correctly predicted elements in the test set and the number of inferred elements in the inference set is considered as the prediction accuracy for grammars.

Element matching recall evaluates the quality of element boundary. As in Fig. 7(b), taking a predicted segment, denoted as B, we compare it with the corresponding element in ground truth denoted as T, the matching recall is $MR = (T \cap B)/T$. If $MR = 0$, it indicates that no corresponding element is found in the predicted segments. The matching recall for the whole test set is the average recall of all segments. To be mentioned, the matching recall of each episode is computed based on the grammar which has higher prediction accuracy. In Tab .2, we list the results of quantitative evaluations for the three programs. NEWS has a prefect prediction accuracy with a matching recall equal to 0.83. This result owes to its structure stability and to the fact that less structural elements are determined. Comparing with NEWS, GAME and TALK have more structural elements determined which leads to lower prediction accuracy. Even though GAME and TALK have similar prediction accuracy, TALK has a lower element matching recall. This can be explained by two reasons. Firstly, the episodes from TALK are taken from the year 2014, the recent programs vary more in structure comparing with earlier ones, such as GAME in 1991 and 1992. Secondly, TALK has a longer duration, i.e., 71.3 minutes in average, therefore, in the uniform segmentation step, with the same number of time stamps, the longer a programs is, the coarser structural information it exhibits. If we increase the number of time stamps, a finer structure will obtained, however at the same time some redundant structural information may also comes out. So it is not wise to utilize a larger number of time stamps for a program with such duration.

8   Conclusion

In this paper, we address the problem of unsupervised structure discovery for recurrent TV programs using grammatical inference. Taking the properties of recurrent programs, we are able to discover the structural elements composing the programs and identify the existence of multiple structures for the programs.

We proposed a three-stage approach, including structural element determination, structure modeling and grammar generation. The use cases on different types of programs demonstrate the effectiveness of our approach in practice. Results reported here mostly hint that structure discovery in recurrent collections using grammatical inference is viable and deserves further attention. The proposed approach can be extended in a number of directions to ameliorate program structuring. Elements determination beyond the repetitiveness may provide hierarchical program structures. At the same time, uniform time stamps in stage of episode discretization may no long be suited, because increased structural elements requires improving structure modeling techniques. Furthermore, enriching grammars with more structural information is also an important tasks.

References

1. Xie, L., Xu, P., Chang, S.F., Divakaran, A., Sun, H.: Structure analysis of soccer video with domain knowledge and hidden Markov models. Pattern Recognition Letters (2004)
2. Bertini, M., Del Bimbo, A., Pala, P.: Content-based indexing and retrieval of TV news. Pattern Recognition Letters (2001)
3. Abduraman, A.E., Berrani, S.A., Merialdo, B.: An unsupervised approach for recurrent TV program structuring. In: European Interactive TV Conference (2011)
4. Qu, B., Vallet, F., Carrive, J., Gravier, G.: Content-based inference of hierarchical structural grammar for recurrent TV programs using multiple sequence alignment. In: IEEE International Conference on Multimedia and Expo (2014)
5. Qu, B., Vallet, F., Carrive, J., Gravier, G.: Using grammar induction to discover the structure of recurrent TV programs. In: International Conferences on Advances in Multimedia (2014)
6. Kijak, E., Gravier, G., Oisel, L., Gros, P.: Audiovisual integration for tennis broadcast structuring. Multimedia Tools and Applications (2006)
7. Eickeler, S., Muller, S.: Content-based video indexing of TV broadcast news using hidden Markov models. In: IEEE International Conference on Acoustics, Speech, and Signal Processing (1999)
8. Hanjalic, A., Lagendijk, R., Biemond, J.: Template-based detection of anchorperson shots in news programs. In: International Conference on Image Processing. IEEE (1998)
9. Gao, X., Tang, X.: Unsupervised video-shot segmentation and model-free anchorperson detection for news video story parsing. IEEE Transactions on Circuits and Systems for Video Technology (2002)
10. Chen, S.C., Shyu, M.L., Chen, M., Zhang, C.: A decision tree-based multimodal data mining framework for soccer goal detection. In: IEEE International Conference on Multimedia and Expo (2004)
11. Ancona, N., Cicirelli, C., Branca, A., Distante, A.: Goal detection in football by using support vector machines for classification. In: IEEE International Joint Conference on Neural Networks (2001)
12. Ben, M., Gravier, G.: Unsupervised mining of audiovisually consistent segments in videos with application to structure analysis. In: IEEE International Conference on Multimedia and Expo (2011)

13. Jacobs, A.: Using self-similarity matrices for structure mining on news video. In: Antoniou, G., Potamias, G., Spyropoulos, C., Plexousakis, D. (eds.) SETN 2006. LNCS (LNAI), vol. 3955, pp. 87–94. Springer, Heidelberg (2006)
14. Li, Z., Ding, B., Han, J., Kays, R., Nye, P.: Mining periodic behaviors for moving objects. In: ACM SIGKDD International Conference on Knowledge Discovery and Data Mining (2010)

FOCUSING PATCH:
Automatic Photorealistic Deblurring
for Facial Images by Patch-Based Color Transfer

Masahide Kawai[1] and Shigeo Morishima[2]

[1] Waseda University
[2] Waseda Research Institute for Science and Engineering,
3-4-1 Okubo Shinjuku-ku Tokyo, Japan, 169-8555

Abstract. Facial image synthesis creates blurred facial images almost without high-frequency components, resulting in flat edges. Moreover, the synthesis process results in inconsistent facial images, such as the conditions where the white part of the eye is tinged with the color of the iris and the nasal cavity is tinged with the skin color. Therefore, we propose a method that can deblur an inconsistent synthesized facial image, including strong blurs created by common image morphing methods, and synthesize photographic quality facial images as clear as an image captured by a camera. Our system uses two original algorithms: patch color transfer and patch-optimized visio-lization. Patch color transfer can normalize facial luminance values with high precision, and patch-optimized visio-lization can synthesize a deblurred, photographic quality facial image. The advantages of our method are that it enables the reconstruction of the high-frequency components (concavo-convex) of human skin and removes strong blurs by employing only the input images used for original image morphing.

Keywords: inconsistent facial image, blur, patch color transfer, patch-optimized visio-lization.

1 Introduction

Deblurring is one of the most important topics in computer vision. In this study, we consider two primary types of blurring: 1) motion blur due to camera shake and 2) morphing blur created by common image morphing methods. Various deblurring techniques have been proposed to remove these types of blurring. Representative techniques include blur kernel estimation methods and nonblind deconvolution methods from a single blurred image [1] or from multiple images captured by a video camera [2]. Both methods can robustly address some types of blurring found in images. However, these methods cannot be applied to facial images that include morphing blur because morphing blur results in *inconsistent regions*, such as the white part of the eye being tinged with the color of the iris or the nasal cavity being tinged with the skin color, owing to alignment errors. Therefore, morphing blur poses a significant challenge to deblurring techniques,

X. He et al. (Eds.): MMM 2015, Part I, LNCS 8935, pp. 155–166, 2015.

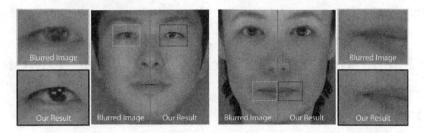

Fig. 1. Comparison between facial images that suffer from morphing blur (given as the blurred image) and the results of the proposed method. The blurred images were created from multiple facial images. Our results provide photographic quality facial images not obtained by the blurred images. From the comparison of the images of the person on the left, the originally strong blur in the eye region is completely removed by our method. Moreover, the comparison of the images of the person on the right indicates that our method can create facial images that include the high-frequency components of the lip by appending the detailed concavo-convex of the lip to the blurred flattened lip. Furthermore, our system is fully automatic.

particularly for facial images. However, if morphing blur can be removed, images created by image morphing methods can then be more effectively applied to image recognition.

To address the need for the deblurring of facial images that include inconsistent regions, we propose a novel method that can deblur an inconsistent facial image, which includes strong blurs, by reconstructing a photographic quality image by using small square images called *patches* derived only from the input images used for the original image morphing. Our system uses two original algorithms: *patch color transfer* and *patch-optimized visio-lization* (advanced visio-lization [3]). These algorithms function as patch-based texture synthesis that can create novel images via reconstruction. Consequently, our system can deblur images even if these images include inconsistent regions. In this paper, we demonstrate that a photorealistic deblurred image can be generated by combining the benefits of patch color transfer and patch-optimized visio-lization, as illustrated in Fig. 1. The main contributions of this paper can be summarized as follows.

1. **Patch color transfer (PCT):** using this algorithm, we can robustly, automatically, and rapidly normalize the facial luminance value. This method has a crucial impact in our deblurring accuracy.
2. **Patch-optimized visio-lization:** using this algorithm, we can automatically create photorealistic facial images by using only the original input images. This approach is an improvement over visio-lization.

2   Related Work

As mentioned previously, there are several methods developed for image deblurring. Li et al. [4] developed a deblurring system that can robustly remove motion blur from a single blurred image by estimating the blur kernel. Unfortunately, any inconsistent regions of an image become highly visible when applying this method to inconsistent images created by image morphing. Moreover, numerous other methods have been proposed. Fergus et al. [5] proposed a system that can remove camera shake from a single photograph. Cai et al. [6] introduced a system to remove motion blurring from a single image by formulating blind blurring as a new joint optimization problem. Levin et al. [7] showed that common maximum a posteriori (MAP) methods involving the estimation of the image and kernel likely fail with sparse priors. Finally, Pan et al. [8] also applied a MAP deblurring algorithm to the blurred-face images on the basis of the predicted salient edges visualized by Poisson reconstruction. However, as with Li et al.'s method, none of the aforementioned methods are applicable to morphing blur that includes inconsistent regions. Furthermore, Cho et al. [9] proposed patch-based deblurring method using captured videos; however, their method cannot be applied to still images because their method requires obtaining sharp patch images from the video sequence.

Therefore, our method focuses not only on motion blur but also on morphing blur. Morphing blur problems cannot be resolved by the aforementioned methods because morphing blur is a strong blur. Our method is considerably more effective at handling strong blurs because it can create a novel image by replacing a blurred image with patches derived only from the input images used for original image morphing.

3   Data Acquisition

An overview of our method and our FOCUSING PATCH system are presented in Fig. 2. The method is primarily composed of two steps (discussed in Sections 4 and 5). First, the luminance values of the input images are normalized to fit a luminance value of the blurred facial image. Second, a deblurred facial image is created using the luminance normalized inputs. In this section, we describe the manner in which blurred input and database images are constructed.

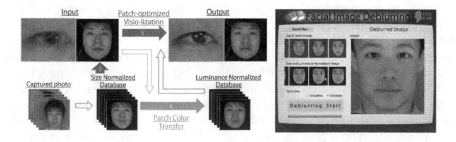

Fig. 2. Overview of our proposed method and our FOCUSING PATCH system

3.1   Input Image

For our method, a strongly blurred facial image is created by image morphing, and it includes several inconsistent regions. There are several methods for creating an illustrative morphing blur. For example, the morphable model [10] and the active appearance model [11] are statistical models that can represent various facial images, and by changing parameters, synthesized facial images can include the same type of blurs as previously mentioned. In this paper, we work with an average face image [12] (*AF image*) created from multiple facial images, which outputs inconsistent images for use here, as examples of blurred facial images. An AF image is created according to the following procedure. Initially, we normalize the position of the eyes, nose, mouth, and outline from 661 feature points by using a feature point detector [13] and radial basis function (RBF) interpolation [14]. Next, we create the AF image by averaging each red, green, and blue (RGB) value from multiple size-normalized inputs per pixel. We consider the AF image as the input and the sizes of all the images are 512×512.

3.2   Database Images

To deblur an input image, we construct the image database. In general, we use those multiple images that were used to create the AF image, as discussed in Section 3.1. Therefore, our method requires the construction of no additional databases. The size of all database images is 512×512.

4   Patch Color Transfer

Our method creates a novel image from multiple images by calculating the RGB distance between the AF and the database images. As a result, most regions of the resulting image are composed of an image from the database with a skin color resembling that of the AF image. To avoid this bias, we must transfer the luminance value in the AF image to those in the facial images from the database. However, it is difficult to transfer the luminance value of a facial image because its luminance gradient exhibits wide local variations, and the skin color changes overall when the lighting environment is changed. Therefore, we propose a novel method called patch color transfer (PCT) that transfers a luminance value in one facial image to another image according to the following procedure. First, the source (AF) and target (size-normalized database) images are separated into multiple patches of 60×60 pixels with a 15-pixel overlap. Next, the luminance value of a patch in the AF patch image is transferred to the corresponding patch in the database image at the same position by adding the average value ($\boldsymbol{Avg}_i$) of the difference between the luminance values of the two images for the patch of the AF image to the patch of the database image on a pixel by pixel basis. The process is repeated for each database image. Because the average luminance difference is added on a pixel by pixel basis, PCT maintains the variation in the luminance gradient for each patch. The details of the described method are given by the following equations:

Left Patch   After liner interploation   Right Patch

Fig. 3. Overviews of the liner interpolation of overlap regions

$$PC_{Out\_xy\_i} = PC_{Tar\_xy\_i} + Avg_i$$
$$(0 \leq x < 60 = x_{max}, 0 \leq y < 60 = y_{max}) \tag{1}$$

where

$$Avg_i = \frac{\sum_{(x,y)\in\Omega_i}(PC_{Src\_xy\_i}) - (PC_{Tar\_xy\_i})}{x_{max} \times y_{max}} \tag{2}$$

$$PC_{Src\_xy\_i} = \{R_{Src\_xy\_i}, G_{Src\_xy\_i}, B_{Src\_xy\_i}\} \tag{3}$$

$$PC_{Tar\_xy\_i} = \{R_{Tar\_xy\_i}, G_{Tar\_xy\_i}, B_{Tar\_xy\_i}\} \tag{4}$$

In Eqs. (1)-(4), PC represents a patch pixel, Out indicates an output, Tar indicates the target, Src indicates the source, N is the patch number per image, i is an index between 0 and N, x is a coordinate value in the horizontal direction of the patch image, y is a coordinate value in the vertical direction of the patch image, Ω_i is the i-th patch image domain (60 × 60), and $R\,(G,B)_{Src\_xy\_i}$ is the R(G,B) value of the (x, y) position in the i-th patch of the source image. Applying the above equation, the luminance value in the i-th patch image of the target is normalized, and all normalized patch images are synthesized into each patch position from the top left to bottom right with a 15-pixel overlap. For the patch syntheses, we take an interpolation of overlap regions to not create artifacts of overlap regions between neighboring patches. Therefore, we use a linear interpolation as shown in Fig. 3. Using this interpolation, we can smoothly represent an image.

To evaluate the PCT method, we compare it with the color transfer (CT) method of Reinhard et al. [15]. CT is a luminance transfer method that can adapt the luminance value of a target image according to a source image's average luminance value and variance by computing the average value and the variance of the target and source images in lab space. Comparisons between the PCT and the CT method are shown in Fig. 4. To compare the accuracy of both methods, we used source images with large luminance gradient variations, where, in the figure, the source image on the left was applied to the first two target images, and that on the right was applied to the final two target images. PCT can accommodate local luminance variations caused by lighting because the luminance values are optimized as patches for each local patch. However, CT cannot accommodate local luminance variations such as specular color because CT uses only a global optimization. Moreover, PCT conducts the computation 2.43 times faster than CT, and PCT is processed in real-time.

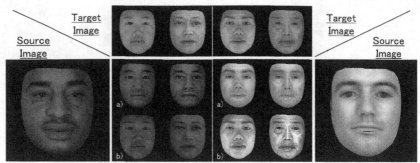

a) Our result: 0.0189 sec /image   b) Reinhard er al.'s result: 0.0460 sec/image

Fig. 4. Comparison between the results of a) our method (PCT) and b) the color transfer (CT) method of Reinhard et al. [15], and a comparison between the performance times of a) and b). The source image on the left was applied to the first two target images, and the source image on the right was applied to the final two target images.

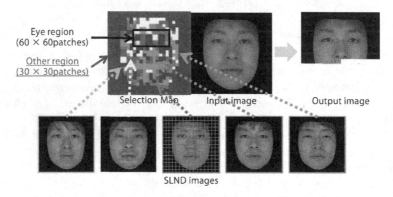

Fig. 5. Overview of the patch-optimized visio-lization method

5   Deblurring the Facial Image

Our deblurring method employing patch-based texture synthesis generates a novel deblurred image from an arbitrary image using size and luminance normalized database images (*SLND images*). Therefore, we use a visio-lization's concept of Mohammed et al. [3] as a patch-based facial image synthesis. The original visio-lization can generate a new hybrid face with patch-based texture synthesis. We modify the original visio-lization method to use it as a deblurring method and apply the modified method called patch-optimized visio-lization to the AF image by using the techniques outlined in Fig. 5. In general, the AF and SLND images are separated into multiple patches of 60 × 60 pixels with a 15-pixel overlap in the eye region and those of 30 × 30 pixels with a 9-pixel overlap in the other regions. Although the original method uses the same patch size throughout, our method can increase the potential for enhancing the representational power of synthesis by employing different patch sizes. Next, the RGB distances between the patches in the AF image and the SLND images are calculated, and the best patch image from the SLND images is selected. In the original method, patch selection is limited to the same position in the AF and SLND images. However, our method selects patches from the neighborhood, but not necessarily from the same positions, in order to obtain the most suitable patches from the SLND images. This flexibility enables the reliable reproduction of the AF patches. The best patch image is selected as the image with the smallest RGB distance, as defined by the following equation:

$$\operatorname*{arg\,min}_{s,m,n} \sum_{(x_m,y_n)\in\Omega_{p\_mn}} \|C_{AF\_x_{m_p}y_{n_p}} - C_{SLND\_s\_x_my_n}\|^2$$

$$(0 \le s \le N), (m_p - 2 \le m \le m_p + 2), (n_p - 2 \le n \le n_p + 2) \qquad (5)$$

where

$$C_{AF\_x_{m_p}y_{n_p}} = \{R_{AF\_x_{m_p}y_{n_p}}, \ G_{AF\_x_{m_p}y_{n_p}}, \ B_{AF\_x_{m_p}y_{n_p}}\} \qquad (6)$$

$$C_{SLND\_s\_x_my_n} = \{R_{SLND\_s\_x_my_n}, \ G_{SLND\_s\_x_my_n}, \ B_{SLND\_s\_x_my_n}\} \qquad (7)$$

Here, N is the total number of SLND images, s is an index between 0 and N, m is the column number of a patch, n is the row number of a patch, m_p is the column number of the present referred patch in the AF image, n_p is the row number of the present referred patch in the AF image, Ω_{p_mn} is the patch image domain of the m-th column and n-th row, and $R(G,B)_{AF\_x_{m_p}y_{n_p}}$ is the R(G,B) values of the (x,y) position of Ω_{p_mn} in the AF image. According to the above equation, the m-th column and n-th row patch image of the s-th image is selected from the SLND images, and all selected patch images are embedded into each patch position from the top-left to bottom-right. With respect to an overlap interpolation, we employ a liner interpolation in same way as Section 4.

6   Result and Comparison with Related Work

To demonstrate our approach, our method was applied to two types of AF images involving high and low resolution SLND images. For this demonstration, we simply created two AF images from five photo images, as described in Section 3, and then applied our method to these AF images, as described in Sections 4 and 5. Fig. 6 provides comparisons between the blurred facial images and our results. As shown in the person on the left of Fig. 6, our method entirely deblurred the AF image. In addition, our method removed the strong blur from the eye regions that results from morphing. As for the cheek region, before applying our method, the cheek surface was flat because the image was averaged. However, after applying our method, the cheek represents a detailed concavo-convex shape. Our method is not only able to remove strong blur but also yields high-frequency components such as this detailed concavo-convex surface.

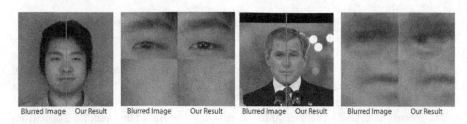

| Blurred Image | Our Result | Blurred Image | Our Result | Blurred Image | Our Result | Blurred Image | Our Result |

Fig. 6. Comparisons between our resulting images and blurred images [16]

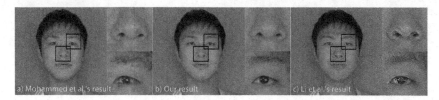

a) Mohammed et al.'s result   b) Our result   c) Li et al.'s result

Fig. 7. Comparisons between the results of (a) Mohammed et al.'s method, (b) the proposed method, and (c) Li et al.'s method

However, our resulting image shown in the person on the right of Fig. 5 is of low-resolution. One possible limitation of our method is that the resolution of the resulting image depends on the resolution of the database images. Consequently, using database images of high-resolution facial images or constructing databases of high-resolution facial images can further improve the performance of our results.

To demonstrate the effectiveness of our proposed method, we compared our method with the visio-lization method of Mohammed et al. [3] and a previous deblurring method of Li et al. [4]. Fig. 7 shows a comparison of the results of (b) the proposed method with those of the methods of (a) Mohammed et al. and (c) Li et al. The individual images correspond to the entire view of the face and detailed views of the nose region and the left eye region. From a comparison of our method with Mohammed et al.'s method, we observe that the newly proposed patch color transfer and patch-optimized visio-lization are significantly more effective for photorealistic synthesis. Moreover, from a comparison of our method with Li et al.'s method, our method can be seen to better improve inconsistent regions, such as the eyes and nose, while the other method amplifies on inconsistent regions. It can be observed that our method is visibly superior to other methods considered.

7   Evaluation

To evaluate the validity of our method, as described in Sections 4 and 5, we conducted an objective evaluation. To evaluate the differences between the AF image and our resulting image regarding facial-underlying high-frequency components, we compared each of these two images with database images that were

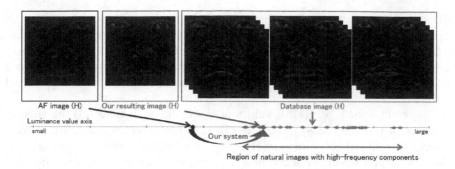

Fig. 8. Comparison of images with high-frequency components and luminance values

natural images captured by a camera. Moreover, we evaluated the images divided into separate groups based on the age and sex of the subject to evaluate our method from various perspectives. We divided images according to sex and in general into four groups according to subject age: 0-9, 10-29, 30-49, and 50-69 for a total of eight groups. According to the above classifications, we created eight AF images each from thirty database images according to the procedure described in Section 3. Our method was then applied to these AF images, as described in Sections 4 and 5.

Next, it is necessary to extract the high-frequency components from the AF image, our resulting image, and the database images. However, the high-frequency components of facial images are disproportionately affected by the lighting environment when the database images were captured and so were the differences in the individual skin color of each. Therefore, we select a single database image to be regarded as a source image (criterion image), and all other database images are luminance-normalized by using PCT. Thereby, the luminance values of the AF image, our resulting image, and the database images are consistent with those of the criterion image, and a justifiable evaluation of the high-frequency components can be conducted. Later, we can obtain each power in a frequency range by applying a Fourier transform algorithm to all images and then extracting the high-frequency components of images via a high-pass filter. An AF image whose high-frequency components have been extracted is denoted as an AF image (H). An equivalently processed output image is denoted as our resulting image (H). An equivalently processed database image is denoted as an database image (H), as illustrated in Fig. 8. The plotted coordinate values at the bottom of Fig. 8 illustrate the summation of each image's luminance value in the high-frequency components range.

Next, the AF (H) and database images (H), or our resulting (H) and database images (H) were separated into multiple patches of 32 × 32 pixels without an overlap, and we computed the root mean square error (RMSE) per patch. The RMSE per patch can be computed by obtaining the difference between the luminance values of the AF patches (H) and database images (H) from the same positions, or those of our resulting (H) and database images (H) from the same positions and then adding the differences of all patch positions. The RMSE is computed per patch, not per pixel, in order to alleviate the position gaps between images by quantizing the images using patches. There are thirty database images for each resulting image; therefore, we computed the RMSE averaged over all 30 database images (H). The RMSE results of the eight groups are listed in Tab. 1.

As shown in Tab. 1, all RMSE values of our resulting images (H) are smaller than those of the AF images (H). Therefore, in each case, our resulting image (H) is close to the database images (H) with natural high-frequency components. As such, the application of our method to an input image is commensurate with appending natural high-frequency components to the input image. Further, from the results of the plotted coordinate values, such as that shown at the bottom of Fig. 8, we determined that our resulting image contains the same degree of high-frequency components as the natural database images and that the AF image contains a lesser degree of high-frequency components. As described, because our method can append natural high-frequency components to the input image, it is effective for adequate facial image deblurring.

Table 1. RMSE values of the average image (H) and our resulting image (H) for the eight groups

No.	RMSE of AF image(H)	RMSE of our resulting image(H)
0-9's women	1.61	1.14
10-29's women	1.55	1.36
30-49's women	1.81	1.51
50-69's women	1.86	1.64
0-9's men	1.41	1.27
10-29's men	1.60	1.37
30-49's men	1.91	1.63
50-69's men	1.80	1.64

8   Application for Super-Resolution of Facial Images

We also attempted to achieve super-resolution (SR) of facial images by using the same methodology of our proposed method as another application of our method. In this attempt, we used low-resolution images (64×64 and 128×128) as input images and used fifty images (512×512) for each image as database images. In general, the original 64×64 (or 128×128) pixel input images were linearly interpolated to correspond to 512×512 pixels. Next, we normalized the size and luminance of the database images based on the interpolated input image, as discussed in Sections 3 and 4. In addition, the resulting SLND images were separated into multiple patches of 15×15 pixels with a 9-pixel overlap, and the best patches were selected according to equation (5). Finally, we synthesized an SR image by embedding selected patches into each patch position from the top-left to bottom-right. By employing a 15×15 pixel patch rather than a

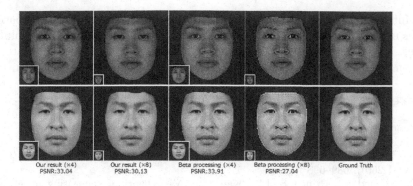

Fig. 9. Results of super-resolution(×4 and ×8) by our method and beta processing [17] and the ground truth [18]

30 × 30 (60 × 60) pixel patch (as was employed in Section 5), we can represent more detailed luminance variation. Moreover, we evaluated our SR results via the peak signal-to-noise ratio (PSNR) based on ground truth images. Fig. 9 shows the results of the application of our method along with the corresponding PSNR values. We also depict SR images created by beta processing (BP) [17] along with its corresponding PSNR values. From Fig. 9, our resulting images show sensitive SR results that favorably compare with the ground truth. As for the results of the 4 times SR images, the PSNR of those created by BP is larger than those created by our method; however, as for results of the 8 times SR images, the PSNR of our method is larger than that of BP. Our 8 times SR results represent highly satisfactory facial images. Therefore, we can conclude that our method is highly satisfactory for the SR of high magnification (8 times and over).

9   Conclusion and Future Work

In this paper, we proposed a novel method for facial image deblurring. Our deblurring system consists of two key algorithms: patch color transfer and patch-optimized visio-lization. These algorithms are effective for the removal of inconsistent regions such as the conditions where the white part of the eye is tinged with the color of the iris and the nasal cavity is tinged with the skin color. Moreover, we have demonstrate that the method can append the high-frequency components of natural images to the resulting images, although the high-frequency components of the input images created by common image morphing have been predominantly removed. Further, via comparison with related work and objective evaluation, we can ensure that our method can deblur strong blur that cannot be removed by previously developed methods. As another application of our method, we attempted to achieve super-resolution by using the same methodology. From these results, we can conclude that our method performs high precision super-resolution, particularly for high magnification.

In future work, we will consider constructing a high-resolution database of images used by patch-based texture synthesis in order to output high-resolution images. The construction of a large and varied database would facilitate the creation of images that are not only deblurred but also high in resolution.

References

1. Shan, Q., Jia, J., Agarwala, A.: High-quality Motion Deblurring from a Single Image. ACM Trans. Graph. 27(3), article 73 (2008)
2. Ben-Ezra, M., Nayar, S.K.: Motion-based Motion dDblurring. IEEE Trans. PAMI 26(6), 689–698 (2004)
3. Mohammed, U., Prince, S.-J.D., Kautz, J.: Visio-lization: generating novel facial images. ACM Trans. Graph. 28(3), article 73 (2009)
4. Xu, L., Jia, J.: Two-Phase Kernel Estimation for Robust Motion Deblurring. In: Daniilidis, K., Maragos, P., Paragios, N. (eds.) ECCV 2010, Part I. LNCS, vol. 6311, pp. 157–170. Springer, Heidelberg (2010)
5. Fergus, R., Singh, B., Hertzmann, A., Roweis, S.-T., Freeman, W.T.: Removing Camera Shake from a Single Photograph. ACM Trans. Graph. 25(3), 787–794 (2006)
6. Cai, J.-F., Ji, H., Liu, C., Shen, Z.: Blind Motion Deblurring from a Single Image Using Sparse Approximation. In: CVPR, pp. 104–111 (June 2009)
7. Levin, A., Weiss, Y., Durand, F., Freeman, W.T.: Understanding and Evaluating Blind Deconvolution Algorithms. In: CVPR, pp. 1964–1971 (June 2009)
8. Pan, J., Hu, Z., Su, Z., Yang, M.-H.: Deblurring Face Images with Exemplars. In: Fleet, D., Pajdla, T., Schiele, B., Tuytelaars, T. (eds.) ECCV 2014, Part VII. LNCS, vol. 8695, pp. 47–62. Springer, Heidelberg (2014)
9. Cho, S., Wang, J., Lee, S.: Video Deblurring for Hand-held Cameras Using Patch-based Synthesis. ACM Trans. Graph. 31(4), article 64 (2012)
10. Blanz, V., Vetter, T.: A Morphable Model for the Synthesis of 3D Faces. In: SIGGRAPH 2099, pp. 187–194 (1999)
11. Cootes, T.F., Edwards, G.J., Taylor, C.J.: Active Appearance Models. In: Burkhardt, H., Neumann, B. (eds.) ECCV 1998. LNCS, vol. 1407, pp. 484–498. Springer, Heidelberg (1998)
12. Jenkins, R., Burton, A.M.: 100% Accuracy in Automatic Face Recognition. Science 25 319(5862), 435 (2008)
13. Irie, A., Takagiwa, M., Moriyama, K., Yamashita, T.: Improvements to Facial Contour Detection by Hierarchical Fitting and Regression. In: 1st ACPR, pp. 273–277 (2011)
14. Noh, J.-Y., Fidaleo, D., Neumann, U.: Animated Deformations with Radial Basis Functions. ACM symposium on VRST, pp. 166–174 (2000)
15. Reinhard, E., Ashikhmin, M., Gooch, B., Shirley, P.: Color Transfer between images. IEEE Computer Graphics and Applications, 34–41 (2001)
16. Kemelmacher-Shlizerman, I., Seitz, S.M.: Beta Processing Dictionary Learning for Coupled Feature Space with Application to Single Image Super Resolution. In: ICCV, pp. 1746–1753 (2011)
17. Li, H., Hairong, Q., Russell, Z.: Beta Process joint Dictionary Learning for Coupled Feature Space with Application to Single Image Super Resolution. In: CVPR, pp. 345–352 (2013)
18. Milborrow, S., Morkel, J., Nicolls, F.: The MUCT Landmarked Face Database. Pattern Recognition Association of South Africa (2010)

Efficient Compression of Hyperspectral Images Using Optimal Compression Cube and Image Plane

Rui Xiao and Manoranjan Paul

School of Computing and Mathematics
Charles Sturt University, Australia
{rxiao,mpaul}@csu.edu.au

Abstract. *Hyperspectral* (HS) images (HSI) provide a vast amount of spatial and spectral information based on the high dimensionality of the pixels in a wide range of wavelengths. A HS image usually requires massive storage capacity, which demands high compression rates to save space with preservation of data integrity. HS image can be deemed as three dimensional data cube where different wavelengths (W) form the third dimension along with X and Y dimensions. To get a better compression result, spatial redundancy of HS images can be exploited using different coders along X, Y, or W direction. This article focuses on taking maximum advantage of HS images redundancy by rearranging HS image into different 3D data cubes and proposes a directionlet based compression scheme constituted the *optimal compression plane* (OCP) for adaptive best approximation of geometric matrix. The OCP, calculated by the spectral correlation, is used to the prediction and determination of which reconstructed plane can reach higher compression rates while minimizing data loss of hyperspectral data. Moreover, we also rearrange the 3D data cube into different 2D image planes and investigate the compression ratio using different coders. The schema can be used for both lossless and lossy compression. Our experimental results show that the new framework optimizes the performance of the compression using a number of coding methods (inclusive of lossless/lossy HEVC, motion JPEG, JPG2K, and JPEG) for HSIs with different visual content.

Keywords: Hyperspectral images, lossless compression, and optimal compression plane (OCP).

1 Introduction

Hyperspectral (HS) images are designed to focus and measure the light reflected by hundreds of narrow and adjacent spectral bands (i.e. wavelengths). HS images may have wide range of wavelengths from hundred nano meter (nm) to thousand micro meter (m) to get different characteristics of objects. The applications of HS imaging are in the fields of military to detect chemical weapon, agriculture to quantify crops, mineralogy to identify different minerals, physics

X. He et al. (Eds.): MMM 2015, Part I, LNCS 8935, pp. 167–179, 2015.
© Springer International Publishing Switzerland 2015

to identify different properties of materials, surveillance to detect different objects and reflections, and environment to measure surface CO2 emissions, to map hydrological formations, and to track pollution levels. A HS camera can capture a HS image with different wavelengths from visual wavelength to far-infrared wavelength. It effectively divides the spectrum into many thin image slices corresponds to wavelength. A commonly used sub-division of wavelengths as follows: *visual* (VIS) from 400 to 700 nano meter (nm), *near infrared* (NIR) from 750 to 1400nm, *shortwave infrared* (SWIR) from 1400 to 3000nm, *medium wave infrared* (MWIR) from 3000 to 8000nm, and *long wave infrared* (LWIR) from 8000 to 15000nm, and *far infrared* from 15000nm to 1000 macro meter (m). Research shows that different materials radiate different reflections in certain range of wavelengths. This characteristic is normally exploited for above mentioned applications. Examples of a HS image and intensity variations in different type objects are shown in Fig. 1 where we can see different slices of a HS image in different wavelength from 400 to 720nm (left) and intensity differences in wavelengths for real and plastic objects.

Fig. 1. A hyperspectral image (left) with different slices correspond to wavelength and an example of intensities (right) for real and plastic objects in different wavelengths

To illustrate a three-dimensional (X, Y, W) hyperspectral data cube, X and Y represent two spatial dimensions usually deemed as successive natural images, and W represents a range of wavelengths, which is defined as a HS image, $H \in R^{x \times y \times w}$ where R represents the real number values for reflectance of a 3D cube with X, Y, and W dimensions. The development of these complex image data challenges in the acquisition, transmission and storage. For example, plant phenomics data required processing 1.2TB per day and high-speed hyperspectral camera for monitoring environmental requires 3GB per second [7]. Such volumes already exceed available transmission bandwidths [14]. Efficient compression techniques are therefore needed before HS images can be integrated into the existing and emerging communication systems.

Recent compression methods can fall broadly into two main categories in terms of information preservation: lossless and lossy compression methods [8]. In lossless compression, the original image can be regenerated from the compressed data with small compression ratio; on the other hand, partially distorted image can be generated from the lossy compression methods although they led higher compression ratios. Some examples of flexible lossy-to-lossless compression techniques are available in the literature [3,4,24]. The compression algorithms based on Discrete Wavelet Transform (DWT) and three dimensional DWT have been presented in [8,20,23]. Huge data blocks are required in these methods that demand a future reduction of the core tensor computations for coding efficiency enhancement and lower the computational load. The compression of multispectral HS image is also considered by using principal component (PC) analysis (PCA) in conjunction with JPEG2000 encoder [3,4,24]. In the PCA-based methods they reduce a number of slices (i.e. 2D images of a certain number of wavelengths) based on the PCA and then apply JPEG2000 on major PCs for efficient rate-distortion performance. Normally, PCA-based compression techniques outperform DWT-based compression techniques for HS images. The reduction of dimensions (i.e., remove some slices) may not be suitable in the compression of HS images because during the PCA it may remove some PCs which are valuable for some specific applications. Some researchers also applied H.264/AVC video coder with the aim of determining its feasibility when applied to HS images compression in [11,18]. They proposed some modifications in the configuration of the coder to exploit different redundancy within HS image for better compression and unmixing applications. So far in our knowledge feasibility analysis of HS image compression using the latest high efficiency video encoder [19] HEVC standard are still deficient in current research.

Recently researchers treat a video as a temporal (along T-axis) collection of two dimensional pictures (formed by XY axes); compress them by exploiting spatial and temporal redundancy in the pictures using different video encoders [9,10,16]. They also define an optimal compression plane (OCP) determination strategy based on the cross-correlation among different directions (i.e., XYT, TXY, TYX) as a preprocessing step to find the best compression plane without applying video coder in each plane exhaustively. The most of the cases the OCP plane can predict the optimal plane which provides the best rate-distortion performance by a coder. In this paper we also treat a HS image as a 3D cube similar to a video as the spectral correlation (band redundancy) is generally high value but not always stronger than spatial correlation [6]. To determine the best 3D cube of a HS image for better compression, we propose an optimal compression cube (OCC) which is the modification of the existing OCP determination strategy [9,10,16]. focusing on the HS image properties in this paper. Then we apply different video and image coding standards such as Motion-JPEG [13,15,22], JPEG [1,12,21], JPG2000 [2,17], and HEVC [16] on the optimal compression cube (OCC) for the best rate-distortion performance. Some compression techniques for example Wavelet transform [5] are not effective in dealing with directional information. Therefore, a directional representation in a 2D dimension $(XY) \times W$, $(YX) \times W$, and $(WX) \times Y$ is also introduced in this paper and investigate the best compression image plane using different image coders. The experimental results show that 2D image plane provides the best rate-distortion performance.

Our contributions in this paper are (i) formulation of an OCC determination strategy based on the HS property, (ii) explore both 3D compression cube and 2D image plane for better compression, (iii) investigate the compressibility of the latest video coding standard HEVC on HS images, and (iv) comprehensive compressibility analysis using different standard image and video coders.

2   Proposed Techniques

Finding efficient geometric representations of images is one of the kernel problems to improving HS image compression. The purpose is to reduce the spatial and spectral correlation simultaneously and then archiving or transmitting optimized compression result. We rearrange a HS image into different 3D data cubes and 2D image planes. Subsequently an OCC determination strategy is applied on 3D data cubes to find the best one based on the cross correlation focusing on the HS property. After finding the best 3D data cube plane we apply different image and video coders for actual rate-distortion performance. A HS image can be reconstructed in 6 directions of 3D data cube, XYW (in this direction normally HS images are captured, stored, and display), YXW, WXY, XWY, WYX, and YWX. Although there are six directions we can form a HS 3D data cube, our experiments using different coders find that two reversed data cubes such as WXY and XWY provide the same rate-distortion performance. Thus, in our experiment we only use three directions (such as XYW, WXY, and WYX) to find the OCC. In addition to 3D formation we also explore directional representation in a 2D dimension $(XY) \times W$, $(YX) \times W$, and $(WX) \times Y$ where two dimensions in bracket form the first dimension and the third dimension forms the second dimension in an image.

The following sub-sections will describe 3D cube formations, 2D image plane formation, OCC formulation, and different encoders used in the proposed technique.

2.1   3D Plane Reconstruction

In the sense of data structure of a HS image which contains extensive redundancy among X (i.e. a spatial dimension), Y (i.e. the other spatial dimension), and W (i.e. the wavelenth dimension). Therefore, we need to determine higher redundancy alone different axis before we compress each HS image. In the first step, we aim to construct 3D cube along the different axis of a HS image. Fig. 2 shows a HS image named as *Vase* in different cubic forms. From the figure, we can easily observe that the spatial and spectral correlations in XYW, WXY and WYX cubes are different. Thus, a specific coder may exploit the redundancy of a cube better way compared to other coders which justifies the rearrangement of a HS image into different 3D cubes.

2.2   2D Plane Reconstruction

As we mentioned earlier some coder may not exploit the directional redundancy in better way, thus we also explore the compressibility of different coders by

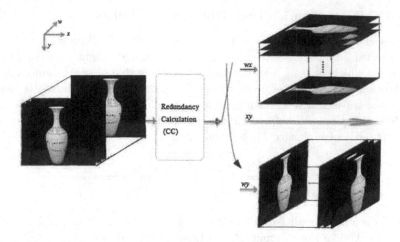

Fig. 2. Different 3D data cubes (XYW, WXY, WYX) from where the proposed optimal compression cube algorithm selects the best cube based on the cross correlation, the proposed OCC algorithm and cross correlation equation is defined in Section 2.3

rearranging a HS image into different 2D image planes. Like 3D cube formation, we can make 2D image plane using six directions. Again we find that two reversed image planes have no difference when we apply different image encoders. Three examples of 2D images of the same HS image, Vase are given in Fig. 3 (for better visualization we could not provide whole images in the figure). In our data set, a HS image has 1040139261 resolutions where $X = 1040$, $Y = 1392$, and $W = 61$. We can form 2D images in different directions of a HS image in $X \times (YW)$ i.e., 104084912(=139261), $Y \times (XW)$ i.e., 139263440(=104061), and $W \times (XY)$ i.e., 611447680(=10401392).

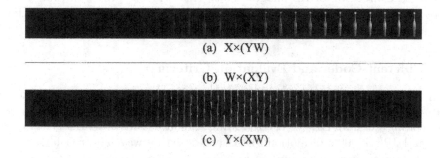

(a) X×(YW)

(b) W×(XY)

(c) Y×(XW)

Fig. 3. Reconstructed 2D images in different planes

2.3   OCC Determination from Different 3D Cubes

Normally when a HS image is captured, slices are organized on a certain wavelength interval. In most cases there are 10 nm. HS images might have two or more bands slices which may or may not be. To obtain value of spatial and spectral redundancies, the average of cross correlation Intra-Band Prediction is used to describe the statistical relationship between neighbouring spectral bands. Then OCC determination will help to choose an optimal cube. The *cross-correlation coefficient* (CC) of *n-th* slice P_n is formulated as follows:

$$CC_n = \frac{\sum_{i,j}((p_n(i,j) - \overline{p_n})(S(i,j) - \overline{S}))}{\sqrt{\sum_{i,j}((p_n(i,j) - \overline{p_n})^2 \sum_{i,j}(S(i,j) - \overline{S})^2}} \tag{1}$$

where S is a median slice of a HIS cube and P_n is the average pixel intensities of *n-th* slice. Unlike the existing OCP scheme [10] we do not calculate the cross correlation coefficient using adjacent slices or frames in the proposed OCC technique. As the property of a HS image is different from the property of a video, we use median slice information to calculate cross correlation coefficient of each slice in different directions. The median slice is the most dominant wavelength of a HS image for major applications and the median slice has the most dominant visual information of a HS image, thus, we use the median slice as the reference slice to calculate CCs. The higher redundancy may reflect in raising the value of CC. We need to determine average CC in each 3D data cube and then the proposed OCC technique find the optimal cube based on the maximum value of average CC from different cubes. The experimental results shown in Fig. 4 reveal that most of the cases XYW cube provides the maximum CC. In the experiment section we also show that the best encoder provides better results in XYW cube in 3D formation in most cases. We also test the rate-distortion performance by different coders based on the predicted direction using the existing OCP technique. The experimental results show that OCP fails to predict the accurate direction for HS images. Thus, this evidence proves the validity of the proposed OCC technique to predict the optimal compression cube direction.

2.4   Different Coder and Evaluation Criteria

The performance comparison of OCC is verified using two most popular image compression methods such as Motion JPEG, JPEG and JPEG2000 and the latest video coding standard HEVC. Image coding standards such as JPEG is based on DCT and JPEG2000 and motion JPEG are based on wavelets transformation. For 3D data cube compression we also use the latest video coding standard with limited motion search to exploit the redundancy between adjacent slices or images. To verify compression performance, we use quantitative and qualitative comparison. The quantitative com parison is based on the PSNR (*peak signal to noise ratio*) which is defined as PSNR $=10.log_{10}(MAX_I^2/MSE)$ where MAX_I is the maximum possible pixel value of the image and MSE is the mean square

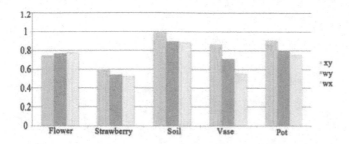

Fig. 4. The correlation coefficient along different axes where xy axes indicates XYW cube, wy axes indicates WYX cube and wx axes indicates WXY cube

errors between the original image and the reconstructed image. We also use subjective evaluation by showing reconstructed images using different cube or planes with different encoders.

3   Experiments

The proposed algorithms have been applied to different HS images such as *Pot*, *Vase*, *Flower*, *Strawberry*, and *Soil*. In this experiment, we use HEVC video encoder along with image encoders to exploit the redundancy between slices. The quantization parameters (QPs) used in HEVC are from 10 and 51 to see wide range different quality and bit rates where QP=10 provides super quality picture with high bit rates and QP=51 provides inferior quality picture with low bit rates. For all JPEG encoders we use 3 to 93 quantization parameters for a wide range of quality and bit rates where 3 provides lower quality and 93 provides higher quality.

3.1   Compression Results Based on 3D Cubes

In this subsection, the OCC framework is validated using image encoders such as JPEG2000, JPEG and HEVC. The PSNR gain is listed based on those coding standards respectively. These examples demonstrate in five HS images: *Pot, Soil, Vase, Flower*, and *Strawberry*.

Fig. 5 shows that the rate-distortion performance in the XYW cube is better than the WXY or WYX cube for all images. This result is also consistence with the proposed OCC strategy where the OCC selects XYW cube based on the CC in this direction (see Fig. 4). Although our OCC technique selects different cube as the optimal cube, the best coder provides better rate-distortion performance in XYW cube for Flower image. However, the rate-distortion performance of the selected cube by OCC does not perform significant inferior. JPEG2000 with XYW cube provides the most distinguishing results compared to other two cubes where XYW direction provides comparable results with other coders, however, it provides significantly lower compression results in other two cubes. HEVC does

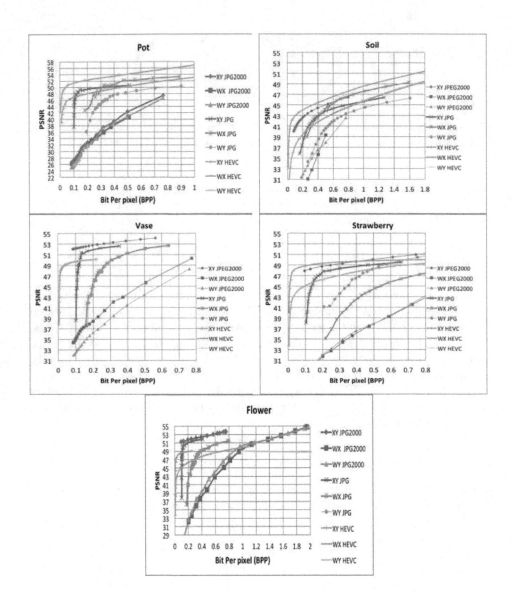

Fig. 5. Rate-distortion curves of five standard hyperspectral images using JPEG, JPEG2000, and HEVC coders on three different 3D arrangements (i) XYW, (ii) WXY, and (iii) WYX

Table 1. The voting result of the optical compression ratio using HEVC video coder and other image video coders JPEG/JPEG2K

Images	Low Bit rates 0 to 0.5bpp (+ 35 PSNR or more)			High bit rate >0.5		
Soil	HEVC	JPEG2000	JPEG	HEVC	JPEG	HEVC
	XY	XY	XY	XY	WX	WX
Straw-berry	JPEG2000	JPEG	HEVC	JPEG2000	JPEG	JPEG
	XY	XY	XY	XY	WY	XY
Vase	JPEG2000	JPEG	HEVC	JPEG2000	JPEG	JPEG
	XY	XY	XY	XY	XY	WY
Pot	HEVC	JPEG	JPEG	HEVC	JPEG	HEVC
	XY	XY	WY	XY	WX	WX
Flower	JPEG2000	JPEG	JPEG	JPEG2000	JPEG	JPEG
	XY	XY	WY	XY	XY	WX

not provide the best rate-distortion performance compared to the other coders, however, it performs reasonably well with different cubes.

Table. 1 summaries the results shown in Fig. 5 where the coding results of HEVC, JPEG2000 and JPEG are listed and compared among 5 images in terms of PSNR and different bit per pixel (BPP). Table.1 votes the best compression ratio between low bit rates and high bit rates. The experiment demonstrates that HEVC encoder achieves better compression result at XYW and WXY cubes at high bite rate, especially when the BPP is higher than 0.5 or PSNR is more than 35 dB. At low bit rates i.e. below 0.5 BPP and above 35 dB PSNR, XYW cube outperforms mainly. JPEG2000 is also a prime encoder of HSI in both low and higher bit rate.

We also investigate the rate-distortion performance using another popular video compression encoder, Motion JPEG to see the best cube for compression. The experimental results (see show Fig. 6) that Motion JPEG along XYW cube also provides better results compared to other directions.

3.2   Compression Results Based on 2D Image Planes

Reconstructed 2D image planes in our experiments demonstrate excellent performance for HS image compression. Corresponding PSNR can reach up to 50 dB at 0.5 BPP. Images well preserve clear edge and patterns information for clear visualization. Rate-distortion curves in Fig. 7 illustrate that the reconstructed image using $X \times (YZ)$ image plane presents noteworthy higher performance comparing to other planes.

In addition to objective result comparison, we also investigate subjective evaluation with the reconstructed images in 2D planes. Fig. 8 demonstrates that reconstructed 2D plane on $Y \times (XW)$ well preserves clear edge and pattern information for better visual perception. Although other two planes $X \times (YW)$

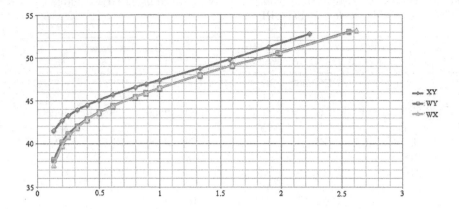

Fig. 6. Using Motion JPEG encoder in XYW, WXY, and WYX cubes for Soil image

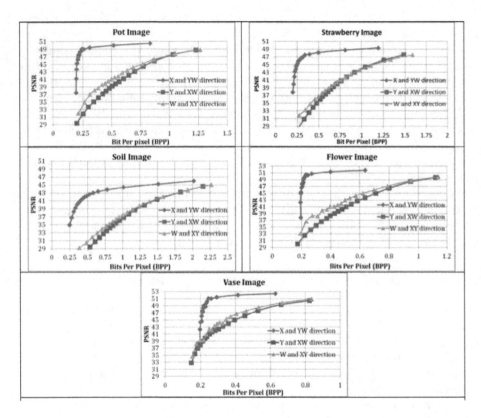

Fig. 7. Rate-distortion performance of five standard hyperspectral images using JPEG coder on three different 2D arrangements (i) X and YW i.e., 1040 84912, (ii) Y and XW i.e., 1392 63440, and (iii) W and XY i.e., 611447680

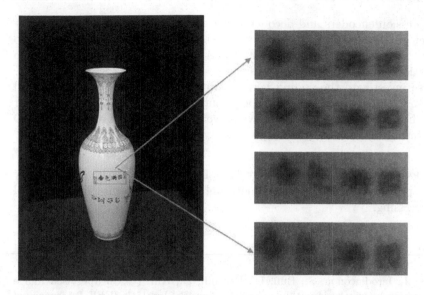

Fig. 8. Reconstructed Vase HS image (shows a small part of the Vase image for clear visualization of patterns and edges) for subjective evaluation in different 2D image planes using JPEG encoder; the left sub-figure shows entire slice of Vase HS image, the right sub-figure shows a small part of an original image, reconstructed image using $X \times YW$ plane, $Y \times WX$, and $W \times XY$ from top row to bottom row respectively

and $W \times (XY)$ compressed by same encoding standard in same quality, partial edge information lost can be detected by normal human vision.

4   Conclusion

There are a number of methods to compress a hyperspectral images using different transformations and feature extractions strategies. However, in the most of cases they reduce dimensions (i.e., reduce reflectance in some wavelength range) of a hyperspectral image for compression purpose. This may not be appropriate for a number of applications as in this process some valuable information might be removed. In this paper we rearrange a hyperspectral image into different 3D cubes and 2D image planes and then apply popular image and video encoders to see the compression performance by exploiting spatial and spectral redundancy without losing specific wavelength information. To do this we propose an optimal compression cube to predict the optimal 3D cube for the best rate-distortion performance. We confirm that the natural cube (i.e., XYW) is the best compression cube. We also confirm that HEVC and JPEG2000 are better encoder where HEVC provides more consistence performance in different directions compared to JPEG2000. Experimental results also show that our proposed optimal compression cube prediction technique can predict the optimal 3D cube successfully in the most cases. The proposed OCC framework is compatible with any compression encoders and decoders since the required pre-processing is independent of the encoding scheme to be used. The OCC framework is easy to implement with low additional computational complexity, only the simple step of calculation of the CCs need to be involved.

We also investigate the coding performance by rearranging a hyperspectal image into 2D image planes. The experimental results show that an encoder provides the best compression results when we form a 2D plane compared to 3D cube. In future we will test the compressibility of a hyperspectral video.

Acknowledgments. We acknowledge that Dr Jun Zhou, Griffith University Australia, provided us datasets and valuable information regarding hyperspectral image processing.

References

1. Byrne, J., Ierodiaconou, S., Bull, D., Redmill, D., Hill, P.: Unsupervised image compression-by-synthesis within a jpeg framework. In: 15th IEEE International Conference on Image Processing, ICIP 2008, pp. 2892–2895. IEEE (2008)
2. Ciżnicki, M., Kierzynka, M., Kopta, P., Kurowski, K., Gepner, P.: Benchmarking jpeg 2000 implementations on modern cpu and gpu architectures. Journal of Computational Science 5(2), 90–98 (2014)
3. Du, Q., Fowler, J.E.: Hyperspectral image compression using jpeg2000 and principal component analysis. IEEE Geoscience and Remote Sensing Letters 4(2), 201–205 (2007)
4. Du, Q., Ly, N., Fowler, J.E.: An operational approach for hyperspectral image compression. In: 2012 IEEE International Geoscience and Remote Sensing Symposium (IGARSS), pp. 1357–1360 (2012)
5. Gao, X., Lu, W., Tao, D., Li, X.: Image quality assessment based on multiscale geometric analysis. IEEE Transactions on Image Processing 18(7), 1409–1423 (2009)
6. Huo, C., Zhang, R., Peng, T.: Lossless compression of hyperspectral images based on searching optimal multibands for prediction. IEEE Geoscience and Remote Sensing Letters 6(2), 339–343 (2009)
7. Jun, Z.: Hyperspectral imaging in environmental informatics. In: Seminar Presentation at Charles Sturt University (April 16, 2014)
8. Karami, A., Yazdi, M., Mercier, G.: Hyperspectral image compression based on tucker decomposition and wavelet transform. In: 2011 3rd Workshop on Hyperspectral Image and Signal Processing: Evolution in Remote Sensing (WHISPERS), pp. 1–4. IEEE (2011)
9. Liu, A., Lin, W., Zhang, F.: Lossless video compression with optimal compression plane determination. In: IEEE International Conference on Multimedia and Expo, ICME 2009, pp. 173–176. IEEE (2009)
10. Liu, A., Lin, W., Paul, M., Zhang, F., Deng, C.: Optimal compression plane for efficient video coding. IEEE Transactions on Image Processing 20(10), 2788–2799 (2011)
11. Liu, G., Zhao, F., Qu, G.: An efficient compression algorithm for hyperspectral images based on a modified coding framework of h. 264/avc. In: IEEE International Conference on Image Processing, San Antonio, TX, USA, pp. 341–344 (2007)

12. Noor, N.R.M., Vladimirova, T.: Investigation into lossless hyperspectral image compression for satellite remote sensing. International Journal of Remote Sensing 34(14), 5072–5104 (2013)
13. Miaou, S.-G., Ke, F.-S., Chen, S.-C.: A lossless compression method for medical image sequences using jpeg-ls and interframe coding. IEEE Transactions on Information Technology in Biomedicine 13(5), 818–821 (2009)
14. Motta, G., Rizzo, F., Storer, J.A.: Hyperspectral data compression. Springer (2006)
15. Mudassar Raza, A.A., Sharif, M., Haider, S.W.: Lossless compression method for medical image sequences using super-spatial structure prediction and inter-frame coding. J. Appl. Res. Technol. 10(4), 618–628 (2012)
16. Paul, M.: Efficient video coding using optimal compression plane and background modelling. IET Image Processing 6(9), 1311–1318 (2012)
17. Rucker, J.T., Fowler, J.E., Younan, N.H.: Jpeg2000 coding strategies for hyperspectral data. In: Proceedings of the 2005 IEEE International Geoscience and Remote Sensing Symposium, IGARSS 2005, vol. 1, p. 4. IEEE (2005)
18. Santos, L., López, S., Callico, G.M., Lopez, J.F., Sarmiento, R.: Performance evaluation of the h. 264/avc video coding standard for lossy hyperspectral image compression. IEEE Journal of Selected Topics in Applied Earth Observations and Remote Sensing 5(2), 451–461 (2012)
19. Sullivan, G.J., Ohm, J., Han, W.J., Wiegand, T.: Overview of the high efficiency video coding (hevc) standard. IEEE Transactions on Circuits and Systems for Video Technology 22(12), 1649–1668 (2012)
20. Tang, X., Pearlman, W.A., Modestino, J.W.: Hyperspectral image compression using three-dimensional wavelet coding. In: Electronic Imaging 2003, pp. 1037–1047. International Society for Optics and Photonics (2003)
21. Vo, D.T., Nguyen, T.Q.: Quality enhancement for motion jpeg using temporal redundancies. IEEE Transactions on Circuits and Systems for Video Technology 18(5), 609–619 (2008)
22. Wang, Z., Chanda, D., Simon, S., Richter, T.: Memory efficient lossless compression of image sequences with jpeg-ls and temporal prediction. In: Picture Coding Symposium (PCS), pp. 305–308. IEEE (2012)
23. Wu, Y.Q., Wu, C.: Hyperspectral remote sensing image compression based on wavelet and support vector regression. Journal of Astronautics 3, 024 (2011)
24. Zhu, W., Du, Q., Fowler, J.E.: Multitemporal hyperspectral image compression. IEEE Geoscience and Remote Sensing Letters 8(3), 416–420 (2011)

Automatic Chinese Personality Recognition Based on Prosodic Features

Huan Zhao, Zeying Yang, Zuo Chen, and Xixiang Zhang

School of Information Science and Engineering,
Hunan University, Changsha, Hunan, P.R. China 410082
chenzuo@hnu.edu.cn

Abstract. Many researches based on the English, French and German language have been done on the relationship between personality and speech with some relevant conclusions. Due to the difference between Chinese and other languages in pronunciation of acoustic characteristics, Chinese personalities and westerners, we put forward the Chinese and his personality prediction research in view. During the study, we collected 1936 speech pieces and their Big Five questionnaires from 78 Chinese. Built models for male and female with arguments of prosodic features such as pitch, intensity, formants and speak rate. Experiments' result shows: (1) the third formant has the same effect as the first two in prediction of personality; (2) combination of pitch, intensity, formants and speak rate as classification parameters can achieve higher classification accuracy(more than 80%) than in single prosodic feature.

Keywords: Automatic personality recognition, Chinese speech, Prosodic features, Personality traits, Big five.

1 Introduction

It is well known that personality is very important in our social activities. Effective assessment of personality is helpful in interpersonal communication, social relations, and career planning. As it is well known, every movement or single speech includes lot of information that can bespontaneous, unconscious, which largely references our personality[1]. Furthermore, our personality can also be referenced through our behavior, appearance[2], or even through videos[3] and audios[4].

In this work, we focus on the speech and personality. A lot of work have already been done on this, with good results obtained. For example, In article [5]and[6], speech audio taken from a specialist have been classified in Big Five based on acoustics and prosodic features. In[7], the relationship between prosodic features and personality were explored. The authors used pitch, energy, first two formants, voiced-time and unvoiced-time as prosodic features. A corpus of 640 speeches was built which is about 10 seconds in average, and personality traits grades was assessed by listeners. Then they used Logistic regression and Support Vector Machines(SVM) to classify the data. Following the procedure described above, they claimed an accuracy of about 70%.

X. He et al. (Eds.): MMM 2015, Part I, LNCS 8935, pp. 180–190, 2015.

Previous works that studied the relationship between prosodic features and personality used a corpus of either English, French or German but not Chinese. When collecting data, they mainly took the thirds assessment as the standard value in personality traits perception. As discussed in [8], there are some differences between Chinese pronunciation and western language (such as English), so was the Chinese personality and western's. Therefore, we put forward the Chinese personality research based on speeches. One of the most important reason this work is proposed is that past researches are mostly using the third person's(listeners) perception results as the prediction standard values in perception of speakers' personality. As we know, the third person's perception results may not rightly response speakers' personality if he is not familiar with the speaker. In order to be able to accurately predict the speaker's personality characteristics, we adopt the speaker to self-assessment as the standard to determine the classification accuracy of the model. In view of the above points, our mainly work is to: (1)Collecting 78 persons' speeches and their own Big Five questionnaire; (2)Taking pitch, energy, first three formants, voiced-time and unvoiced-time as prosodic features to test the effect of personality recognition; (3)Building personality predictions' model for male and female, respectively using an SVM.

The rest of this paper introduces theories and principles of our assumption, experiments and results. At the end of the paper, we make a summary and some prospects about future works.

2   Speech and Personality

2.1   Assessment of Personality

Personality is *individual's characteristic patterns of thought, emotion, and behavior together with the psychological mechanisms-hidden or not-behind those patterns*[9]. Many experts built lots of personality assessment models for the importance of personality traits. The famous models are: Big Five[10], Sixteen Personality Factor Questionnaire(16PF)[11], Eysencks personality theory[12] and so on. In these models, Big Five has wide applicability[13] and high credibility[14] than other models. So we take it as personality standard. Big Five is the result of vocabulary method research. Researches found that personality can be covered just by the following 5 traits:

Extraversion(Ex): high extraversion means more enthusiastic, sociable, energetic, optimistic and adventurous etc.

Neuroticism(Ne): includes characteristics such as: anxiety, hostility, repression, impulsion and frailty etc.

Openness(Op): people has high score in this trait will be more imaginative, wisdom, aesthetic, creativity and affettuoso and vice versa.

Agreeableness(Ag): features are trust, altruism, frank and modesty etc.

Conscientiousness(Co): people with high score in conscientiousness indicates he/she is fairer, more dependable, self-disciplined and dutiful than lowers.

Questionnaire of Big-Five that we use in this work includes 60 questions. Each question has five answers: "strongly disagree", "disagree", "neutral", "agree" and "strongly agree", which can be transformed into scores from 1 to 5. As every personality trait owns 12 questions in this inventory, every trait's total score is ranging from 12 to 60.

2.2 Speech and Personality

In 1927, Edward Sapair presented that there is some relationship between personality and voices[15]. Then papers about using vocal behavior to percept personality have sprung up, and most of them are based on prosodic features and personality traits.

Prosodic features mainly include pitch, energy and rate. Pitch is used to indicate the ears' feeling about vibration of objects and is decided by frequency of sound wave. Generally, the higher the frequency of the vibration, the higher pitches; so is the lower vibration frequency. The energy of voice is measured by intensity, which can express average energy of voice. Voice time interval can indicate speak rate.

In previous researches, some of them take a kind of prosodic features as characteristic to explore. In[16], pitch is taken as the parameter to percept personality and the result indicates that a higher pitch represents a more active personality for him or a more exoscopic personality for her. But most of previous researches take two or more prosodic features as target. In [17], five assumptions are used to explore the relationship between competence and benevolence of personality and prosodic features. And its conclusion shows that high speed and high pitch suggests high competence and vice versa. But high speed and low pitch variation indicates low benevolence ratings. Many papers have explored topic of competence and prosodic features[7][18] and most of them got a simple result: higher speed leads to higher competence ratings. Moreover, in some other articles, voices volume[19], tone quality and speech's pause time[20] have also been taken in to consideration.

The above researches mainly concentrated in the late of the 20th century. In recent years, the study of personality and prosodic features took more characteristics in consideration. In [21][22], authors take pitch spectrum, pitch grade, rate and intensity as prosodic features. In the aspect of personality, self-assessment was adopted to evaluate the sincerity, excitability, ability, etc. In a research about excitability[23], the parameters include not only pitch spectrum, pitch variation range, rate, but also eye movement, blinking and eyebrow movement.

As mentioned before, people made a lot of research on the relationship between the personality and speech, and achieved rich results. But all of these studies used corpus based on English, German, French or other western language and their objects are westerns too. In[8], authors discuss the pronunciation difference between Chinese and English. So due to the difference between Chinese and other western language in pronunciation, in this work, we try to use pitch, energy and rate as prosodic features to predict Chinese's personality.

3   Approach

The APR(Automatic Personality Recognition) approach proposed in this article mainly includes four steps: (1)collecting corpus for the experiments; (2)extracting low-level short-term and estimate long-term prosodic features from the speeches; (3)matching prosodic parameter and personality traits; (4)building statistical model of personality traits based on prosodic characteristics and test its performance.

3.1   Building Corpus

As the limit of corpus, we have to build a corpora for experiments. There are some principles in collecting speech pieces: firstly, subjects' mother language must be Chinese, secondly, they should have complete language ability, and can use Chinese to express their own ideas and things fluently.

3.2   Extracting Parameters

After collecting the data, we can extract features. This can be separated into 2 steps: short-term features extraction and long-term features extraction. As long-terms used in this experiment can be counted from short-terms, the accuracy of short-term is very important.

In this work, the short-term features extracted are first three formants, intensity, pitch, voiced time and unvoiced time. The formants, which are *"the spectral peaks of the sound spectrum of the voice"*[23] was defined by Gunnar Fant. Intensity is used to describe energy of the speech clips. Pitch represents the main correlation between tone and intonation. Pitch, rate and energy are the most important characteristics of prosody, so these four features are used in our experiments[7].

To extract these features, we used a wildly applied software–Praat[24]. The features are extracted from 25ms analysis windows at step of 10ms. Every clip's feature-parameters can be stored in a matrix $F = (f_0, f_1, \cdots, f_8)$, the first row($f_0$) represents time(from 0 to end of the clip), vectors from f_1 to f_8 correspond to pitch, intensity and first three formants and its bandwidth. After getting these three features, we can separate voiced and unvoiced time bites from the speeches and stored in matrix $T = (t_0, t_1)$.

The long-term features can be estimated from the data got above. Maximum, minimum and average are used to describe the dynamic range, and standard deviation is a measure of data average dispersion degree. These four statistical features are enough for pitch, intensity and formants, but for time we add another dimensionality: voiced-time/unvoiced-time.

So far, we have built our data set which includes two matrix(one for male, another for female) like this $D = (d_0, \cdots, d_n)$, n=33. d_0 to d_4 indicate the classes of personality traits. Each trait is separated into three classes; d_5 to d_8 belong to pitch features; d_9 to d_{12} are the intensity parameters; d_{13} to d_{24} are formant parameters; the last 9 vectors are time parameters.

3.3   Studying and Classifying

In the experiments, Support Vector Machines(SVM)–a binary classifier was used to train and classify the corpus. As we have three classes, is necessary to improve original SVM. So first, we use the one-vs-all method to translate three classes into binary, then use SVM again.

The purpose of SVM is finding a separating hyperplane $\boldsymbol{w}^T\boldsymbol{x} + b$ so that the points belong to different classes can space on the either side of hyperplane. The points closest to the hyperplane is called support vectors and the distance from them to separating hyperplane $|\boldsymbol{w}^T\boldsymbol{P} + b|/||\boldsymbol{w}||$ must be the maximum. Therefore the goal is find w and b values in the classifier,we can write it as:

$$max_{w,b}\{min_n(sign \cdot (\boldsymbol{w}^T\boldsymbol{x} + b)) \cdot \frac{1}{||\boldsymbol{w}||}\} \tag{1}$$

It's difficult to solve this problem directly, unless we convert it to another form. A method called Lagrange multipliers has be widely used to solve this type problems. Hence, the function above can be writen as:

$$max_\alpha[\sum_{i=1}^{m} \alpha - \frac{1}{2}\sum_{i,j=1}^{m} sign^{(i)} \cdot sign^{(j)} \cdot a_i \cdot a_j \langle x^{(i)}, x^{(j)}\rangle] \tag{2}$$

with constraints:

$$\alpha \geq 0 \quad and \quad \sum_{i=1}^{m}\alpha_i \cdot sign^{(i)} = 0 \tag{3}$$

Hereto, the main work of SVM is solving these α[25][26]. There are many methods that can solve this optimization problem. But to train our data, the Sequential Minimal Optimization(SMO), which is published by John Platt in 1996[27] is been chosen. "SMO breaks this large quadratic programming(QP) problem into a series of smallest possible QP problems. These small QP problems are solved analytically, which avoids using a time-consuming numerical QP optimization as an inner loop". Thus, SMO is one of the fastest method used to solve these kind of problems. So using it in our experiment for training our model is best.

k-fold cross validation[28] is also used in the experiment. We split the data D into 10(as stated in[28] "10 folds can get a better than the more expensive leave-one out cross validation") equal subsets $D = \{D_1, D_2, \cdots, D_{10}\}$ at random, every time $D_i(i \in \{1, 2, \cdots, 10\})$ is used for test, and the other 9 subsets are used to train the model. So each fold of training and classifying is independent and the end result is more convincing.

4   Experiments and Results

In this section, experiments and its results are mainly described in three subsections: (1)data collection; (2)personality traits recognition performance comparison when taken first three formants as features; (3)personality traits recognition performance comparison when taking pith, intensity, formants and time as prosodic features. In (2) and (3), correctly classified rate was used to measure prosodic features performance in personality traits recognition.

4.1   Data Collection Results

As mentioned before, we interviewed 78 Chinese, whose mother language are Chinese, and whose are able to communicate in Chinese in general situations. Of all the subjects, there are 38 males and the others are females. For each of subject, firstly we got his(or her) NEO-FFI which contains 60 questions. After that, we recorded their speech clips which are spoken in dispassionate by letting him(or her) describe some thing or just answer specific questions. About 25 dispassionate clips are taken from every participant and most clips are in 10s to 15s. So finally the corpus was composed by 1936 speech clips 975 bites belonging to males and the other 961 chips belonging to females, as shown in Fig. 1.

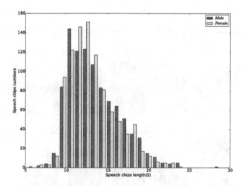

Fig. 1. Distribution of speech clips length. In this chart, male's mean length of speech clips is 13.455s, and female's is 13.158s. The number of speech clips, whose length between 9s-15s is 682 and 731 for male and female, respectively.

Figure 2 shows the NEO-FFI scores of each subject. For Ag and Ex, their score distribution is around the average 36; and the other three traits' average score is about 40. Personality is hard to measure in quantitative because of it selfs characteristic, and not needed. Therefore in the experiments below, personality traits will be divided into three grades based on the traits score.

4.2   Formants Predict Personality

Formant is a prosodic feature, which refers to vibration frequency of pronunciation and intonation. People were aware of the importance of first two formants, but the importance of the third formant in phonetics were enlightened in papers such as [7] and[29]. In this work, the relationship between personality traits were judged through classification results of men's and women's different traits based on first three formants, which are taken as prosodic features.

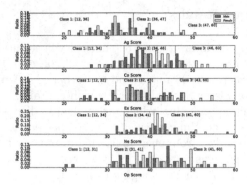

Fig. 2. NEO-FFI score distribution

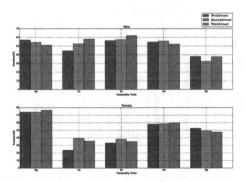

Fig. 3. The personality traits classification result based on single formant

For two subgraphs in figure 3, the x axis represents the distribution of each personality characteristics, the y axis represents classification accuracy. In every characteristic, the three rectangles stand for the first three formants classification accuracy. For male, we get a best accuracy of 61.95% for Ex with is obtained by using the third formant. But using the first formant we get an accuracy of 56.31% and an accuracy of 57.33% for the second formant. These accuracy results means Ex is easy to classify. The lowest recognition accuracy is from Op trait. Its three formants classification accuracy are only 38.05%, 32.41% and 37.64%, respectively. In general for the two personality traits Ag and Op, the first formant classification effect is the best; For personality traits Co and Ex, the third formant has the highest classification accuracy. For the other personality trait the second formant effect is slightly better than the other two. For female, the best accuracy result is obtained from Ag, which accuracy rates are 73.98%, 73.99%, 76.48%. That is about 20% higher than the accuracy rates obtained for male. The Co and Ex have the poor effect in classification accuracy, which are both less than 35%. In the picture, although each single feature recognition

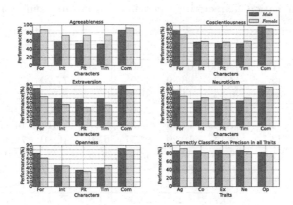

Fig. 4. Personality recognition results based on prosodic features(Formants(For), Intensity(Int), Pitch(Pit), Time(Tim), Combination(Com))

rate is low, it can be known that any formant can play similar role when used as prosodic features in personality traits recognitions. Although the role of the third formant in the phonetics is not well-known, we can apply the third formant to personality traits recognition based on prosodic features.

4.3   Prosodic Predicts Personality

In the above section, we explore the formant's influence in personality traits. Pitch, sound intensity, voiced time, unvoiced time, the first three formants, and the combinations of them are used to explore the impact on the classification of personality traits. As shown in figure 4, the first 5 subgraphs represents 5 different personality traits. The abscissas of the 5 subgraphs represent formant, intensity, pitch, timing and the combination of them. The ordinates are classification accuracy rate. For different personality traits, prosodic features as the basis of classification results in different identification accuracy rate. In general, when using single prosodic feature, formant can get better effect than other threes, especially for Op. Also for Op no matter which single prosodic feature(or their combination), its recognition rate is lower than the other four personality traits. As shown in the figures, either male or female, the recognition rates when based on single prosodic feature, are lower than the recognition rates obtained by combining those features. For Ag, when using a single feature, the average recognition rate of men and women are 58.1% and 74.71% respectively; 84.62% and 87.62% in combination. In each other personality traits, also it can be seen that combined characteristics receive a higher classification accuracy rate than a single characteristic with a gain of about 20% accuracy. This means each prosodic features can reflect one or more aspects of our personality.

When it comes to classification, there are some mistakes more or less. The confusion matrix(see table 1) will show classification more directly and concretely when using the combined prosodic features. The table is divided into two parts,

Table 1. Combination characteristics classification results' confusion matrix(%)

Traits		Male			Female		
		1	2	3	1	2	3
Ag	1	96.00	0	4.00	87.33	12.67	0
	2	4.22	86.89	8.89	0	77.20	22.80
	3	6.95	11.79	81.26	0	8.72	91.28
Co	1	97.00	1.00	2.00	92.00	4.00	4.00
	2	13.43	77.14	9.43	9.07	76.17	14.77
	3	5.14	10.29	84.57	6.80	16.80	76.40
Ex	1	92.00	5.00	3.00	90.67	5.33	4.00
	2	6.78	89.04	4.17	9.87	77.07	13.06
	3	11.33	12.34	76.33	9.56	14.09	76.32
Ne	1	89.71	8.29	2.00	91.82	7.27	0.91
	2	15.40	83.40	1.20	31.33	67.34	1.33
	3	19.20	15.20	65.60	30.63	8.12	61.26
Op	1	81.60	10.80	7.60	85.00	8.00	7.00
	2	15.11	80.22	4.67	10.96	80.43	8.61
	3	17.09	11.64	71.28	12.57	20.86	66.57

men and female, and then divided into the five personality traits(Ag, Co, Ex, Ne, Op) according to Big Five personality model. Each of the personality trait is partitioned into three categories according to the NEO-FFI score. Every value in the table represents the classification accuracy. For example, the value 96% in third row and third column is said that for male's Ag trait, when using the model trained in SVM to classify speeches belonging class one, 96% of them are identified correctly. And the remaining 4% are classified as class three incorrectly, and so on. For each personality trait, their recognition rates are showing a phenomenon that the first class performs best, the second class performs better than the third.

5   Conclusions

The main work of this paper is the recognition of personality in Chinese speeches prosodic features. We use the three main prosodic features: pitch, energy and speed rate with the personality traits, which are got from NEO-FFI. After data collection, the first three formants have been analyzed for personality recognition. Even though the importance of the third formant is still unclear, one of the finding of this paper is that the third formant has the same effect in automatic personality recognition as the first two. In another personality traits recognition experiment, a comparison between single character and combination characters has been made. This experiment shows that combination's accuracy rate is higher than the singles in average by 20%. Also by combining prosodic features for, we classification can get a better recognition rate. As other papers are based on different data set, comparing our work with theirs can hardly get convincing.

As mentioned before, we present our personality traits not only in speech but also in many other ways. So exploring prosodic features with other characteristics such as facial movements, body language[30] even eye movements is very meaningful. As China is a very large country, the research about automatic personality recognition based on the local language is also meaningful.

References

[1] Barbero, C., Zovo, P.D., Gobbi, B.: A flexible context aware reasoning approach for iot applications. In: 2011 12th IEEE International Conference on Mobile Data Management (MDM), vol. 1, pp. 266–275. IEEE (2011)

[2] Uleman, J.S., Saribay, S.A., Gonzalez, C.M.: Spontaneous inferences, implicit impressions, and implicit theories. Annu. Rev. Psychol. 59, 329–360 (2008)

[3] Mayer, R.E.: Multimedia learning. Cambridge University Press (2009)

[4] Olivola, C.Y., Todorov, A.: Elected in 100 milliseconds: Appearance-based trait inferences and voting. Journal of Nonverbal Behavior 34(2), 83–110 (2010)

[5] Polzehl, T., Moller, S., Metze, F.: Automatically assessing personality from speech. In: 2010 IEEE Fourth International Conference on Semantic Computing (ICSC), pp. 134–140. IEEE (2010)

[6] Polzehl, T., Moller, S., Metze, F.: Automatically assessing acoustic manifestations of personality in speech. In: 2010 IEEE Spoken Language Technology Workshop (SLT), pp. 7–12. IEEE (2010)

[7] Mohammadi, G., Vinciarelli, A.: Automatic personality perception: Prediction of trait attribution based on prosodic features. IEEE Transactions on Affective Computing 3(3), 273–284 (2012)

[8] Haiying Li, Y.W.: Comparative study on the phonetic features chinese, english and japanese. Social Science Forum 9, 176–180 (2009)

[9] Allport, G.W.: Personality: A psychological interpretation (1937)

[10] Komarraju, M., Karau, S.J., Schmeck, R.R., Avdic, A.: The big five personality traits, learning styles, and academic achievement. Personality and Individual Differences 51(4), 472–477 (2011)

[11] Cattell, R.B., Eber, H.: Sixteen personality factor questionnaire (16pf). Institute for Personality and Ability Testing, Champaign, Illinois, USA (1972)

[12] Eysenck, S.B., Eysenck, H.J., Barrett, P.: A revised version of the psychoticism scale. Personality and Individual Differences 6(1), 21–29 (1985)

[13] Hattie, J.: Visible learning: A synthesis of over 800 meta-analyses relating to achievement. Routledge (2013)

[14] John, O.P., Naumann, L.P., Soto, C.J.: Paradigm shift to the integrative big five trait taxonomy. Handbook of Personality: Theory and Research 3, 114–158 (2008)

[15] Sapir, E.: Speech as a personality trait. American Journal of Sociology, 892–905 (1927)

[16] Argyle, M.: Bodily communication. Routledge (2013)

[17] Ray, G.B.: Vocally cued personality prototypes: An implicit personality theory approach. Communications Monographs 53(3), 266–276 (1986)

[18] Collier, G.J.: Emotional expression. Psychology Press (2014)

[19] Lindzey, G., Gilbert, D., Fiske, S.T.: The handbook of social psychology. Oxford University Press (2003)

[20] Burgoon, J.K., Guerrero, L.K., Floyd, K.: Nonverbal communication. Allyn & Bacon, Boston (2010)

[21] Schmitz, M., Krüger, A., Schmidt, S.: Modelling personality in voices of talking products through prosodic parameters. In: Proceedings of the 12th International Conference on Intelligent User Interfaces, pp. 313–316. ACM (2007)

[22] Trouvain, J., Schmidt, S., Schröder, M., Schmitz, M., Barry, W.J.: Modelling personality features by changing prosody in synthetic speech (2008)

[23] Fant, G.: Acoustic theory of speech production: with calculations based on X-ray studies of Russian articulations, vol. 2. Walter de Gruyter (1971)

[24] Boersma, P.: Praat, a system for doing phonetics by computer. Glot International 5(9/10), 341–345 (2002)

[25] Bishop, C.M., et al.: Pattern recognition and machine learning, vol. 1. Springer, New York (2006)

[26] Harrington, P.: Machine Learning in Action. Manning Publications Co. (2012)

[27] Platt, J., et al.: Sequential minimal optimization: A fast algorithm for training support vector machines (1998)

[28] Kohavi, R., et al.: A study of cross-validation and bootstrap for accuracy estimation and model selection. IJCAI 14, 1137–1145 (1995)

[29] Mohammadi, G., Vinciarelli, A., Mortillaro, M.: The voice of personality: mapping nonverbal vocal behavior into trait attributions. In: Proceedings of the 2nd International Workshop on Social Signal Processing, pp. 17–20. ACM (2010)

[30] Pianesi, F., Mana, N., Cappelletti, A., Lepri, B., Zancanaro, M.: Multimodal recognition of personality traits in social interactions. In: Proceedings of the 10th International Conference on Multimodal Interfaces, pp. 53–60. ACM (2008)

Robust Attribute-Based Visual Recognition
Using Discriminative Latent Representation

Yuqi Wang, Yunfei Gong, and Qiang Liu

School of Software, Tsinghua University, Beijing 100084, China
{wangyuqi10,gongyf07}@gmail.com, liuqiang@mail.tsinghua.edu.cn

Abstract. Recent work in visual recognition have addressed attribute-based classification. However, semantic attributes that are designed and labeled by humans generally contain some noise, and have weak learnability for classifiers and discrimination between categories. As a fine supplement to semantic attribute, data-driven attribute learned from training data suffers from the ineffectiveness in novel category classification with no or few samples. In this paper, we introduce the Discriminative Latent Attribute (DLA) as a mid-level representation, which has connection with both visual low-level feature and semantic attribute through matrix factorization. Furthermore, we propose a novel unified formulation to efficiently train category-DLA matrix and attribute classifiers together, which makes DLA more learnable and more discriminative between categories. Our experiments show the effectiveness and robustness of our approach which outperforms the state-of-the-art approach in zero-shot learning task.

Keywords: object recognition, latent attribute, transfer learning.

1 Introduction

Visual attributes have been advocated in the computer vision community in the past few years. The term "attribute" is first proposed to describe images containing unfamiliar objects in [1,2]. As an intermediate high-level representation extracted from visual low-level feature, attribute is widely used for image recognition [1-8], image retrieval [20-24] and video action recognition [9,10]. Due to its cross-category generalization property, attribute is often used to recognize novel categories with no training examples [2]. This problem is referred to as zero-shot learning [2,11,12] in general.

Attribute is often referred to as semantic attribute, e.g. color, shape, parts. It is designed by humans to describe objects according to the knowledge bases provided by domain experts [1,2]. However, it has three shortcomings. Firstly, semantic attributes are labeled by humans in general (few works like [17, 18] automatically discover attributes from knowledge source from web). Thus, absolutely relying on semantic attributes suffers too much from noise [8]. Secondly, semantic attributes designed by humans may not be suitable for attribute classifiers due to lack of feedback in design phase. Thirdly, the discrimination between categories in semantic attribute space is not sufficient. In this case, some previous works [6,7,10,19] use non-semantic

X. He et al. (Eds.): MMM 2015, Part I, LNCS 8935, pp. 191–202, 2015.

data-driven attribute as a fine supplement for semantic attribute. However, data-driven attribute just relies on training data. For novel categories with no or few samples, it is nearly impossible to obtain high-quality data-driven attribute representation of novel category without any additional prior knowledge.

Motivated by above observations, in this paper, we introduce the Discriminative Latent Attribute (DLA). Figure 1 shows an illustration of DLA. DLA is a mid-level representation between semantic attribute and visual feature of image. To our knowledge, it's the first attempt to represent image with the latent attributes which are different from semantic attributes. In connection between DLA and semantic attribute, we obtain DLA from semantic attribute representation using matrix factorization. Benefiting from matrix factorization, DLA captures the latent knowledge of semantic attribute and can deal with novel category classification in a similar way of semantic attribute. In other words, unlike data-driven attribute, DLA has exactly the same generalization property as semantic attribute. We provide the approaches to transfer knowledge from known categories to novel categories and to deal with zero-shot learning based on DLA. Besides, DLA can also overcome the three shortcomings of semantic attributes. Take full advantage of visual low-level feature that also connects to DLA, DLA can suppress noise in semantic attribute effectively.

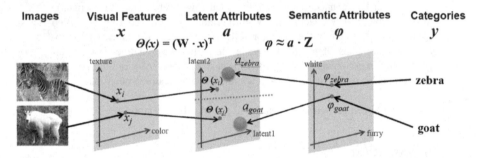

Fig. 1. Illustration of the proposed Discriminative Latent Attribute (DLA)

The main contributions of this paper are summarized as follows:

We introduce the Discriminative Latent Attribute (DLA), which has connection with both visual low-level feature and semantic attribute as an intermediate representation. DLA captures the latent knowledge of semantic attribute and is more robust to noise.

We propose a novel unified formulation to train category-DLA matrix and attribute classifiers. The formulation makes DLA more learnable for attribute classifiers and more discriminative to categories than semantic attribute.

Based on DLA, we provide the knowledge transfer approaches to deal with zero-shot learning tasks and demonstrate the generalization property of DLA by comprehensive experiments.

The rest of this paper is organized as follows. Section 2 reviews related work. Section 3 describes the framework and our DLA model. Section 4 describes how to transfer the knowledge based on DLA model. Section 5 shows the experimental results and analysis. Finally, we summarize conclusions and discuss the future work.

2   Related Work

Some previous work [3,8] used latent variables in the model, but they treated semantic attributes as latent variables directly. In image retrieval field, [23] also used latent variables as weak attributes, but they obtain them from the construction of latent tree in which latent variables are used to describe the relation of some group of weak attributes. Different from previous works, our work uses latent attributes as an image representation which have similar meanings as semantic attributes.

Many researchers have paid attention to attribute-based knowledge transfer due to the generalization property of attributes. C. H. Lampert et al. [2] firstly proposed direct attribute prediction (DAP) model and indirect attribute prediction (IAP) model to solve zero-shot learning task. [16] proposed a function to measure the compatibility between an image and its category label. Instead of attribute classifiers, the function is the intermediate used for knowledge transfer. Similar to [16], [4] proposed the topic model based framework for zero-shot learning. Above approaches are based on manually-designed semantic attributes, which is a difficult task for humans.

3   Learning Latent Attribute Representation

We firstly define the symbols and the problem of known category classification.

Known Category Classification: Given a training set $S = \{(x_i, y_i), i = 1, 2, \ldots M\}$, in which $x_i \in X$ is the visual low-level feature of an image and $y_i \in Y$ is the category label associated with x_i, and a category-attribute matrix Φ, in which each row represents a category and each column represents a semantic attribute, our goal is to learn a function $f : X \rightarrow Y$.

3.1   The Framework

To solve this classification problem, we introduce Discriminative Latent Attribute (DLA) space shown in Figure 1. Suppose there are k categories, and the dimension of DLA space is d. We define the category-DLA matrix as $A \in R^{k \times d}$, in which the rows $\{A_i, i = 1, .., k\}$ correspond to k categories. We infer the image label following three steps:

Mapping Image: For an input image, map its visual low-level feature to DLA by attribute classifiers $\Theta_i(x) = [\Theta_1(x), \ldots, \Theta_d(x)]$, in which $\Theta_i(x) \in R$ is the ith attribute classifier's output, d is the dimension of DLA space.

Mapping Category: For known categories, use the category-DLA matrix A trained in the model. For all novel categories, map the novel category-attribute matrix Φ' to DLA space as A'. The mapping approaches (detailed in section 4) is based on matrix factorization technique, including minimizing $\|\Phi' - A'Z\|_F^2$, in which Z represents the semantic attribute latent feature matrix (Z can also be seen as the transform matrix from DLA space to semantic attribute space).

Output Category: Choose one nearest neighbor (1NN) category (row of A) for input image in DLA space by Euclidean distance: arg min$\| A_i - \Theta(x)\|$.

According to above framework, we should train the attribute classifiers $\Theta_i(x)$, the category-DLA matrix A and the transform matrix Z in the model formulation.

3.2   Formulation

As mentioned in Section 1, DLA should be discriminative between categories, learnable for attribute classifiers and capture the latent knowledge of semantic attribute.

In order to make it more discriminative and learnable, the separation between every two categories (rows of A) should be larger, and the distance between every image and its category in DLA space should be smaller. Similar to [7], we define the separation ρ and the distance ϵ respectively as follows:

$$\rho = \left\| A_{y_1} - A_{y_2} \right\|, \quad y_1, y_2 \in Y \tag{1}$$

$$\epsilon = \left\| A_{y_i} - \Theta(x_i) \right\|, \quad \{x_i, y_i\} \in S \tag{2}$$

in which A_{y_1}, A_{y_2} are two categories' representations in DLA space (two rows of A), $\Theta(x)$ is the image's representation in DLA space (the output of attribute classifiers). We use linear SVM as attribute classifier in consideration of efficiency. So we can define $\Theta(x)$ as:

$$\Theta(x) = (W \cdot x)^T \tag{3}$$

in which each row of W correspond to the parameters of one linear SVM.

In order to capture the latent knowledge of semantic attribute, we use matrix factorization technique to analyze the factor of manually-defined category-attribute trix Φ. Suppose the category and the semantic attribute have their own latent features, the dot product of the two latent features represent the value of this attribute in this category. We define category latent matrix as A, in which each row of A corresponds to one category, and define semantic attribute latent matrix as Z (Z can also be considered as a transform matrix from DLA space to semantic attribute space), in which each column of Z corresponds to one semantic attribute. If we strictly follow the constraint, we will have $\Phi = A \times Z$. But just like many other models that use matrix factorization technique, relax the strict constraint will result in high performance. So we define the reconstruction error as follows:

$$\delta = \left\| \Phi - AZ \right\|_F^2 \tag{4}$$

We consider ρ, ϵ, δ together, and we need larger ρ, smaller ϵ, and smaller δ. So we put them together and consider the object function without any constraints:

$$\min_{A,Z,W} \sum_{\{x_i,y_i\}_{i=1}^M} \left\| A_{y_i} - (W \cdot x_i)^T \right\|_2^2 - \gamma_1 \sum_{y_1,y_2 \in Y} \left\| A_{y_1} - A_{y_2} \right\|_2^2$$

$$+ \gamma_2 \left\| \Phi - AZ \right\|_F^2 \tag{5}$$

in which γ_1 and γ_2 are weight parameters for ρ and δ. This is the major principle of our model formulation. Then we consider adding some constraints to A, Z and W respectively.

A represents the category-DLA matrix. It should be less redundant in order to capture knowledge from more aspects of semantic attribute. So we add orthogonal constraint to A. Just like PCA or other models, the strict orthogonal constraint will result in low performance. So we relax the orthogonal constraint as follows:

$$\min_A \|A^T A - I\|_F^2 \tag{6}$$

W is the parameters of linear SVM and Z is the sematic attribute latent matrix in matrix factorization. We add the Frobenius norm constraint similar to the one used in SVM and matrix factorization. Thus, we add the three constraints to object function and get the final one as follows:

$$\min_{A,Z,W} \sum_{\{x_i,y_i\}_{i=1}^M} \left\|A_{y_i} - (W \cdot x_i)^T\right\|_2^2 - \gamma_1 \sum_{y_1,y_2 \in Y} \left\|A_{y_1} - A_{y_2}\right\|_2^2$$

$$+\gamma_2 \|\Phi - AZ\|_F^2 + \beta_1 \|A^T A - I\|_F^2 + \beta_2 \|W\|_F^2 + \beta_3 \|Z\|_F^2 \tag{7}$$

Unlike most previous work which firstly trained attribute classifiers, we propose a unified formulation to train the attribute classifiers W, the category-DLA matrix A and the transform matrix Z together. The unified formulation make it convenient for them to optimize each other. To solve Eq. (7), repeat following steps: holding A, W fixed, find the best transform matrix Z; holding W, Z fixed, find the best category-DLA matrix A; holding A, Z fixed, find the best attribute classifiers W. A local optimum can be obtained using coordinate descent.

3.3   Efficiency and Scalability

During the optimization of our formulation, we have iterations in the SGD phase of A, and in the global coordinate descent. Benefitting from the property of coordinate descent, we don't have to get the optimum solution of A in each global iteration. We just set max iterations to 20 for the SGD phase of A. In addition, the global coordinate descent is also efficient so that we need only 3 to 5 iterations to get the global solution. Our model is approximately linear to the scale of training set as the number of total iterations is less than 100. Compared with non-linear kernel-based SVM models, our model is significantly fast. For example, on a 4-core Intel I7 3.4GHz PC, it just takes about half an hour to train our model on AWA (contains about 30000 images) dataset, while it takes more than 20 hours to train the non-linear SVM.

Different from the fixed dimension of manually-designed semantic attribute, we can set arbitrary dimension of DLA. It allows us to apply DLA to more scenarios. Besides, even low-dimension DLA can achieve the comparable experimental results for high-dimension DLA, which we will demonstrate in Section 5.

4   Knowledge Transfer by Latent Attribute

In this section, we focus on the problem of zero-shot learning task.

Zero-Shot Learning: Given training data defined in Section 3, the disjoint category-attribute matrix Φ', our goal is to learn a function $f : X' \rightarrow Y'$, in which X' is visual low-level feature, Y' is category label in disjoint categories with regard to training categories.

In zero-shot scenario, we only have the novel category-attribute matrix Φ' as prior information about novel categories. As mentioned in Section 2, most previous work only transfer attribute classifiers learned from training set to obtain the semantic attribute representations of images. Considering the intermediate DLA space, we should transfer both attribute classifiers and semantic attribute latent feature matrix. Attribute latent feature is the latent feature of semantic attribute. Since semantic attributes in novel categories are the same as the ones in known categories, attribute latent feature matrix can also be transferred to novel categories. We use the matrix W and Z trained in known categories. Reviewing the framework proposed in Section 3, since we have W and Z, we merely need the category-DLA matrix A'. Similar to Eq. (7), for novel categories, we follow the same principle to have larger ρ and smaller δ. The object function is as follows:

$$\min_{A'} -\gamma_1 \sum_{y_1, y_2 \in Y'} \left\| A'_{y_1} - A'_{y_2} \right\|_2^2 + \gamma_2 \left\| \Phi' - A'Z \right\|_F^2 \qquad (8)$$

We can use the Stochastic Gradient Descent (SGD) to solve this equation again. Since W, Z, and A' are obtained, we follow the framework described in Section 3, and we can classify novel categories.

5   Experiments

In this section, we conduct experiments of three tasks to evaluate the effectiveness and robustness of the proposed Discriminative Latent Attribute (DLA). The experimental setup will be described in Section 5.1. In Section 5.2, we focus on the problem of known category classification. In Section 5.3, experiments of zero-shot tasks will be conducted.

5.1   Experimental Setup

Datasets. We conducted experiments on the two well-known datasets, Animal with Attributes (AwA) dataset [2], and a-Pascal/a-Yahoo (aP/aY) dataset [1]. AwA contains 30475 images of 50 animal categories. Each category is annotated with 85 semantic attributes with both binary and continuous value. The a-Pascal dataset contains 6340 training images and 6355 testing images. There are 20 object categories and each image is annotated with 64 attributes with binary value. The a-Yahoo dataset contains 2644 images of 12 disjoint object categories with the same 64 semantic attributes.

Visual Features. On AwA, the same as [2], we used the pre-computed 10940-dimensional feature vectors that contain HSV color histograms, SIFT, rgSIFT, PHOG, SURF, and local self-similarity histograms. For our model, we further used PCA to get a 1912-dimensional feature. On aP/aY, the same as [1], we used the pre-computed 9751-dimensional feature vectors that contain information on color, texture, edge orientation, and HoG extracted from the objects' bounding boxes. We further used PCA to get a 1998-dimensional feature in the same way.

Category-Attribute Matrix. Our model relies on category-level attribute annotation. On AwA, we just used the provided category-attribute matrix with binary value. On aP/aY, since there are only annotations for each image, we first calculated the average attribute value of all images from the same category and got a category-attribute matrix with continuous value and then changed it to binary matrix.

Noisy Labels. In our experiments, we assumed that the manually-labeled category-attribute matrix is ground truth without any noise. In order to evaluate the performance of DLA under different noisy situations, we added some noise (5%, 10%, 20%, 30%, 40% and 50%) into category-attribute matrix by randomly picking one attribute and modifying it from 1 to 0 or 0 to 1.

5.2   Known Category Classification

In this task, we verify the multi-class classification performance on the 40 training categories of AwA and 20 categories of a-Pascal.

Baseline. We choose two baselines in this task. The first one is multi-class SVM with χ^2 kernel based on low-level feature. The same as [2], we set the bandwidth as 0.2 times median distance. The second one is the DAP model, which is based on semantic attribute and performs better than the basic attribute-based SVM method[1].

On both datasets, we selected different amount (15, 20, 30, 50 and 50%[2]) of images per category for training and validation, 25 images per category for testing. Using cross-validation, we tune the parameters we used as follows: $\gamma_1 \in \{10,50,100,200\}$, $\gamma_2 \in \{0.5,1,2,5,10\}$, $\beta_1 \in \{5\gamma_2, 10\gamma_2\}$, $\beta_2, \beta_3 \in \{0.05, 0.1, 0.5, 10, 20\}$. We conducted the experiments with all configures of noise and different dimensions of our DLA model.

Firstly, we demonstrate the robustness performance of DLA in Table 1. We only use DAP as baseline here because accuracy of SVM is irrelevant to the noise. The DLA (85) or DLA (64) means the DLA model with 85 or 64 dimension which is the same dimension as semantic attribute. With 50% training data and all configurations of noise, accuracy of DLA model has much smaller range than accuracy of DAP model on both datasets. On AwA, although DAP has higher accuracy with configurations of 20%, 30% and 40% noise, DLA has higher mean accuracy (39.71%) than DAP's (38.99%). On a-Pascal, DLA has obvious advantage with 47.26% in mean, about 5% higher than DAP's 42.51%. Hence, in this aspect DLA is more robust than semantic attribute.

---

[1] The method represents an image with attributes and uses SVM based on the attribute representation. According to our experimental results, the method has lower performance than DAP.

[2] When we select 50% images per category on a-Pascal which is heavily biased toward category "person", we only used the first 200 images of category "person" for training and testing in consideration of the balance of categories.

Table 1. Multi-class accuracy in known category classification task with 50% training data

	AwA		aP/aY	
Noise	DAP	DLA(85)	DAP	DLA(64)
0%	35.26	**39.72**	38.11	**47.29**
5%	38.07	**39.68**	39.94	**47.29**
10%	38.13	**39.74**	41.26	**47.27**
20%	**40.13**	39.76	43.63	**47.24**
30%	**41.50**	39.75	43.85	**47.14**
40%	**41.05**	39.68	46.07	**47.32**
50%	38.78	**39.67**	44.72	**47.24**

Figure 2 shows the mean accuracy of our DLA model and two baselines with different amount of training data. DLA model uses 1NN to output category label. However, the DLA representation can also be used for SVM to output category. Using SVM increases the complexity of the model, but can improve the performance when a large number of training data are provided. Thus, DLA+SVM performs best with 50% data for training and has similar performance as DLA with other amount of training data. On both datasets, DLA model significantly outperforms the two baselines.

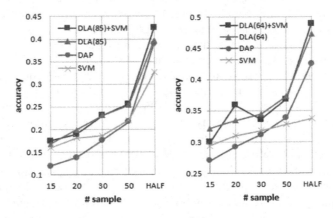

Fig. 2. Accuracy of known category classification on AwA (left) and a-Pascal (right)

We also conducted experiments for DLA with different dimensions and the mean accuracy is shown in Table 2. As we see, the dimensions affects the performance little in this scenario. The maximum difference between three dimensions is within 0.1% with the same number of training data. In most cases, high-dimension DLA performs better than low-dimension DLA although the improvement is little. This observation indicates that semantic attribute is redundant so that our DLA model can efficiently achieve better performance even with half dimensions of semantic attribute. In other words, DLA can also reduce redundant in semantic attribute.

Table 2. Multi-class accuracy of DLA with different dimensions in known category classification task. We report the mean accuracy with all configurations of noise.

	AwA			aP/aY		
# sample	40D	85D	200D	30D	64D	200D
15	**16.83**	16.76	16.80	32.17	32.17	**32.17**
20	19.93	19.91	**19.87**	33.34	**33.43**	33.37
30	23.03	**23.07**	22.99	34.43	34.46	**34.46**
50	25.33	25.34	**25.36**	37.26	37.29	**37.31**
50%	39.69	39.71	**39.71**	47.27	47.26	**47.32**

5.3   Novel Category Classification

In this section, we demonstrate that our DLA model has exactly the same generalization property as semantic attribute. Moreover, it is more robust than semantic attribute. On AwA, we used 10 disjoint categories provided by dataset as novel categories. On aP/aY, we used 12 disjoint categories of a-Yahoo as novel categories. We used 90% images from 40 known categories as training data on AwA and used the default split training data (about 50%) on a-Pascal. We tuned parameters of DLA with the same strategy described in Section 5.2.

DLA ($Z = I$). In order to further evaluate the performance of DLA, we introduce this advanced baseline. In Eq. (7), we fix the transform matrix Z to identity matrix I. Thus the latent attribute matrix A is similar to semantic attribute matrix Φ. And in the optimization phase, we do not need to optimize Z. This advanced baseline can benefit from the unified formulation which makes the representation more learnable for attribute classifiers. So we treat it as advanced baseline to evaluate the performance of DLA without the factor of the unified formulation.

In zero-shot scenario, there is no training data in novel categories. The only prior knowledge for novel category is the category-attribute matrix which describes semantic attributes of each category. In order to get the novel category-DLA matrix, we optimized Eq. (8) using the same parameters as the ones tuned in the training phase of DLA model.

Baseline. We choose two baselines. The first one is the advanced baseline DLA ($Z=I$) mentioned above. The second one is the DAP model with the same configurations as [2]. According to previous reported results, among semantic attribute based approaches, DAP is the state-of-the-art approach. Under the 0% noise situation (using the original category-attribute matrix), the DAP model in our experiments achieves accuracy of 39.1% on AwA and 17.8% on aP/aY that is similar to the performance reported in [5] (41.1% on AwA and 19.1% on aP/aY).

The accuracy of zero-shot learning with different configurations of noise is shown in Figure 3. We analyze the comparison of DLA and two baselines with the same dimension first (Left 1 and Left 3). On AwA, DLA outperforms two baselines with all noisy situations. Moreover, even DLA ($Z=I$) outperforms DAP benefitting from the unified formulation. With 0% noise, DLA and the two baselines have the similar accuracy (about 1% higher than DAP). With the increasing noise, the accuracy of all the models decrease. However, the robustness of DLA model makes sense and the difference between DLA and DAP increase. With 20% noise, DLA achieves accuracy of 33.0% which is significantly (7.5%) higher than DAP. When the noise further increases, the accuracy of all the models decrease rapidly. With the 50% noise which seems to be completely chaos, all the models perform nearly as chance classification (10% on AwA and 8.33% on aP/aY). On aP/aY, the trends of DLA and two baselines are similar to the ones on AwA, but there are some different observations. The maximum of difference between DLA and DAP is 5.1% with 0% noise. As the noise increase, the difference decrease monotonously. Another observation is that the accuracy of 0% noise on AwA (40.0%) is obviously higher than the accuracy on aP/aY (22.9%). We inferred that there is more noise (about 20%) in the labeled semantic attributes (assumed as ground truth) on aP/aY than AwA. Thus if we shift the configuration from 0% noise to 20% noise on aP/aY, we get the similar trends of accuracy as the ones on AwA.

The comparison of DLA with different dimensions can be seen in Figure 3 (Left 2 and Left 4). Unlike known category classification task, accuracy of high-dimension DLA is visibly higher than accuracy of low-dimension DLA. On AwA, the accuracy of 40-dimension DLA is slightly lower with less than 30% noise. 85-dimension DLA and 200-dimension DLA performs quite the same. On aP/aY, 200-dimension DLA outperforms the other two absolutely. 64-dimension DLA outperforms 30-dimension DLA with less than 30% noise configurations. With high dimension, DLA achieves better performance.

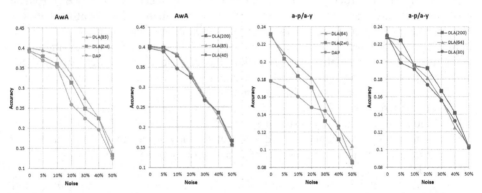

Fig. 3. Accuracy of zero-shot learning with different configurations of noise on AwA (left two images) and aP/aY(right two images)

With almost all noise configurations on AwA and 5% to 40% noise configurations on aP/aY, our DLA model significantly outperforms DAP model. Therefore our DLA model is more robust than DAP.

6   Conclusion

In this paper, we introduce the Discriminative Latent Attribute (DLA), which has connection with both visual low-level feature and semantic attribute as a mid-level representation. DLA captures the basic knowledge of semantic attribute and is more robust. We proposed a novel unified formulation to train DLA model efficiently by joining attribute classifiers and matrix factorization. The formulation makes DLA more learnable for attribute classifiers and more discriminative to categories than semantic attribute. In addition, DLA has almost the same generalization property as semantic attribute. Based on DLA model, we proposed the knowledge transfer approaches to deal with the zero-shot learning tasks. We conducted known and novel category classification experiments respectively on the two well-known datasets, Animals with Attributes (AwA) and a-Pascal/a-Yahoo (aP/aY). The experimental results shows the robustness, learnability, discriminability and generalization of our model. In known category classification task, it significantly improve the performance of traditional classification. In the zero-shot learning task, it outperforms the DAP model with all configurations of noise.

We can continue our future work in two directions. The first one is to research how to reduce the effect of noise in image-level attributes. The second one is to analyze attribute with uncertain values and to find the most significant attributes for each image or category.

References

1. Farhadi, A., Endres, I., Hoiem, D., Forsyth, D.: Describing objects by their attributes. In: CVPR (2009)
2. Lampert, C.H., Nickisch, H., Harmeling, S.: Learning to detect unseen object classes by between-class attribute transfer. In: CVPR (2009)
3. Wang, Y., Mori, G.: A discriminative latent model of object classes and attributes. In: Daniilidis, K., Maragos, P., Paragios, N. (eds.) ECCV 2010, Part V. LNCS, vol. 6315, pp. 155–168. Springer, Heidelberg (2010)
4. Yu, X., Aloimonos, Y.: Attribute-based transfer learning for object categorization with zero/One training example. In: Daniilidis, K., Maragos, P., Paragios, N. (eds.) ECCV 2010, Part V. LNCS, vol. 6315, pp. 127–140. Springer, Heidelberg (2010)
5. Lampert, C.H., Nickisch, H., Harmeling, S.: Attribute-Based Classification for Zero-Shot Visual Object Categorization. TPAMI (2014)
6. Rastegari, M., Farhadi, A., Forsyth, D.: Attribute discovery via predictable discriminative binary codes. In: Fitzgibbon, A., Lazebnik, S., Perona, P., Sato, Y., Schmid, C. (eds.) ECCV 2012, Part VI. LNCS, vol. 7577, pp. 876–889. Springer, Heidelberg (2012)
7. Yu, F.X., Cao, L., Feris, R.S., Smith, J.R., Chang, S.F.: Designing category-level attributes for discriminative visual recognition. In: CVPR (2013)
8. Vahdat, A., Mori, G.: Handling uncertain tags in visual recognition. In: ICCV (2013)
9. Ma, Z., Yang, Y., Xu, Z., et al.: Complex event detection via multi-source video attributes. In: CVPR (2013)
10. Liu, J., Kuipers, B., Savarese, S.: Recognizing human actions by attributes. In: CVPR (2011)
11. Larochelle, H., Erhan, D., Bengio, Y.: Zero-data Learning of New Tasks. In: AAAI (2008)
12. Palatucci, M., Pomerleau, D., Hinton, G.E., Mitchell, T.M.: Zero-shot Learning with Semantic Output Codes. In: NIPS (2009)
13. Choi, J., Rastegari, M., Farhadi, A., Davis, L.S.: Adding unlabeled samples to categories by learned attributes. In: CVPR (2013)
14. Wah, C., Belongie, S.: Attribute-Based Detection of Unfamiliar Classes with Humans in the Loop. In: CVPR (2013)

15. Sharma, G., Jurie, F., Schmid, C.: Expanded parts model for human attribute and action recognition in still images. In: CVPR (2013)
16. Akata, Z., Perronnin, F., Harchaoui, Z., Schmid, C.: Label-embedding for attribute-based classification. In: CVPR (2013)
17. Berg, T.L., Berg, A.C., Shih, J.: Automatic attribute discovery and characterization from noisy web data. In: Daniilidis, K., Maragos, P., Paragios, N. (eds.) ECCV 2010, Part I. LNCS, vol. 6311, pp. 663–676. Springer, Heidelberg (2010)
18. Rohrbach, M., Stark, M., Szarvas, G., Gurevych, I., Schiele, B.: What helps where and why semantic relatedness for knowledge transfer. In: CVPR (2010)
19. Yang, Y., Shah, M.: Complex events detection using data-driven concepts. In: Fitzgibbon, A., Lazebnik, S., Perona, P., Sato, Y., Schmid, C. (eds.) ECCV 2012, Part III. LNCS, vol. 7574, pp. 722–735. Springer, Heidelberg (2012)
20. Douze, M., Ramisa, A., Schmid, C.: Combining attributes and fisher vectors for efficient image retrieval. In: CVPR (2011)
21. Kumar, N., Belhumeur, P.N., Nayar, S.K.: FaceTracer: A search engine for large collections of images with faces. In: Forsyth, D., Torr, P., Zisserman, A. (eds.) ECCV 2008, Part IV. LNCS, vol. 5305, pp. 340–353. Springer, Heidelberg (2008)
22. Siddiquie, B., Feris, R.S., Davis, L.S.: Image ranking and retrieval based on multi-attribute queries. In: CVPR (2011)
23. Yu, F.X., Ji, R., Tsai, M.H., Ye, G., Chang, S.F., Weak, S.F.: attributes for large-scale image retrieval. In: CVPR (2012)
24. Zhang, H., Zha, Z.J., Yang, Y., Yan, S., Gao, Y., Chua, T.S.: Attribute-augmented Semantic Hierarchy. In: MM (2013)

An Analysis of Time Drift in Hand-Held Recording Devices

Mario Guggenberger, Mathias Lux, and Laszlo Böszörmenyi

Institute of Information Technology,
Alpen-Adria-Universität Klagenfurt,
9020 Klagenfurt am Wörthersee, Austria
{mg,mlux,lb}@itec.aau.at

Abstract. Automatic synchronization of audio and video recordings from events like music concerts, sports, or speeches, gathered from heterogeneous sources like smartphones and digital cameras, is an interesting topic with lots of promising use-cases. There are already many published methods, unfortunately none of them takes time drift into account. Time drift is inherent in every recording device, resulting from random and systematic errors in oscillators. This effect leads to audio and video sampling rates deviating from their nominal rates, effectively leading to different playback speeds of parallel recordings, with deltas measured up to 60 ms/min. In this paper, we present experiments and measurements showing that time drift is an existing problem that cannot be ignored when good quality results are demanded. Therefore, it needs to be taken care of in future synchronization methods and algorithms.

Keywords: Audio, video, multimedia, crowd, events, synchronization, clock drift, time drift.

1 Introduction

Digital cameras, smartphones and tablets are ubiquitous devices that many people carry with them all day. They make it incredibly easy to record good quality audio or video at all kinds of events, e.g. concerts, sports, or speeches. Synchronizing all those recordings from an event opens up various interesting use-cases like detecting key moments by looking at the frequency of concurrent recordings, temporal stitching of clips to get a complete and continuous coverage of a whole event, creating vivid videos by switching between different perspectives or showing different shots side-by-side, improving presentation quality by picking the best audio and video tracks from concurrent recordings, or reconstructing 3D scenes from recordings of different angles. Many approaches for automatic synchronization have already been proposed, and a recent overview of methods is presented in [2]. Unfortunately, all devices have an inherent error of time, which results in slightly different recording speeds across devices, effectively separating them into different time scales. We first ran into this problem in private amateur projects, e.g. recording concerts with multiple consumer cameras from different

X. He et al. (Eds.): MMM 2015, Part I, LNCS 8935, pp. 203–213, 2015.
© Springer International Publishing Switzerland 2015

perspectives or recording talks with separate audio and video recording devices, later during our work on an automatic audio synchronization software [5], and most recently while trying to synchronize the Jiku Mobile Video Dataset [12], a dataset with crowd-sourced smartphone video recordings. We were not able to get satisfactory results, neither with an automatic fingerprint-based method, nor by doing it manually. Parallel recordings were often out of sync, resulting in audio echoes, whose intensity varied between hardly noticeable to totally unacceptable. This problem, called drift, has also been identified in [2]. Drift by itself is not a new phenomenon, but it has been mostly ignored in the multimedia community. It has been covered in other areas, e.g. in network delay measurements [11] and for the identification of physical network devices through fingerprinting [14].

Drift in electronic devices is an error in the oscillators that drive, coordinate and synchronize the digital circuitries. Due to random and systematic influences, no oscillator runs at its specified nominal frequency. This frequency is specified in hertz (Hz), an SI unit based upon the second as defined by the Coordinated Universal Time (UTC) world time scale [3]. Since oscillators are also integral parts of clocks, this means that no clock, even when perfectly set exactly to UTC time at some point, can keep the correct time infinitely long. Even more important, oscillators are responsible to accurately time audio sampling rates and video frame rates during recording and playback. They also drive the system clock of a device which is additionally used to e.g. stamp recordings with the capture time. In case of recordings, this boils down to system and user errors in timestamps denoting the moment *when* a recording was captured, and system errors in timing of *how long* the capture process was running. With accurate clocks, synchronization would be as easy as reading the timestamps and aligning the recordings on a common timeline. All of the currently published synchronization methods handle only the unreliable timestamps and ignore the drift altogether. This concerns mainly recordings from uncontrolled environments, respectively consumer and prosumer equipment used by amateurs; professional environments usually avoid the problem by feeding all devices with a common master clock signal at recording time.

With this paper, we want to highlight a problem to the community that, in our opinion, it is not aware of, but that we think it should consider. This paper is not intended to provide concrete solutions, albeit we discuss suggestions on how the problem can be tackled. In the remainder of this paper, we continue to give a more detailed introduction into the issue. We present a method to measure drift with a precision that is sufficient for multimedia use cases and to underline our claim. We present experiments that provide proof that drift is a recent and pressing problem in current devices that needs to be handled by synchronization methods to achieve good quality results. We then continue with suggestions on how to detect and how to remove or compensate drift, and finish with our conclusion.

1.1 Oscillators

Oscillators exhibit a number of frequency/period instabilities and are usually specified in terms of short term, long term and environmental frequency stability [1]. Short term instabilities are known as jitter and degrade audio quality, but are not of our concern. Responsible for the drift problem are the long term and environmental instabilities that are a result of the initial error in crystal manufacturing, aging, and dependencies from temperature, vibration, and power supply [1,13]. Temperature usually has the biggest impact. The error ϵ in accuracy is measured in parts per million (ppm) and called *drift*. One ppm can be regarded as one microsecond per second, meaning that a clock with an oscillator specified at -10 ppm looses $10\,\mu s/s$ from UTC time, amounting to almost one second per day. According to various service manuals of smartphones and tablets, commonly used oscillators are either crystal oscillators (XO), temperature controlled XOs (TCXO), or voltage controlled TCXOs (VCTCXO). XOs are usually specified with an accuracy of ±10 ppm to ±100 ppm, TCXOs are more accurate at ±1 ppm which is why they are often used to drive audio components. In comparison, atomic oscillators are accurate to at least 0.001 ppm, UTC is accurate to ≈ 0.00000001 ppm [16] and even more precise oscillators exist.

1.2 Time Drift

Drift can be regarded as a deviation of a value in a time series from an ideal time series. In our case, the ideal time series is a timebase itself, e.g. UTC time, and the clock drift is a deviation from it, where the drift factor is another function over time which the drift accumulates by. Clock drift leads to deviations in the audio playback and recording sample rates in devices, that result in pitch shifts in, and bandwidth changes of the transmitted signals, but also impact the runtime of played or recorded files. It also leads to changes in video frame rates. This change of runtime is what we call the time drift. We specify it in ppm, but for better understanding in the multimedia domain, it is sometimes more intuitive to specify it in milliseconds per minute (ms/min) or per hour (ms/h).

1.3 Synchronization

Synchronization of two or more audio/video recordings is the act of mapping each moment in each recording to a common timeline such that all captured moments, which simultaneously happened at recording time, are mapped to the same time instant on the timeline. This mapping can essentially be divided into two steps: (i) compensating the drift in all recordings, and (ii) positioning the recordings on a timeline. Drift compensation is the process of removing or altering the drift inherent in all recordings to establish a mapping to the common time over their whole runtime. When the common synchronization time is UTC, we call it absolute drift compensation and all recordings need to be fit to that time. An alternative is relative drift compensation, where the inherent time in recordings of one device is taken as the common time and only recordings from other devices need to be compensated. This is still a simplification as it assumes that the drift of a recording device is constant, which we will later show is wrong, but the variance in the drift is a much smaller problem than the existence of the drift itself. The result of such a mapping is a timeline of synchronized recordings.

2　Measurements

To achieve high precision and absolute measurements, we chose to use GPS as the reference time source, since it is the cheapest and easiest way to obtain a high precision time signal. While GPS has been shown to be accurate up to ≈ 3 ns [9], the datasheet of our particular Garmin GPS receiver [4] specifies an accuracy of 1 µs, which is sufficient for our measurements. The GPS receiver emits a one pulse per second (PPS) signal that we fed into one channel of an audio capture interface connected to a computer workstation. The PPS signal was adjusted to audio line-level through a simple voltage divider circuit. The second channel of the audio device was fed with a 440 Hz sine wave test signal played back on the device to be measured. To obtain absolute measurement results, we used the audio signal analysis software Spectrum Lab[1] which measures the computer workstation's drift by analyzing the GPS signal and removing it from the analysis result of the input signal from the measured device. With this setup, we could measure the playback drift factor $d_p = 440/f_p$, where f_p is the measured frequency output from the measured playback device. Since the hardware audio codec of a device C derives the playback sample rate from the same clock signal as the recording sample rate, we can assume that both sample rates are equal, thus both drift factors are equal and $d_p^C \equiv d_r^C$. The drift factor can be converted to ppm by calculating $\epsilon = (d - 1) \times 10^6$. For quick estimates of inherent drift in multimedia devices, we have published an Android app that is capable of on-the-fly measurements built upon this method [6].

2.1　Inter-device Drift

To investigate the drift between different devices, we measured 16 devices in an isolation booth of a recording studio at constant room temperature by playing the test signal from a laptop computer and recording it in parallel by each device for 90 minutes. We then filtered the recordings with a band-pass filter of 40 Hz width around the center at 440 Hz to remove environmental noise, and deduced their recording drift d_r from the average frequency over the whole length. We additionally measured the absolute drift of the laptop and removed it from the device measurements to convert them to absolute drifts from UTC. It is important to note that the drifts cannot be added but must be multiplied to get correct results. To calculate the absolute drift d_t^D of device D, the playback drift of the notebook d_p^N must be inserted into the formula $d_t^D = d_p^N \times d_r^D$.

---

[1] http://www.qsl.net/dl4yhf/spectra1.html

Table 1. Drift measurements (ppm) of various hand-held recording devices (Tablet, Smartphone, MP3 Player, Video Camera, Audio Recorder)

Device	Type	audio	video
Samsung Galaxy Note 10.1	T	17.03	13.70
Samsung Galaxy SII	S	273.93	272.46
Samsung Galaxy Spica	S	-15.13	n/a
iRiver H120	M	93.98	n/a
iRiver H320	M	57.96	n/a
Canon HF10	V	n/a	6.45
Acer Iconia A200	T	13.39	-554.19
Apple iPad 2 Wi-Fi	T	13.73	11.96
Apple iPod touch 4G	M	416.76	413.64
M-Audio Microtrack 24/96	A	-40.42	n/a
LG Nexus 4 (rev. 10)	S	6.74	3.39
LG Nexus 4 (rev. 11)	S	4.12	1.88
Asus Nexus 7 2012 Wi-Fi	T	2.92	2.15
Sony PCM-M10	A	8.61	n/a
Editor UA-5	A	-3.21	n/a
Zoom R16	A	23.46	n/a

The results are listed in Table 1. It is interesting to note that two devices, marketed as exceptionally high-end, suffer from huge drift making compensation utterly important, even for short clips. Another interesting observation is that almost all devices run too fast and just a small fraction too slow. We repeated the measurement where possible, this time recording the test signal on video instead of pure audio. The results in Table 1 show us that the drifts generally decreased, which is typical for the increased temperature of the oscillators resulting from higher computational demands and lit screens. The most interesting point, however, is the immensely different drift of the Acer tablet, leading to the assumption that this particular device does use different time sources for audio and video recordings, against the usual practice of timing video frames with the audio clock. This leads to the conclusion that video drift cannot be assumed to be the same as audio drift and needs to be determined separately. Experiments on selected devices have also shown that the drift of the user-visible clocks of devices, taken offline to avoid time synchronization over a network, equals the drift measured in the audio signal.

2.2   Intra-device Drift

The next series of measurement answers the question how big the drift variance between devices from the same make, model and production batch are. For this measurement, we took five Nexus 5 smartphones and eight Nexus 7 tablets bought at the same time in the same store. We therefore assume each type coming from the same production batch. We measured all of them on our measurement workstation by playing back the test signal, and again conducted a second run recording the signal on video for 20 minutes with the same method as described in the previous subsection. Results are listed in Table 2, and again show a similar decrease of video drift as before. They also show a variance of a few ppm between devices. The most interesting point is that their drift is generally pretty low and uniform, making these devices good choices for parallel recordings and bypassing the need of drift compensation if the individual recordings are kept reasonably short.

Table 2. Drift measurements of 5 LG Nexus 5 smartphones and 8 Asus Nexus 7 (2013) tablets from the same production batches with their means (λ) and standard deviations (σ)

Device	No	audio drift		video drift	
		ppm	ms/h	ppm	ms/h
Nexus 5	1	6.89	25	5.17	19
	2	4.51	16	1.41	5
	3	7.36	26	4.30	15
	4	7.82	28	5.01	18
	5	5.61	20	4.25	15
	λ	6.44	23	4.03	15
	σ	1.36	5	1.52	5
Nexus 7	1	5.11	18	0.88	3
	2	8.80	32	2.66	10
	3	8.88	32	3.66	13
	4	6.98	25	2.30	8
	5	8.77	32	3.82	14
	6	8.48	31	5.04	18
	7	5.89	21	1.21	4
	8	6.62	24	4.14	15
	λ	7.44	27	2.97	11
	σ	1.49	5	1.46	5

2.3 Temperature Influence

We already indicated that the temperature is the major influence on oscillator drift rates. Figure 1 shows a plot of the room temperature in our office and the drift of our measurement workstation as calculated from GPS time, recorded over five winter days. It clearly shows that these two variables have a strong inverse correlation and that temperature is our primary concern. The two sudden drops in temperature and corresponding spikes in drift result from venting our office by opening the windows at an outside temperature of about 0 °C. Plotting the 5-minute moving average drift over time from the devices measured in the previous subsection as shown in Figure 2 demonstrates the warm-up phase of the devices during which the drift is not linear. It takes about 10 minutes for the devices to reach their working temperature and for the drift to change into a linear progression. This implicates that recordings of short clips with cool-down pauses in between will have a higher inherent average drift rate than long running recordings. Finally, we measured a few devices under extreme conditions to discover how much impact the temperature is expected to have in the worst case. We put them into a freezer and cooled them down to −20 °C, then moved them onto a heater with an air temperature of +50 °C and waited until the maximum drift was reached. We think this covers almost all situations in which recordings with consumer hand-held devices are made, and it is also the range in which crystal oscillators have an almost linear dependency between drift and temperature [13]. The results are shown in Table 3.

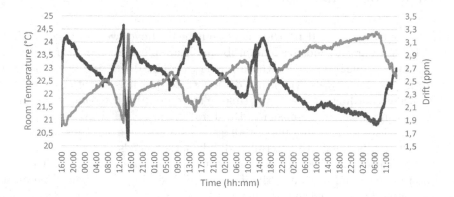

Fig. 1. Impact of room temperature (blue) on the computer workstation's drift (green) over five days

Table 3. The influence of extreme temperature on device drift rates

Device	−20 °C		+50 °C	
	ppm	ms/h	ppm	ms/h
iRiver H120	71.27	257	94.37	340
Asus Nexus 7 (2012)	0.49	2	4.95	18
Samsung Nexus S	11.04	40	20.54	74

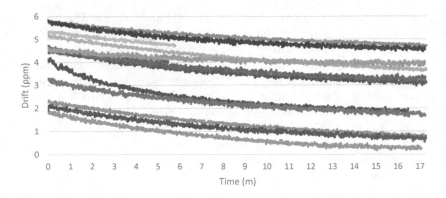

Fig. 2. 5-minute moving averages of drifts in devices recording HD video. The warm-up phase is clearly visible during the first minutes until they reach their working temperature and the drift stabilizes.

3   Drift Compensation

With the experiments and measurements outlined above, we have shown that time drift is a current and existent problem. Examples like the Acer Android tablet and the Apple iPod touch, whose video recordings drift apart by 60 ms/min (amounting to almost 3.5 s/h), clearly demonstrate that it is essential to remove drift to keep recordings in sync over time. This removal, drift compensation, is basically a contraction or expansion of the duration of each affected audio and/or video signal, leading to a mapping of all streams onto the same time scale.

The first step is always to determine the drift that needs to be compensated. Except for directly measuring recording devices as shown in Section 2, it is also possible to detect and measure drift if only the recordings are given. The most obvious method is manual measurement, which is easy for a pair of recordings, but gets tedious the more recordings are involved. It can be done by manually establishing synchronization points at the start and the end of an overlapping interval of two recordings, and calculating the difference of the lengths of the corresponding intervals in both recordings. Automatic synchronization methods can also be changed to incorporate drift handling. Approaches based on the correlation of feature series, e.g. [2], can consider using dynamic time warping [10] instead of cross-correlation, which we have already shown to work in a demo application [5]. Approaches that yield multiple discrete synchronization points between pairs of recordings, e.g. fingerprinting methods like in [7], can employ a method similar to the manual approach above, by applying linear regression to the relative time offsets between the synchronization points and calculating the slope of the fitted line.

The next two subsections cover suggestions for drift compensation of audio and video streams, we are however not going into detail as this is out of scope of this paper. To compensate drift in multimedia audio/video streams, we suggest to split them into their elementary streams and process them separately with the appropriate method.

3.1 Audio

Audio drift can be easily removed by resampling. Changing the duration of an audio signal by resampling usually has the effect of a pitch change, which is often unwanted. To avoid the effect, elaborate pitch-preserving time stretching algorithms have been proposed [15]. In the case of drift, however, a pitch change is already introduced by the drifted recording process and inherent in the signal, but it is usually too small to be noticed by the human ear. Relative drift compensation by resampling would again shift the pitch of the signal and not necessarily remove or decrease it. Absolute compensation by resampling completely removes the pitch shift from the recording, reconstructs the originally recorded signal, and is therefore the optimal choice.

3.2 Video

Video resampling is a much more complex problem. In comparison to audio, a video stream has less temporal samples (frames) but the complexity comes from their twodimensionality which requires handling of motion between consecutive frames. Simple video resampling methods like frame averaging or frame repetition/skipping incur highly visible video quality degradations. Methods based on motion compensation produce better results but are much more complex and computationally expensive [8], and artifacts still remain visible to the trained eye. We propose that video streams should be left untouched whenever possible to preserve quality. Due to the low number of frames per second, videos can be allowed a little bit of drift which goes completely undetected if it stays below the frame presentation interval. This could be compensated by shifting the complexity from image processing to drift minimization. Examples are skipping or repeating frames in particular sections of low movement or uniform colors, e.g. black frames at a nighttime concert when the lights go off, or by intelligently cutting between different recordings to lose or catch up drifted time.

4 Conclusion

In this paper we have shown that time drift is a real problem that needs to be taken care of in the domain of multimedia synchronization. This especially concerns recordings that are crowd-sourced and/or crawled from Internet sites or social networks, where control over the recording devices being used is not possible. For coordinated amateur productions, where there is control over the devices, there is at least the possibility to select devices with minimized relative drift to avoid post-processing.

Drift is inherent in recording devices, and therefore inherent in all audio and video recordings, and makes the exact synchronization of recordings more complex than current synchronization methods suggest. Depending on the signal's content and the consumer's experience, an audio drift of 10 ms may or may not be noticeable, but a drift of 300 ms after five minutes is unacceptable, even for video frames. It not only ruins the experience of a human consumer, but also makes any kind of post-processing error-prone.

We have suggested ways of detecting and dealing with drift, but questions are still open on how much drift is acceptable, and how much it negatively impacts the performance of content-based synchronization methods. We know that synchronization methods that synchronize recording upon a single point in time do not yield satisfying results for drifted recordings as the synchronization gets lost with increasing distance to the synchronization point. We hypothesize that synchronization methods that either directly incorporate drift handling, or are used on recordings where drift has been compensated as a pre-processing step, yield better results, and leave that also open for further discussion.

Acknowledgments. This work was supported by Lakeside Labs GmbH, Klagenfurt, Austria, and funding from the European Regional Development Fund (ERDF) and the Carinthian Economic Promotion Fund (KWF) under grant 20214/22573/33955. We also thank Wolfgang Büscher for his great work on the software Spectrum Lab and offering it for free.

References

1. Cardinal Components Inc. Clock Oscillator Stability (No. A.N. 1006)
2. Casanovas, A.L., Cavallaro, A.: Audio-visual events for multi-camera synchronization. Multimedia Tools and Applications (March 2014)
3. Allan, D.W., Ashby, N., Hodge, C.C.: Agilent AN 1289 The Science of Timekeeping
4. Garmin International, Inc., Olathe, Kansas. GPS 18x Technical Specifications (October 2011)
5. Guggenberger, M., Lux, M., Böszörményi, L.: Audioalign - synchronization of a/v-streams based on audio data. In: 2012 IEEE International Symposium on Multimedia (ISM), pp. 382–383 (December 2012)
6. Guggenberger, M., Lux, M., Böszörményi, L.: Clockdrift: A mobile application for measuring drift in multimedia devices. In: Proceedings of the 22st ACM International Conference on Multimedia, MM 2014, ACM, New York (2014)
7. Kennedy, L., Naaman, M.: Less talk, more rock: Automated organization of community-contributed collections of concert videos. In: Proceedings of the 18th International Conference on World Wide Web, WWW 2009, pp. 311–320. ACM, New York (2009)
8. Kuo, C.-H., Chang, L.-C., Liu, Z.-W., Liu, B.-D.: System level design of a spatio-temporal video resampling architecture. In: IEEE International Symposium on Circuits and Systems, ISCAS 2008, pp. 2797–2800 (May 2008)
9. Lombardi, M.A.: The use of gps disciplined oscillators as primary frequency standards for calibration and metrology laboratories. Measure: The Journal of Measurement Science 3(3), 56–65 (2008)
10. Müller, M.: Information Retrieval for Music and Motion. Springer-Verlag New York, Inc., Secaucus (2007)

11. Moon, S.B., Skelly, P., Towsley, D.: Estimation and removal of clock skew from network delay measurements. In: Proceedings of the Eighteenth Annual Joint Conference of the IEEE Computer and Communications Societies, INFOCOM 1999, vol. 1, pp. 227–234. IEEE (March 1999)
12. Saini, M., Venkatagiri, S.P., Ooi, W.T., Chan, M.C.: The jiku mobile video dataset. In: Proceedings of the 4th ACM Multimedia Systems Conference, MMSys 2013, pp. 108–113. ACM, New York (2013)
13. Semtech. Improving the Accuracy of a Crystal Oscillator (AN1200.07) (January 2009)
14. Sharma, S., Hussain, A., Saran, H.: Experience with heterogenous clock-skew based device fingerprinting. In: Proceedings of the 2012 Workshop on Learning from Authoritative Security Experiment Results, LASER 2012, pp. 9–18. ACM, New York (2012)
15. Zoelzer, U. (ed.): Dafx: Digital Audio Effects. John Wiley & Sons, Inc., New York (2002)
16. Thomas, C.: Stability and accuracy of international atomic time tai. In: Tenth European Frequency and Time Forum, EFTF 1996 (IEE Conf. Publ. 418), pp. 520–527 (March 1996)

A Real-Time People Counting Approach in Indoor Environment

Jun Luo[1], Jinqiao Wang[2], Huazhong Xu[1], and Hanqing Lu[2]

[1] School of Automation, Wuhan University of Technology, Wuhan, China
[2] National Laboratory of Pattern Recognition,
Institute of Automation, Chinese Academy of Sciences, Beijing, China
junjing2218@gmail.com, wutxhz@163.com, {jqwang,uhq}@nlpr.ia.ac.cn

Abstract. Due to complex background information, shadow and occlusions, it is difficult to count people accurately. In this paper, we propose a fast and robust human counting approach in indoor space. Firstly, we use foreground object extraction to remove background information. In order to get both moving people and stationary people, we designed a block-updating way to update the background model. Secondly, we train a multi-view head-shoulder model to find candidate people, and an improved k-means clustering is proposed to locate the position of each people. Finally, a temporal filter with frame-difference is used to refine the counting results and detect noise, such as double-count, random disturbance. An indoor people dataset is recorded in the classroom of our university. Experiments and comparison show the promise of the proposed approach.

Keywords: People counting, block-updating, improved k-means clustering, temporal refinement.

1 Introduction

Counting the number of human indoors is a challenging problem that has lots of practical applications, such as building security, room resources adjustment, market research, and intelligent building etc. In indoor environments, as a moving pixel extraction, the traditional background subtraction method is limited because not all humans' bodies are moving. What's more, when people get together, we will get a large blob with several objects inside. These blobs can not provide the object level information and are hard to segment. Actually, most of the time the number of human remain stable in indoor spaces. In other words, Although occlusions often occur, the room is at a dynamic stability state. Therefore, people counting in indoor environment is a challenging topic.

2 Related Work

A direct top-down view can avoid most occlusions in people counting, Teixeira and Savvides [5] proposed a lightweight method for localizing and counting people in indoor spaces using custom-built camera installed on the ceiling. Li et al. [2]

X. He et al. (Eds.): MMM 2015, Part I, LNCS 8935, pp. 214–223, 2015.
© Springer International Publishing Switzerland 2015

improved the accuracy rate of people counting by analyzing across multiple cameras. The temporal continuity of objects is an important reason to achieve stable pedestrian counting. Zhao and Nevatia [8] treated problem of segmenting individual humans in crowded as a model-based Bayesian segmentation problem. They presented an efficient Markov chain Monte Carlo (MCMC) method to get the solution. Wang et al. [6] built a spatio-temporal group context model to model the spatio-temporal relationships between groups, formulate the problem of pedestrian counting as a joint maximum a posteriori (MAP) problem. Zhang and Chen [7] used group tracking to compensate weakness of multiple human segmentation, which can handle complete occlusion. Chan and Vasconcelos [1] using dynamic texture, segmented the scene into different regions with different motions, extracted various features from each segment. A Gaussian process is used for estimating the pedestrian count for each segment. Most of the previous works focused on how to get an accurate counting result at the expense of real-time counting. Moreover, different from counting outdoors, the assumption that people would be in motion for a substantial amount of time is invalid, the number of human indoor changes little during most periods. The key problem of counting indoor is how to get a stable and accurate number of human in these periods.

3   Proposed Framework

As shown in Fig.1, our people counting method mainly includes three modules: foreground object extraction, head-shoulder detection and temporal refinement. Firstly, through the background subtraction, we will get blobs and corresponding gradient maps with human bodies inside. Then, a multi-view head-shoulder model is trained for head-shoulder detection. A head detection and a head-shoulder detection can find the general position of human, vice versa, results of head-shoulder detection will help update the background model with a block-updating method. After that, the clustering method is used for obtaining the counting results. Finally, the frame-difference method is utilized to estimate occlusions and refine the counting result.

3.1   Foreground Object Extraction

Traditional background subtraction approaches segment video frames into stationary pixels and moving pixels. The background model is built based on stationary pixels as a reference of moving pixels. Similarly, we also can consider that frames are constituted by pixels on and off human bodies. We can build and update the background model with no-human pixels to extract pixels on human bodies. When the room is empty, we capture a frame to build a initial background model. Then, we update the background model for each region based on the result of head-shoulder detection. Blobs without heads and shoulders inside will be updated into the background model. To simplify the process, we put blobs into

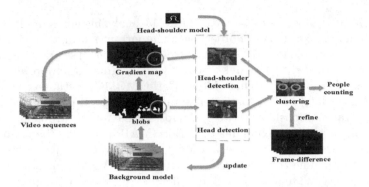

Fig. 1. Overall framework of people counting in indoor environment

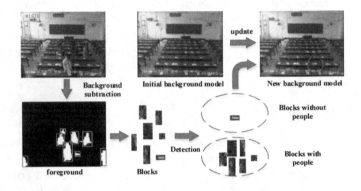

Fig. 2. Block-updating method

some blocks based on the height and width of blobs and blobs are marked by the coordinate of center points of blocks. We update these blocks instead of blobs.

Fig.2 shows the process of block-updating. As shown in Fig.2, because a book was moved in the frame, it became a blob. The foreground was divided into some blocks. The block with the book inside was updated into the background model based on the result of detection. This block-updating way can be described as follows:

$$B_k = B_{k-1} \oplus P b_j \tag{1}$$

Here B_{k-1} and B_k represent the last background model and the current model respectively. b_j is one block in the foreground. $\oplus$ means update the block b_j to the corresponding area of background model. P is the probability of being updated. We get the probability P by:

$$P = \frac{n}{N} \tag{2}$$

where n is the number of frames one block doesn't contain human bodies. N is the maximum of n and set to 25. Because we always update the no-human pixel into the back ground model, both moving people and stationary people can be extracted. As shown in Fig.3, the blob caused by light is removed after several frames.

(a)   (b)   (c)   (d)

Fig. 3. An example of the block-updating process. (a)The frame when turning on the light; (b)the blob of light bulb is extracted; (c)the area with light bulb is updated to background model after several frames;(d)the blob of light bulb disappeared.

3.2 Head-shoulder Detection

Some blobs we extracted through background subtraction algorithm may contain several people. Zhao and Nevatia [8] proposed a fast and efficient algorithm to locate people in a crowd, focusing on the boundary of the foreground. Similar to [8], head candidates and head-shoulder candidates are detected from foreground images and gradient maps respectively.

Head Candidates. Assuming that heads of humans are visible in a crowd most of the time. Head candidates need meet two conditions: it is the local vertical peak of the boundary, there are enough foreground pixels under it [9]. The result of head detection will be used as candidates for head-shoulder detection.

Head-shoulder Detection. We sample a lot of head-shoulder images from surveillance video and train a set of generic head-shoulder models. Because of the location of cameras, we mainly collect images from two representative categories: back view and side view. Through edge extraction and manually refining, we get a set of profile images as samples.

The training of head-shoulder model is similar to [4]. The training process is a collection of the probabilities of the key points' appearances in the image. But our method process samples more directly and can avoid the error brought by key points selecting. We firstly resize all the samples to a fixed size and superimpose them into a model. The model is divided into some smaller blocks. It's important to note that the top of human head is selected as the reference point. The reference

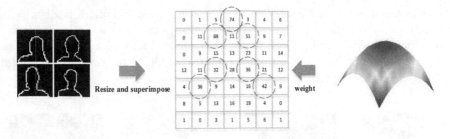

Fig. 4. A toy example of Head-shoulder model

point locates on the upper center of the model, all the head points in samples are aligned to this reference point. As shown in Fig.4, we accumulate the number of edge points in each block. We will get a general distribution of edge points of samples. Then we select some blocks with higher values as the positions of key points. A Parzen Windows method is used to assign every pixel a weight in each block. Assuming the window function obeys a Gaussian distribution, the weight is computed as follows:

$$w(x) = \frac{n}{N}\varphi\left(\frac{x - x_c}{h_N}\right) \tag{3}$$

where n represents the number of edge points in each selected block, N is the number of edge points in samples, x is the coordinate of arbitrary point in a block, x_c is coordinate of the center pixel in the block, h_N is the variance of the Gaussian distribution, φ is the window function, i.e. the Gaussian distribution in our case.

With head-shoulder detection, we can obtain the candidates position of human. We choose K-means clustering with max and min distance to decide the people count. The cluster centers are filtered by checking if there are enough head-shoulder candidates inside. The number of rest cluster centers is treated as the result of counting. Uncertainty of initial center points limits the convergence speed and effect of K-means. Here the sequence of head candidates is utilized as initial center points to make K-means faster and better.

Given a head detect sequence $A = \{\alpha_j\}_{j=1}^M$ and a head-shoulder detect sequence $B = \{\beta_j\}_{j=1}^N$, $A \cup B$ is the cluster set, and A is initial center points too. The goal is to find cluster center sequence $C = \{\gamma_j\}_{j=1}^K$. In order to reduce the impact of occlusions on counting result, we defined a maximum cluster radius R_{max} and a minimum cluster radius R_{min}. R_{min} and R_{max} are set to 1000 and 5000 respectively in our experiments. The algorithm is given in $Algorithm.1$. The normal K-means algorithm will be used to process the rest points in B.

Algorithm 1. Dynamic programming for K-means clustering

Input: Head sequences $A = \{\alpha_j\}_{j=1}^{M}$. head-shoulder sequence $B = \{\beta_j\}_{j=1}^{N}$.
Output: cluster center sequence C.
1: **for** $i = 1; i \leq N; i++$ **do**
2:     **for** $j = 1; j \leq M; j++$ **do**
3:         **if** $distance(\alpha_j, \beta_j) < R_{max}$ **then**
4:             delete β_j;
5:             $R_{max} = distance(\alpha_j, \beta_j)$;
6:         **else**
7:             $R_{max} = R_{max}$;
8:         **end if**
9:         **if** $distance(\alpha_j, \beta_j) > R_{min}$ **then**
10:             $C \leftarrow \alpha_j$;
11:         **else**
12:             delete β_j;
13:         **end if**
14:     **end for**
15: **end for**
16: **return** C;

3.3   Temporal Refinement

Except for enter and exit, occlusions is the main cause for change in the count number. We consider reasons of the change as two representative states: One is that the counting number increased as the occlusion disappeared. Another one is the counting number decrease when the occlusion happen. We can find that whatever state in indoor environment, the moving pixels exist. So whether the change is caused by occlusions can be estimated by moving pixels. We mainly consider two cases of occlusions, moving people occluded by stationary people and stationary people occluded by moving people.

As shown in Fig.5, there are two stationary people and four moving people in the frame. A stationary people is occluded by a moving people. we use frame-difference method to get the set of moving pixels. According to different contour of bodies, we divide them into different sequences. When the counting number decreases, we estimate occlusions according to the location of disappeared cluster center. For the case of stationary people occluded by moving people, the position of cluster center will be surrounded by moving pixels in the result of frame-difference. Our judgement is described as follows:

$$I(x_i, y_i) = \begin{cases} 1 \text{ if } \begin{cases} min(x_j) \leq x_i \leq max(x_j) \\ min(y_j) \leq y_i \leq max(y_j) \end{cases} \\ 0 \text{ else} \end{cases} \tag{4}$$

where x_j and y_j is the coordinate of moving pixel in one sequence. x_i and y_i is the coordinate of disappeared cluster center. If $I(x_i, y_i)$ equal to 1, we consider that $I(x_i, y_i)$ is occluded. The counting result remains the same. For the case of moving people occluded by stationary people, occlusions will be estimated by the disappearance of cluster center corresponding moving pixels. Furthermore, if the number of moving pixels is less than a threshold (set to 50 in our experiments) in one frame, we think the state in this indoor environment is steady and keep the counting result the same.

(a)   (b)   (c)

Fig. 5. Temporal refinement for people counting.(a)occlusion is occurring;(b)some information of head-shoulder disappeared;(c)moving pixels.

4   Experimental Results

We test our approach on the dataset collected by existing cameras installed on the back wall of classrooms in our university. The dataset reflect all states in these classroom within twenty-four hours. We select videos from two groups respectively, which named *video*1, *video*2.

We compare our approach with [8] and [3] by comparing the counting result on *video*1 and *video*2. The result of [8] is named "single frame", and [3] is "Adaptive model". As shown in Fig.5, because of the limitation of traditional background subtraction in indoor environment, the foreground with some stationary persons inside can't be extracted by [8]. So the counting results of [8] are below the groundtruth during some periods. As shown in Fig.5 (a), in indoor environment, "Adaptive model" [3] is easier to be affected by occlusions and disturbance of noise than our approach. The result of our approach is smoother. Although sometimes the counting result is incorrect, it can get closer to the groundtruth slowly after a number of frames. Because we use spatial temporal information to refine the counting result, the result of our approach falls behind the groundtruth most of the time. If the state in the classroom remains stable for a long time, the refinement on counting result will be constant. Fig.5 (b) shows the same conclusion on *video*2. Fig.6 shows some frames with counting results in our experiments.[1] The evaluation criterion is as follows:

$$RMSE = \sqrt{\frac{1}{n}\sum_{k=1}^{n}\left(num_g(i) - num_t(i)\right)^2} \tag{5}$$

---

[1] http://www.nlpr.ia.ac.cn/iva/homepage/jqwang/Demos.htm

where n is the total number of frames in the tested video. $num_g(i)$ is the ground truth of the ith frame. and $num_t(i)$ is the test result of the ith frame. The comparison with [8] and [3] is given in $Table.2$. Comparison in $Table.2$ shows our approach adapts better to the indoor environment.

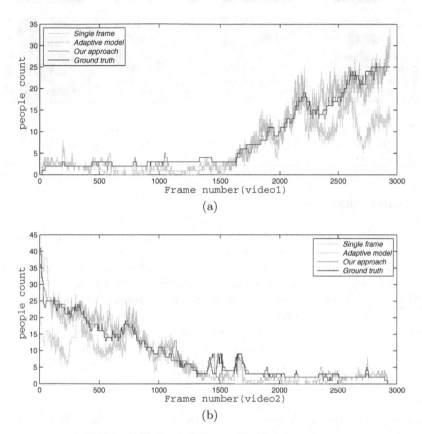

Fig. 6. Counting result on $video1$ and $video2$

Table 1. The comparison results

video name	RMSE		
	Our approach	Single frame [8]	Adaptive model [3]
Classroom1	1.0225	4.9142	2.1847
Classroom2	1.3879	4.9618	2.7618
Classroom3	1.2613	4.9252	2.0137

Fig. 7. Counting result in *video*1 and *video*2, the number of human was marked in green on the top right corner of images, and the clustering result was marked on the head of human in red points.

5 Conclusion

In this article, we propose a new method for people counting in indoor environment. We extract foreground objects and present a block-updating way to update the background model, which is fit for the indoor environment. The head-shoulder detection is the key problem in our approach, because it is linked closely with block-updating and clustering. So we trained a generic head-shoulder model and combined with improved K-means clustering to detect human bodies. Based on the spatial temporal information between frames, we refine the counting result and get a stable and accurate number of human in a number of frames. Compared with [8] and [3], our approach achieves a better performance than state-of-the-art approaches in the indoor environment.

Acknowledgements. This work was supported by 863 Program (2014AA015104), and National Natural Science Foundation of China (61273034, and 61332016).

References

1. Chan, A.B., Vasconcelos, N.: Counting people with low-level features and bayesian regression. IEEE Transactions on Image Processing 21(4), 2160–2177 (2012)
2. Li, J., Huang, L., Liu, C.: People counting across multiple cameras for intelligent video surveillance. In: 2012 IEEE Ninth International Conference on Advanced Video and Signal-Based Surveillance (AVSS), pp. 178–183 (2012)
3. Liu, J., Wang, J., Lu, H.: Adaptive model for robust pedestrian counting. In: Lee, K.-T., Tsai, W.-H., Liao, H.-Y.M., Chen, T., Hsieh, J.-W., Tseng, C.-C. (eds.) MMM 2011 Part I. LNCS, vol. 6523, pp. 481–491. Springer, Heidelberg (2011)

4. Sun, C., Zou, Q., Fu, W., Wang, J.: Multiple hypotheses based spatial-temporal association for stable pedestrian counting. In: Huet, B., Ngo, C.-W., Tang, J., Zhou, Z.-H., Hauptmann, A.G., Yan, S. (eds.) PCM 2013. LNCS, vol. 8294, pp. 803–810. Springer, Heidelberg (2013)
5. Teixeira, T., Savvides, A.: Lightweight people counting and localizing in indoor spaces using camera sensor nodes. In: First ACM/IEEE International Conference on Distributed Smart Cameras, ICDSC 2007, pp. 36–43. IEEE (2007)
6. Wang, J., Fu, W., Liu, J., Lu, H.: Spatio-temporal group context for pedestrian counting. IEEE Transactions on Circuits and Systems for Video Technology, 1–11 (2014)
7. Zhang, E., Chen, F.: A fast and robust people counting method in video surveillance. In: 2007 International Conference on Computational Intelligence and Security, pp. 339–343. IEEE (2007)
8. Zhao, T., Nevatia, R.: Bayesian human segmentation in crowded situations. In: Proceedings of the 2003 IEEE Computer Society Conference on Computer Vision and Pattern Recognition, vol. 2, pp. II-459. IEEE (2003)
9. Zhao, T., Nevatia, R., Lv, F.: Segmentation and tracking of multiple humans in complex situations. In: Proceedings of the 2001 IEEE Computer Society Conference on Computer Vision and Pattern Recognition, CVPR 2001, vol. 2, pp. II-194. IEEE (2001)

Multi-instance Feature Learning Based on Sparse Representation for Facial Expression Recognition

Yuchun Fang and Lu Chang

School of Computer Engineering and Science, Shanghai University
{ycfang,changlu}@shu.edu.cn

Abstract. Usually, sparse representation is adopted to learn the intrinsic structure in label spaces to fulfil recognition tasks. In this paper, we propose a feature learning scheme based on sparse representation and validate its effectiveness taking facial expression recognition as a multi-instance learning problem. By introducing the sparse constraint with l_1 sparse regularization, the proposed model learns the instance-specific feature based on label variance information. In this paper, we propose two schemes for denoting the label variance in multi-instance facial expression recognition. Experimental analysis shows that the sparse constraint is useful in feature learning when label variance is properly expressed and utilized. We successfully obtain the stable structure in the feature spaces with the sparse representation based on multi-instance feature learning.

Keywords: Feature learning, Multi-instance learning, Sparse representation, Facial expression recognition.

1 Introduction

In computer vision and pattern recognition research, the problem of classifying a single visual instance into multi-valued label is treated as supervised learning based on separate feature representation and classifier design. Though the instance is corresponding to different values of a single label, it is usually used in the classification part instead of the representation part. Taking expression recognition as an example illustrated in the left of Figure. 1, for a given face image, a single representation is extracted for a multi-class classifier [1]. If introducing the label information into the feature extraction or selection stage, it is expected to learn the multi-instance features that can further decrease dimensionality of representation and promote precision of the classifier.

When it comes to feature learning for recognition task, there are several general methods. The most common one is to learn weights based on statistics of defined criterion in feature space [1,2]. Evidence from psychology research reveals that facial expressions may be better recognized from localized facial regions [4]. Hence, another instance feature learning way is to learn the local combination of sub-facial regions or sub-feature space [5]. Machine learning methods such

X. He et al. (Eds.): MMM 2015, Part I, LNCS 8935, pp. 224–233, 2015.

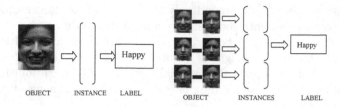

Fig. 1. Single instance vs. multi-instance expression recognition

as boosting [6,7] and manifold [8] are also employed for better presentation of features. No matter how we extract instance specific information, to incorporate as much as correlation among features is very important.

Inspired by the above analysis, we propose a multi-instance feature learning model for facial expression analysis as illustrated in the right of Figure. 1. We propose a feature learning model based on the sparse constraint of l_1 regularization to form a multi-instance representation corresponding to multi-values of the label. Normally, the convergence of the label learning of the Sparse Representation Classification (SRC) is strongly supported by the assumption that samples of the same category could form the best linear combination for each other [9,10]. While in the case of feature spaces, there usually exists no obvious inter-dimension linear combination in feature space. To handle this problem, we propose two simple methods to denote the variance of label value that can joint together with the sparse constraint in a non-negative way to realize feature learning. With the learned adaptability of the feature to each value of label, we obtain a much lower dimensional multi-instance feature representation.

To validate the proposed multi-feature learning framework, we analyze the performance variation of the learned feature with respect to the ordered value of sparseness. We find that low value of sparseness corresponds to highly effective feature components. With the learned lower dimensional multi-instance features, we obtain higher precision.

2 Multi-instance Feature Learning with Sparse Representation

With the constraints of sparseness, a lot of improved results are obtained in machine vision and pattern recognition [9,10]. Instead of using the sparse constraints for label learning, we explore to employ sparseness constraints in feature learning tasks.

2.1 Multi-instance Feature Learning

Multi-instance learning has been used in many applications as a general framework of machine learning [11]. In multi-instance learning, the training set is denoted with many bags, each of which contains many instances. Multi-instance learning aims at forming concept labels. Correspondingly, multi-instance feature learning serves to explore effective representations that are appropriate for each bag.

In this paper, we take one bag with multiple-instances in facial expression recognition. As shown in the right of Figure. 1, taking any general feature representation as a bag, we utilize the inter-label variation to learn multiple representations for an object. The learning is realized with introducing of sparseness constraints in feature space.

2.2   Sparse Representation

SRC has been used successfully in face recognition [9,10]. By introducing the sparse constraint in the optimization problem of least square fitting, the obtained value of sparseness is corresponding to the label of the objects. The convergence of the optimization is ensured by the fact that the feature vector of face image of each subject can be linearly well combined with feature vector of other images from the same subject. Hence, it is possible to come up with a solution with sparse representation in label spaces. To employ sparse representation in feature learning, a different learning framework needs to be designed.

2.3   Sparseness for Multi-instance Feature Learning

Motivated by SRC for its compressed information annotation, we adopt the sparse constraints to learn instance specific representations. In the feature space, denote the label variance set of the i-th value of a given label containing N_i column vectors of M dimension as $\left\{Z_1^{(i)}, \cdots, Z_{N_i}^{(i)}\right\}$ $(i = 1, \cdots, C)$, the combination of all $M \times N_i$ matrix $A^{(i)} = \left[Z_1^{(i)}, \cdots, Z_{N_i}^{(i)}\right]$ forms a $N \times M$ sample matrix across all values of label as $A = \left[A^{(1)}, \cdots, A^{(C)}\right]^T$, in which $N = \sum N_i$. Denote the label vector as $y = \left[\overbrace{l_1, \cdots, l_1}^{N_1}, \cdots, \overbrace{l_C, \cdots, l_C}^{N_C}\right]^T$, the sparse solution in feature space as M-dimensional vector x, the linear combination with sparse constraints can be denoted in Equation (1).

$$x = \arg\min_x \|x\|_1$$
$$s.t. Ax = y, x \geqslant 0 \tag{1}$$

We adopted the l_1 sparse regularization item due to its tractable solution in most cases. Since the learning is aimed at forming a feature evaluation, the sparse value is constrained to be non-negative.

To make the learned sparseness value meaningful for instance learning, we constrain $z_j^{(i)}$ to bear instance correlation information. Hence, the proposed multi-instance feature learning model is based on the assumption that the values of the given label are correlated. Taking facial expression as an example, expression is the label with multiple values such as sad, happy and fear etc. The appearance of expressions can be regarded as a non-rigid transform from a neutral expression.

3 Multi-instance Feature Learning for Expression Recognition

In our opinion, the facial expression analysis can be modeled as a multi-instance problem. It is expected that for the multiple values of expression label, corresponding instance could be learned in the format as feature. More importantly, the values of facial expression are highly correlated, since the appearance of any expression is just a non-rigid transformation of the raw neutral expression. Hence, we can deduce very useful correlation information to realize the proposed feature learning model in Section 2.3.

3.1 Multi-instance Facial Expression Recognition

Two strategies are proposed by us to learn feature representation in facial expression recognition. The first one is denoted as IL-I as shown in Equation (2), in which each sample for learning is the raw feature $f_j^{(i)}$ translated with the mean $\overline{f^{(i)}}$

$$z_j^{(i)} = f_j^{(i)} - \overline{f^{(i)}} \tag{2}$$

The second strategy IL-II is denoted in Equation (3). Each sample takes any neutral appearance as references.

$$z_j^{(i)} = f_j^{(i)} - NE_k^{(i)} \tag{3}$$

For both IL-I and IL-II, we actually set the $C = 2$ in Equation (1) by taking samples from one expression as positive and samples from other expressions as negative, and learn the feature for each expression separately. Hence, the learned instance can directly serve to solve the expression verification problem.

3.2 Feature and Classifier

For the raw feature $f_j^{(i)}$, we take the very popular Uniform Local Binary Pattern (ULBP) feature [2] in experiments, which is very popular in face recognition as well as in facial expression recognition. The ULBP feature contains a sequence of histograms in blocked sub-images. Each histogram is a statistic of one of the K-by-L squares coded by the LBP operator. The LBP operator is decided by the radius of neighborhoods and the sampling density in the neighborhood. By taking some very popular parameter setting, a 3304 dimensional feature is employed.

Since our major concentration is feature learning, complex classifiers such as Neural Network or Support Vector Machine are not used to avoid the unfairness in comparison that may be caused by extra performance added by these classifiers. We simply use the distance classifier to show the effectiveness of the proposed feature learning model. For each class, we pick the mean of training samples as the compared object. Two normal metrics used are L1 and Chi2.

4   Experimental Analysis

To validate the proposed multi-instance feature learning model, we perform facial expression recognition to analyze the performance of the model and the affection of parameter settings such as the face preprocessing, usage of label information.

4.1   Experimental Database and Setting

The experiments are conducted in the most comprehensive facial expression database of Cohn-Kanade (CK) [12]. The database consists of expressional sequences from subjects of varied gender, race and age with expression changing from neutral (NE) to surprise (SU), sadness (SA), angry (AN), disgust (DI), fear (FE), happy (HA) or contempt (CO) (denoted in abbreviation for identifying hereafter). The images are preprocessed with two strategies into resolution 160*140 as illustrated in Figure. 2. Strategy FPP-I aligned the global face through bi-cubic interpolation based on parameters obtained with least square fitting with 2 eye centers, nose tip and mouth center. Strategy FPP-II aligned local facial area with triangle texture mapping based on 68 facial feature points. With FPP-I, the facial area along with some information outside face contour is kept, while with FPP-II only the area inside facial contour is kept with the expression transformed drastically.

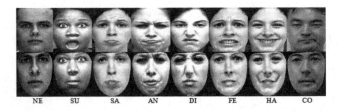

Fig. 2. Preprocessing: Top row are processed with FPP-I, bottom row with FPP-II

As in most other literatures, we select first 3 frames as neutral and last 3 frames as specific expression from 327 sequences with over 6 frames. A total of 1962 images are used in our experiments. The statistics of the database in our experiments is summarized in Table 1. According to Section 3.3, we learn the two types of multi-instance features based on Strategy IL-I and IL-II respectively for facial expression recognition with samples listed in Table 1. Based on Equation (1), for Strategy IL-I, the number of positive samples is just the number of images in the database as shown in Table 1, and the number of negative samples is the number of total samples minus the number of corresponding positive samples. In comparison, there are much more samples for strategy IL-II. Since there are at least 3 images labeled as Neutral for each image with non-neutral expression for a subject, the intersection between each pair of non-neutral and neutral images from the same subject end up to a large amount of training samples.

Table 1. The comparison results of different combination of feature space and classifier

Expression	# of sequence	# of images	IL-I		IL-II	
			# of pos.	# of neg.	# of pos.	# of neg.
SU	83	249	249	1713	2376	7191
SA	28	84	84	1878	945	8622
AN	45	135	135	1827	1332	8235
DI	59	177	177	1785	1728	7839
FE	25	75	75	1887	837	8730
HA	69	207	207	1755	2052	7515
CO	18	54	54	1908	297	9270
NE	327	981	981	981	-	-

4.2 Effectiveness of Multi-instance Feature Learning

All sessions of feature learning with sparse representation lead to a convergent optimization. The learned instance is a sparse vector that indicates the relevance of each dimension for corresponding expression. In this paper, we simply select subsets of features according to the order of value of sparseness. Since each learned instance can provide verification information about the corresponding expression, we adopt the Equal Error Rate (EER) to observe the variation of precision with respect to the size of subsets. The subsets are created in both ascending and descending way according to the value of sparseness. Example results for expression AN are listed in Figure. 3 and Figure. 4 for learning strategy IL-I and IL-II respectively. The baseline is the EER of the raw feature without learning information integrated.

By observing the variation of ascending curves in the top rows of the Figure. 3 and Figure. 4, we notice that with the increase of dimension, the errors tend to be stable soon with very slow dimensional feature. More promisingly, the errors approach the baseline or even lower than the baseline in all cases shown in the top rows of the Figure. 3 and Figure. 4. While for the descending curves, the errors of curves in bottom rows of the Figure. 3 and Figure. 4 are also decrease with the increase of feature dimension, but the minimum EER happens when the dimension approaches 3304. By Figure. 3 and Figure. 4, we can also notice that the conclusion holds for all two compared metrics L1 and Chi2, and also for all two compared face preprocessing strategy FPP-I and FPP-II. In general, Chi2 is a little better than L1, FPP-II is a little better than FPP-I. Though Figure. 3 and Figure. 4 only show the results of AN verification, similar results can be obtained for all the other 7 expression labels.

The observed results for comparing EER in feature learning according to descending or ascending sparseness values prove several important conclusions about our study. First, by the change of EER with dimension according to descending sparseness value, the features corresponding to higher sparseness values are less valuable for expression recognition. Second, the effective feature can be obtained by selecting subsets according to the ascending order of the sparseness

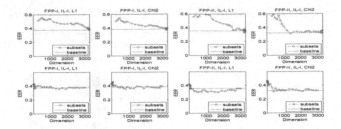

Fig. 3. Variation of EER of ANGER verification with regards to dimension of learned feature with IL-I for FPP-I and FPP-II, L1 and Chi2: top row for ascending manner and bottom row descending manner with the sparseness values

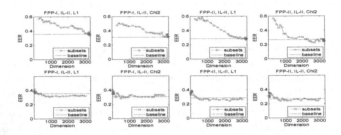

Fig. 4. Variation of EER of ANGER verification with regards to dimension of learned feature with IL-II for FPP-I and FPP-II, L1 and Chi2: top row for ascending manner and bottom row for descending manner with the sparseness values

value. With feature dimension much lower than the raw feature, the baseline can be reached or surpassed. Hence, the proposed multi-instance feature learning model is very effective and valuable.

4.3   Comparisons for Facial Expression Recognition

In this section, we mainly evaluate the facial expression recognition performance with the confusion matrices as in other literatures. Based on the analysis in Section 4.2, we pick the Chi2 metrics and FPP-II for both IL-I and IL-II and list the experimental results in the Table 2.(b) and Table 3.(b) respectively, in which the second row is the minimum dimension of feature for the required precision for each expression. The results without instance learning (i.e. 3304 dimensional ULBP feature) are also listed in the Table 2.(a) and Table 3.(a). By comparing the Table 2.(a), Table 2.(b) and Table 3.(a), Table 3.(b), we can tell that with much lower dimension of features, the performance of the proposed multi-instance learning with IL-I and IL-II is very close to or better than the single instance cases in facial expression recognition.

Table 2. Confusion matrix comparison for IL-I without (a) and with (b) instance learning

(a)

%	SU	SA	AN	DI	FE	HA	CO	NE
DIM				3304				
SU	83.2	0.8	0.0	0.0	0.8	0.0	3.2	12.0
SA	0.0	61.9	0.0	0.0	0.0	0.0	0.0	38.1
AN	0.0	8.8	70.6	0.0	0.0	0.0	10.3	10.3
DI	0.0	1.1	1.1	89.9	0.0	0.0	3.4	4.5
FE	0.0	0.0	0.0	0.0	52.6	18.4	15.8	13.2
HA	0.0	0.0	0.0	0.0	7.7	91.4	1.0	0.0
CO	0.0	0.0	0.0	0.0	0.0	0.0	96.3	3.7
NE	0.6	1.6	3.5	1.0	0.2	0.2	10.8	82.1

(b)

%	SU	SA	AN	DI	FE	HA	CO	NE
DIM	800	2700	1500	700	1700	1100	100	2500
SU	83.2	1.6	0.0	0.0	0.8	0.8	6.4	7.2
SA	0.0	61.9	0.8	0.0	0.0	0.0	0.8	0.0
AN	0.0	7.4	72.1	0.0	0.0	0.0	11.8	8.8
DI	0.0	1.1	1.1	88.8	0.0	0.0	5.6	3.4
FE	0.0	0.0	0.0	0.0	52.6	18.4	15.8	13.2
HA	0.0	0.0	0.0	0.0	6.7	91.4	1.9	0.0
CO	0.0	0.0	0.0	0.0	0.0	0.0	100	0.0
NE	0.6	1.4	3.5	1.0	0.0	0.2	10.8	82.5

5   Conclusions

In this paper, we propose a feature learning model based on the sparse constraint to learn multi-instance representation for facial expression recognition. By taking advantage of the correlation among values of labels, we propose to utilize two strategies IL-I and IL-II to train the sparse representation based multi-instance feature learning model. Experiments discover that when picking features by ascending order of the learned sparse value, we can obtain a group of low-dimensional feature subsets that are of close or better expression verification results compared to the single instance situation. The result of the proposed feature learning model can be regarded as a multi-instance representation under a uniform learning framework. The learned features are adaptive instances corresponding to the label value, which results in more compressed representation without losing precision.

Acknowledgments. The work is funded by the National Natural Science Foundation of China (No.61170155 & No.61371149).

Table 3. Confusion matrix comparison for IL-I without (a) and with (b) instance learning

(a)

%	SU	SA	AN	DI	FE	HA	CO
DIM				3304			
SU	93.6	2.4	0.0	0.0	0.8	0.0	3.2
SA	9.5	85.7	0.0	0.0	4.8	0.0	0.0
AN	0.0	10.3	79.4	0.0	0.0	0.0	10.3
DI	0.0	1.1	1.1	94.4	0.0	0.0	3.4
FE	5.3	0.0	0.0	0.0	60.5	18.4	15.8
HA	0.0	0.0	0.0	0.0	7.7	91.4	1.0
CO	0.0	0.0	0.0	0.0	0.0	0.0	100

(b)

%	SU	SA	AN	DI	FE	HA	CO
DIM	2300	2100	1900	2600	2300	2300	1700
SU	93.6	2.4	0.0	0.0	0.8	0.0	3.2
SA	7.1	85.7	0.0	0.0	4.8	0.0	2.4
AN	0.0	7.4	80.9	0.0	0.0	0.0	11.8
DI	0.0	1.1	1.1	94.4	0.0	0.0	3.4
FE	5.3	0.0	0.0	0.0	60.5	18.4	15.8
HA	0.0	0.0	0.0	0.0	5.8	94.2	0.0
CO	0.0	0.0	0.0	0.0	0.0	0.0	100

References

1. Shan, C., Gong, S., McOwan, P.W.: Facial expression recognition based on local binary patterns: A comprehensive study. IVC 27, 803–816 (2009)
2. Ahonen, T., Hadid, A., Pietikainen, M.: Face description with local binary patterns: Application to face recognition. TPAMI 28, 2037–2041 (2006)
3. Liao, C.T., Chuang, H.J., Duan, C.H., Lai, S.H.: Learning spatial weighting for facial expression analysis via constrained quadratic programming. PR 46, 3103–3116 (2013)
4. Wang, S., Liu, Z., Lv, S., Lv, Y., Wu, G.,, P.: A natural visible and infrared facial expression database for expression recognition and emotion inference. Mul 12, 682–691 (2010)
5. Zhong, L., Liu, Q., Yang, P., Liu, B., Huang, J., Metaxas, D.N.: Learning active facial patches for expression analysis. In: CVPR, pp. 2562–2569 (2012)
6. Yang, P., Liu, Q., Metaxas, D.N.: Boosting coded dynamic features for facial action units and facial expression recognition. In: CVPR, pp. 1–6 (2007)
7. Zhao, G., Pietikäinen, M.: Boosted multi-resolution spatiotemporal descriptors for facial expression recognition. PRL 30, 1117–1127 (2009)
8. Xiao, R., Zhao, Q., Zhang, D., Shi, P.: Facial expression recognition on multiple manifolds. PR 44, 107–116 (2011)
9. Wright, J., Yang, A.Y., Ganesh, A., Sastry, S.S., Ma, Y.: Robust face recognition via sparse representation. TPAMI 31, 210–227 (2009)

10. He, R., Zheng, W.S., Hu, B.G.: Maximum correntropy criterion for robust face recognition. TPAMI 33, 1561–1576 (2011)
11. Zhang, M.L., Zhou, Z.H.: Improve multi-instance neural networks through feature selection. NPL 19, 1–10 (2004)
12. Kanade, T., Cohn, J.F., Tian, Y.: Comprehensive database for facial expression analysis. In: FGR, pp. 46–53 (2000)

Object Detection in Low-Resolution Image via Sparse Representation

Wenhua Fang[1], Jun Chen[1,2], Chao Liang[1,2], Xiao Wang[1],
Yuanyuan Nan[1], and Ruimin Hu[1,2]

[1] National Engineering Research Center for Multimedia Software,
School of Computer, Wuhan University, Wuhan, 430072, China
[2] Research Institute of Wuhan University in Shenzhen, China

Abstract. We propose a novel object detection framework in extreme
Low-Resolution (LR) images via sparse representation. Object detection
in extreme LR images is very important for some specific applications
such as abnormal event detection, automatic criminal investigation from
surveillance videos. Object detection has achieved much progress in computer vision, but it is still a challenging task in LR image, because traditional discriminative features in high resolution usually disappear in
low resolution. The precision of the detector in LR will decrease by a
large margin. Our model uses sparse coding of part filters to represent
each filter as a sparse linear combination of shared dictionary elements.
The main contribution of this paper: 1) the object detection framework
in extreme LR is proposed by detecting objects in reconstructed HR image; 2) a mapping function from LR patches to High-Resolution (HR)
patches will be learned by a local regression algorithm called sparse support regression, which can be constructed from the support based of the
LR-HR dictionary; 3) a novel feature extraction method is proposed to
accelerate by extracting visual features from HR dictionary atoms. Our
approach has produced better performance for object detection than
state-of-the-art methods. Testing our method from INRIA and PASCAL
VOC 2007 datasets has revealed similar improvements, suggesting that
our approach is suitable for general object detection applications.

1 Introduction

Object detection is one of the most challenging problems in computer vision. It
is difficult due to the significant amount of variation between images belonging
to the same object category. Other factors, such as changes in viewpoint and
scale, illumination, partial occlusions and multiple instances further complicate
the problem of object detection. Object detection is the problem of finding the
positions of all concerned objects in an image. More specifically, the goal is to
find the bounding box for each object. One common approach is to use a sliding
window to scan the image exhaustively in scale-space, and classify each window
individually [2, 17, 14, 1, 19].According to Dalal [3], detection performance in
INRIA dataset is 90% when the sliding window is (64 × 128) and resolution of

X. He et al. (Eds.): MMM 2015, Part I, LNCS 8935, pp. 234–245, 2015.
© Springer International Publishing Switzerland 2015

(a) Human detection on INRIA dataset

(b) Car Detection on surveillance dataset

Fig. 1. Red and green bounding boxes represent missing objects and detecting objects respectively

image is 640 × 480 and the size of object is similar to (64 × 128). However, the performance dropped significantly to 40% when the sliding window is (16 × 32). Because the proper size of the pedestrian in the image is unknown previously, it is prone to missing when the resolution of the object is low. Objects in LR are always missed in the traditional detection method [3]as shown in Fig. 1. Moreover, object detection in the LR image is also important as well as in the HR image, such as in criminal investigation and public security field. Additionally, object detection is much more difficult in LR image than in HR one because traditional discriminative features in high resolution cannot be extracted in low resolution. The performance of the object detectors, including the state-of-the-art method [8, 15], in LR will decrease by a large margin.

Traditional multi-scale object detection methods in LR images are resizing the detection windows or/and images [4]. And magnifying the image from LR image by linear interpolation will lose the high frequency details such as edges, which are discriminative feature for classification. For retaining the critical detailed information during the magnification, the sparse representation in Super Resolution, is used to reconstruct the image and then the object detection methods are performed in the image. In this paper, we improve the perceptual image quality from LR to HR, among which the resolution enhancement technology is called super-resolution. We can get high-resolution and high quality image with more details economically, not by imposing higher requirements on hardware devices and sensors. We use a manifold regularized regression framework for

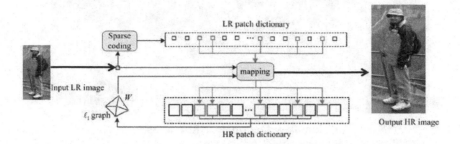

Fig. 2. Flowchart of the proposed method. Note that the red patches denote the sparse support domain of the input LR patch on the dictionary, and we use the sparse graph of the HR sparse support domain to guide the construction of the mapping function.

super-resolution as shown in Fig. 2. The sparse representation is relaxed for LR to HR sparse support domain regression, which is flexible in using the information of local training samples. Note that image patches have regular structures where accurate estimation of pixel values via regression is possible. Accordingly, the proposed method has more power and flexibility to describe different image patterns. In addition, the proposed method simultaneously considers the manifold regularization, thus capturing the intrinsic geometrical structure of the dictionary. We use a mapping function from low resolution (LR) patches to high-resolution (HR) patches will be learned by a local regression algorithm called sparse support regression, which can be constructed from the support bases of the LR-HR dictionary.

Then, we can also use the two important components (feature and classifier) in the object detection from HR we can get from the above. First, features capture the most discriminative information of pedestrians. Second, a classifier decides whether a candidate window shall be detected as enclosing a pedestrian and SVM (Support vector machine) is often used. The connection between features and classifier components is usually achieved using manual parameter configuration. The HOG feature is individually designed with its parameters manually tuned given the linear SVM classifier [3]. There are two main contributions of this paper: 1) a object detection framework in extreme low resolution is proposed by detecting objects in reconstructed high resolution image; 2) a mapping function from low resolution patches to high resolution patches will be learned by a local regression algorithm called sparse support regression, which can be constructed from the support based of the LR-HR dictionary. Our low resolution detector produced better performance for pedestrian detection than state-of-the-art methods. Testing our method on pedestrians from the INRIA database revealed similar improvements, suggesting that our approach is suitable for general object detection applications.

2   Sparse Representation for Object Detection

Sparse representation is usually used to reconstruct the image in super resolution and is rarely used for object detection. Xiaofeng Ren et al[15] firstly adopted sparse representation to extract the histogram of sparse codes feature for object detection, which outperformed the famous method [8]. But they did not consider the object detection in LR. In our work, the object detection framework is presented on the reconstructed image via sparse representation in this section. The framework is divided into two phrases. Firstly, the HR image is reconstructed from LR by specific dictionary learning. Secondly, the optimized object detection method is performed in the HR dictionary.

2.1   Sparse Representation Model

Given a set of image patches $Y = [y_1, \cdots, y_n]$, finding a dictionary $D = [d_1, \cdots, d_m]$ and an associated sparse code matrix $X = [x_1, \cdots, x_n]$ by minimizing the reconstruction error

$$\min_{D,X} ||Y - \mathrm{DX}||_F^2 s.t. \forall_i, ||x_i||_0 \leq K \tag{1}$$

where x_i are the columns of X, the zero-norm $|| \cdot ||_0$ counts the non-zero entries in the sparse code x_i, and K is a predefined sparsity level.

2.2   LR-HR Pairwise Dictionary

Given a set of LR and HR training image patch pairs, $\{(x_1, y_1), \cdots, (x_N, y_N)\} \subset R^d \times R^D$, d and D are the dimensions of one LR and one HR patch respectively. Define $X = [x_1, \cdots, x_N]$ and $Y = [y_1, \cdots, y_N]$, each column of which is a patch sample. Thus the matrixes X and Y can be viewed as the LR and HR patch dictionaries respectively.

Considering that the manifold assumption (two manifolds spanned by the feature spaces of the LR and HR patches are locally similar) may not be tenable, we learn a much more stable LR-HR mapping in the support domain for super-resolution. Thus it can be transformed to a regression problem.

Our another important goal is to encode the geometry of the HR patch manifold, which is much more credible and discriminated compared with that of the LR one [9], and preserve the geometry for the reconstructed HR patch space [12]. This will ensure that the local geometric structure of the reconstructed HR patch manifold is consistent with that of the original IIR one. Based on the above discussions, our MSSR (Manifold regularized Sparse Support Regression)algorithm for image super-resolution should be equipped with two properties: 1) The shared support of each LR patch and HR patch has an explicit regression relationship; 2) The local geometrical information on the original HR patch dictionary is preserved. In the following part, we will describe how we formulate MSSR with these two desired properties.

2.3   Sparse Support Regression

Instead of assuming that each pair of HR and LR patches [10] has the same sparse representation, in our proposed MSSR method, this strong regularization of sparse representation is relaxed for sparse support regression, and the sparse coefficient vectors of one LR and HR patch pair share the same support, i.e., the same indices of nonzero elements.

Given a set of LR and HR training patches (dictionary pairs), $\{(x_1, y_1), \cdots, (x_N, y_N)\} \subset R^d \times R^D$ for an unseen LR patch x_t, we try to learn a mapping function $f(x, P) = Px$, from the LR patch to the HR one to minimize the following regularized cost function for the regression,

$$\varepsilon(P) = \sum_{i \in S} (Px_i - y_i)^2 + \alpha \|P\|_H^2 \qquad (2)$$

where α is a regularization parameter, P is a $D \times d$ matrix to be learned, $\|P\|_H^2$ is the induced norm of f in the reproducing kernel Hilbert space (RKHS) space H, and S is the support of the coding coefficients θ^* of the unseen patch x_t on LR training patches X:

$$\theta^* = \arg\min_{\theta} \|x_t - X\theta\|_2 + \lambda_1 \|\theta\|_1 \qquad (3)$$

Thus, $S = \text{support}(\theta^*)$. In Eq. (2), $\|\theta\|_1$ denotes the l_1 norm of θ and the parameter λ_1 balances the coding error of x_t and the sparsity of θ The solution of Eq. (2) can be achieved by convex optimization methods referring to [13]. The support of one vector is referring to the indices of nonzero elements in the vector. Defining X_s and Y_s as $X_s = \{x_i | i \in S\}$ and $Y_s = \{y_i | i \in S\}$ respectively and using Fibonacci norm to represent the smoothness of H, we can rewrite Eq. (1) as the following matrix form:

$$\varepsilon(P) = \|PX_S - Y_S\|_F^2 + \alpha \|P\|_F^2 \qquad (4)$$

2.4   Mining the Geometry on HR Patch Dictionary

This section targets on the second property, which is to preserve the local geometrical information on the HR patch dictionary. Note that the neighborhood relation, which guides the formulation of sparseness, is defined on the manifold rather than the Euclidean space.

Researchers have proposed various methods to measure the similarity between data points [16, 11], e.g., pair-wise distance based similarity and reconstruction coefficient based similarity. Since the former is suitable for discriminant analysis problems, such as recognition and clustering. Alternatively, reconstruction coefficient based similarity is datum-adaptive, and thus more suitable for image super-resolution. LLE is one of the representative works for reconstruction coefficient similarity estimation. It calculates the coefficient for each data through k-NN searching, thus k sparsity. The performance of LLE graph will decrease rapidly when the datas are non-uniformly sampled from underlying manifold, and this situation is very common in practice.

Recently, some researchers have demonstrated that the sparse structure of one manifold can be explored by the l_i graph [16], resulting in many benefits for machine learning and image processing problem. Let y_i be the i -th HR patch , which is under consideration now. We want to identify its neighbors on the smooth manifold rather than the entire Euclidean space. On the smooth patch manifold space, the patch can be well sparsely approximated by a linear combination of a few nearby patches. Thus, it has a sparse representation over the support domain Y_s. For any HR patch y_i , it can be sparsely approximate by the data matrix Y_s except y_i :

$$W_i^* = \arg\min_{W_i} \|y_i - Y_S W_i\|_2 + \lambda_2 \|W_i\|_1 \tag{5}$$

where W_i denotes the i -the column of the matrix W whose diagonal elements are zeros, and λ_2 is the parameter balancing the coding error of y_i and the sparsity of W_i

2.5 MSSR Objective Function and Optimization

We preserve the geometry relation represented by W for the reconstructed HR patch manifold. When LR patch is transformed to the HR patch, we try to preserve geometry constraint from W for $f(X_S, P)$ It can be gained by minimizing,

$$\sum_{i \in S} \|P x_i - P X_S W_i\|_2^2 = \|P X_S - P X_S W\|_F^2 = \|P X_S (I - W)\|_F^2 \tag{6}$$

where I is an identity matrix.

Considering both of the two properties we want to engage, the objective function of our proposed MSSR is defined as:

$$O_{MSSR} = \|P X_S - Y_S\|_F^2 + \alpha \|P\|_F^2 + \beta \|P X_S (I - W)\|_F^2 \tag{7}$$

where β is a regularization parameter. Using matrix properties $tr(AB) = tr(BA)$, $\|A\|^2 = tr(AA^T)$ and $tr(A) = tr(A^T)$, we have

$$\begin{aligned} O_{MSSR} &= tr\left\{(P X_S - Y_S)(P X_S - Y_S)^T\right\} + \alpha tr(P P^T) \\ &\quad + \beta tr(P X_S (I - W)(I - W)^T X_S^T P^T) \\ &= tr(P X_S X_S^T P^T - P X_S Y_S^T - Y_S X_S^T P^T + Y_S X_S^T) \\ &\quad + \alpha tr(P P^T) + \beta tr(P X_S G X_S^T P^T) \end{aligned} \tag{8}$$

where $G = (I - W)(I - W)^T$. In order to minimize the objective Eq. (7), we would like to take the derivative of O_{MSSR} with respect to P and set it to zero, i.e., we have the following equation

$$\begin{aligned} \frac{\partial O_{MSSR}}{\partial P} &= 2 P X_S X_S^T - 2 Y_S X_S^T + 2\alpha P + 2\beta P X_S G X_S^T = 0 \\ &\Rightarrow P(X_S X_S^T + \alpha I + \beta X_S G X_S^T) = Y_S X_S^T \\ &\Rightarrow P = Y_S X_S^T (X_S X_S^T + \alpha I + \beta X_S G X_S^T)^{-1} \end{aligned} \tag{9}$$

Following [20], we perform a back projection for the super-resolved HR image of the proposed MSSR method to satisfy the global reconstruction constraint.

Note that our method is similar to the Local Learning based Regression (LLR) method proposed in [18], which is also trying to learn a mapping between LR and HR patches. However, there are essential differences between LLR and the proposed method: 1) LLR learns the LR-HR mapping in the local space of K nearest neighbors, which uses a fixed number of nearest neighbors through the feature space, while the proposed method adaptively selects the neighbors without any predefined neighborhood size, and reveals the mapping relationship between the LR and HR patch in the sparse support domains; 2) LLR does not take into account the geometric structure of the patch manifold that plays an important role in the choice of example patches, while the proposed method aims to preserve the geometric structure of the original HR patch manifold space for that of the reconstructed HR one, thus well revealing the similar local geometric structure manifold of LR and HR patch spaces and enhancing the learning performance.

2.6   Object Detection in Reconstructed High Resolution Image

Object detection can be directly in the reconstructed high resolution image. In order to accelerate the detection, we can divide the detection into on-line and off-line sections. Since the high resolution image is reconstructed from high resolution dictionary, the feature extraction can be shift from the reconstructed image to dictionary atoms. In the reconstruction stage, the patch from reconstructed image is linearly combined by dictionary atoms. The feature extraction of the patch can be also combined by that of dictionary atoms. Denoting the feature pyramid of an image I as Φ, and $I = [P_1, \cdots, P_N]$, and D_j in $D = [D_1, \cdots, D_K]$ is the atom of D (Dictionary), we have $\Psi * P_i \approx \Psi * (\sum_j \alpha_{ij} D_j) = \sum_j \alpha_{ij}(\Psi * D_j)$, where $*$ denotes the convolution operator. Concretely, we can recover individual part fter responses via sparse matrix multiplication (or lookups) with the activation vector replacing the heavy convolution operation as shown in Eq. (10)

$$
\begin{bmatrix} \Psi * P_1 \\ \Psi * P_2 \\ \vdots \\ \vdots \\ \vdots \\ \vdots \\ \Psi * P_N \end{bmatrix} \approx \begin{bmatrix} \alpha_1 \\ \alpha_2 \\ \vdots \\ \vdots \\ \vdots \\ \vdots \\ \alpha_N \end{bmatrix} \begin{bmatrix} \Psi * D_1 \\ \Psi * D_2 \\ \vdots \\ \Psi * D_K \end{bmatrix} = AM \tag{10}
$$

We can use different detection algorithms detect object in the reconstructed high-resolution images. In Dalal'method [3], first, HOG features capture the most discriminative information of pedestrians. Second, a SVM classifier decides whether a candidate window shall be detected as enclosing a object. The detection performance achieved 80%, but the calculation process is complex.

The state-of-the-art deformable models [8] have handled the articulation of human parts such as legs, head and torso and allowed objects parts to articulate with constraint. In this paper, we compare object detection performance in the original image with these algorithms in reconstruction image. We chose the most representative object detection algorithm, because our framework applies to any object detection algorithms. It can get the similar effect on the other algorithms and our approach retain the high-frequency features of the edge at maximum extent, and these features are most discriminative for classification.

Dollar et al. [5] proposed an extension of where Haar-like feature are computed over multiple channels of visual data [CHNFTRS], including LUV color channels, grayscale, gradient magnitude and gradient magnitude quantized by orientation (implicitly computing gradient histograms), providing a simple and uniform framework for integrating multiple feature types. The gradient feature is used Dalal's, Felzenszwalb's and Dollar's object method, because the gradient is high-frequency features. So our approach is effective absolutely.

3 Experiment

We use both the INRIA Person Dataset [3] and the PASCAL2007 challenge dataset [6] for validating our proposed method. we randomly choose 1000 images from the test dataset and downsample the images by 2 and 4 times to construct the new test dataset. For INRIA, we compare our method with the HOG model. For PASCAL2007, we compare our method with the state-of-the-art DPM system [7].

INRIA Person Datasets. It contains 1774 pedestrian positive examples and 1671 negative images without people. The pedestrian annotations were scaled into a fixed size of 64×128 windows, which include a margin of 16 pixels around the pedestrians. The data set was divided into two, train and test; and the image resolution is 640×480. We get 320×240 resolution images as LR obtained from sampling the original image. In our experiments, train images contain both LR and HR objects. For each cascade level, the SVM classifier was trained using all the positive examples and $N_n = 500$ negative examples generated by boostrapping. Detection on the INRIA pedestrian data set is challenging since it includes subjects with a wide range of variations in pose, clothing, illumination, background, and partial occlusions. **PASCAL2007 dataset (comp3)** includes 20 object classes in a total of 9963 images, widely used as the standard benchmark for general object detection. There are large variations across classes in terms of the consistency of shape, appearance, viewpoint or occlusion.

Sparsity level and dictionary size. Do we need a sparsity level $K > 1$? This is an intriguing question and illustrates the difference between reconstructing signals (what sparse coding techniques are designed for) and extracting meaningful structures for recognition. Fig. 3 (a) shows the average precision on INRIA when we change the sparsity level along with the dictionary size using 5x5 patches. We observe that when the dictionary size is small, a patch cannot be well represented with a single codeword, and $K > 1$ (at least 2) seems to help. However,

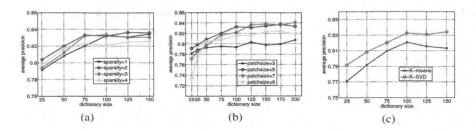

Fig. 3. Investigating the use of sparse codes on INRIA. (a) Average precision (AP) of sparsity level vs dictionary size; sparsity=1 works well when the dictionary is large. (b) Patch size vs dictionary size; larger patches do code richer information but requires larger dictionaries. (c) Dictionary learning with K-SVD works better than K-means.

when the dictionary size grows and includes more structures in its codes, the K = 1 curve catches up, and performs very well. Therefore we use K = 1 in all the following experiments, which makes the HOG features behave indeed like histograms using a sparse code dictionary.

Patch size and dictionary size. Our approach retain the edge information better and can capture richer structures using larger patches. Fig. 3(b) shows the average precision as we change both the patch size and the dictionary size. It is encouraging to see that indeed the average precision greatly increases as we use larger patches (along with larger dictionary size). While 3×3 codes barely show an edge over HOG, 5× 5 and 7×7 codes work much better, and the trend continues beyond 200 codewords. 9×9 patches, however, may be too large for our setting and do not perform well. The ability to code and make use of larger patches shows the merits of our feature design and K-SVD learning comparing to the spherical k-medoids, which had considerable trouble with larger patches and observed decreases in accuracy going beyond the small size 3x3.

K-SVD vs K-means. With K = 1, one can also use K-means to learn a dictionary (after normalizing the magnitude of each patch). Fig. 3(c) compares the detection accuracy with K-SVD vs K-means dictionaries on 5x5 patches. K-SVD dictionaries have a clear advantage over K-means, probably because the reconstruction coefficient in sparse coding allows for a single codeword to model more appearances including the change of sign.

Fig. 4 shows that the performance of our method is comparable to the state-of-art approaches. We compare ours experiments results with HOG+SVM and HOG+SVM+Linear interpolation on INRIA dataset and PASCAL VOC 2007dataset. The xaxis corresponds to false positives per window (FPPW) and Per-Image (FPPI) , the y-axis corresponds to the miss rate, and we plot the detection error tradeoff curves on a log-log scale for INRIA dataset and PASCAL 2007 dataset.

Detection Results. Our proposed method outperform the HOG in INRIA dataset, improving the mAP from 80.2% to 84.5%. In PASCAL VOC 2007

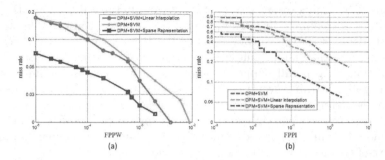

Fig. 4. (a)INRIA Per-Window results (b) PASCAL VOC 2007 results

dataset, our methods has improved the mAP from 30.1% to 33%, compared with the DPM.

Running Time. The consuming time of our method is consisting of offline and online sections. The offline section is for constructing the dictionary from learning the nearly 100,000 patches. Here we only consider the online section. For a 300*300 image, the online computation of our method takes ~30ms on an desk computer (single-core). The HOG and DPM take the ~100ms and ~9 seconds respectively.

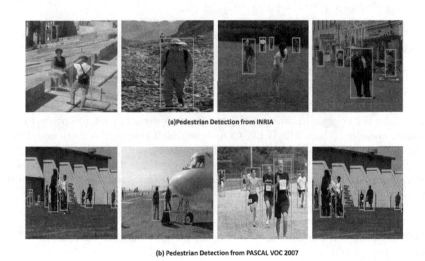

Fig. 5. Detection results. Red box is missing detection of DPM.

4   Conclusions

In this work we proposed a novel object detection framework in extremely low resolution image using sparse representation and divided the object detection into online and off-line stages. In the off-line stages, the patch based feature extraction on dictionary atoms was conducted. And in the online stage, the final visual feature is combined linearly by above patch based feature. Our work is first clearly demonstrate the advantages of LR-HR for general object detection. Our studies show that large structures in large patches, when are captured in a large dictionary, generally improve object detection, calling for future work on designing and learning even richer representation. We design a novel sparse regression algorithm for object detection which can enhance the learning performance. It is experimentally shown that the proposed MSSR methods can produce more faithful details and higher objective quality in comparison to the other state-of-the-art object detection approaches.

Acknowledgement. The research was supported by the National Nature Science Foundation of China (61303114, 61231015, 61170023), the Specialized Research Fund for the Doctoral Program of Higher Education (20130141120024), the Technology Research Project of Ministry of Public Security (2014JSYJA016), the Fundamental Research Funds for the Central Universities (2042014kf0250), the China Postdoctoral Science Foundation funded project (2013M530350), the major Science and Technology Innovation Plan of Hubei Province (2013AAA020), the Key Technology R&D Program of Wuhan (2013030409020109), the Guangdong-Hongkong Key Domain Break-through Project of China (2012A090200007), and the Special Project on the Integration of Industry, Education and Research of Guangdong Province (2011B090400601).

References

[1] Aytar, Y., Zisserman, A.: Tabula rasa: Model transfer for object category detection. In: 2011 IEEE International Conference on Computer Vision (ICCV), pp. 2252–2259. IEEE (2011)

[2] Cevikalp, H., Triggs, B.: Efficient object detection using cascades of nearest convex model classifiers. In: 2012 IEEE Conference on Computer Vision and Pattern Recognition (CVPR), pp. 3138–3145. IEEE (2012)

[3] Dalal, N., Triggs, B.: Histograms of oriented gradients for human detection. In: IEEE Computer Society Conference on Computer Vision and Pattern Recognition, CVPR 2005, vol. 1, pp. 886–893. IEEE (2005)

[4] Dollár, P., Belongie, S., Perona, P.: The fastest pedestrian detector in the west. In: BMVC, vol. 2, p. 7. Citeseer (2010)

[5] Dollár, P., Tu, Z., Perona, P., Belongie, S.: Integral channel features. In: BMVC, vol. 2, p. 5 (2009)

[6] Everingham, M., Van Gool, L., Williams, C.K., Winn, J., Zisserman, A.: The pascal visual object classes (voc) challenge. International Journal of Computer Vision 88(2), 303–338 (2010)

[7] Felzenszwalb, P., McAllester, D., Ramanan, D.: A discriminatively trained, multiscale, deformable part model. In: IEEE Conference on Computer Vision and Pattern Recognition, CVPR 2008, pp. 1–8. IEEE (2008)

[8] Felzenszwalb, P.F., Girshick, R.B., McAllester, D.: Cascade object detection with deformable part models. In: 2010 IEEE conference on Computer vision and pattern recognition (CVPR), pp. 2241–2248. IEEE (2010)

[9] Jiang, J., Hu, R., Han, Z., Huang, K., Lu, T.: Efficient single image super-resolution via graph embedding. In: 2012 IEEE International Conference on Multimedia and Expo (ICME), pp. 610–615. IEEE (2012)

[10] Jiang, J., Hu, R., Han, Z., Huang, K., Lu, T.: Position-patch based face hallucination via locality-constrained representation. In: 2012 IEEE International Conference on Multimedia and Expo (ICME), pp. 212–217. IEEE (2012)

[11] Jiang, J., Hu, R., Wang, Z., Han, Z., Dong, S.: Manifold regularized sparse support regression for single image super-resolution. In: 2013 IEEE International Conference on Acoustics, Speech and Signal Processing (ICASSP), pp. 1429–1433. IEEE (2013)

[12] Jiang, J., Hu, R., Wang, Z., Han, Z.: Noise robust face hallucination via locality-constrained representation. Transaction on Multimedia 16(5), 1268–1281 (2014)

[13] Lee, H., Battle, A., Raina, R., Ng, A.Y.: Efficient sparse coding algorithms. Advances in Neural Information Processing Systems 19, 801 (2007)

[14] Rahtu, E., Kannala, J., Blaschko, M.: Learning a category independent object detection cascade. In: 2011 IEEE International Conference on Computer Vision (ICCV), pp. 1052–1059. IEEE (2011)

[15] Ren, X., Ramanan, D.: Histograms of sparse codes for object detection. In: 2013 IEEE Conference on Computer Vision and Pattern Recognition (CVPR), pp. 3246–3252. IEEE (2013)

[16] Shen, B., Si, L.: Non-negative matrix factorization clustering on multiple manifolds. In: AAAI (2010)

[17] Sun, M., Savarese, S.: Articulated part-based model for joint object detection and pose estimation. In: 2011 IEEE International Conference on Computer Vision (ICCV), pp. 723–730. IEEE (2011)

[18] Tang, Y., Yan, P., Yuan, Y., Li, X.: Single-image super-resolution via local learning. International Journal of Machine Learning and Cybernetics 2(1), 15–23 (2011)

[19] Vijayanarasimhan, S., Grauman, K.: Efficient region search for object detection. In: 2011 IEEE Conference on Computer Vision and Pattern Recognition (CVPR), pp. 1401–1408. IEEE (2011)

[20] Yang, J., Wright, J., Huang, T.S., Ma, Y.: Image super-resolution via sparse representation. IEEE Transactions on Image Processing 19(11), 2861–2873 (2010)

A Novel Fast Full Frame Video Stabilization
via Three-Layer Model

Wei Long[1], Jie Yang[1], Dacheng Song[1], Xiaogang Chen[2], and Xiangjian He[3]

[1] Institute of Image Processing and Pattern Recognition,
Shanghai Jiao Tong University, Shanghai, China
{canoexp,jieyang}@sjtu.edu.cn, 576762979@qq.com
[2] University of Shanghai for Science and Technology, Shanghai, China
xg.chen@live.com
[3] School of Computing and Communications, University of Technology, Sydney, Australia
xiangjian.He@uts.edu.au

Abstract. Video stabilization is an important video enhancement technology which aims at removing undesired shaking from input videos. A challenging task in stabilization is to inpaint the missing pixels of undefined areas in the motion-compensated frames. This paper describes a new video stabilization method. It adopts a multi-layer model to improve the efficiency of the video stabilization. The undefined areas can be inpainted in real-time. Compared with traditional methods, our proposed algorithm only need maintain a single updated mosaic image for video completion, while previous methods require to store all neighboring frames and then registered with the current frame. The experimental results demonstrated the effectiveness of the proposed approach.

Keywords: Video stabilization, motion compensation, video completion, global motion estimation.

1 Introduction

Recently, with the pervasion of inexpensive low-end video cameras, and with the increasing popularity of video sharing on websites, we have witnessed a dramatic increase in amount of captured videos. When using a hand-held or vehicle-mounted video camera to capture a scene, the output is most likely not what we desired; it is rather a distorted representation of the view, where a lot of unexpected movement exists. A video stabilization process tries to remove this unintentional camera movement, and thus improve the quality of the input video.

Generally, there are two steps should be included to remove the jitter of input video, that is: (1) estimating the latent smooth camera path, and (2) calculating the stabilized frames according to this new path. Most existing methods [4,5] apply a full-frame 2D transformation to the input frame to obtain the output stabilized output frames. Ideally, we need the dense 3D structure of the scene to produce image from a novel view. However, obtaining the 3D structure from 2D images is extremely challenging due to

X. He et al. (Eds.): MMM 2015, Part I, LNCS 8935, pp. 246–256, 2015.
© Springer International Publishing Switzerland 2015

the computational complexities. There are several methods [17,18,19] have been proposed to tackle this problem, which mostly depend on image-based rendering to generate new images of a scene as seen along the smooth camera path. However, these algorithms are still limited to still scenes. For the dynamic scenes the estimation model could have many outliers.

In this paper, we propose a three-layer model for fast video stabilization and video completion. In the beginning, the global image transformation parameters are estimated between current frame and its previous frame. Next, unexpected motion is estimated and eliminated based on the transformation parameters. At last, motion compensated image may have undefined regions, thus we propose a new method to perform the video completion. Our proposed scheme only requires to store previous frame and the updated mosaic image for video stabilization and video completion. It is much faster than existing methods and suitable for real time full frame processing.

2   Related Work

The goal of the video stabilization is to eliminate unexpected jitters occurring on vehicle-mounted or unsteady handheld cameras. Although this problem has been studied over decades [1,2], great progress has been made in recent years using visual fixation [3], motion trajectory analysis [14], video completion [5,6], etc.

Since camera shake is one of the most common causes of image blur, we only focus on recovering the undesired jitters in this paper. Therefore, the image deblurring or video deblurring is out of the scope in this paper.

The steps for video stabilization typically containing three successive steps: camera motion estimation, motion filtering (motion smoothing), and motion compensation [7]. The video stabilization algorithms can be distinguished by the methods adopted in these stages. In some cases, video completion is considered in the final step [5,6],[24], and camera motion estimation is calculated in the first step. The motion estimation could be implemented by many methods, such as feature-based matching [2],[4],[6],[8], entire frame based matching [7],[9] and signature-based estimation [10,11]. Signature-based estimation methods are very fast but only compensate horizontal and vertical translations. Other two kind of methods are more adaptive for global image transformation such as rotation and similarity transformation [12,13], affine transformation [6],[9] and homographic transformation [2], etc.

Motion filter methods estimate the intentional camera motions and eliminate the unwanted jitters. It can be implemented either by cumulative transformation [9],[16] or smoothing transformations between the neighboring frames [13]. However, smoothing camera motion will produce undefined regions in stabilized video. Some motion filter methods [4],[13],[15] consider the tradeoff between camera motion smoothness and the amount of undefined areas by user specified parameters. But they did not provide a solution to rectify undefined areas.

Filling in the missing pixels of undefined areas in stabilized image is called video completion [5], which may be achieved by mosaicking [6],[9] or by motion inpainting method [5]. However, most of these approaches are computational expensive and difficult for real-time calculation. They have to store and register all neighboring frames.

Recently, 3D stabilization methods [17,18,19,20,21] attempt to recover true camera motion and scene structures via structure from motion (SFM) systems, in order to fully handle general scene structures and camera motion. Stabilization is subsequently done by smoothing the camera path in 3D and synthesizing a new video stream based on the smoothed path. To avoid the dependency on structure from motion techniques, the method [22] directly smoothes the 2D feature trajectories which based on the observation that they approximately lie in a low-dimensional subspace over any short period of time. However, these methods are generally too slow for stabilization purposes, taking hours to process a video with a few hundred frames.

Another literature [23] described a method to detect and correct the lighting and shaking artifacts of home videos. The lighting artifacts are corrected by appropriately adjusting the luminance. And in order to remove shaking, image mosaic is first employed with edge blending technique, then employed Bezier-curve based blending of motion trajectory to perform motion compensation.

3   Camera Global Motion Estimation

The geometrical transformation between the neighboring frames can be generally modeled as:

$$[x' \quad y' \quad 1]^T = \Phi [x \quad y \quad 1]^T \tag{1}$$

where $[x, y, 1]$ and $[x', y', 1]$ are the pixel coordinate in previous and current frame in homogeneous form respectively. If the translation is considered, the parameter Φ can be defined as:

$$\Phi = \begin{bmatrix} 1 & 0 & t_x \\ 0 & 1 & t_y \\ 0 & 0 & 1 \end{bmatrix} \tag{2}$$

Alternatively, we can define the motion as affine transformation, and then Φ is defined as:

$$\Phi = \begin{bmatrix} A & B & t_x \\ C & D & t_y \\ 0 & 0 & 1 \end{bmatrix} \tag{3}$$

Besides, the similarity transform is also quite common in practice, we can define Φ as:

$$\Phi = \begin{bmatrix} a\cos\theta & a\sin\theta & t_x \\ a\sin\theta & a\cos\theta & t_y \\ 0 & 0 & 1 \end{bmatrix} \tag{4}$$

In our implementations, we tested the similarity transform based motion compensation model as well as the affine transformation based model. Without loss of generality, in the following discussions, we assume that the similarity transformation is adopted.

In our implementation, we extracted SIFT feature points from consecutive frames to estimate global camera motion. The corresponding feature points are registered to each other, and then employ RANSAC approach to accurately estimate the similarity transformation matrix entries of Φ .

| (a) | (b) | (c) |

Fig. 1. Illustration of motion direction estimation (a) current frame (b) the next frame (c) the estimated motion

Using above equation, global camera motion direction can be calculated through the two neighboring frames, as shown in Fig.1. In the Fig.1(c), the black points denote original point of previous frame, and white lines denote the motion directions and magnitudes of the next consecutive frames from the previous one.

4   Video Stabilization

For the purpose of efficiently process video stabilization and video completion, we have been proposed a three-layer model that shows in Fig.2. The whole workflow of our model is illustrated in Fig.2. The correspondence parameters of this model, that is $\{T, P, \Phi, N, Q\}$ with form of 3×3 similarity transformation matrix. To the first layer $L1$, the original shaking video stream is inputted. To the second layer $L2$, video registration across two consecutive frames is processed. From diagram of video stabilization model, we assume that first frame registered to all of other remaining frames in input video stream. Therefore, the registration matrix N is an identity matrix; we assign identity matrix I to registration matrix N, i.e. $N = I$. This means that all frames in $L2$ are supposed to be aligned with the first input frame. Therefore, the transformations between consecutive frames are all identity. As we can see in the experimental results section, this registration assumption can deal with video stabilization properly. To the third layer $L3$, the stabled inpainted video stream is outputted. Undesired jitters are eliminated, video completion is achieved. For the convenience of description, the individual layer of the model with video stream are denoted three layer images as f , g and h respectively. Subscript t and $t-1$ indicates current frame and previous frame in timeline. From above analysis, the final transformation matrix needed to be estimated becomes $\{T, P, \Phi, I, Q\}$ since we let $N = I$.

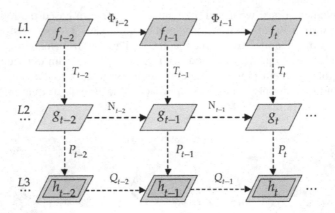

Fig. 2. Illustration of our proposed model for video stabilization

Since transformation parameter T_1 and P_1 are set to I from the model, therefore, both h_1 and g_1 are equal to f_1. As shown in Fig. 2, the transformation parameters between two consecutive frames in $L1$ are denoted by Φ, which is estimated by the method presented in global motion estimation aforementioned. Because all frames in $L2$ are aligned with f_1, the transformation parameter T_t which maps input frame f_t to g_t actually aligns f_t with f_1. From Fig. 2 we can easily induce equality $N_{t-1}T_{t-1} = T_t\Phi_{t-1}$. Therefore,

$$T_t = N_{t-1}T_{t-1}\Phi_{t-1}^{-1} \tag{5}$$

The jitters could be eliminated when mapping input frames to $L2$ layer. But the intentional camera motions are also removed and the output window always covers f_1 region. To remain the intentional camera motions, we warp aligned frame from g_t to h_t. Actually, the camera jitters contribute to the high-frequency components of this sequence. Hence, when we adopted a scheme that based on the latest K parameters $T_t, T_{t-1}, \ldots, T_{t-K+1}$ this influence can eliminated. The means of these parameters, $\dfrac{1}{K}\sum\limits_{i=t-K+1}^{t} T_i$, reflects the smoothed camera motion. Therefore, we warp g_t to the stabilized output frame h_t by its inverse form, *i.e.* $P_t = \left(\dfrac{1}{K}\sum\limits_{i=t-K+1}^{t} T_i\right)^{-1}$. It is necessary to further define a parameter as:

$$E_t = P_t \cdot T_t \tag{6}$$

Thus, the satisfying stabilized output frame h_t can be directly warped by E_t from the input frame f_t.

(a)   (b)   (c)

Fig. 3. Illustration of example for different K values. (a) K=2   (b)   K=10 (c) K=30.

It is not difficult to see that $E_t = I$ when $K = 1$, the annoying jitters in the input video are not reduced. Larger K produces smoother camera motion, but it may introduce more undefined regions in output frame, an example illustration shown in Fig.3. Empirically, we set $K = 10$ in our all experiments in this paper, and it is also suitable for most cases. To estimate the transform parameter P_t, we actually calculate the mean of K latest parameters $\{a, \theta, t_x, t_y\}$ firstly, which are decomposed from transform parameter T. Then, based on the mean parameter and the inversed of P_t, a transformation matrix is constructed. After stabilization, for the undefined regions in stabilized output image, we employ video completion method to fill in it.

5   Video Completion

In order to estimate the transformation matrix Q shown in Fig.2 at $L3$ between two successive frames, we can induce their relations through multi-layer model. In layer $L1$ the frame f_{t-1} could be aligned with h_t either by $E_t \cdot \Phi_{t-1}$ or $Q_{t-1} \cdot E_{t-1}$. Therefore, we can easily obtained equality below:

$$Q_{t-1} \cdot E_{t-1} = E_t \cdot \Phi_{t-1} \tag{7}$$

$$Q_{t-1} = E_t \cdot \Phi_{t-1} \cdot E_{t-1}^{-1} \tag{8}$$

When estimated Q_{t-1} is done, previous output frame h_{t-1} can be aligned with h_t. Therefore, the undefined missing pixels in h_t can be filled up by pixels in h_{t-1} with coordinate transformation parameter Q_{t-1}^{-1}, or by pixels in h_{t-2} with parameter $Q_{t-2}^{-1} Q_{t-1}^{-1}$ and so on.

However we could define an expanded image to h by adding margin, like the yellow regions in Fig. 2 (output frame is centered in the expanded image h). Previous frame's pixels which might be used for filling up undefined area in later output frames are then inherited in the expanded image. Actually the filled expanded image is the mosaicking image formed by latest frames. This process can be automatically updated by the input frame. Such mosaicking method is basically different from previous methods and the construction process is more efficient.

Therefore, the undefined missing pixels in current output expanded frame can be efficiently filled up by pixels in previous mosaic image h_{t-1}. As an example of illustration, comparison is shown in Fig.6. In Fig.6, the bottom row shows output frame, compared without video completion and with video completion technique. Alternatively, we can also adopt cross-weighting method to reduce artifacts in the mosaic image by image blending. The blend weight is depended on the distance to the boundary [9].

In general, the whole procedure for our proposed methods of video stabilization and video completion is concluded as follows:

(1) Estimate the global camera motion between both current frame and previous frame of the input video.

(2) Construct an expanded image which has larger size than input frame, warping input frame to the expanded image by parameter E_t.

(3) All undefined missing pixels in current expanded image areas are filled up by pixels in previous expanded image according to parameter Q_{t-1}^{-1}. These pixels in the output window of current expanded image are stored into final stabilized video.

In the algorithm, only previous expanded image is stored in memory for video completion. The latest K parameters T are used to estimate stabilized camera motion. Based on our observation, 10% percent of the width of the input frame is enough for the margin width in the expanded image.

6   Experimental Results

The performance of the proposed method is evaluated with different video sequences covering different type of scenes. Our method was implemented in C++ and tested by many videos with annoying jitters, and its efficacy has been approved. Our method can be achieved in real time. To the video with a resolution of 640×360 pixels, the whole procedure including stabilization and video completion only takes 40.316ms per frame, on computer with Inter(R) Xeon(R) CPU 2.40 GHz without any hardware acceleration.

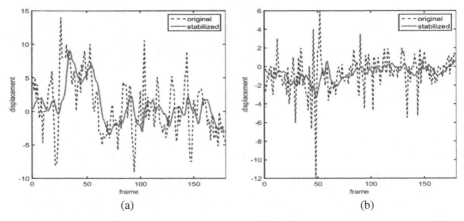

(a)   (b)

Fig. 4. Illustration of the input video motion and the stabilized video motion between successive frames. (a) The horizontal translations (b) The vertical translations.

We test our algorithm on several real-life video sequences acquired with hand held video device without any image stabilization. Fig. 4 demonstrates the horizontal and vertical translations curves between successive frames, with an input video (1) shown in Fig.5 and its output stabilized video respectively. It is easily observed that high-frequency components caused by jitters are greatly reduced by our proposed method. The stabilized camera motion follows a smoother path. Additionally, camera scaling and rotation are also smoothed by the proposed method.

We tested Matsushita [5], Tico [6], Litvin [9], and our algorithm on 7 video sequences (see Fig. 5). Video (3),(4),(5),(6) and (7) are used in [20]. The first two video

Fig. 5. Illustration of snapshots of the evaluation videos

Fig. 6. Illustration of video completion (a) previous frame (b) consecutive frame next previous frame (c) stabilization output without video completion (d) stabilization output with our video completion

clips are acquired with a hand-held device, and the resolution is 288x216 pixels. Video (3)-video (5) are with a resolution of 640x360, and the last two clips are with a resolution of 1280x720 pixels. The computational cost is recorded and compared. We run the algorithms of Matsushita [5], Tico [6], Litvin [9], and our algorithm on the same computer. The computation time is compared and shown in Fig.7.

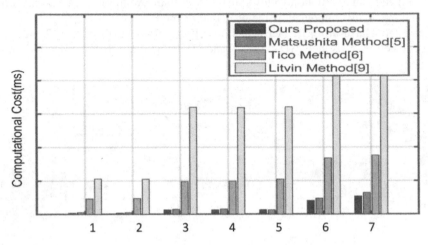

Fig. 7. Computational cost comparison between Matsushita [5], Tico [6], Litvin [9], and our method

In Fig. 6, we demonstrate the video completion of two consecutive input frames. The top-left image is the previous input frame, top-right image corresponds to the next input frame, bottom-left is the corresponding output frame without video completion, and bottom-right is the output frame with video completion.

In the bottom row of Fig.6, the white box in the expanded image denotes the valid output frame. Compared with Fig.6 (c) and Fig.6 (d), the undefined missing pixels in the expanded area of output frame, there are more than ten pixels are translated after video stabilization. The undefined missing pixels are well painted and they are indistinguishable as the neighboring pixels. The stabilized frames are shown in Fig.8.

Fig. 8. Illustration snapshot of video stabilization results Top row: snapshot of input frame; Bottom row: snapshot of stabilization output frame

7   Conclusions and Discussion

In this paper, we propose a novel framework of multi-layer model for full frame real time video stabilization. From the results of demonstration aforementioned section, and based on the updated mosaic image, the video stabilization and video completion process is implemented efficiently. It costs less computationally expensive than traditional methods, and produces satisfying results. And its effectiveness has been validated on different video clips.

Our method is based on global motion estimation which may become unstable, e.g., when a moving object covers large amount of image area. However, local motion estimation also has limitations. It may generate wrong results for fast moving objects. As a result, the composed mosaicking images could have outliers. Our future work is to further improve the quality of video completion and motion compensation algorithm.

References

1. Ratakonda, K.: Real-time digital video stabilization for multi-media applications. In: IEEE Proceedings of the 1998 IEEE International Symposium on Circuits and Systems, vol. 4, pp. 69–72 (1998)
2. Censi, A., Fusiello, A., Roberto, V.: Image stabilization by features tracking. In: IEEE Proceedings of the 9th International Conference on Image Analysis and Processing, Venice, Italy, pp. 665–667 (September 1999)
3. Kurz, C., Thormahlen, T., Seidel, H.: scene-aware video stabilization by visual fixation. In: IEEE Conference for Visual Media Production, CVMP 2009, pp. 1–6 (2009)
4. Lee, K., Chuang, Y., Chen, B.Y., Ouhyoung, M.: Video Stabilization using Robust Feature Trajectories. In: IEEE 12th International Conference on Computer Vision, ICCV 2009, pp. 1397–1404 (2009)
5. Matsushita, Y., Ofek, E., Tang, X., Shum, H.: Full-frame Video Stabilization. In: IEEE Computer Society Conference on Computer Vision and Pattern Recognition, CVPR 2005, pp. 50–57 (2005)
6. Hu, R., Shi, R., Shen, F., Chen, W.: Video Stabilization Using Scale-Invariant Features. In: IEEE Proceedings of 11th International Conference on Information Visualization, IV 2007, pp. 871–877 (2007)
7. Tico, M., Vehvilainen, M.: Constraint motion filtering for video stabilization. In: IEEE International Conference on Image Processing, pp. 569–576 (2005)
8. Marcenaro, L., Vernazza, G., Regazzoni, C.: Image stabilization algorithms for video-surveillance applications. In: IEEE Proc. of International Conference on Image Processing, ICIP 2001, vol. 1, pp. 349–352 (2001)
9. Litvin, A., Konrad, J., Karl, W.: Probabilistic Video Stabilization using Kalman Filtering and Mosaicking. International Society for Optics and Photonics: Electronic Imaging 5022, 663–674 (2003)
10. Bombini, L., Cerri, P., Grisleri, P., Scaffardi, S., Zani, P.: An evaluation of monocular image stabilization algorithms for automotive applications. In: IEEE Intelligent Transportation Systems Conference, ITSC 2006, pp. 1562–1567 (2006)

11. Broggi, A., Grisleri, P., Graf, T., Meinecke, M.: A software video stabilization system for automotive oriented applications. In: IEEE 61st Vehicular Technology Conference, VTC 2005, pp. 2760–2764 (2005)
12. Xu, L., Lin, X.: Digital image stabilization based on circular block matching. IEEE Transactions on Consumer Electronics, 566–574 (2006)
13. Chang, H., Lai, S., Lu, K.: A robust real-time video stabilization algorithm. Journal of Visual Communication and Image Representation 17, 659–673 (2006)
14. Niskanen, M., Silven, O., Tico, M.: Video Stabilization Performance Assessment. In: IEEE International Conference on Multimedia and Expo, ICME 2006, pp. 405–408 (2006)
15. Zhang, G., Hua, W., Qin, X., Shao, Y., Bao, H.: Video stabilization based on a 3D perspec-tive camera model. J. The Visual Computer Visual Computer 25, 997–1008 (2009)
16. Kwon, O., Shin, J., Paik, J.: Video Stabilization Using Kalman Filter and Phase Correlation Matching. Image Analysis and Recognition, 141–148 (2005)
17. Bhat, P., Zitnick, C., Snavely, N., Agarwala, A., Agrawala, M., Cohen, M., Curless, B., Kang, S.: Using photographs to enhance videos of a static scene. In: Proceedings of the 18th Eurographics conference on Rendering Techniques, pp. 327–338 (2007)
18. Buehler, C., Bosse, M., McMillan, L.: Non-metric image based rendering for video stabilization. In: Proceedings of the 2001 IEEE Computer Society Conference on Computer Vision and Pattern Recognition, CVPR 2001, pp. 609–614 (2001)
19. Fitzgibbon, A., Wexler, Y., Zisserman, A.: Image-based rendering using image-based priors. International Journal of Computer Vision 63, 141–151 (2005)
20. Liu, F., Gleicher, M., Jin, H., Agarwala, A.: Contentpreserving warps for 3d video stabilization. ACM Transactions on Graphics 28, Article: 44 (2009)
21. Liu, F., Gleicher, M., Wang, J., Jin, H., Agarwala, A.: Subspace video stabilization. ACM Transactions on Graphics 30(1), 4 (2011)
22. Zhou, Z., Jin, H., Ma, Y.: Plane-Based Content Preserving Warps for Video Stabilization. In: IEEE Conference on Computer Vision and Pattern Recognition, CVPR 2013, pp. 2299–2306 (2013)
23. Yan, W., Kankanhalli, M.: Detection and removal of lighting & shaking artifacts in home videos. In: Proceedings of the Tenth ACM International Conference on Multimedia, pp. 107–116 (2002)
24. Song, D., Chen, X., Yang, J.: Real-time full frame video stabilization. In: Proceeding of the International conference on Applied Informatics and Communication, ICAIC 2012 (2012)

Cross-Modal Self-Taught Learning
for Image Retrieval

Liang Xie, Peng Pan*, Yansheng Lu, and Sheng Jiang

School of Computer Science and Technology,
Huazhong University of Science and Technology,
Wuhan, China, 430074
whutxl@hotmail.com, {panpeng,lys,jwt}@mail.hust.edu.cn

Abstract. In recent years, cross-modal methods have been extensively studied in the multimedia literature. Many existing cross-modal methods rely on labeled training data which is difficult to collect. In this paper we propose a cross-modal self-taught learning (CMSTL) algorithm which is learned from unlabeled multi-modal data. CMSTL adopts a two-stage self-taught scheme. In the multi-modal topic learning stage, both intra-modal similarity and multi-modal correlation are preserved. And different modalities have different weights to learn the mutli-modal topics. In the projection stage, soft assignment is used to learn projection functions. Experimental results on Wikipedia articles and NUS-WIDE show the effectiveness of CMSTL in both cross-modal retrieval and image hashing.

Keywords: cross-modal retrieval, image retrieval, image hashing, self-taught learning.

1 Introduction

Over last decades, with the advance of computer network and multimedia technologies, we have witnessed a massive explosion of multimedia content on the web. Large amounts of multi-modal data, such as images and texts, are generated, shared and accessed on social websites, e.g., Flickr, Wikipedia and YouTube. Classical uni-modal approach [1] is not able to deal with these multi-modal data.

Cross-modal approach which analyzes the correlation of heterogeneous modalities, has been extensively studied in the multimedia literature [2,3,4,5,6]. They can solve the retrieval of heterogeneous data, e.g., using text query to retrieve images. Since the cross-modal correlation is beneficial for bridging the "semantic gap", the performance of uni-modal retrieval is likely to be improved by cross-modal approach [3,7]. In many previous studies on cross-modal retrieval [3,4,5], class labels are used to construct the cross-modal correlation. If an image and a text belong to the same class, then they are considered as relevant. However, labeling the training data is usually labor intensive and expensive, which makes the

---

* Corresponding author.

X. He et al. (Eds.): MMM 2015, Part I, LNCS 8935, pp. 257–268, 2015.

labeled data difficult to be collected. Another disadvantage of using class labels for cross-modal learning is that classes may be limited in some domain. For example, 'apple' may not be connected to 'computer' if labeled data is about fruit.

In this paper we propose a novel method: cross-modal self-taught learning (CMSTL) for cross-modal image retrieval. Unlike previous methods which analyze cross-modal correlation according to class labels, CMSTL analyzes the latent correlation according to the co-occurrence of heterogenous data. CMSTL does not require any labeled training data, it only needs multi-modal documents for training. Multi-modal data are usually tend to co-occur in the same documents on many social websites. Therefore, CMSTL is more practical than previous methods in the real world.

CMSTL adopts the two-stage self-taught scheme [8]. Since self-taught scheme have two stages: unsupervised learning and supervised learning, methods based on this scheme can both benefit from effective unsupervised and supervised approaches. In the first stage, an effective hierarchical multi-modal approach is proposed to generate latent topic. Intra-modal topics of each modality are first generated from intra-modal similarity. Then they are all combined to a uniform multi-modal topic space. These two generations are combined in a joint objective function from which the final multi-modal topics can be optimized. In the generation of multi-modal topics, different modalities have different weights. It makes our method more adaptive than traditional methods which treat all modalities equally. In the second stage, a soft supervised projection is used, and all modality are projected to the latent topic space via kernel least square regression (KLSR). We test our method on two real-world datasets: Wikipidia articles [3] and NUS-WIDE [9]. The experimental results show the effectiveness of our method in cross-modal retrieval. We further extend CMSTL to image hashing and the results show that our cross-modal method improves uni-modal image retrieval.

The rest of this paper is organized as follows. In section 2 we discuss the related work. In section 3 we describe the framework of our CMSTL. Section 4 shows the experimental results of cross-modal retrieval and image hashing on two datasets. Finally we conclude in Section 5.

2   Related Work

In recent years, many efforts have been devoted to the cross-modal multimedia retrieval. Most cross-modal methods focus on learning an uniform space where different modalities are correlated. One type of methods analyze latent correlation, which is based on the co-occurrence of multi-modal data, to construct the uniform space. In [2], multimedia correlation space (MMCS) is constructed from multi-modal data. However, its main limitation is the lack of out-of-sample generalization. New queries must be first mapped to their nearest neighbors in the training set. [10] uses canonical correlation analysis (CCA) and cross-modal factor analysis (CFA) in the context of audio-image retrieval. Both CCA and CFA learn a latent space where two modalities are correlated. Kernel CCA (KCCA) is proposed in [11] to extract translation invariant semantics of text documents written in multiple languages. [12] also uses KCCA to model correlation between web images and corresponding text captions.

Another type of cross-modal methods learn the cross-modal correlation space from class labels. In [3], both images and texts are represented by the posteriors of class labels. Logistic regression is used to project documents into the probabilistic space of class. In [4], a semantic generation model is proposed for cross-modal retrieval, the correlation of different modalities is described by their generation from semantic labels. In [13], a joint graph regularized heterogeneous metric learning (JGRHML) algorithm is proposed to improve the semantic metric, which is learned through label propagation. [5] proposes the latent semantic cross-modal ranking (LSCMR) to discriminatively learn a latent low-rank embedding space by structural large margin learning. LSCMR is trained from supervised ranking examples, which are determined by the class labels shared among documents. These methods have the limitation that their performance depends on the class labels. If the class labels are not sufficient to describe the cross-modal correlation well, their performance may be affected.

Recently some methods are proposed to solve the problem of large-scale multimedia retrieval, including cross-modal retrieval. Spectral hashing [14] and self-taught hashing [8] are two representative methods for large-scale uni-modal retrieval. Spectral Hashing uses a subset of thresholded eigen-vectors of the graph Laplacian as hashing codes. Self-taught hashing adopts the self-taught scheme which is similar to our methods, the difference is that it uses hard assignment to obtain the binary codes, which may cause the loss of semantic information. Cross-modality similarity-sensitive hashing (CMSSH) [15] uses the supervised similarity learning method boosting to embed data into the hamming space. Multimodal latent binary embedding (MLBE) learns the hashing codes in a probabilistic framework. Both CMSSH and MLBE rely on training data labeled by classes, and they cannot work while the labels are missing. Cross-view hashing (CVH) [16] requires predefined cross-modal similarities of the training data. But if the cross-modal similarity matrix is set to identity matrix, CVH can be learned from unlabeled multi-modal data.

3   The Description of Our Framework

In this section we describe the framework of cross-modal self-taught learning (CM-STL), which contains two learning stages: multi-modal topic learning (MMTL) and projection function learning. MMTL learns latent topics from multi-modal training data, then projection function is learned to represent test data from each modality by these topics.

3.1   Learning Multi-modal Latent Topics

MMTL learns latent topics which can correlate different modalities, from unlabeled multi-modal training data. Suppose there are N multi-modal documents

$D_1, \ldots, D_N$. Each document D_n contains M modalities, and $D_n = \{x_n^1, \ldots, x_n^M\}$, where x_n^m is the feature of m-th modality. In our experiments, we only consider two modalities: image and text, thus $M = 2$ and D_n is an image-text pair in fact. But our methods can also be applied to more modalities ($M > 2$).

For each modality, we construct its intra-modal similarity graph A_m, which is defined as:

$$A_{ij}^m = \begin{cases} sim_m(x_i^m, x_j^m), & x_i^m \text{ and } x_j^m \text{ are } k \text{ nearest neighbors} \\ 0, & otherwise \end{cases} \tag{1}$$

where $sim_m(x_i^m, x_j^m)$ is the similarity of x_i^m and x_j^m.

To learn the latent topics from multi-modal documents, we first generate intra-modal topic matrix $F_m \in \mathbb{R}^{N \times T_m}$ for each modality. T_m is the number of intra-modal topics. F_m can preserve the latent information of the m-th modal features, it is obtained by minimizing the following Graph Laplacian regularizer:

$$\sum_{k=1}^{T_m} \sum_{i=1}^{N} \sum_{j=1}^{N} A_{ij}^m \left(\frac{f_{ik}^m}{d_{ii}^m} - \frac{f_{jk}^m}{d_{jj}^m} \right) = Tr\left(F_m^T L_m F_m\right) \tag{2}$$

where f_{ik}^m is an element of F_m, and d_{ii}^m is the sum of i-th row of A_m. $L_m = I - D_m^{-1/2} A_m D_m^{-1/2}$, I is the identity matrix, D_m is the diagonal matrix and its diagonal element is d_{ii}^m. $Tr(\cdot)$ denotes the trace operator.

After we obtain intra-modal latent semantic matrices $F_m|_{m=1}^{M}$, we use them to generate the multi-modal latent topic matrices $F \in \mathbb{R}^{N \times T}$. T is the number of multi-modal topics. F combines the latent information of all modalities, and each F_m generates the final F by minimizing the following function:

$$\|F - F_m W_m\|_F^2 \tag{3}$$

where $\|\cdot\|_F^2$ denotes the Frobenius norm. $W_m \in \mathbb{R}^{T_m \times T}$ is the weight matrix for the generation of F.

In sum, the multi-modal topic matrix F is hierarchically generated. At first, intra-modal topic matrices $F_m|_{m=1}^{M}$ are generated by intra-modal similarity. Then multi-modal topic matrix F is generated from all $F_m|_{m=1}^{M}$.

We optimize the hierarchical generation in a joint framework. By combining (2) and (3), we arrive at the following objective function:

$$\min_F \sum_{m=1}^{M} \left(Tr\left(F_m^T L_m F_m\right) + \alpha_m^2 \|F - F_m W_m\|_F^2 \right)$$
$$s.t. \qquad F_m^T F_m = I, \ m = 1, \ldots, M$$
$$F^T F = I$$
$$\sum_{m=1}^{M} \alpha_m = 1 \tag{4}$$

where α_m is the weight parameter, it represents the importance of m-th modality for the generation of F, and we can easily find that (4) is convex respect to α_m.

By setting the derivative of (4) w.r.t. W_m to zero, we have:

$$W_m = F_m^T F \tag{5}$$

Substituting W_m in (4), the objective function becomes:

$$\sum_{m=1}^{M} \left(Tr\left(F_m^T L_m F_m \right) + \alpha_m^2 Tr\left(I - F^T F_m F_m^T F \right) \right) \tag{6}$$

We adopt an alternating optimization to solve (6). More specifically, we alternatively update F, F_m and α_m to optimize the objective function.

1) Optimizing F: We fix F_m and α_m, then (6) can be reformulated as:

$$\max_{F} Tr\left(F^T \sum_{m=1}^{M} \left(\alpha_m^2 F_m F_m^T \right) F \right)$$

$$s.t. \qquad F^T F = I \tag{7}$$

It is obviously that (7) is an eigenvalue problem, and we can obtain F by eigendecomposition of $\sum_{m=1}^{M} \left(\alpha_m^2 F_m F_m^T \right)$.

2) Optimizing F_m: We fix F and α_m. According to the trace property: $Tr\left(F^T F_m F_m^T F \right) = Tr\left(F_m^T F F^T F_m \right)$, (6) can be transformed to:

$$\min_{F_m} Tr\left(F_m^T C_m F_m \right)$$

$$s.t. \quad F_m^T F_m = I \tag{8}$$

where

$$C_m = L_m - \alpha_m^2 F^T F \tag{9}$$

We can also find that F_m is learned by solving the eigenvalue problem of (8),

3) Optimizing α_m: F and F_m are fixed, by using Lagrange multiplier, we can obtain:

$$\alpha_m = \frac{1/Tr\left(I - F^T F_m F_m^T F \right)}{\sum_{i=1}^{M} 1/Tr\left(I - F^T F_i F_i^T F \right)} \tag{10}$$

The whole alternating optimization process is illustrated in Algorithm 1. In the implementation of this algorithm, we initialize F_m by solving the eigenvalue problem of (2), and $\alpha_m|_{m=1}^{M}$ are set to the same. Since the objective function is lower bounded by 0 and it will keep decreasing in each step, its convergence is guaranteed. One advantage of our topic learning is that the importance of different modality for generating the latent topic is different, while previous cross-modal methods such as CCA treat all modalities equally. Thus our topic learning methods is more adaptive. Another advantage is that the hierarchical generation preserves not only intra-modal similarity, but also multi-modal correlation which is seldom considered in previous methods.

Algorithm 1. The learning process of MMLSA

Input:

$A_m|_{m=1}^M$

Output:

F

1: Compute $L_m|_{m=1}^M$;
2: Initialize $F_m|_{m=1}^M$ and $\alpha_m|_{m=1}^M$;
3: **while** $t < T$ **do**
4:     Update F by solving the eigenvalue problem of (7);
5:     Update $F_m|_{m=1}^M$ by solving the eigenvalue problem of (8);
6:     Update $\alpha_m|_{m=1}^M$ according to (10);
7:     t=t+1;
8: **end while**

3.2 Learning Projection Functions

Heterogenous data in training documents are correlated by multi-modal topics. However, for new data out of the training set, we still have to learn the explicit projection from each modality to the topic space. In the previous self-taught scheme [8], F is converted into binary codes via thresholding, and then binary topics can be treated as class labels. At last, classifiers which project documents into the latent topic space, are trained via some classification methods, such as support vector machine (SVM). However, using binary codes may lose some semantic information. Generally, it's better to use a soft assignment for latent topics.

In this paper we use a probabilistic soft assignment to learn the projection functions. Previous methods [3,17] have shown the effectiveness of image retrieval in probabilistic space. Therefore, the probabilistic representation for latent topics should be better than binary codes. To obtain a probabilistic space, we use Gaussian mixture on F to learn S probabilistic topics, and predict $Y = [y_1^T, \ldots, y_N^T]^T$. y_n is the posterior vector of f_n and it is predicted by the Gaussian mixture model.

Y is the final topic representation for training documents, we can use supervised methods to learn the function to project new documents into this topic space. Since Y is not binary, classification methods cannot work in this condition, and we use kernel least square regression (KLSR) instead. The objective function of KLSR is described as follows:

$$\min_{P_m} \|K_m P_m - Y\|_F^2 + \lambda Tr \left(P_m^T K_m P_m \right) \tag{11}$$

where K_m is the kernel function of the m-th modal training features. P_m is the projection weight for the m-th modality. $\lambda > 0$ is the regularization parameter. By setting the derivative of (11) w.r.t P_m to zero, we can easily get the weight matrix:

$$P_m = (K_m + \lambda I)^{-1} Y \tag{12}$$

In order to effectively compute the inverse. We perform singular singular value decomposition (SVD) to obtain a pseudo-inverse of $K_m + \lambda I = U_m \Lambda V_m^T$, which is computed by $(K_m + \lambda I)^{-1} = V_m \bar{\Lambda} U_m^T$, where $\bar{\Lambda}$ is defined as:

$$\bar{\Lambda}_{ii} = \begin{cases} 0, & \text{if } \Lambda_{ii} < \epsilon \\ \Lambda_{ii}^{-1}, & \text{otherwise} \end{cases} \tag{13}$$

where $\epsilon = 1$ is a threshold. Then the weight matrix is computed by:

$$P_m = V_m \bar{\Lambda} U_m^T Y \tag{14}$$

Algorithm 2. The process of cross-modal self-taught learning

Input:
   $A_m|_{m=1}^M, K_m|_{m=1}^M$
Output:
   $P_m|_{m=1}^M$
1: Compute F according Algorithm (1);
2: Train the Gaussian Mixture Model and predict the posterior matrix Y on F;
3: **for** each modality m **do**
4:     Do the SVD: $K_m + \lambda I = U_m \Lambda V_m^T$;
5:     Compute $\bar{\Lambda}$ according to (13);
6:     Compute P_m according to (14) ;
7: **end for**

The self-taught learning process is described in Algorithm 2. Given a new document from the m-th modality, we first compute its kernel vector k, then we can compute its topic representation according to $y = kP_m$. Note that some elements in y may not between 0 and 1, thus we have to normalize y to make it be a probabilistic vector. Softmax function is used to normalize y, for each element y_i in y, it is normalized by:

$$\bar{y}_i = \frac{\exp(y_i)}{\sum\limits_{j=1}^{S} \exp(y_j)} \tag{15}$$

Finally, we use the normalized vector $\bar{y} = [\bar{y}_1, \dots, \bar{y}_T]$ to represent new document for retrieval.

4   Experiments

4.1   Datasets and Features

In this paper, two real world multi-modal image datasets: Wikipedia articles [3] and NUS-WIDE [9] are used for evaluation. These two datasets are both split to training set and test set. All methods are only learned from training set, and their performance is evaluated on test set.

Wikipedia dataset was assembled from the "Wikipedia feature articles". It contains 2,866 multi-modal documents (image-text pairs), and each of them is labeled with exactly one of 10 semantic concepts which can be used as the ground truth. Documents share the same concept are regarded as relevant. 2,173 of the image-text pairs in Wikipedia dataset are chosen as training set, and the rest 693 are used as test set. 10-D LDA features are extracted for texts, and images are represented by 128-D SIFT BoVWs [1].

NUS-WIDE dataset contains 269,648 multi-modal documents, each multi-modal documents is also an image-text pair and text in NUS-WIDE refers to the associated social tags. Each image-text pairs are labeled by 81 concepts that can be used for evaluation. We prune the original NUS-WIDE to form a new dataset consisting of 203,597 image-text pairs by keeping the images that have at least one tag and one concept. Then this dataset is split to 5,090 training set and 198,507 test set. 1000-D binary features are used for tags, and 500-D SIFT BoVWs are used for images [2].

4.2 Results of Cross-Modal Retrieval

In this subsection we compare our method with several representative methods for cross-modal retrieval. CFA[10], CCA[18], and KCCA[12] are used for comparison. We also show the performance of a baseline method MMLT+SVM, where the latent topics is first learned by Algorithm 1, and then the self-taught scheme described in [8] is used to learn projection functions for each modality respectively. In our methods, the topic dimension of $F_m|_{m=1}^{M}$, F, and Y are all set to the same, they are set to 8 on Wikipedia dataset and 32 on NUS-WIDE dataset. In the computing of intra-modal similarity matrices $A_m|_{m=1}^{M}$, histogram intersection distance is used for both image and text features. The nearest neighbors for $A_m|_{m=1}^{M}$ is set to 500 on Wikipedia dataset, and 1000 on NUS-WIDE dataset. In all kernel based methods, the histogram intersection kernel which is same to histogram intersection distance, is used for both image and text features. The dimension of latent space in all methods are set to the same, 8 in Wikipedia dataset and 32 in NUS-WIDE dataset. Normalize correlation (NC) distance [3] are used for all methods in cross-modal retrieval.

We adopt the non-interpolated mean average precision (MAP) to evaluate the performance of cross-modal retrieval. Given a query and the rank list of retrieval set, the average precision (AP) is defined as:

$$AP = \frac{1}{R} \sum_{i=1}^{N} pre(i)rel(i) \tag{16}$$

---

[1] All features can be downloaded from http://www.svcl.ucsd.edu/projects/
crossmodal/

[2] All features can be downloaded from http://lms.comp.nus.edu.sg/research/
NUS-WIDE.htm

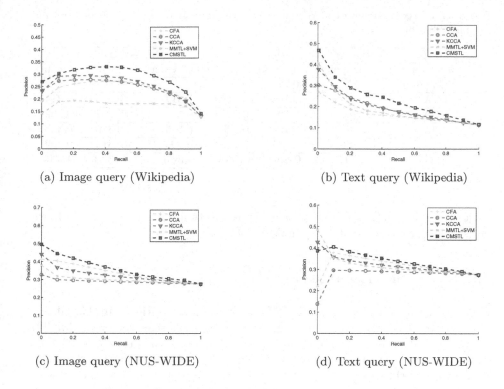

(a) Image query (Wikipedia)

(b) Text query (Wikipedia)

(c) Image query (NUS-WIDE)

(d) Text query (NUS-WIDE)

Fig. 1. The PR curves of cross-modal retrieval on two datasets

where N is the size of retrieval set. R is the number of relevant documents in the retrieved set, $pre(i)$ is the precision of top i retrieved documents. $rel(i) = 1$ if the i-th retrieved documents is relevant to query, otherwise $rel(i) = 0$. The MAP score is the mean of AP scores from all the queries. Besides MAP, the retrieval performance is also measure by Precision-Recall (PR) curve.

We evaluate two types of cross-modal retrieval on test set. In image query, images in test set are used as queries and texts in test set form retrieval set. In text query, test texts are queries and test images form retrieval set. Table 1 shows the MAP scores of cross-modal retrieval. It should be noted that the results of SCM have been reported in [3], thus its results on NUS-WIDE is null. We can find our CMSTL performs best on both two datasets. It performs even better than SCM which is learned from labeled training data on Wikipedia dataset. In usual methods learned from the training data labeled by ground truth should obtain better performance. CMSTL performs better than KCCA, CCA and CFA which treat all modalities equally, which confirms that it is worth weighting different modalities for the learning of latent topics. CMSTL also obtains higher MAP scores than MMTL+SVM, which demonstrates the advantage of our soft

Table 1. The comparison of MAP scores for cross-modal retrieval. The best results are marked in bold.

Datasets	Wikipedia			NUS-WIDE		
Methods	Image Query	Text Query	Mean	Image Query	Text Query	Mean
SCM[3]	0.277	0.226	0.252	-	-	-
CFA	0.245	0.166	0.210	0.313	0.315	0.314
CCA	0.249	0.193	0.221	0.291	0.292	0.292
KCCA	0.262	0.196	0.229	0.326	0.320	0.323
MMLT+SVM	0.181	0.181	0.181	0.305	0.313	0.309
CMSTL	**0.295**	**0.234**	**0.265**	**0.366**	**0.344**	**0.355**

projection scheme. The PR curves on two datasets are shown in Figure 1, we can see they are consistent with the MAP scores.

4.3   Results of Image Hashing

We have shown the superiority of CMSTL in cross-modal retrieval. In this subsection we show its performance on image hashing for large-scale retrieval. CMSTL can be easily extended for image hashing by thresholding the probabilistic latent vector $\bar{y}$. For the i-th topic, we obtain its threshhold θ_i by computing the mean of all values of this topic in the retrieval set. If $\bar{y}_i > \theta_i$, we set it to 1, otherwise we set it to 0. Generally cross-modal correlation is benefit to image retrieval. Thus CMSTL should be effective in image hashing.

We compare CMSTL based hashing (CMSTLH) to several representative hashing methods, including laplacian co-hashing (LCH)[19], spectral hashing (SH) [14] [3], self-taught hashing (STH) [8] [4], cross-view hashing (CVH) [16], where CVH is a cross-modal hashing method. For all hashing methods, their code length is set to 16 on Wikipedia and 32 on NUS-WDE, which are optimal on two datasets respectively. We use MAP50 to measure the performance, MAP50 is similar to MAP described in section 4.2, the difference is that MAP50 is compute by the top 50 relevant documents, it can be compute by setting $N = 50$ in (16). The performance is also evaluated on test set. In each retrieval, an image is used as query and other images in test set form retrieval set.

Table 2 shows the MAP scores on Wikipedia and NUS-WIDE. We can observe that CMSTL obtains the highest MAP scores on both two datasets. Although CVH is a cross-modal method, it does not always performs better than uni-modal methods. These results confirm that our method is also suit to image hashing. Since our methods is not specially designed for hashing, the improvement of CMSTLH is not large. Figure 2 shows the PR curve of image hashing, the results are consistent with Table 2.

---

[3] http://www.cs.huji.ac.il/~yweiss/SpectralHashing/

[4] http://www.dcs.bbk.ac.uk/~dell/publications/dellzhang_sigir2010_suppl.html

Table 2. The comparison of MAP50 scores for image hashing. The best results are marked in bold.

Datasets	LCH	SH	STH	CVH	CMSTLH
Wikipedia	0.382	0.263	0.374	0.385	**0.397**
NUS-WIDE	0.545	0.353	0.542	0.511	**0.569**

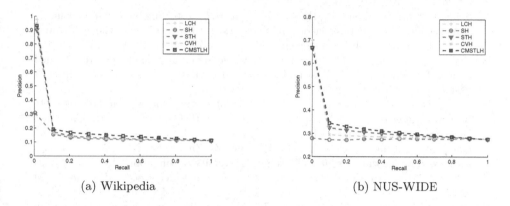

(a) Wikipedia   (b) NUS-WIDE

Fig. 2. The PR curves of image hashing on two datasets

5 Conclusion

In this paper we propose a cross-modal self-taught topic learning (CMSTL) algorithm which contains two stages: multi-modal topic learning (MMTL) and projection function learning. In MMTL, hierarchial generation is used to obtain multi-modal topics. Different modalities have different weights in the generation process. Then MMTL optimizes the intra-modal similarity and multi-modal correlation jointly. In projection learning stage, the soft assignment for topics is used. Topic matrix is converted to a probabilistic form and KLSR is used to learn the projections function. Experimental results on two real world image dataset demonstrate the effectiveness of CMSTL in cross-modal retrieval. We further extend CMSTL to image hashing and find that CMSTL can also improve the hashing performance.

References

1. Datta, R., Joshi, D., Li, J., Wang, J.Z.: Image retrieval: Ideas, influences, and trends of the new age. ACM Computing Surveys (CSUR) 40(2), 5 (2008)
2. Yang, Y., Xu, D., Nie, F., Luo, J., Zhuang, Y.: Ranking with local regression and global alignment for cross media retrieval. In: Proceedings of the 17th ACM International Conference on Multimedia, pp. 175–184. ACM (2009)
3. Rasiwasia, N., Pereira, J.C., Coviello, E., Doyle, G., Lanckriet, G.R., Levy, R., Vasconcelos, N.: A new approach to cross-modal multimedia retrieval. In: Proceedings of the International Conference on Multimedia, pp. 251–260. ACM (2010)

4. Xie, L., Pan, P., Lu, Y.: A semantic model for cross-modal and multi-modal retrieval. In: Proceedings of the 3rd ACM Conference on International Conference on Multimedia Retrieval, pp. 175–182. ACM (2013)

5. Lu, X., Wu, F., Tang, S., Zhang, Z., He, X., Zhuang, Y.: A low rank structural large margin method for cross-modal ranking. In: Proceedings of the 36th International ACM SIGIR Conference on Research and Development in Information Retrieval, pp. 433–442. ACM (2013)

6. Song, J., Yang, Y., Yang, Y., Huang, Z., Shen, H.T.: Inter-media hashing for large-scale retrieval from heterogeneous data sources. In: Proceedings of the 2013 International Conference on Management of Data, pp. 785–796. ACM (2013)

7. Hwang, S.J., Grauman, K.: Learning the relative importance of objects from tagged images for retrieval and cross-modal search. International Journal of Computer Vision 100(2), 134–153 (2012)

8. Zhang, D., Wang, J., Cai, D., Lu, J.: Self-taught hashing for fast similarity search. In: Proceedings of the 33rd International ACM SIGIR Conference on Research and Development in Information Retrieval, pp. 18–25. ACM (2010)

9. Chua, T.-S., Tang, J., Hong, R., Li, H., Luo, Z., Zheng, Y.: NUS-WIDE: a real-world web image database from National University of Singapore. In: Proceedings of the ACM International Conference on Image and Video Retrieval, p. 48. ACM (2009)

10. Li, D., Dimitrova, N., Li, M., Sethi, I.K.: Multimedia content processing through cross-modal association. In: Proceedings of the Eleventh ACM International Conference on Multimedia, pp. 604–611. ACM (2003)

11. Vinokourov, A., Cristianini, N., Shawe-Taylor, J.S.: Inferring a semantic representation of text via cross-language correlation analysis. In: Advances in Neural Information Processing Systems, pp. 1473–1480 (2002)

12. Hardoon, D., Szedmak, S., Shawe-Taylor, J.: Canonical correlation analysis: An overview with application to learning methods. Neural Computation 16(12), 2639–2664 (2004)

13. Zhai, X., Peng, Y., Xiao, J.: Heterogeneous Metric Learning with Joint Graph Regularization for Cross-Media Retrieval. In: AAAI (2013)

14. Weiss, Y., Torralba, A., Fergus, R.: Spectral hashing. In: Advances in Neural Information Processing Systems, pp. 1753–1760 (2009)

15. Bronstein, M.M., Bronstein, A.M., Michel, F., Paragios, N.: Data fusion through cross-modality metric learning using similarity-sensitive hashing. In: IEEE Conference on Cumputer Vision and Pattern Recognition (CVPR), pp. 3594–3601. IEEE (2010)

16. Kumar, S., Udupa, R.: Learning hash functions for cross-view similarity search. In: IJCAI Proceedings-International Joint Conference on Artificial Intelligence, vol. 22(1), p. 1360 (2011)

17. Rasiwasia, N., Moreno, P., Vasconcelos, N.: Bridging the gap: Query by semantic example. IEEE Transactions on Multimedia 9(5), 923–938 (2007)

18. Hotelling, H.: Relations between two sets of variates. Biometrika, 321–377 (1936)

19. Zhang, D., Wang, J., Cai, D., Lu, J.: Laplacian co-hashing of terms and documents. In: Gurrin, C., He, Y., Kazai, G., Kruschwitz, U., Little, S., Roelleke, T., Rüger, S., van Rijsbergen, K. (eds.) ECIR 2010. LNCS, vol. 5993, pp. 577–580. Springer, Heidelberg (2010)

Multimedia Social Event Detection in Microblog

Yue Gao[1], Sicheng Zhao[2], Yang Yang[1], and Tat-Seng Chua[1]

[1] School of Computing, National University of Singapore, Singapore
[2] School of Computer Science and Technology, Harbin Institute of Technology, Harbin, China
{dcsgaoy,dcsyangy,dcscts}@nus.edu.sg, zsc@hit.edu.cn

Abstract. Event detection in social media platforms has become an important task. It facilities exploration and browsing of events with early plans for preventive measures. The main challenges in event detection lie in the characteristics of social media data, which are short/conversational, heterogeneous and live. Most of existing methods rely only on the textual information while ignoring the visual content as well as the intrinsic correlation among the heterogeneous social media data. In this paper, we propose an event detection method, which generates an intermediate semantic entity, named *microblog clique* (MC), to explore the highly correlated information among the noisy and short microblogs. The heterogeneous social media data is formulated as a hypergraph and the highly correlated ones are grouped to generate the MCs. Based on these MCs, a bipartite graph is constructed and partitioned to detect social events. The proposed method has been evaluated on the Brand-Social-Net dataset. Experimental results and comparison with state-of-the-art methods demonstrate the effectiveness of the proposed approach. Further evaluation has shown that the use of the visual content can significantly improve the event detection performance.

Keywords: Event detection, Microblog clique, Live data, Multimedia.

1 Introduction

Social media platforms [11], such as Twitter[1], Facebook[2], and Sina Weibo[3], have become important real-time information resources and host a huge amounts of user contributed content (UGC). The rapid development of social media platforms has led to continuously increasing data, which plays an important role in information sharing and diffusion. An example is the Super Bowl 2013, which attracts up to 24 million tweets in total and the number of tweets about just the blackout is over 231k per minute. The users can employ these platforms to report real-life events which may spread quickly and widely across the entire social network. This live information in social media streams requires effective technique for organization and management.

Events in social media refer to observable occurrences of people, places, times and activities [8]. As introduced in [16], an event can be regarded as a single episode of a large story and event detection can benefit the social media content analysis and enable powerful event browsing.

---

[1] https://twitter.com/
[2] https://www.facebook.com/
[3] http://www.weibo.com

X. He et al. (Eds.): MMM 2015, Part I, LNCS 8935, pp. 269–281, 2015.
© Springer International Publishing Switzerland 2015

In recent years, event detection in social media platforms has attracted extensive research attention [14,15]. Most of existing event detection methods consider only the textual context and the social connection while ignoring the visual context which has been growing in importance in social media. It is noted that detecting events in social media is challenging due to the following three characteristics. First, social media posts tend to be short and conversational in nature. Thus, the contents and vocabularies used in these posts tend to change rapidly. Under this circumstance, a single post may not be adequate to reflect meaningful content, and exploration of highly correlated posts on the same topic becomes an urgent requirement. Second, the content of social media posts has become increasingly heterogeneous and multimedia. The social posts may contain not only the text and image, but also time-stamp, location, social connection, user preference, and other metadata. Our recent investigation shows that about 30% of microblog posts now contain images and this number is still increasing. Therefore, visual content becomes more important in these days. Third, social media content comes in the form of social media streams. The amount of social media data is not only enormous but also continuously growing in every minute. These live data make it a hard task to detect new events and to handle increasingly large scale data.

In this paper, we aim to detect events from social media posts. To address the problems of short and conversational posts, we propose to generate *microblog cliques* (MCs), which is a group of highly correlated microblogs. MCs help to enrich the single post content and tackle the data sparseness issue. To address the heterogeneity of microblog data, we propose to jointly employ the textual and visual content in microblog for analysis, which can substantially explore the intrinsic correlation among the heterogeneous data. Figure 1 presents the framework of the proposed event detection method.

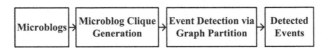

Fig. 1. The framework of the proposed event detection method in microblog

The proposed event detection method can be briefly described as follows. Given a set of microblogs, we first construct a heterogeneous microblog hypergraph, in which the distance between two microblogs is measured by multiple facets, including textual content, visual content, location, time-stamp and the user connection. The heterogeneous hypergraph is then partitioned into small sub-graphs, which are denoted as the MCs. Each MC comprises a group of highly-correlated microblogs, such as near-duplicates or reposted microblogs. We summarize each MC by selecting several representative microblogs. These MCs are then used to construct an MC graph, and the K-way segmentation on the MC graph is conducted using the transfer cut to generate the K events for the given microblogs. In our method, the Bayesian information criteria is employed to select the optimal event number.

The proposed event detection method has been evaluated on the Brand-Social-Net dataset [6], which includes 20 saga events of different types, i.e., the saga story. Experimental results and comparison with state-of-the-art methods demonstrate the effectiveness of the proposed method.

The remainder of this paper is organized as follows. Section 2 reviews related work on event detection in social media platforms. The detailed algorithms, including MC generation and event detection, are elaborated in Section 3 and Section 4, respectively. Section 5 presents the experimental results, followed by conclusions and discussions for further work in Section 6.

2 Related Work

In this section, we briefly review the related work in event detection in social media platforms. Given new incoming data, the similarity between the new data and the existing events are computed first and the event with the maximal similarity is selected. When all the similarities are below a predefined threshold, it will be considered as a new event. A modified TF/IDF and time-based threshold are employed in [2] to measure the relevance between events and documents, in which an auxiliary dataset is used to estimate the IDF due to the fact that the future documents are unknown. An incremental IDF is introduced in [20] which considers a time window and a decay factor to measure the similarity between documents and events. Fung et al. [5] explored the word appearance as the binomial distribution, and the word burst is identified by a heuristic with thresholds. The frequency domain of textual content has also been investigated. The Wavelet-based signal processing has been introduced in [18] to detect events, in which the cross correlation between the word appearance is measure by using the Wavelet-based feature.

Reuter and Cimiano [12] proposed an event classification method to deal with incremental data in social media streams. In their method, a candidate retrieval step is first performed to gather related events by using the capture time, upload time, geographic location, tags, titles and the description. Then the similarities between the document and each event for the top returned retrieved events are measured based on nine features, including the temporal information, geographical information and textual information. The probabilities of the documents belonging to the event or belonging to a new event can then be computed by a trained Support Vector Machine. A threshold is empirically selected using a gradient descent method on a split of training data. Becker et al. [4] introduce the learning similarity metrics to identify events in social media streams, in which the event identification task is formulated as a clustering problem. In this method, each event is denoted by a document cluster, and the scalable clustering is evaluated using normalized mutual information and B-cubed [3]. Considering the different information in social media documents, such as the textual feature and the location data, different similarities are combined in an ensemble-clustering procedure. To classify new data into existing events, a group of training samples are first selected from labeled data and the logistic regression and SVM are employed as the classifier, which shows the best performance in experimental results, i.e., CLASS-LR and CLASS-SVM.

It is noted that most of the existing methods are based on the textual content associated with the time-stamp. With the increasing amount of multimedia content in social media streams, such as images and videos, it is important to further explore the roles of visual context in microblogs for event analysis.

3   Microblog Clique Generation

Most of the social media posts are short and conversational. Therefore, it is difficult to explore useful information from the limited and noisy content of one single microblog. On the other hand, most of the microblogs are not alone due to the conversational nature of social media. For instance, the highly correlated reposts and/or comments can be exploited as a valuable resource for enriching the original microblog post. Under this circumstance, we propose to generate a middle level object, termed *microblog clique* (MC), to represent the grouped microblogs. Here an MC is a set of highly correlated microblogs, which are all related to the same topic in a short time window. Each MC is a combination of several relevant microblogs, which is more informative. In this way, MCs can be used as basis to explore a set of microblogs as a basic unit instead of a single microblog. Figure 2 illustrates the workflow of the proposed MC generation method. To formulate the relationship among microblogs, a heterogeneous microblog hypergraph is constructed and the hierarchical bi-partition on the hypergraph is conducted under the constraint of Bayesian information criteria.

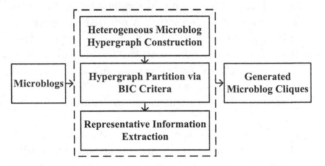

Fig. 2. The framework for *microblog clique* generation

3.1   Microblog Hypergraph Construction

Considering the multi-modal data in microblogs, such as the textual information, visual information, social connection, and location, a hypergraph is a good structure to formulate the microblog relationship. Hypergraph [22] is able to handle heterogeneous data and has been extensively employed in many data mining and information retrieval tasks [9,7] due to its superiority in high-order relationship modeling. Given a group of microblogs $M = \{m_1, m_2, \ldots, m_n\}$, a microblog hypergraph $\mathcal{G}_H = \{\mathcal{V}, \mathcal{E}, \mathbf{W}\}$ is constructed. In $\mathcal{G}_H$, each vertex denotes one microblog and there are n vertices in total. To generate the edges $\mathcal{E}$ linking different vertices, the heterogeneous data for microblogs are employed to measure the distance between each two microblogs.

- **Textual information:** The textual information of each microblog is described by TF-IDF, and the cosine similarity is employed to measure the pairwise microblog textual distance.
- **Visual content:** Given two microblogs with images, the visual content distance can be measured. Here a spatial pyramid image feature [19] is extracted for each image, which is highly discriminative on spatial layout and local information. The dense

sift feature is extracted for each image and a visual dictionary size of 1,024 is learnt. The spatial pyramid structure includes three levels, i.e., 1×1, 2×2 and 4×4, and a 21,504-D feature is generated for each image. A 200-D feature is further extracted by using PCA as the visual feature for each image.

- **Location:** The geographical similarity between two microblogs (if available) is measured by using the Harversine-formula [13].

- **Social connection:** When the two microblogs share the same owner or the two corresponding owners are connected in the social media platform, such as the follower/followee relation, these microblogs are close in social space. We measure the social similarity between two microblogs by:

$$s_s\left(m_i, m_j\right) = \begin{cases} 1, & u_i = u_j; \\ 0.5, & u_i \Leftrightarrow u_j; \\ 0, & otherwise. \end{cases} \tag{1}$$

where $u_i \Leftrightarrow u_j$ indicates u_i and u_j are connected in the social media platform.

- **Temporal information:** When two microblogs are posted within a short time gap, there is a high probability that the two microblogs are related. Here the temporal similarity between two microblogs m_i and m_j is measured by:

$$s_t\left(m_i, m_j\right) = 1 - \frac{\left(|t_i - t_j|\right)}{\tau}, \tag{2}$$

where t_i and t_j are the time-stamps of m_i and m_j respectively, and τ is a normalized factor.

Each microblog m_i is regarded as the centroid, and the top N nearest microblogs are selected to generate an edge based on the textual information and the visual content, respectively. For the location information and the temporal information, each microblog m_i is connected with its neighbors with a geo-distance threshold and a time-distance threshold, respectively. For the social connection, each user generates one edge, which connects all the microblogs of this user and the related users. Figure 3 illustrates the hypergraph construction procedure.

The incidence matrix $\mathbf{H}$ of the microblog hypergraph $\mathcal{G}_H$ is generated by:

$$\mathbf{H}\left(v, e\right) = \begin{cases} 1 \text{ if } v \in e \\ 0 \text{ if } v \notin e \end{cases}. \tag{3}$$

The vertex degree of a vertex $v \in \mathcal{V}$ is defined by:

$$d\left(v\right) = \sum_{e \in \mathcal{E}} w\left(e\right) \mathbf{H}\left(v, e\right), \tag{4}$$

and the edge degree of an edge $e \in \mathcal{E}$ is defined by:

$$\delta\left(e\right) = \sum_{v \in \mathcal{V}} \mathbf{H}\left(v, e\right). \tag{5}$$

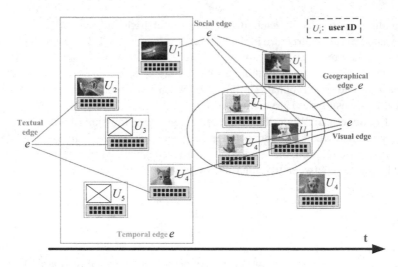

Fig. 3. Illustration for microblog hypergraph construction

3.2 Microblog Hypergraph Segmentation

Given the microblog hypergraph, we aim to generate the MCs, which are groups of microblogs with the same topics. We partition the hypergraph using the hypergraph cut approach, which has been widely investigated in recent years.

Let S and $\bar{S}$ denote the two-way partition of $\mathcal{G}_H$. The hypergraph cut can be defined as

$$Cut_H\left(S, \bar{S}\right) := \sum_{e \in \partial S} w\left(e\right) \frac{|e \cap S| \, |e \cap \bar{S}|}{d\left(e\right)}, \tag{6}$$

where ∂S is the hyperedge boundary and it is defied by:

$$\partial S := \left\{ e \in E | e \cap S \neq \emptyset, e \cap \bar{S} \neq \emptyset \right\}.$$

The two-way normalized hypergraph partition can be defined as:

$$NCut_H\left(S, \bar{S}\right) := Cut_H\left(S, \bar{S}\right) \left(\frac{1}{vol\left(S\right)} + \frac{1}{vol\left(\bar{S}\right)} \right), \tag{7}$$

where $vol\left(S\right) = \sum_{v \in S} d\left(v\right)$ and $vol\left(\bar{S}\right) = \sum_{v \in \bar{S}} d\left(v\right)$ are the volume of S and $\bar{S}$ respectively.

Following [22], the normalized cut can be relaxed as a real valued optimization task, which can be solved by using the eigenvector for the smallest non-zeros eignevector of the hypergraph Laplacian, i.e., $\Delta = \mathbf{I} - \mathbf{D}_v^{-\frac{1}{2}} \mathbf{H} \mathbf{W} \mathbf{D}_e^{-1} \mathbf{H}^T \mathbf{D}_v^{-\frac{1}{2}}$.

In this way, the input microblogs $M = \{m_1, m_2, \ldots, m_n\}$ are partitioned into two parts, and the two-way partition will be conducted in each new partition. This procedure continues until the optimal results are achieved. Here we employ the Bayesian information criteria (BIC) [17] to evaluate the partition results, which is used to determine whether to accept the two-way partitions or not.

3.3 Bayesian Information Criteria

To identify an optimal partition, we should measure the representability of different partition results. Here the Bayesian information criteria (BIC) [17] is employed to evaluate the representation of a model, i.e., the selected representative microblogs from each partition.

Given a group of partitions $P=\{P_1, P_2, \ldots, P_m\}$ for the data $M=\{m_1, m_2, \ldots, m_n\}$, the BIC value is calculated by:

$$BIC = \delta(M) - \frac{N_p}{2} \log n, \tag{8}$$

where N_p is the number of parameters, which can be regarded as the feature dimension for microblog description, $\delta(M)$ is the log-likelihood of the microblog for the partition P with the maximum likelihood, and n is the number of microblogs for processing.

In our experiments, the maximum likehood estimate for the variance can be calculate by:

$$\widehat{\theta}^2 = \frac{1}{n-m} \sum_i d(m_i, c_{m_i})^2, \tag{9}$$

where $d(m_i, c_{m_i})$ is the distance between m_i and the corresponding representative microblog c_{m_i}.

The log-likelihood of the microblog data can be measured by:

$$\delta(M) = \sum_i \left(\frac{1}{\sqrt{2\pi\widehat{\theta}}^{N_p}} - \frac{1}{2\widehat{\theta}^2} \left\| d(m_i, c_{m_i})^2 \right\| + \log \frac{n_i}{n} \right), \tag{10}$$

where n_i is the number of microblogs in the corresponding partition of m_i.

Here we take the bi-partition as an example. The to-be-measured partitions are $\{P_0\}$ and $\{P_1, P_2\}$, where $P_0 = P_1 + P_2$ indicates that P_1 and P_2 are the partition results of P_0. The BIC values for these two partition results are calculated, and the partition result with higher BIC value is employed as the final result.

3.4 The MC Representation

With the microblog hypergraph partition results, all the input microblogs can be divided into a group of clusters. These sets of microblogs are regarded as the MCs, which are used in the next event detection procedure. For each MC, we use the combination of the textual content and visual content of all its microblogs to represent the MC. After the duplicate textual and visual content removing, the enriched textual content and combined images are employed for MC description. In comparison with the use of a single microblog, MC provides enriched information from a set of highly relevant microblogs, which can generate more meaningful content for microblog analysis.

4 Event Detection with MC

The generated MCs can be regarded as the intermediate level semantic entities for the event detection task. The task is formulated to explore the relations and infer events between different MCs, $MC = \{MC_1, MC_2, \ldots, MC_p\}$ and the corresponding microblogs $M = \{m_1, m_2, \ldots, m_n\}$. The objective here is to further partition the MCs and the microblogs into event clusters. We formulate MCs and the corresponding microblogs in a bipartite graph $\mathcal{G}_B = \{X, Y, B\}$, where the vertex set $X = MC \cap M$, the vertex set $Y = MC$, and B is the across-affinity matrix to link X and Y. B is defined as follows:

$$B_{ij} = \begin{cases} \eta & if\ x_i \in M\ and\ x_i \in y_i \\ e^{-\gamma d_{ij}} & if\ x_i \in MC\ and\ y_j \in MC \\ 0 & otherwise \end{cases} , \qquad (11)$$

where η and γ are two parameters to balance the inner-MC correlation and the between-MC smoothness; d_{ij} is the distance between two MCs, which can be calculated by using the combination of textual distance and visual distance, if the images are available. To partition MCs, the transfer cut method [10] is again employed here, which can be summarized as follows.

Given the bipartite graph $\mathcal{G}_B$ and the number of required partition numbers K, we first generate $\mathbf{D}_X = diag\,(\mathbf{B1})$ and $\mathbf{D}_Y = diag\,(\mathbf{B}^T\mathbf{1})$. As $|X|$ is much larger than $|Y|$, we first focus on the smaller bipartite graph $\mathcal{G}_{BY} = \{Y, W_Y\}$, which only contains the MC vertices and $\mathbf{W}_Y = \mathbf{B}^T\mathbf{D}_X^{-1}\mathbf{B}$. The graph Laplacian of $\mathcal{G}_{BY}$ can be calculated by $\mathbf{L}_Y = \mathbf{D}_Y - \mathbf{W}_Y$. The K bottom eigenpairs $\{\lambda_i, \mathbf{v}_i\}_1^K$ of $\mathcal{G}_{BY}$ can be obtained. As proved in [10], the bottom K eigenpairs $\{\xi_i, \mathbf{f}_i\}_1^K$ can be calculated by:

$$\begin{aligned} 0 &\leq \xi_i \leq 1 \\ \xi_i\,(2 - \xi_i) &= \lambda_i \\ \mathbf{u}_i &= \tfrac{1}{1-\xi_i}\mathbf{D}_X^{-1}\mathbf{B}\mathbf{v}_i \\ \mathbf{f}_i &= \left(\mathbf{u}_i^T, \mathbf{v}_i^T\right)^T \end{aligned} \qquad (12)$$

Then $\{\mathbf{f}_1, \mathbf{f}_2, \ldots, \mathbf{f}_K\}$ can be used for spectral clustering [21] on the bipartite graph $\mathcal{G}_B$, and K microblog clusters can be obtained. Due to the noise in microblogs, some small clusters are formed but they are disregarded as noise and only those clusters with more than 2% of microblogs are selected as the detected events. It is nontrivial to select an optimal K value. Here we further employ BIC to evaluate the selection of K. We assume that the number of existing events is K_0 which is initialized as 0 at the beginning, and the largest number of events with new incoming data is no more than $K_0 + n_{new}/t_m$, where t_m is a threshold to determine the minimal microblog requirement for an event which is set as 50 in our experiment. The bipartite graph will be partitioned $n_{new}/t_m + 1$ times, and the partition result with the highest BIC value is selected as the event detection output. Here we assume that $\{\Gamma_1, \Gamma_2, \ldots, \Gamma_K\}$ are the K detected events in the last procedure. The description for each Γ_i is based on the MC, where a MC selection is conducted to find key MCs for the event. Here the weight for each MC is measured by the importance, such as the number of microblogs, reposts, and comments. Then top n_s MCs are selected, which is set as 3 in our experiments.

5 Experiments

5.1 Experimental Settings

The Testing Dataset. In the experiments, we employ the Brand-Social-Net dataset [6]. This dataset consists of 3 million microblogs with 1.3 million images from Sina Weibo on June and July, 2012. Each microblog contains the text description, the image if available, the owner information, posting time, geo location and user connections on Sina Weibo. The number of users is 1 million. This dataset contains 20 saga events, such as "Windows 8 Preview", "Chongqing Auto Expo", and "Honda Elysion". These saga events happened during June and July, 2012, and the number of relevant microblogs for each saga event ranges from hundreds to thousands. Given the microblogs of each saga event, event detection is conducted to explore the sub-events in these saga events.

Compared Methods. To evaluate the proposed event detection method, the following methods are employed for comparison.

- The Candidate-Ranking method [12] (CR). The Candidate-Ranking method first retrieves several promising events and the probability of the incoming document for these events or a new event can be measured by an SVM classifier.
- The Candidate-Ranking method [12] with visual content (CR+V). We further implement the Candidate-Ranking method by incorporating visual content analysis.
- The CLASS-SVM method [4] (CS). CS is an incremental clustering method which employs SVM as the classifier to identify whether a new document belongs to an existing event or a new event.
- The CLASS-SVM method [4] with visual content (CS+V). We further implement the CLASS-SVM method with visual content analysis.
- The proposed method, denoted by Proposed.
- The proposed method without visual content (Proposed-V). In this method, the visual content of microblogs are not taken into consideration.
- The proposed method without MC, i.e., Proposed-MC. In this method, the MC generation process is removed.

Evaluation Criteria. Event detection is conducted on all related data. To evaluate the event detection performance, two types of ground truth are manually annotated, i.e., the summarized ground truth, consisting of tens of microblogs and reflecting most of the main content in the data, and the top-ranking content, consisting of 10 microblogs which are the most important content in the data. Three students were employed to manually select the summarized ground truth and the top-ranking content from all related microblogs.

We adopt the following performance evaluation measures. **Recall** is to measure the data coverage of the generated events, **Precision** aims to evaluate the event detection accuracy, and **F-Measure** is a joint measure of Precision and Recall.

Average normalized modified retrieval rank (ANMRR) [1] is a rank-based measure, which considers the ranking information of microblogs. A lower ANMRR value indicates better performance, i.e., relevant microblogs rank at top positions. In our experiments, the selected top-ranking microblogs are regarded as the positive samples, and ANMRR is to evaluate the ranking results. A lower ANMRR value indicates that the important microblogs are listed at the top positions.

5.2 Comparison with the State-of-the-Art

We first compare the proposed method with the state-of-the-art methods, i.e., CR [12] and CS [4]. The average performance of different methods on event detection is presented in Figure 4.

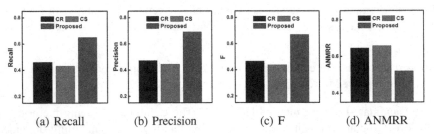

(a) Recall   (b) Precision   (c) F   (d) ANMRR

Fig. 4. The performance comparison on static event detection among different methods

From Figure 4, we observe that the proposed method can achieve better results in comparison with CR and CS in the event detection task. The proposed method achieves an improvement of 40.9%, 46.1%, 43.4%, and 19.4% in terms of Recall, Precision, F, and ANMRR, respectively as compared to CR, and an improvement of 50.7%, 55.5%, 53.0%, and 20.9% as compared to CS. These results demonstrate the effectiveness of the proposed method on event detection.

The better results are benefited from the proposed intermediate semantic level, i.e., the MCs, which can jointly explore the highly related microblogs to address the inadequate information issue. In our method, the visual content has been investigated in both the MC generation and the event detection procedures, which also contributes a lot to the event detection performance. In next two subsections, we will elaborate the effects of visual content and MC on event detection.

5.3 On Visual Content

We evaluate the influence of visual content in event detection, in which we compare the performance of CR, CS and the Proposed method with/without visual content. Figure 5 presents the experimental results, which show that the use of visual content can significantly improve the event detection performance.

In comparison with the textual content, the visual content has shown its superiority on information spreading in social media platforms. The visual content in microblogs has been forward to be able to enrich the short and conversational textual data.

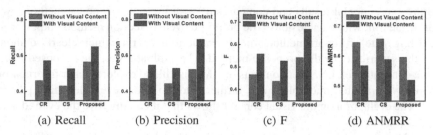

(a) Recall   (b) Precision   (c) F   (d) ANMRR

Fig. 5. The performance comparison of with/without visual content on static event detection

5.4 On MC Performance

As an intermediate concept level representation, MC aims to enrich the microblog information from a small group of highly correlated microblogs. Here we compare the performance of the proposed method with Proposed-MC, which removes the MC generation step for event detection. Experimental results are shown in Figure 6, which indicates that the event detection performance degrades without MC. The use of MC achieves an improvement of 18.2%, 23.2%, 20.6%, and 14.8% in terms of Recall, Precision, F, and ANMRR, respectively. These results indicate that the proposed MC is effective and essential for the event detection task.

The advantage of MC comes from its intermediate concept level representation, which is beyond that can be conveyed in a single microblog. An MC is composed of a group of highly correlated microblogs, which may share similar textual content, visual content, closed geographical information and connected owners, such as the reposted microblogs and the corresponding comments. These microblogs can reinforce each other with the continuous data such as the new content from the reposts and/or the comments, which can address the information sparseness issue in microblogs.

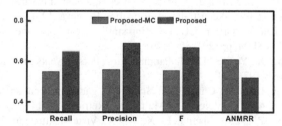

Fig. 6. The performance comparison of with/without MC on static event detection

6 Conclusion

In this paper, we proposed an event detection method in microblog. An intermediate concept level, i.e., *microblog clique*, is introduced to explore the highly correlated microblogs to enrich event representation. To tackle the heterogeneous data in microblogs, the microblogs are formulated in a hypergraph structure and hierarchical bi-partition is conducted to generate MCs. A bipartite graph is then constructed using the MCs and the corresponding microblogs, and the bipartite graph partition is performed to detect events. The proposed method has been evaluated on the Brand-Social-Net dataset.

From the experimental results and comparisons with the state-of-the-art methods, we can draw the following conclusions. The proposed event detection method outperforms the existing state-of-the-art methods on all evaluation criteria, which clearly demonstrates the superiority of the proposed method. The evaluation on the proposed intermediate concept level, i.e., MC, confirms that MC is able to explore richer information from highly correlated microblogs and further leads to better event detection performance. The evaluation on the visual content shows that it can improve the event detection performance too.

To address the event detection task in social media platforms, there are still several difficult tasks. First, most existing methods directly combine multi-modal data in social media posts, such as the textual and visual content. It is noted that these heterogeneous data may have high correlation, and how to jointly investigate the multi-modal data in microblogs requires further attention. Second, the social network can infer important latent information for social events behind the microblogs and the users, which is another future research topic.

Acknowledgements. This research is supported by the Singapore National Research Foundation under its International Research Centre @ Singapore Funding Initiative and administered by the IDM Programme Office.

References

1. Description of core experiments for mpeg-7 color/texture descriptors. In: Standard ISO/MPEGJTC1/SC29/WG11 MPEG98/M2819 (1999)
2. Allan, J., Papka, R., Lavrenko, V.: On-line new event detection and tracking. In: ACM SIGIR (1998)
3. Amigó, E., Gonzalo, J., Artiles, J., Verdejo, F.: A comparison of extrinsic clustering evaluation metrics based on formal constraints. Information Retrieval 12(4)
4. Becker, H., Naaman, M., Gravano, L.: Learning similarity metrics for event identification in social media. In: WSDM, pp. 291–300 (2010)
5. Fung, G.P.C., Yu, J.X., Yu, P.S., Lu, H.: Parameter free bursty events detection in text streams. In: VLDB, pp. 181–192 (2005)
6. Gao, Y., Wang, F., Luan, H., Chua, T.-S.: Brand data gathering from social media streams. In: Proceedings of ACM Conference on Multimedia Retrieval (2014)
7. Gao, Y., Wang, M., Zha, Z.-J., Shen, J., Li, X., Wu, X.: Visual-textual joint relevance learning for tag-based social image search. IEEE Transactions on Image Processing 22(1), 363–376 (2013)
8. Hearst, M.: Search user interfaces. Cambridge University Press (2009)
9. Huang, Y., Liu, Q., Zhang, S., Metaxas, D.: Image retrieval via probabilistic hypergraph ranking. In: CVPR (2010)
10. Li, Z., Wu, X.M., Chang, S.F.: Segmentation using superpixels: A bipartite graph partitioning approach. In: CVPR, pp. 789–795 (2012)
11. Naveed, N., Gottron, T., Kunegis, J., Alhadi, A.C.: Searching microblogs: coping with sparsity and document quality. In: Proceedings of CIKM, pp. 183–188 (2011)
12. Reuter, T., Cimiano, P.: Event-based classification of social media streams. In: Proceedings of the 2nd ACM International Conference on Multimedia Retrieval (2012)

13. Reuter, T., Cimiano, P., Drumond, L., Buza, K., Schmidt-Thieme, L.: Scalable event-based clustering of social media via record linkage techniques. In: ICWSM (2011)
14. Ritter, A., Etzioni, O., Clark, S., et al.: Open domain event extraction from twitter. In: KDD, pp. 1104–1112. ACM (2012)
15. Rozenshtein, P., Anagnostopoulos, A., Gionis, A., Tatti, N.: Event detection in activity networks. In: KDD, pp. 1176–1185. ACM (2014)
16. Sayyadi, H., Hurst, M., Maykov, A.: Event detection and tracking in social streams. In: WSDM (2009)
17. Schwarz, G.: Estimating the dimension of a model. Ann. Statist. 6, 461–464 (1978)
18. Weng, J.S., Lee, B.S.: Event detection in twitter. In: ICWSM (2011)
19. Yang, J., Yu, K., Gong, Y., Huang, T.: Linear spatial pyramid matching using sparse coding for image classification. In: CVPR, pp. 1794–1801 (2009)
20. Yang, Y., Pierce, T., Carbonell, J.G.: A study on retrospective and on-line event detection. In: ACM SIGIR (1998)
21. Yang, Y., Yang, Y., Shen, H.T., Zhang, Y., Du, X., Zhou, X.: Discriminative nonnegative spectral clustering with out-of-sample extension. IEEE Transactions on Knowledge and Data Engineering 25(8), 1760–1771 (2013)
22. Zhou, D., Huang, J., Schokopf, B.: Learning with hypergraphs: Clustering, classification, and embedding. In: NIPS (2007)

A Study on the Use of a Binary Local Descriptor and Color Extensions of Local Descriptors for Video Concept Detection

Foteini Markatopoulou[1,2], Nikiforos Pittaras[1], Olga Papadopoulou[1], Vasileios Mezaris[1], and Ioannis Patras[2]

[1] Information Technologies Institute (ITI), CERTH, Thermi 57001, Greece
{markatopoulou,npittaras,olgapapa,bmezaris}@iti.gr
[2] Queen Mary University of London, Mile end Campus, UK, E14NS
i.patras@qmul.ac.uk

Abstract. In this work we deal with the problem of how different local descriptors can be extended, used and combined for improving the effectiveness of video concept detection. The main contributions of this work are: 1) We examine how effectively a binary local descriptor, namely ORB, which was originally proposed for similarity matching between local image patches, can be used in the task of video concept detection. 2) Based on a previously proposed paradigm for introducing color extensions of SIFT, we define in the same way color extensions for two other non-binary or binary local descriptors (SURF, ORB), and we experimentally show that this is a generally applicable paradigm. 3) In order to enable the efficient use and combination of these color extensions within a state-of-the-art concept detection methodology (VLAD), we study and compare two possible approaches for reducing the color descriptor's dimensionality using PCA. We evaluate the proposed techniques on the dataset of the 2013 Semantic Indexing Task of TRECVID.

Keywords: Video feature extraction, concept detection, concept-based video retrieval, binary descriptors.

1 Introduction

Concept-based video annotation and indexing is a very important task for the multimedia analysis field and a significant part of applications such as video retrieval, video event detection and video hyperlinking [27], [17]. A typical video concept detection system consists of three main modules: the video decomposition module, where video sequences are segmented into shots and each shot is represented by e.g. one or more characteristic keyframes/images; the feature extraction module, where features (e.g. local image descriptors, motion descriptors) are extracted from the visual information and encoded into a descriptor vector; and, finally the learning module, which employs machine learning algorithms in order to solve the problem of associating image descriptor vectors and concept labels.

X. He et al. (Eds.): MMM 2015, Part I, LNCS 8935, pp. 282–293, 2015.

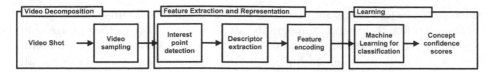

Fig. 1. Block diagram of a typical concept detection system

In this work we focus on the feature extraction process. Scale Invariant Feature Transform (SIFT) [16] and Speeded Up Robust Features (SURF) [2] are probably the two local descriptors that are most-widely used for this task. However, they are non-binary descriptors, which makes them not so suitable for applications requiring the transmission of descriptor vectors. For example, when considering a mobile application where pictures are taken with a mobile device and local descriptors from these pictures need to be sent to a server for further processing, then it is very important that the local descriptors are as compact as possible, to minimize transmission requirements [8]. ORB (Oriented FAST and Rotated BRIEF) [22] is a binary local descriptor, which was originally proposed for similarity matching between local image patches. We examine ORB in the task of video concept detection, and we show that it constitutes a viable alternative to the non-binary descriptors currently used in this task, while its compact size and low storage needs make this descriptor appealing for mobile applications. Subsequently, inspired by two color extensions of SIFT [24], namely RGB-SIFT and OpponentSIFT, we define the corresponding color extensions for the two other local descriptors considered in this work (SURF, ORB), and we show that this relatively straightforward way of introducing color information is in fact a generic methodology that works similarly well for different local descriptors. In addition, we present a different way of performing Principal Component Analysis (PCA) [28] for feature reduction, which improves the results of SIFT/SURF/ORB color extensions when combined with VLAD encoding. Our experiments were performed on the TRECVID 2013 Semantic Indexing (SIN) dataset [19], which consists of a development set and a test set (approximately 800 and 200 hours of internet archive videos for training and testing, respectively).

The rest of this paper is organized as follows: Section 2 reviews related work, focusing on local image descriptors. Section 3 discusses how the binary ORB descriptor can be used for video concept detection. Section 4 introduces the color extensions of SURF and ORB, while Section 5 discusses two possible approaches of employing PCA for color descriptors. Section 6 presents our experiments and results, and finally Section 7 summarizes our main conclusions.

2   Related Work

Figure 1 summarizes a typical concept detection system. The video decomposition module uses shot segmentation algorithms in order to divide the initial video sequence into shots, and then possibly also single-out a subset of the visual information (e.g. keyframes, tomographs [26]) to be used for further processing. Then, the feature extraction module deals with the extraction of meaningful feature vectors to represent each piece of visual information. A variety of visual, textual and audio features can be extracted to this end; a review of different types of features can be found in [27]. In large-scale video concept detection, typically local image features are utilized, being extracted from representative keyframes or similar 2D image structures [26]. Two of the most popular local descriptors are SIFT [16] and SURF [2]. Both of them extract features that are invariant to rotation, scale and illumination variations, while SURF extraction is somewhat less computationally-demanding (SURF is two times faster than SIFT according to [2]). SIFT and SURF construct vectors of floating-point values (which are often quantized to integers in the range [0,255]). For many modern applications, though, e.g. concept detection on mobile devices, small-sized yet discriminative descriptors are very important in order to extract, store and transmit them efficiently (e.g. send local descriptors to a server for performing concept detection). Binary local descriptors are an attractive alternative to non-binary descriptors such as SIFT and SURF, generating binary strings which can be computed efficiently while also requiring lower storage space. BRIEF [5], ORB [22], BRISK [15], and FREAK [1] are some examples of binary local descriptors that have been proposed for similarity matching between local image patches. They are all based on calculating the differences between pairs of pixel intensity values within an image patch; what distinguishes them is the pattern they follow in order to perform these pair-wise pixel comparisons. Studies show that ORB is among the most accurate binary descriptors for image matching [6]. The possibility of using ORB in image classification was also briefly examined in [12].

The above mentioned non-binary and binary local descriptors are intensity-based: they are applied to grayscale images (e.g. an RGB image is firstly converted to grayscale), and the extracted features are calculated from the pixel intensity values. Two color variants of SIFT, namely RGB-SIFT and OpponentSIFT, that increase the illumination invariance, the discriminative power and also make the descriptor invariant to light color changes were proposed in [24]. Methods that consider the color information in order to improve the SURF descriptor have also been proposed, but were examined only on the image matching problem [11], [10], [9]. For example, [10] calculates a color local kernel histogram in the neighborhood of each keypoint and concatenates it with the original SURF descriptor that has been extracted from the pixel intensity values of the same neighborhood. In [12], the extraction of ORB from all three color channels of the RGB color space was considered.

For the purpose of visual concept detection, local descriptors extracted from different patches of one image are subsequently aggregated into a global image representation, a process known as feature encoding. The most popular encoding in the last years has been the Bag-of-Words (BoW) [21]. Fisher vector (FV) [20], Super Vector (SV) [30] and VLAD (Vector of Locally Aggregated Descriptors) [13] are three state-of-the-art encodings that significantly outperform the

BoW [25] [7]. FV encoding describes the difference between the distribution of features for an image and the distribution fitted to the features of all the training data. VLAD [13] is a fast approximation of FV that performs somewhat worse but is more compact and faster to compute [14], which makes it a good compromise. SV [30] works in the same lines, however requires larger codebooks than VLAD and FV in order to exhibit similar levels of accuracy, which increases the memory and computation requirements. The three latter encodings are high-dimensional and their dimensionality is affected by the dimensionality of the local descriptors they encode, thus dimensionality reduction approaches such as PCA are widely used for making the image representation more compact prior to learning/classification. Dimensionality reduction can be performed at two stages: local descriptors can be reduced prior to the encoding, and then the final encoding can also be further compacted [14].

Finally, for learning the associations between the image representations and concept labels, algorithms such as Logistic Regression (LR) and Support Vector Machines (SVM) are typically trained separately for each concept, on ground-truth annotated corpora. Then, when a new unlabeled video shot arrives, the trained concept detectors will return confidence scores that show the belief of each detector that the corresponding concept appears in the shot. This baseline learning process can be further improved in different ways, e.g. by taking into account concept correlations instead of training each detector independently [18].

3   Using a Binary Local Descriptor for Concept Detection

ORB [22] is a binary local image detector and descriptor that presents similar discriminative power with SIFT and SURF in image matching problems, it has similar properties such as invariance in rotation, scale and illumination, but at the same time is more compact and faster to be computed. A 256-element binary ORB vector requires 256 bits to be stored; in contrast, an integer-quantized 128-element SIFT vector requires 1024 bits. In addition, according to [22], ORB is an order of magnitude faster than SURF to compute, and more than two orders of magnitude faster than SIFT.

There is not a single way for introducing binary descriptors in the visual concept detection pipeline. [12] did so by considering the BoW encoding, and proposed a modified K-means algorithm (the "K-majority" algorithm) for generating the codebook (vocabulary) of BoW, that would result in a binary codebook. To illustrate the modifications, Algorithm 1 presents the steps of the original K-means clustering algorithm. In order to create a binary codebook, [12] used the Hamming distance in Step 2 of the Construction stage and also in the Assignment stage, while in Step 3 they used their "K-majority" voting method in order to calculate a binary cluster center.

In this work we claim that a binary descriptor (ORB) can be used for the video concept detection in the same way as its non-binary counterparts. Specifically, let us assume that I is a set of images and x_i $i = 1, ..., N$ are ORB descriptors extracted from I, where $x_i \in \{0, 1\}^d$. N is the total number of extracted local

Algorithm 1. Steps of K-means algorithm

Codebook construction:
1. Randomly initialize a set of K cluster centers w_k
2. For each descriptor vector x_i, compute index k_i of the cluster centre nearest to x_i
3. Update the cluster centers w_k
4. Repeat steps 2 and 3 until convergence
Word assignment:
Given a new local descriptor vector x', assign it to the nearest cluster w_k

---

descriptors and d is the dimension of the ORB descriptor. From these binary descriptors, we generate a floating-point codebook of K visual codewords $w_k \in \mathbb{R}^d$, $k = 1, ..., K$, using a standard K-means. The distances between the binary ORB descriptors and the codewords (Construction: Step 2 and Assignment stage of Algorithm 1) are calculated by the L2 norm. In Step 3 of Algorithm 1, averaging is also performed as in the original K-means (calculating the mean of a set of vectors).

Assigning binary local descriptors to a binary codebook using the hamming distance, as in [12], is faster than assigning them to a floating-point codebook using the L2 distance. However, considering that for the concept detection problem the time needed for the assignment to codebook is negligible compared to other processes of the pipeline (e.g. feature encoding, classification), more important is what leads to a more discriminative codebook that improves the concept detection accuracy. We report results of comparing these two codebook creation strategies (that of [12] and the one described in this section) in Section 6.

4   Color Extensions of Binary and Non-binary Local Descriptors

Based on the good results of two color extensions of SIFT, namely RGB-SIFT and OpponentSIFT [24], we examine the impact of using the same methodology for introducing color information to other descriptors (SURF, ORB). Our objective is to examine if this is a methodology that can benefit different local descriptors and is therefore generally applicable.

Let d denote the dimension of the original local descriptor (typically, d will be equal to 64 or 128 for SURF and 128 or 256 for ORB). Figure 2 summarizes the process of extracting RGB-SURF, RGB-ORB, OpponentSURF and OpponentORB descriptors. An RGB image has three 8-bit channels (for red, green and blue). The original non-color local descriptors are calculated on 8-bit grayscale images, so they first transform the RGB image to grayscale. In contrast to this, our RGB-SURF/ORB (Fig. 2:(a)) apply the original SURF or ORB descriptors directly to each of the three R, G, B channels and for each keypoint extract three d-element feature vectors. These are finally concatenated into one $3 \cdot d$-element feature vector, which is the RGB-SURF or RGB-ORB descriptor vector.

Similarly, our OpponentSURF/ORB (Fig. 2:(b)) descriptors firstly transform the initial RGB image to the opponent color space [24]. We refer to the transformed channels as O_1, O_2 and O_3. O_3 is the luminance channel, i.e. the one

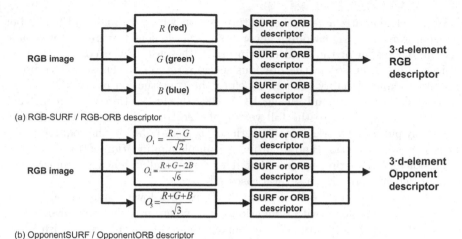

(a) RGB-SURF / RGB-ORB descriptor

(b) OpponentSURF / OpponentORB descriptor

Fig. 2. Block diagram of color SURF/ORB descriptor extraction, where d denotes the dimension of the original local descriptor

that the original SURF/ORB descriptors use. The other two channels (O_1 and O_2) capture the color information, where O_1 is the red-green component and O_2 is the blue-yellow component. Following the transformation, a normalization step that converts the ranges of each channel within the [0,255] range is employed, as in [24]. Similarly with RGB-SURF/ORB, the original SURF or ORB descriptors are then applied separately to each transformed channel and the final $3 \cdot d$-element feature vectors are the concatenation of the three feature vectors extracted from the three channels.

5   Reducing the Dimensionality of Local Color Descriptors

State-of-the-art encoding methods generate high-dimensional vectors that make difficult the training of machine learning algorithms. For example, while the

---

Algorithm 2. Algorithm for channel-PCA

Input: The number of color channels c ($c = 3$ in our color descriptors); the dimension d of each channel of the color descriptor (normally 128 or 256); the desired dimension l' of the reduced feature vector (the full feature vector will be reduced from $l = c \cdot d$ to l'); the complete feature matrix A that will be used for learning the projection matrices

Projection Matrix calculation: Calculate c projection matrices of size $d \times p_i$ according to:

for $i = 1$ to c **do**

  1. Perform eigenvalue decomposition of the covariance matrix corresponding to the features of the current channel (i.e., corresponding to a part of the features in A)

  2. Select the number of principal components p_i to retain for this channel ($\sum_{i=1}^{c} p_i = l'$)

  3. Form the channel's projection matrix using only the first p_i principal components

end for

Dimensionality reduction using channel-PCA: Given a new feature vector x', transform the features of each color channel using the corresponding projection matrix, and concatenate the transformed feature vectors for all channels

BoW model generates a k-element feature vector, where k equals to the number of visual words, VLAD encoding generates a $k \cdot l$-element feature vector (where l is the dimension of the local descriptor; in the case of the color extensions of descriptors discussed in the previous section, $l = 3 \cdot d$). Thus, it is common to employ dimensionality reduction before the construction of VLAD vectors, on local descriptors, mainly using PCA [28]. In this section we explain that directly applying PCA to the full vector of color descriptors, as implied from previously published works (e.g. [7]; termed "typical-PCA" in the sequel), is not the only possible solution, and we propose a simple modification of this descriptor dimensionality reduction process that it experimentally shown to improve the concept detection results in several cases.

PCA projects linearly l-dimensional features to a lower-dimensional feature space. Given a matrix A with dimension $l \times n$, where n is the number of observations, if we want to perform dimensionality reduction (from l to l') with PCA, the reduced matrix A' will be $A' = E^T \cdot A$, where E is the projection matrix (of dimension $l \times l'$) and T denotes the transpose of a matrix.

PCA aims to find those directions in the data space that present high variance. When PCA is applied directly to the entire vector of one of the color extensions of (binary or non-binary) local descriptors, if one or two of the three color channels of the descriptor exhibit lower diversity than the others, then these risk being under-represented in the reduced dimensionality space. To avoid this, we propose performing PCA separately for each color channel and consider an equal number of principal components from each of them, to create three projection matrices that correspond to each of the three channels, instead of one projection matrix that corresponds to the complete descriptor vector. The three reduced single-channel descriptor vectors that can be obtained for a color descriptor using the aforementioned projection matrices are finally concatenated in a reduced color-descriptor vector. Algorithm 2 summarizes the proposed channel-PCA algorithm.

6   Experiments

6.1   Experimental Setup

Our experiments were performed on the TRECVID 2013 Semantic Indexing (SIN) dataset [19], which consists of a development set and a test set (approximately 800 and 200 hours of internet archive videos for training and testing, respectively). We evaluate our system on the test set using the 38 concepts that were evaluated as part of the TRECVID 2013 SIN Task, and we follow the TRECVID methodology for the evaluation of the results [19].

For experimenting with all methods, one keyframe was initially extracted for each video shot and was scaled to 320×240 pixels prior to feature extraction. For some of our final experiments, we also extracted two visual tomographs [26] from each shot. Regarding feature extraction, we followed the experimental setup of [7] and we used the toolbox that its authors have published. More specifically, we used the dense SIFT descriptor, that accelerates the original SIFT descriptor, in combination with the Pyramid Histogram Of visual Words (PHOW) approach [4]. PHOW is a simple modification of dense SIFT that uses more than one square regions at different scale levels in order to extract features. For SURF and ORB we used their implementations included in OpenCV, and further extended these implementations with the corresponding color variants that we introduced in Section 4. The same square regions at different scale levels of the PHOW approach were used as the image patches that were described by ORB and SURF. We calculated 128-SIFT, 128-SURF and 256-ORB grayscale descriptors; then, each color extension of a descriptor resulted in a color descriptor vector three times larger than that of the corresponding original descriptor, as explained in Section 4. All the local descriptors were compacted (to 80 dimensions for SIFT, SURF and their color extensions, following the recommendations of [7] and [14]; to 80 dimensions for grayscale ORB and to 256 dimensions for ORB color extensions) using PCA and were subsequently aggregated using the VLAD encoding. Similarly with the authors of [7], we divided each image into the same 8 regions using spatial binning and we used sum pooling to combine the encodings from different regions. As a result of the above process, a VLAD vector of 163840 elements for SIFT, SURF or grayscale ORB and of 524288 elements for ORB color extensions was extracted for each image (by image we mean here either a keyframe or a visual tomograph). These VLAD vectors were compressed into 4000-element vectors by applying a modification of the random projection matrix [3]. These reduced VLAD vectors served as input to the Logistic Regression (LR) classifiers that we used. Following the *cross validated committees* methodology of [17], we trained five LR classifiers per concept and per local descriptor (SIFT, ORB, RGB-ORB etc.), and combined the output of these five by means of late fusion (averaging). When different descriptors were combined, again late fusion was performed by averaging of the classifier output scores. In all cases, the final step of concept detection was to refine the calculated detection scores by employing the re-ranking method proposed in [23].

6.2   Results and Discussion

Tables 1, 2 and 3 present the results of our experiments in terms of Mean Extended Inferred Average Precision (MXinfAP) [29], which is an approximation of the Mean Average Precision (MAP) suitable for the partial ground truth that accompanies the TRECVID dataset [19].

In Table 1 we examine the performance of the original grayscale ORB descriptor in concept detection, when used in conjunction with a binary codebook

Table 1. Performance (MXinfAP, %) for ORB, when the binary codebook proposed in [12] and when a floating-point codebook is used. In parenthesis we show the relative improvement w.r.t. the binary codebook.

Descriptor	Binary codebook [12]	Floating-point codebook (no PCA)	Floating-point codebook (PCA 80)
ORB	4.52	10.36 (+129.2%)	11.43 (+152.9%)

(as in [12]) and a floating-point one (as in Section 3). In both cases, VLAD encoding is employed. We can see that the binary codebook proves ineffective; the floating-point one outperforms it by more than 129%. We also compacted the ORB descriptors to 80 dimensions using PCA before encoding them, which further increased MXinfAP to 11.43%. Based on this result, in all subsequent experiments with ORB and its extensions a floating-point codebook was used. In addition, grayscale ORB was compacted to 80 dimensions.

In Table 2 we evaluate the different local descriptors and their color extensions considered in this work, as well as combinations of them. First, comparing the original ORB descriptor with the other two non-binary descriptors (SIFT, SURF), we can see that ORB performs rather similarly to its non-binary counterparts (more precisely, its MXinfAP is a bit worse). This performance is achieved despite ORB and its extensions being much more compact than SIFT and SURF, as seen in the second column of Table 2. Second, concerning the methodology for introducing color information to local descriptors, we can see that the combination of the original SIFT descriptor and the two known color SIFT variants that we examine ("SIFT combination" in Table 2) outperforms using the original SIFT descriptor

Table 2. Performance (MXinfAP, %) for the different descriptors, when typical and channel-PCA for dimensionality reduction is used. In parenthesis we show the relative improvement w.r.t. the corresponding original grayscale local descriptor for each of the SIFT, SURF and ORB color variants.

Descriptor	Descriptor size in bits	Keyframes, typical-PCA	Keyframes, channel-PCA	Boost(%) w.r.t typical-PCA
SIFT	1024	14.22	14.22	-
RGB-SIFT	3072	14.97 (+5.3%)	14.5 (+2.0%)	-3.1%
OpponentSIFT	3072	14.23 (+0.1%)	14.34 (+0.8%)	+0.8%
SIFT combination	-	**19.11 (+34.4%)**	**19.24 (+35.3%)**	+0.7%
SURF	1024	14.68	14.68	-
RGB-SURF	3072	15.71 (+7.0%)	15.99 (+8.9%)	+1.8%
OpponentSURF	3072	14.7 (+0.1%)	15.26 (+4.0%)	+3.8%
SURF combination	-	**19.4 (+32.2%)**	**19.48 (+32.7%)**	+0.4%
ORB	256	11.43	11.43	-
RGB-ORB	768	13.02 (+13.9%)	13.58 (+18.8%)	+4.3%
OpponentORB	768	12.61 (+10.3%)	12.73 (+11.4%)	+1.0%
ORB combination	-	**17.38 (+52.1%)**	**17.45 (+52.7%)**	+0.4%
SIFT/SURF combination	-	**22.4**	**22.35**	-0.2%
SIFT/ORB combination	-	**21.32**	**21.46**	+0.7%
SURF/ORB combination	-	**21.56**	**21.74**	+0.8%
SIFT/SURF/ORB combination	-	**23.00**	**23.01**	0.0%

Table 3. Performance (MXinfAP, %) for different combinations of descriptors, (a) when features are extracted only from keyframes, (b) when horizontal and vertical tomographs described by SIFT, RGB-SIFT and OpponentSIFT are also examined, (c) when the Label Powerset algorithm is also applied [18]

Descriptor	(a) Keyframes (channel-PCA)	(b) Keyframes+ Tomographs	(c) Keyframes+ Tomographs+LP
SIFT combination	19.24	20.28	21.35
SURF combination	19.48	19.74	20.92
ORB combination	17.45	17.83	19.92
SIFT/SURF/ORB combination	**23.01**	**24.49**	**25.58**

alone by 34.4% (35.3% for channel-PCA). The similar combinations of the SURF color variants with the original SURF descriptor, and of the color variants of ORB with the original ORB descriptor, are shown in Table 2 to outperform the original SURF and ORB by 32.2% and 52.1%, respectively (which increase to 32.7% and 52.7% for channel-PCA). These results show that the relatively straightforward way we used for introducing color information to SURF and ORB, based on the similar SIFT extensions, is in fact generally applicable to heterogeneous local descriptors.

To analyse the influence of PCA on the vectors of local color descriptors, we also compared in Table 2 the channel-PCA of section 5 with the typical approach of applying PCA directly on the entire color descriptor vector. In both cases PCA was applied before the VLAD encoding, and in applying channel-PCA we kept the same number of principal components from each color channel (e.g. for RGB-SIFT, which is reduced to $l' = 80$ using typical-PCA, we set $p_1 = p_2 = 27$ for the first two channels and $p_3 = 26$ for the third color channel; $p_1 + p_2 + p_3 = l'$). According to the relative improvement figures reported in the last column of Table 2, performing the proposed channel-PCA in several cases improves the concept detection results, compared to the typical-PCA alternative, without introducing any additional computational overhead.

Another observation from Table 2 is that the concept detection performance increases when pairs of local descriptors (including their color extensions) are combined (i.e., SIFT/SURF, SIFT/ORB and SURF/ORB combinations), which shows a complementarity in the information that the different local descriptors capture. The best overall results among the experiments of Table 2 are achieved when all the local descriptors and their color variants are combined (last row of this table), reaching a MXinfAP of 23.01%.

Finally, in Table 3 we report experiments with two literature techniques that can further benefit the combination of SIFT, SURF and ORB. Specifically, we experiment with video tomographs [26] (for simplicity these are described using only SIFT and its two color extensions) and a two-layer stacking architecture that captures concept correlations using the Label Powerset (LP) algorithm in the second layer [18]. In all experiments of this table, for the color variants of SIFT, SURF and ORB, channel-PCA was used. The results of Table 3 indicate that introducing some form of motion information (through tomographs) and considering the correlations among concepts (through LP) can give an additional 11.2% relative improvement to the best results reported in Table 2 (MXinfAP increased from 23.01% to 25.58%).

7   Conclusions

In this work we showed that a binary local descriptor (ORB) can perform sufficiently well, compared to its non-binary counterparts, in the video concept detection task. We also showed that a methodology previously used for defining two color variants of SIFT is a generic one that is also applicable to descriptors such as ORB and SURF. We proposed a different way of employing PCA for dimensionality reduction of color descriptors that are used in combination with VLAD (channel-PCA). Finally, we quantified the impact of combining the above techniques (e.g. combination of binary and non-binary color descriptors) and other previously proposed methods (tomographs, LP) to a concept detection system.

Acknowledgements. This work was supported by the European Commission under contracts FP7-287911 LinkedTV and FP7-600826 ForgetIT.

References

1. Alahi, A., Ortiz, R., Vandergheynst, P.: Freak: Fast retina keypoint. In: IEEE Int. Conf., CVPR 2012, pp. 510–517 (2012)
2. Bay, H., Ess, A., Tuytelaars, T., Gool, L.V.: Speeded-up robust features (surf). Computer Vision and Image Understing 110(3), 346–359 (2008)
3. Bingham, E., Mannila, H.: Random projection in dimensionality reduction: Applications to image and text data. In: 7th ACM SIGKDD Int. Conf. on Knowledge Discovery and Data Mining, pp. 245–250. ACM, NY (2001)
4. Bosch, A., Zisserman, A., Muoz, X.: Image classification using random forests and ferns. In: IEEE Int. Conf. ICCV 2007, Rio de Janeiro, pp. 1–8 (2007)
5. Calonder, M., Lepetit, V., Ozuysal, M., Trzcinski, T., Strecha, C., Fua, P.: BRIEF: Computing a Local Binary Descriptor Very Fast. IEEE Transactions on Pattern Analysis and Machine Intelligence 34(7), 1281–1298 (2012)
6. Canclini, A., Cesana, M., Redondi, A., Tagliasacchi, M., Ascenso, J., Cilla, R.: Evaluation of low-complexity visual feature detectors and descriptors. In: 18th Int. Conf. on Digital Signal Processing (DSP), pp. 1–7 (2013)
7. Chatfield, K., Lempitsky, V., Vedaldi, A., Zisserman, A.: The devil is in the details: an evaluation of recent feature encoding methods. In: British Machine Vision Conference, pp. 76.1–76.12. British Machine Vision Association (2011)
8. Chen, D.M., Makar, M., de Araújo, A.F., Girod, B.: Interframe coding of global image signatures for mobile augmented reality. In: DCC, pp. 33–42 (2014)
9. Chu, D.M., Smeulders, A.W.M.: Color invariant SURF in discriminative object tracking. In: Kutulakos, K.N. (ed.) ECCV 2010 Workshops, Part II. LNCS, vol. 6554, pp. 62–75. Springer, Heidelberg (2012)
10. Fan, P., Men, A., Chen, M., Yang, B.: Color-SURF: A surf descriptor with local kernel color histograms. In: IEEE Int. Conf. on Network Infrastructure and Digital Content, pp. 726–730 (2009)
11. Fu, J., Jing, X., Sun, S., Lu, Y., Wang, Y.: C-surf: Colored speeded up robust features. In: Yuan, Y., Wu, X., Lu, Y. (eds.) Trustworthy Computing and Services. CCIS, vol. 320, pp. 203–210. Springer, Heidelberg (2013)
12. Grana, C., Borghesani, D., Manfredi, M., Cucchiara, R.: A fast approach for integrating ORB descriptors in the bag of words model. In: SPIE, vol. 8667, pp. 866709–866709-8 (2013)
13. Jegou, H., Douze, M., Schmid, C., Perez, P.: Aggregating local descriptors into a compact image representation. In: IEEE on Computer Vision and Pattern Recognition (CVRP 2010), San Francisco, CA, pp. 3304–3311 (2010)
14. Jegou, H., Perronnin, F., Douze, M., Sanchez, J., Perez, P., Schmid, C.: Aggregating local image descriptors into compact codes. IEEE Transactions on Pattern Analysis and Machine Intelligence 34(9), 1704–1716 (2012)

15. Leutenegger, S., Chli, M., Siegwart, R.: Brisk: Binary robust invariant scalable keypoints. In: IEEE Int. Conf. ICCV 2011, pp. 2548–2555 (2011)
16. Lowe, D.G.: Distinctive Image Features from Scale-Invariant Keypoints. Int. Journal of Computer Vision 60(2), 91–110 (2004)
17. Markatopoulou, F., Moumtzidou, A., Tzelepis, C., Avgerinakis, K., Gkalelis, N., Vrochidis, S., Mezaris, V., Kompatsiaris, I.: ITI-CERTH participation to TRECVID 2013. In: TRECVID 2013 Workshop, Gaithersburg, MD, USA (2013)
18. Markatopoulou, F., Mezaris, V., Kompatsiaris, I.: A comparative study on the use of multi-label classification techniques for concept-based video indexing and annotation. In: Gurrin, C., Hopfgartner, F., Hurst, W., Johansen, H., Lee, H., O'Connor, N. (eds.) MMM 2014, Part I. LNCS, vol. 8325, pp. 1–12. Springer, Heidelberg (2014)
19. Over, P., Awad, G., Michel, M., Fiscus, J., Sanders, G., Kraaij, W., Smeaton, A.F.: Trecvid 2013 – an overview of the goals, tasks, data, evaluation mechanisms and metrics. In: Proceedings of TRECVID 2013, NIST, USA (2013)
20. Perronnin, F., Sánchez, J., Mensink, T.: Improving the fisher kernel for large-scale image classification. In: Daniilidis, K., Maragos, P., Paragios, N. (eds.) ECCV 2010, Part IV. LNCS, vol. 6314, pp. 143–156. Springer, Heidelberg (2010)
21. Qiu, G.: Indexing chromatic and achromatic patterns for content-based colour image retrieval. Pattern Recognition 35, 1675–1686 (2002)
22. Rublee, E., Rabaud, V., Konolige, K., Bradski, G.: ORB: An efficient alternative to SIFT or SURF. In: IEEE Int. Conf. on Computer Vision, pp. 2564–2571 (2011)
23. Safadi, B., Quénot, G.: Re-ranking by local re-scoring for video indexing and retrieval. In: 20th ACM Int. Conf. on Information and Knowledge Management, UK, pp. 2081–2084. ACM, NY (2011)
24. Van de Sande, K.E.A., Gevers, T., Snoek, C.G.M.: Evaluating color descriptors for object and scene recognition. IEEE Transactions on Pattern Analysis and Machine Intelligence 32(9), 1582–1596 (2010)
25. Van de Sande, K.E.A., Snoek, C.G.M., Smeulders, A.W.M.: Fisher and vlad with flair. In: IEEE Conference on Computer Vision and Pattern Recognition (2014)
26. Sidiropoulos, P., Mezaris, V., Kompatsiaris, I.: Video tomographs and a base detector selection strategy for improving large-scale video concept detection. IEEE Transactions on Circuits and Systems for Video Technology 24(7), 1251–1264 (2014)
27. Snoek, C.G.M., Worring, M.: Concept-Based Video Retrieval. Foundations and Trends in Information Retrieval 2(4), 215–322 (2009)
28. Witten, I., Frank, E.: Data Mining Practical Machine Learning Tools and Techniques, 2nd edn. Morgan Kaufmann, San Francisco (2005)
29. Yilmaz, E., Kanoulas, E., Aslam, J.A.: A simple and efficient sampling method for estimating ap and ndcg. In: 31st ACM SIGIR Int. Conf. on Research and Development in Information Retrieval, pp. 603–610. ACM, USA (2008)
30. Zhou, X., Yu, K., Zhang, T., Huang, T.S.: Image classification using super-vector coding of local image descriptors. In: Daniilidis, K., Maragos, P., Paragios, N. (eds.) ECCV 2010, Part V. LNCS, vol. 6315, pp. 141–154. Springer, Heidelberg (2010)

Content-Based Image Retrieval
with Gaussian Mixture Models

Christian Beecks, Merih Seran Uysal, and Thomas Seidl

Data Management and Exploration Group
RWTH Aachen University
Germany
{beecks,uysal,seidl}@cs.rwth-aachen.de

Abstract. Among the various approaches of content-based image modeling, generative models have become prominent due to their ability of approximating feature distributions with arbitrary accuracy. A frequently encountered generative model for the purpose of content-based image retrieval is the Gaussian mixture model which facilitates the application of various dissimilarity measures. The question of which dissimilarity measure provides the highest retrieval performance in terms of accuracy and efficiency is still an open research question. In this paper, we propose an empirical investigation of dissimilarity measures for Gaussian mixture models based on high-dimensional local feature descriptors. To this end, we include a unifying overview of state-of-the-art dissimilarity measures applicable to Gaussian mixture models along with an extensive performance analysis on a multitude of local feature descriptors. Our findings will help to guide further research in the field of content-based image modeling with Gaussian mixture models.

Keywords: Content-based image retrieval, Gaussian mixture models, Kullback-Leibler Divergence, Signature-based distance functions.

1 Introduction

Modeling image contents for the purpose of *content-based image retrieval*, which denotes the process of retrieving images from an image database that share similar inherent characteristics with respect to a given query image, is a significant challenge. Many content-based approaches model the inherent characteristic properties of images via features in a multi-dimensional feature space and define image dissimilarity by quantizing and matching these features according to a visual vocabulary. The probably most prominent approaches follow either the bag-of-visual-words model [25] which utilizes a common visual vocabulary or the more generic signature-based model [24] which makes use of image-specific visual vocabularies. Among the numerous variants of modeling image contents, generative models are considered to be of high quality since they allow to approximate feature distributions with arbitrary accuracy.

X. He et al. (Eds.): MMM 2015, Part I, LNCS 8935, pp. 294–305, 2015.

A frequently encountered generative model for the purpose of content-based image retrieval is the *Gaussian mixture model* [2,9,12,15,19,23,27,28]. It comprises a finite linear combination of *Gaussian probability distributions*, also denoted as *normal distributions*, in order to approximate feature distributions of images. There are many reasons for utilizing normal distributions [6]: First, they are very tractable analytically. Second, they have familiar symmetrical bell shapes. Third, they approximate a large variety of probability distributions according to the central limit theorem.

Given Gaussian mixture models as content-based image representations, the definition of an appropriate dissimilarity measure is of crucial importance. While some approaches [9,12,27] follow the idea of approximating the Kullback-Leibler Divergence between Gaussian mixture models, recent advances [2,3] propose the utilization and adaptation of signature-based distance functions, such as the Earth Mover's Distance [24] and Signature Quadratic Form Distance [5] for the dissimilarity estimation among Gaussian mixture models.

Although the aforementioned dissimilarity measures have been investigated within the context of content-based image retrieval, it remains unclear which type of dissimilarity measure provides the highest retrieval performance in terms of accuracy and efficiency over high-dimensional Gaussian mixture models. In this paper, we thus subject the state-of-the-art dissimilarity measures applicable to Gaussian mixture models to an extensive empirical performance analysis. Our objective consists in supporting novel insights into content-based image modeling and retrieval with Gaussian mixture models.

This paper is structured as follows. Section 2 introduces Gaussian mixture models for content-based image modeling. Section 3 then proposes a survey of dissimilarity measures approximating the Kullback-Leibler Divergence among Gaussian mixture models, whereas Section 4 proposes signature-based distance functions for the comparison of Gaussian mixture models. The results of our empirical performance analysis are reported in Section 5, before we conclude our paper with an outlook on future work in Section 6.

2   Image Modeling via Gaussian Mixture Models

The most frequently encountered approach to represent images is carried out by means of a *feature space*. A feature space $(\mathbb{F}, \delta)$ comprises a set of features $\mathbb{F}$, for instance SIFT descriptors [18], and is endowed with a function $\delta : \mathbb{F} \times \mathbb{F} \to \mathbb{R}$ for the comparison of two features. Frequently but not necessarily, the function δ is supposed to be a metric distance function such as the Euclidean distance L_2. Given a feature space $(\mathbb{F}, \delta)$, each image $\mathcal{I}$ is represented by means of *features* $f_1^{\mathcal{I}}, \ldots, f_n^{\mathcal{I}} \in \mathbb{F}$ which reflect the characteristic image properties and are additionally assigned to real-valued weights indicating their importance. Defining these weights via multiple normal distributions leads to the definition of a Gaussian mixture model.

Definition 1 (Gaussian mixture model). *Let $(\mathbb{F}, \delta)$ be a multi-dimensional Euclidean feature space., i.e. $\mathbb{F} = \mathbb{R}^d$ and $\delta = L_2$. A Gaussian mixture model $g : \mathbb{F} \to \mathbb{R}$ is defined as:*

$$g(x) = \sum_{i=1}^{n} \pi_i \cdot \mathcal{N}_{\mu_i, \Sigma_i}(x), \ where$$

$$\mathcal{N}_{\mu_i, \Sigma_i}(x) = \frac{1}{\sqrt{(2\pi_i)^d |\Sigma_i|}} e^{-\frac{1}{2}(x-\mu_i)\Sigma_i^{-1}(x-\mu_i)^T},$$

with prior probabilities $\pi_i \in \mathbb{R}^+$ such that $\sum_{i=1}^{n} \pi_i = 1$, means $\mu_i \in \mathbb{R}^d$, and covariance matrices $\Sigma_i \in \mathbb{R}^{d \times d}$ for $1 \leq i \leq n$.

As can be seen in Definition 1, a Gaussian mixture model is a finite linear combination of normal distributions. It thus provides an expressive yet compact way of generatively describing a distribution of features $f_1^{\mathcal{I}}, \ldots, f_n^{\mathcal{I}} \in \mathbb{R}^d$ arising from an individual image $\mathcal{I}$. In fact, each normal distribution $\mathcal{N}_{\mu_i, \Sigma_i}$ over a feature space $(\mathbb{R}^d, \delta)$ is completely defined by its mean $\mu_i \in \mathbb{R}^d$ and its covariance matrix $\Sigma \in \mathbb{R}^{d \times d}$, thus the storage required to represent a Gaussian mixture model g is dependent on the dimensionality $d \in \mathbb{N}$ of the multi-dimensional Euclidean feature space and the number of components of g.

Based on Gaussian mixture models as image representations, we will continue with summarizing the Kullback-Leibler Divergence and its approximations for Gaussian mixture models in the following section.

3   Approximations of the Kullback-Leibler Divergence

In this section, we provide a short survey of approximations of the *Kullback-Leibler Divergence* [17] for the comparison of Gaussian mixture models. The Kullback-Leibler Divergence is one of the most frequently encountered dissimilarity measures which can be solved analytically between normal distributions. It is defined between two normal distributions $\mathcal{N}_{\mu^a, \Sigma^a}$ and $\mathcal{N}_{\mu^b, \Sigma^b}$ as follows [12]:

$$KL(\mathcal{N}_{\mu^a, \Sigma^a}, \mathcal{N}_{\mu^b, \Sigma^b}) =$$
$$\frac{1}{2} \left(\log(\frac{|\Sigma^b|}{|\Sigma^a|}) + \mathrm{tr}(\Sigma^{b-1} \cdot \Sigma^a) + (\mu^a - \mu^b) \cdot \Sigma^{b-1} \cdot (\mu^a - \mu^b)^T - d \right),$$

where $\mathrm{tr}(\Sigma) = \sum_{i=1}^{d} \Sigma[i,i]$ denotes the trace of a matrix $\Sigma \in \mathbb{R}^{d \times d}$.

This closed-form expression of the Kullback-Leibler Divergence between normal distributions is further utilized in order to approximate the Kullback-Leibler Divergence between Gaussian mixture models. In fact, there exists no closed-form expression of the Kullback-Leibler Divergence between Gaussian mixture models. Nonetheless, the past decade has yield a number of approximations which are investigated in the remainder of this section.

One approximation of the Kullback-Leibler Divergence between Gaussian mixture models is the matching-based *Goldberger approximation* [9]. The idea of this approximation is to match each mixture component of the fist Gaussian mixture model to one single mixture component of the second Gaussian mixture model. The matching is determined by the Kullback-Leibler Divergence between the mixture components. The formal definition is given below.

Definition 2 (Goldberger approximation). *Let* $(\mathbb{F} = \mathbb{R}^d, \delta)$ *be a feature space and* $g_a = \sum_{i=1}^{n} \pi_i^a \cdot \mathcal{N}_{\mu_i^a, \Sigma_i^a}$ *and* $g_b = \sum_{j=1}^{m} \pi_j^b \cdot \mathcal{N}_{\mu_j^b, \Sigma_j^b}$ *be two Gaussian mixture models. The Goldberger approximation of the Kullback-Leibler Divergence* $\mathrm{KL_{Goldberger}}$ *between* g_a *and* g_b *is defined as:*

$$\mathrm{KL_{Goldberger}}(g_a, g_b) = \sum_{i=1}^{n} \pi_i^a \cdot \left(\mathrm{KL}(\mathcal{N}_{\mu_i^a, \Sigma_i^a}, \mathcal{N}_{\mu_{\rho(i)}^b, \Sigma_{\rho(i)}^b}) + \log \frac{\pi_i^a}{\pi_{\rho(i)}^b} \right),$$

where $\rho(i) = \arg \min_{1 \leq j \leq m} \{\mathrm{KL}(\mathcal{N}_{\mu_i^a, \Sigma_i^a}, \mathcal{N}_{\mu_j^b, \Sigma_j^b}) - \log \pi(\mathcal{N}_{\mu_j^b, \Sigma_j^b})\}$ *denotes the matching function between single mixture components* $\mathcal{N}_{\mu_i^a, \Sigma_i^a}$ *and* $\mathcal{N}_{\mu_j^b, \Sigma_j^b}$ *of the Gaussian mixture models* g_a *and* g_b, *respectively.*

As presented in the definition above, the approximation $\mathrm{KL_{Goldberger}}(g_a, g_b)$ is defined as the sum of the Kullback-Leibler Divergences $\mathrm{KL}(\mathcal{N}_{\mu_i^a, \Sigma_i^a}, \mathcal{N}_{\mu_{\rho(i)}^b, \Sigma_{\rho(i)}^b})$ plus the logarithm of the quotient of their prior probabilities $\log \frac{\pi_i^a}{\pi_{\rho(i)}^b}$ between matching mixture components $\mathcal{N}_{\mu_i^a, \Sigma_i^a}$ and $\mathcal{N}_{\mu_{\rho(i)}^b, \Sigma_{\rho(i)}^b}$, multiplied by the prior probabilities π_i^a.

The Goldberger approximation works well empirically [12] when the mixture components are far apart without significant overlap [9] but it performs poorly in the presence of a few mixture components with low prior probabilities [13]. For this reason, Goldberger et al. [9] furthermore propose another approximation that is based on the *unscented transform* [16]. The idea of this approximation is similar to the Monte Carlo simulation. Instead of taking a large independent and identically distributed sampling, the unscented transform approach deterministically defines a sampling which reflects the mixture components of a Gaussian mixture model. The *unscented transform approximation* of the Kullback-Leibler Divergence is formalized as in the following.

Definition 3 (Unscented transform approximation). *Let* $(\mathbb{F} = \mathbb{R}^d, \delta)$ *be a feature space and* $g_a = \sum_{i=1}^{n} \pi_i^a \cdot \mathcal{N}_{\mu_i^a, \Sigma_i^a}$ *and* $g_b = \sum_{j=1}^{m} \pi_j^b \cdot \mathcal{N}_{\mu_j^b, \Sigma_j^b}$ *be two Gaussian mixture models. The* unscented transform *approximation of the Kullback-Leibler Divergence* $\mathrm{KL_{unscented}}$ *between* g_a *and* g_b *is defined as:*

$$\mathrm{KL_{unscented}}(g_a, g_b) = \frac{1}{2 \cdot d} \cdot \sum_{i=1}^{n} \pi_i^a \cdot \sum_{k=1}^{d} \log \frac{g_a(f_{i,k})}{g_b(f_{i,k})},$$

such that the sample points $f_{i,k} = \mu_i^a \pm \sqrt{d \cdot \lambda_{i,k}} \cdot e_{i,k} \in \mathbb{F}$ *reflect the mean and variance of single components* $\mathcal{N}_{\mu_i^a, \Sigma_i^a}$, *while* $\lambda_{i,k} \in \mathbb{R}$ *and* $e_{i,k} \in \mathbb{F}$ *denote the i-th eigenvalue and eigenvector of* Σ_i^a, *respectively.*

Definition 3 shows how the unscented transform is utilized to approximate the Kullback-Leibler Divergence among Gaussian mixture models. Based on $2 \cdot d$-many sampling points $f_{i,k} \in \mathbb{F}$, the integral of the Kullback-Leibler Divergence is approximated through the corresponding sums. In comparison to the Monte Carlo simulation, this approximation provides a high accuracy by only using a comparatively small sampling size. It thus shows faster computation time than the approximation by the Monte Carlo simulation but slower computation time than the Goldberger approximation.

Whereas the aforementioned approximations of the Kullback-Leibler Divergence are originally defined within the context of content-based image retrieval, the next approximation is investigated in the context of acoustic models for speech recognition. The *variational approximation* [12] is defined below.

Definition 4 (Variational approximation). *Let* $(\mathbb{F} = \mathbb{R}^d, \delta)$ *be a feature space and* $g_a = \sum_{i=1}^{n} \pi_i^a \cdot \mathcal{N}_{\mu_i^a, \Sigma_i^a}$ *and* $g_b = \sum_{j=1}^{m} \pi_j^b \cdot \mathcal{N}_{\mu_j^b, \Sigma_j^b}$ *be two Gaussian mixture models. The* variational approximation *of the Kullback-Leibler Divergence* $\mathrm{KL}_{\mathrm{variational}}$ *between* g_a *and* g_b *is defined as:*

$$\mathrm{KL}_{\mathrm{variational}}(g_a, g_b) = \sum_{i=1}^{n} \pi_i^a \cdot \log \left(\frac{\sum_{i'=1}^{n} \pi_{i'}^a \cdot e^{-\mathrm{KL}(\mathcal{N}_{\mu_i^a, \Sigma_i^a}, \mathcal{N}_{\mu_{i'}^a, \Sigma_{i'}^a})}}{\sum_{j=1}^{m} \pi_j^b \cdot e^{-\mathrm{KL}(\mathcal{N}_{\mu_i^a, \Sigma_i^a}, \mathcal{N}_{\mu_j^b, \Sigma_j^b})}} \right).$$

Unlike other approaches, the variational approximation $\mathrm{KL}_{\mathrm{variational}}(g_a, g_b)$ also takes into account the Kullback-Leibler Divergences of the mixture components within the Gaussian mixture model g_a and those between the mixture components of both Gaussian mixture models g_a and g_b.

The aforementioned approximations of the Kullback-Leibler Divergence between Gaussian mixture models differ in the way the components, i.e. normal distributions, of the Gaussian mixture models are related. In fact, none of these approximations relate all components of both Gaussian mixture models with each other. How this is done by means of signature-based distance functions is investigated in the following section.

4   Signature-Based Distance Functions for the Comparison of Gaussian Mixture Models

In this section, we show how to compare Gaussian mixture models with signature-based distance functions. These distance functions enable the comparison of differently structured feature representations, such as feature histograms and feature signatures, by utilizing a ground-distance between individual components of the feature representations. In order to apply signature-based distance functions to Gaussian mixture models, whose mixture components are normal distributions, we propose the Kullback-Leibler Divergence as ground-distance due to its ability to take into account both means and covariances of the corresponding normal distributions. We briefly explain the main ideas together with formal definitions of state-of-the-art signature-based distance functions applied to Gaussian mixture models in the remainder of this section. We will use the shorthand notation $g_a = \{(\pi_i^a, \mathcal{N}_i^a)\}_{i=1}^{n}$ instead of $g_a = \sum_{i=1}^{n} \pi_i^a \cdot \mathcal{N}_{\mu_i^a, \Sigma_i^a}$ where appropriate.

The *Hausdorff Distance* [11] is a matching-based approach that measures the maximum nearest neighbor distance, i.e. Kullback-Leibler Divergence, among normal distributions of both Gaussian mixture models. Its formal definition adapted to Gaussian mixture models is given below.

Definition 5 (Hausdorff Distance). *Let* $(\mathbb{F} = \mathbb{R}^d, \delta)$ *be a feature space and* $g_a = \{(\pi_i^a, \mathcal{N}_i^a)\}_{i=1}^n$ *and* $g_b = \{(\pi_j^b, \mathcal{N}_j^b)\}_{j=1}^m$ *be two Gaussian mixture models. The* Hausdorff *Distance* $\mathrm{HD}_{\mathrm{KL}}$ *between* g_a *and* g_b *is defined as:*

$$\mathrm{HD}_{\mathrm{KL}}(g_a, g_b) = \max\{\mathrm{h}(g_a, g_b), \mathrm{h}(g_b, g_a)\}, \quad where$$

$$\mathrm{h}(g_a, g_b) = \max_{(\pi^a, \mathcal{N}^a) \in g_a} \min_{(\pi^b, \mathcal{N}^b) \in g_b} \{\mathrm{KL}(\mathcal{N}^a, \mathcal{N}^b)\}.$$

Intuitively, two Gaussian mixture models have a small distance value if each normal distribution of one Gaussian mixture model has a close counterpart in the other Gaussian mixture model with respect to the Kullback-Leibler Divergence. The Hausdorff Distance disregards the prior probabilities of the Gaussian mixture models and thus does not favor any normal distributions. Taking into account prior probabilities can be achieved by modifying the auxiliary function h as done by the *Perceptually Modified Hausdorff Distance* [22]. Its formal definition adapted to Gaussian mixture models is given below.

Definition 6 (Perceptually Modified Hausdorff Distance). *Let* $(\mathbb{F} = \mathbb{R}^d, \delta)$ *be a feature space and* $g_a = \{(\pi_i^a, \mathcal{N}_i^a)\}_{i=1}^n$ *and* $g_b = \{(\pi_j^b, \mathcal{N}_j^b)\}_{j=1}^m$ *be two Gaussian mixture models. The* Perceptually Modified Hausdorff Distance $\mathrm{PMHD}_{\mathrm{KL}}$ *between* g_a *and* g_b *is defined as:*

$$\mathrm{PMHD}_{\mathrm{KL}}(g_a, g_b) = \max\{\mathrm{h}_w(g_a, g_b), \mathrm{h}_w(g_b, g_a)\}, \quad where$$

$$\mathrm{h}_w(g_a, g_b) = \frac{\displaystyle\sum_{(\pi^a, \mathcal{N}^a) \in g_a} \pi^a \cdot \min_{(\pi^b, \mathcal{N}^b) \in g_b} \left\{\frac{\mathrm{KL}(\mathcal{N}^a, \mathcal{N}^b)}{\min\{\pi^a, \pi^b\}}\right\}}{\displaystyle\sum_{(\pi^a, \mathcal{N}^a) \in g_a} \pi^a}.$$

The Perceptually Modified Hausdorff Distance allows to suppress the contribution of noisy mixture components, i.e. normal distributions with low prior probabilities, in a self-adjusting manner. It thus attributes the distance computation to those matching pairs of normal distributions with higher prior probabilities in order to obtain a more stable distance value.

The *Signature Matching Distance* [4] follows the idea of modeling the distance between two Gaussian mixture models by making use of the symmetric difference of pairs of matching normal distributions. This distance is generic in the sense that it can be applied to any matching m $\subseteq g_a \times g_b$, such as the *nearest neighbor matching* [21], *distance ratio matching* [21], *inverse distance ratio matching* [4], or *distance weight ratio matching* [4], and to any cost function c evaluating the quality of a matching. Its formal definition adapted to Gaussian mixture models is given below.

Definition 7 (Signature Matching Distance). *Let* $(\mathbb{F} = \mathbb{R}^d, \delta)$ *be a feature space and* $g_a = \{(\pi_i^a, \mathcal{N}_i^a)\}_{i=1}^n$ *and* $g_b = \{(\pi_j^b, \mathcal{N}_j^b)\}_{j=1}^m$ *be two Gaussian mixture models. The* Signature Matching Distance $\mathrm{SMD}_{\mathrm{KL}}$ *between* g_a *and* g_b *with respect to a matching* m, *a cost function* c, *and parameter* $1 \geq \lambda \in \mathbb{R}^{\geq 0}$ *is defined as:*

$$\mathrm{SMD}_{\mathrm{KL}}(g_a, g_b) = \mathrm{c}(\mathrm{m}_{g_a \to g_b}) + \mathrm{c}(\mathrm{m}_{g_b \to g_a}) - 2\lambda \cdot \mathrm{c}(\mathrm{m}_{g_a \leftrightarrow g_b}).$$

Intuitively, this distance function is based on normal distributions that match in the direction from g_a to g_b and vice versa. In addition, pairs of normal distributions which match in both directions can be excluded by increasing the value of parameter λ. This attributes the distance computation to less similar mixture components. For further details regarding the Signature Matching Distance, we refer to the work of Beecks et al. [4].

The *Earth Mover's Distance* [24] is a transformation-based approach measuring the cost of transforming one Gaussian mixture model into another one. In fact, it can be interpret as a specific matching-based approach. Its formal definition adapted to Gaussian mixture models is given below.

Definition 8 (Earth Mover's Distance). *Let* $(\mathbb{F} = \mathbb{R}^d, \delta)$ *be a feature space and* $g_a = \sum_{i=1}^n \pi_i^a \cdot \mathcal{N}_{\mu_i^a, \Sigma_i^a}$ *and* $g_b = \sum_{j=1}^m \pi_j^b \cdot \mathcal{N}_{\mu_j^b, \Sigma_j^b}$ *be two Gaussian mixture models. The* Earth Mover's Distance EMD_{KL} *between* g_a *and* g_b *is defined as a minimum cost flow of all possible flows* $F = \{f | f : \mathbb{F} \times \mathbb{F} \to \mathbb{R}\}$ *as follows:*

$$\text{EMD}_{\text{KL}}(g_a, g_b) = \min_F \left\{ \frac{\sum_{i=1}^n \sum_{j=1}^m f(\mu_i^a, \mu_j^b) \cdot \text{KL}(\mathcal{N}_{\mu_i^a, \Sigma_i^a}, \mathcal{N}_{\mu_j^b, \Sigma_j^b})}{\min\{\sum_{i=1}^n \pi_i^a, \sum_{j=1}^m \pi_j^b\}} \right\},$$

subject to the constraints: $\forall \mu_i^a, \mu_j^b \in \mathbb{F} : f(\mu_i^a, \mu_j^b) \geq 0, \forall \mu_i^a \in \mathbb{F} : \sum_{j=1}^m f(\mu_i^a, \mu_j^b) \leq \pi_i^a, \forall \mu_j^b \in \mathbb{F} : \sum_{i=1}^n f(\mu_i^a, \mu_j^b) \leq \pi_j^b,$ *and* $\sum_{i=1}^n \sum_{j=1}^m f(\mu_i^a, \mu_j^b) = \min\{\sum_{i=1}^n \pi_i^a, \sum_{j=1}^m \pi_j^b\}.$

The Earth Mover's Distance is defined as a solution of an optimization problem, which is optimal in terms of the minimum cost flow, subject to the constraints defined above. These constraints guarantee a feasible solution, i.e. all flows are positive and do not exceed the corresponding limitations given by the prior probabilities of both Gaussian mixture models.

Unlike the aforementioned distance functions which compute a distance value based on the normal distributions that match with respect to certain criteria, the succeeding *Signature Quadratic Form Distance* [5] is a correlation-based approach which relates all normal distributions with each other. This is done by transforming the Kullback-Leibler Divergence $\text{KL}(\mathcal{N}^a, \mathcal{N}^b)$ between two normal distributions $\mathcal{N}^a$ and $\mathcal{N}^b$ into a similarity function $s_{\text{KL}}(\mathcal{N}^a, \mathcal{N}^b)$, for instance through the *Gaussian kernel* which is one of the best-known positive definite kernels [8]. The formal definition of the Signature Quadratic Form Distance adapted to Gaussian mixture models is given below.

Definition 9 (Signature Quadratic Form Distance [5]). *Let* $(\mathbb{F} = \mathbb{R}^d, \delta)$ *be a feature space,* $g_a = \{(\pi_i^a, \mathcal{N}_i^a)\}_{i=1}^n$ *and* $g_b = \{(\pi_j^b, \mathcal{N}_j^b)\}_{j=1}^m$ *be two Gaussian mixture models, and* $s_{\text{KL}} : \mathbb{R}^{\mathbb{F}} \times \mathbb{R}^{\mathbb{F}} \to \mathbb{R}$ *be a KL-based similarity function. The* Signature Quadratic Form Distance $\text{SQFD}_{s_{\text{KL}}}$ *between* g_a *and* g_b *is defined as:*

$$\mathrm{SQFD}_{s_{\mathrm{KL}}}(g_a, g_b) = \sqrt{\langle g_a, g_a \rangle_{s_{\mathrm{KL}}} - 2 \cdot \langle g_a, g_b \rangle_{s_{\mathrm{KL}}} + \langle g_b, g_b \rangle_{s_{\mathrm{KL}}}},$$

with similarity correlation $\langle g_a, g_b \rangle_{s_{\mathrm{KL}}} = \sum_{i=1}^{n} \sum_{j=1}^{m} \pi_i^a \cdot \pi_j^b \cdot s_{\mathrm{KL}}(\mathcal{N}_i^a, \mathcal{N}_j^b)$.

Intuitively, the Signature Quadratic Form Distance can be thought of as the length of the difference of the two Gaussian mixture models. In addition to transforming the Kullback-Leibler Divergence into a similarity function, there exists also a closed-form solution for estimating the Gaussian kernel directly between two normal distributions with diagonal covariance matrices. This leads to the following special form of the Signature Quadratic Form Distance [2] for Gaussian mixture models, as shown in the definition below.

Definition 10 (Signature Quadratic Form Distance [2]). *Let* $(\mathbb{F} = \mathbb{R}^d, \delta)$ *be a feature space,* $g_a = \{(\pi_i^a, \mathcal{N}_i^a)\}_{i=1}^{n}$ *and* $g_b = \{(\pi_j^b, \mathcal{N}_j^b)\}_{j=1}^{m}$ *be two Gaussian mixture models with diagonal covariance matrices, and* $k_{\mathrm{Gauss}}(x, y) = e^{-\frac{\|x-y\|^2}{2\sigma^2}}$ *be the Gaussian kernel for any* $x, y \in \mathbb{R}^d$ *and parameter* $\sigma \in \mathbb{R}^{\geq 0}$. *The* Signature Quadratic Form Distance $\mathrm{SQFD}_{\mathrm{GMM}}$ *between* g_a *and* g_b *is defined as:*

$$\mathrm{SQFD}_{\mathrm{GMM}}(g_a, g_b) = \sqrt{\mathrm{E}\langle g_a, g_a \rangle_{k_{\mathrm{Gauss}}} - 2 \cdot \mathrm{E}\langle g_a, g_b \rangle_{k_{\mathrm{Gauss}}} + \mathrm{E}\langle g_b, g_b \rangle_{k_{\mathrm{Gauss}}}},$$

where the expected similarity correlation $\mathrm{E}\langle g_a, g_b \rangle_{k_{\mathrm{Gauss}}}$ *is defined between* g_a *and* g_b *as* $\mathrm{E}\langle g_a, g_b \rangle_{k_{\mathrm{Gauss}}} = \sum_{i=1}^{n} \sum_{j=1}^{m} \pi_i^a \cdot \pi_j^b \cdot \left(\prod_{i=1}^{d} \frac{e^{-\frac{1}{2} \cdot \frac{(\mu_i^a - \mu_i^b)^2}{\sigma^2 + (\sigma_i^a)^2 + (\sigma_i^b)^2}}}{\frac{1}{\sigma}\sqrt{\sigma^2 + (\sigma_i^a)^2 + (\sigma_i^b)^2}} \right).$

As can be seen in Definition 10, the Signature Quadratic Form Distance is defined between two Gaussian mixture models without utilizing the Kullback-Leibler Divergence. As we will see in the next section, this approach will yield a higher retrieval performance than conventional KL-based approaches.

5 Performance Analysis

In this section, we study the retrieval performance of the proposed dissimilarity measures among Gaussian mixture models in terms of accuracy and efficiency. The performance analysis was carried out based on the Holidays database [14] which comprises 1,491 holiday photos corresponding to a large variety of scene types. It was designed to test the robustness with respect to rotation, viewpoint, and illumination changes and provides a solid ground truth including 500 selected queries for benchmarking content-based image retrieval approaches.

Gaussian mixture models comprising ten mixture components with diagonal covariance matrices where generated for each image individually by extracting local feature descriptors with the Harris Laplace detector [20] and clustering them with the expectation maximization algorithm [7]. The color descriptor software [26] was used to extract the local feature descriptors and the *WEKA* framework [10] was utilized to cluster the extracted descriptors. A more detailed explanation

Table 1. Mean average precision values of the Kullback-Leibler Divergence approximations on the Holidays database. The highest values are highlighted for each descriptor.

descriptor	$KL_{Goldberger}$ map	$KL_{unscented}$ map	$KL_{variational}$ map
rgbhistogram	0.450	**0.501**	0.463
opponenthist.	0.445	0.451	**0.461**
huehistogram	0.480	**0.579**	0.495
nrghistogram	0.353	**0.490**	0.377
transf.colorhist.	0.286	**0.687**	0.500
colormoments	0.555	**0.612**	0.577
col.mom.inv.	0.526	**0.684**	0.607
sift	0.063	**0.581**	0.141
huesift	0.028	**0.177**	0.092
hsvsift	0.046	–	**0.109**
opponentsift	0.063	–	**0.129**
rgsift	0.049	–	**0.108**
csift	0.048	–	**0.106**
rgbsift	0.059	–	**0.126**

of the utilized *pixel-based histogram descriptors*, *color moment descriptors*, and *gradient-based SIFT descriptors* is given in the work of van de Sande et al. [26].

Table 1 summarizes the mean average precision values of the Goldberger approximation $KL_{Goldberger}$, unscented transform approximation $KL_{unscented}$, and variational approximation $KL_{variational}$. As can be seen in that table, the unscented transform approximation generally shows higher mean average precision values than the other approaches except on high-dimensional gradient-based SIFT descriptors. The high dimensionality of those descriptors causes the unscented transform approximation to fail due to algorithmic reasons [2].

Table 2 summarizes the mean average precision values of the Hausdorff Distance HD_{KL}, Perceptually Modified Hausdorff Distance $PMHD_{KL}$, Signature Matching Distance SMD_{KL}, Earth Mover's Distance EMD_{KL}, and Signature Quadratic Form Distance based on the KL-based similarity function $SQFD_{s_{KL}}$ (cf. Definition 9) as well as that one based on the closed-form expression $SQFD_{GMM}$ (cf. Definition 10). As for the Signature Matching Distance, the inverse distance ratio matching [4] is used and the highest mean average precision values for parameters $\varepsilon \in \{0.1, 0.2, \ldots, 1.0\}$ and $\lambda \in \{0.0, 0.05, \ldots, 1.0\}$ are reported. Regarding the Signature Quadratic Form Distance $SQFD_{s_{KL}}$, the Gaussian kernel with parameter σ is utilized to transform the Kullback-Leibler Divergence into a similarity function. For both variants of the Signature Quadratic Form Distance, i.e. $SQFD_{s_{KL}}$ and $SQFD_{GMM}$, we report the highest mean average precision values for the parameters $\sigma \in \mathbb{R}^{>0}$. As can be seen in Table 2, the Signature Quadratic Form Distance based on the closed-form expression $SQFD_{GMM}$ outperforms the other signature-based approaches except on Gaussian mixture models based on the color moment invariants descriptor.

To sum up, the results indicate that approximations of the Kullback-Leibler Divergence among Gaussian mixture models provide competitive results on

Table 2. Mean average precision values of the signature-based distance functions on the Holidays database. The highest values are highlighted for each descriptor.

descriptor	HD_{KL} map	$PMHD_{KL}$ map	SMD_{KL} map ε λ			EMD_{KL} map	$SQFD_{s_{KL}}$ map σ		$SQFD_{GMM}$ map σ	
rgbhistogram	0.426	0.461	0.438	1.0	0.7	0.418	0.463	2.13	**0.654**	0.43
opponenthist.	0.429	0.460	0.441	0.4	0.55	0.420	0.458	2.23	**0.680**	0.45
huehistogram	0.453	0.502	0.476	0.9	0.4	0.468	0.483	1.38	**0.670**	0.23
nrghistogram	0.377	0.392	0.376	0.9	0.05	0.369	0.387	1.96	**0.643**	0.39
transf.colorhist.	0.463	0.521	0.512	1.0	0.0	0.500	0.491	0.34	**0.630**	0.23
colormoments	0.536	0.589	0.589	1.0	0.0	0.518	0.441	5.73	**0.598**	5.73
col.mom.inv.	0.523	0.591	**0.582**	1.0	0.5	0.562	0.412	532.73	0.508	53.27
sift	0.434	0.457	0.458	0.3	0.4	0.458	0.406	69.70	**0.653**	174.25
huesift	0.438	0.476	0.455	0.2	0.85	0.467	0.410	57.30	**0.703**	200.54
hsvsift	0.447	0.483	0.470	0.2	0.4	0.479	0.407	195.26	**0.675**	390.53
opponentsift	0.455	0.483	0.479	0.4	0.55	0.478	0.409	197.24	**0.675**	295.86
rgsift	0.438	0.477	0.470	0.2	0.75	0.474	0.409	103.93	**0.680**	285.80
csift	0.440	0.472	0.471	0.4	0.3	0.472	0.412	393.40	**0.682**	295.05
rgbsift	0.439	0.468	0.468	0.3	0.85	0.470	0.406	67.00	**0.663**	301.49

Table 3. Computation time values in milliseconds needed to perform a single distance computation between two Gaussian mixture models with ten mixture components

approach	rgbhistogram (45d)	sift (128d)	csift (384d)
$KL_{Goldberger}$	0.016	0.031	0.032
$KL_{unscented}$	2.856	22.59	196.327
$KL_{variational}$	0.041	0.046	0.048
HD_{KL}	0.031	0.032	0.032
$PMHD_{KL}$	0.032	0.033	0.038
SMD_{KL}	0.032	0.037	0.032
EMD_{KL}	0.125	0.112	0.125
$SQFD_{s_{KL}}$	0.048	0.047	0.063
$SQFD_{GMM}$	0.101	0.251	0.719

low-dimensional local feature descriptors. On high-dimensional local feature descriptors, signature-based distance functions are able to outperform the former ones. Superior retrieval performance in terms of accuracy is achieved by using the closed-form expression of the Signature Quadratic Form Distance.

Let us now investigate the efficiency of the proposed dissimilarity measures. Table 3 shows the computation time values needed to perform a single distance computation between two Gaussian mixture models comprising 10 mixture components. The approaches have been implemented in Java 1.6 and evaluated on a single-core 3.4 GHz machine. As can be seen in Table 3, all approaches except the unscented transform approximation $KL_{unscented}$ are computed between two Gaussian mixture models in less than one millisecond. As for signature-based distance functions, the Signature Quadratic Form Distance based on the closed-form expression $SQFD_{GMM}$ requires the highest computation time.

To conclude, the retrieval performance analysis shows that most of the evaluated approaches are applicable to Gaussian mixture models for the purpose of content-based image retrieval. It also shows that none of these approaches is able to compete with the Signature Quadratic Form Distance based on the closed-form expression in terms of accuracy.

6   Conclusions and Future Work

In this paper, we have investigated Gaussian mixture models for content-based image retrieval. We have included a short survey of dissimilarity measures approximating the Kullback-Leibler Divergence and of those following the signature-based approach. Although both kinds of approaches are suitable for the comparison of Gaussian mixture models, signature-based distance functions are to be preferred in combination with high-dimensional local feature descriptors. The conducted empirical performance evaluation points out that the Signature Quadratic Form Distance outperforms existing approaches in terms of accuracy. As future work, we intend to extend this research to further investigate the analytical properties of the proposed approaches. In addition, we aim at examining their scalability and usability to further domains, such as audio and video fields.

Acknowledgments. This work is partially based on the work of Beecks [1] and is funded by DFG grant SE 1039/7-1 and by the Excellence Initiative of the German federal and state governments.

References

1. Beecks, C.: Distance-based similarity models for content-based multimedia retrieval. PhD thesis, RWTH Aachen University (2013)
2. Beecks, C., Ivanescu, A.M., Kirchhoff, S., Seidl, T.: Modeling image similarity by gaussian mixture models and the signature quadratic form distance. In: Proc. of the IEEE Int. Conf. on Computer Vision, pp. 1754–1761 (2011)
3. Beecks, C., Ivanescu, A.M., Kirchhoff, S., Seidl, T.: Modeling multimedia contents through probabilistic feature signatures. In: Proc. of the ACM Int. Conf. on Multimedia, pp. 1433–1436 (2011)
4. Beecks, C., Kirchhoff, S., Seidl, T.: Signature matching distance for content-based image retrieval. In: Proc. of the ACM Int. Conf. on Multimedia Retrieval, pp. 41–48 (2013)
5. Beecks, C., Uysal, M.S., Seidl, T.: Signature quadratic form distance. In: Proc. of the ACM Int. Conf. on Image and Video Retrieval, pp. 438–445 (2010)
6. Casella, G., Berger, R.: Statistical Inference. Duxbury Press (2001)
7. Dempster, A.P., Laird, N.M., Rubin, D.B.: Maximum likelihood from incomplete data via the em algorithm. Journal of the Royal Statistical Society. Series B (Methodological) 39(1), 1–38 (1977)
8. Fasshauer, G.E.: Positive definite kernels: past, present and future. Dolomites Research Notes on Approximation 4, 21–63 (2011)
9. Goldberger, J., Gordon, S., Greenspan, H.: An efficient image similarity measure based on approximations of kl-divergence between two gaussian mixtures. In: Proc. of the IEEE Int. Conf. on Computer Vision, pp. 487–493 (2003)

10. Hall, M., Frank, E., Holmes, G., Pfahringer, B., Reutemann, P., Witten, I.H.: The weka data mining software: an update. ACM SIGKDD Explorations Newsletter 11(1), 10–18 (2009)
11. Hausdorff, F.: Grundzüge der Mengenlehre. Von Veit (1914)
12. Hershey, J., Olsen, P.: Approximating the kullback leibler divergence between gaussian mixture models. In: Proc. of the IEEE Int. Conf. on Acoustics, Speech and Signal Processing, vol. 4, pp. 317–320 (2007)
13. Huo, Q., Li, W.: A dtw-based dissimilarity measure for left-to-right hidden markov models and its application to word confusability analysis. In: Int. Conf. on Spoken Language Processing, pp. 2338–2341 (2006)
14. Jegou, H., Douze, M., Schmid, C.: Hamming embedding and weak geometric consistency for large scale image search. In: Forsyth, D., Torr, P., Zisserman, A. (eds.) ECCV 2008, Part I. LNCS, vol. 5302, pp. 304–317. Springer, Heidelberg (2008)
15. Jeong, S., Won, C.S., Gray, R.M.: An adaptive color image retrieval framework using gauss mixtures. In: Proc. of the IEEE Int. Conf. on Image Processing, pp. 945–948 (2008)
16. Julier, S., Uhlmann, J.: A general method for approximating nonlinear transformations of probability distributions. Robotics Research Group, Department of Engineering Science, University of Oxford, Tech. Rep (1996)
17. Kullback, S., Leibler, R.A.: On information and sufficiency. The Annals of Mathematical Statistics 22(1), 79–86 (1951)
18. Lowe, D.G.: Distinctive image features from scale-invariant keypoints. International Journal of Computer Vision 60(2), 91–110 (2004)
19. Łuszczkiewicz-Piątek, M.: Which color space should be chosen for robust color image retrieval based on mixture modeling. In: Choras, R.S. (ed.) Image Processing and Communications Challenges 5. AISC, vol. 233, pp. 55–64. Springer, Heidelberg (2014)
20. Mikolajczyk, K., Schmid, C.: Scale & affine invariant interest point detectors. International Journal of Computer Vision 60(1), 63–86 (2004)
21. Mikolajczyk, K., Schmid, C.: A performance evaluation of local descriptors. IEEE TPAMI 27(10), 1615–1630 (2005)
22. Park, B.G., Lee, K.M., Lee, S.U.: Color-based image retrieval using perceptually modified hausdorff distance. EURASIP Journal of Image and Video Processing, 4:1–4:10 (2008)
23. Permuter, H.H., Francos, J.M., Jermyn, I.H.: Gaussian mixture models of texture and colour for image database retrieval. In: Proc. of the IEEE Int. Conf. on Acoustics, Speech and Signal Processing, pp. 569–572 (2003)
24. Rubner, Y., Tomasi, C., Guibas, L.J.: The earth mover's distance as a metric for image retrieval. International Journal of Computer Vision 40(2), 99–121 (2000)
25. Sivic, J., Zisserman, A.: Video google: A text retrieval approach to object matching in videos. In: Proc. of the IEEE Int. Conf. on Computer Vision, pp. 1470–1477 (2003)
26. van de Sande, K.E.A., Gevers, T., Snoek, C.G.M.: Evaluating color descriptors for object and scene recognition. IEEE TPAMI 32(9), 1582–1596 (2010)
27. Vasconcelos, N.: On the complexity of probabilistic image retrieval. In: Proc. of the IEEE Int. Conf. on Computer Vision, pp. 400–407 (2001)
28. Xing, X., Zhang, Y., Gong, B.: Mixture model based contextual image retrieval. In: Proc. of the ACM Int. Conf. on Image and Video Retrieval, pp. 251–258 (2010)

Improving Interactive Known-Item Search in Video with the Keyframe Navigation Tree

Marco A. Hudelist[1], Klaus Schoeffmann[1], and Qing Xu[2]

[1] Klagenfurt University, Universitaetsstr. 65-67, 9020 Klagenfurt, Austria
[2] School of Computer Science and Technology, Tianjin University, Tianjin, China
{Klaus.Schoeffmann,MarcoAndrea.Hudelist}@aau.at, qingxu@tju.edu.cn

Abstract. In this paper we propose the Keyframe Navigation Tree (KNT) as navigational aid in video for interactive search. The KNT is a hierarchical visualization of keyframes that can compactly represent the content of a video with different levels of details. It can be used as an alternative, or in addition, to a common seeker-bar of a video player. Through a user study with 20 participants we show that the proposed navigation approach not only allows significantly faster interactive search in video than a common video player, but also requires significantly less effort (also less mental and physical load) and is much more enjoyable to use.

1 Introduction

Known-Item search (KIS) in video is a search situation where someone knows of a specific video segment, and knows it is contained in a video or video collection, wants to find it, but does not know where it is located. In interactive KIS scenarios the user employs a video browsing tool [30] - or an interactive video retrieval tool - to find the desired video content. Interactive KIS had been evaluated as a task in TRECVID [23] for several years (from 2010 through 2012, when it was stopped due to insignificant progress) and is still an active evaluation scenario in the Video Browser Showdown (VBS) since 2012 [27]. The evaluation of the VBS [2,25] has shown that at interactive KIS tasks it is especially important to give the user flexible content navigation means and rich insight into the video content. More specifically, in VBS 2012 a simple but highly interactive tool could easily outperform another highly sophisticated video retrieval tool (e.g., with different forms of content analysis as well as concept detection) [2]. This simple interactive tool was actually an enhanced video player with no content analysis but high focus on interactive use and rich content insight, e.g. through parallel video player windows and synchronized navigation in different parts of the video [7].

It has been shown in the literature that displaying keyframes of shots as navigational means – through common lists [20], enhancements of the timeline/seeker-bar [8,31], or special arrangements [4,9,16] – can improve content navigation performance in video. The keyframes provide a good representation of the content and can aid the user at search tasks in videos. It has also been shown that users are

X. He et al. (Eds.): MMM 2015, Part I, LNCS 8935, pp. 306–317, 2015.

able to recognize thumbnails of keyframes even if they are presented at moderately small size [15] or with wrong aspect ratio [1].

However, the navigation by common keyframes has also several limitations: (i) presenting keyframes as thumbnails along the timeline limits [8,31] content insight at a glance to a few keyframes only, while the total number of keyframes/shots in a video is often in the several hundreds, (ii) special arrangements or video surrogates like the *VideoTrees* [16] or *Table-of-Video-Content* [9] need a lot of screen space due to the special structure, while the thumbnails are typically very small – and hard to recognize – to not produce overlappings, and finally (iii) presenting many small thumbnails can quickly exaggerate to too high mental load for the user as confirmed by the study of [26].

In this work we focus on improving interactive known-item search on tablet computers with touch interaction. We propose a video browsing tool that uses the *Keyframe Navigation Tree* (KNT) for the purpose of video content navigation. More precisely, it consists of a hierarchical set of navigation areas that show different levels of detail of the keyframes in a synchronized way. The keyframes are extracted with a novel keyframe selection method, which was recently published [19], and then visualized in a compact way with different levels of detail. Our approach is inspired by the idea of the best performing video browsing tool of the VBS 2012 competition [7] (which is basically a video player with slight extensions) as well as previous work on video navigation through extended seeker-bars [21,28]. Moreover, it considers latest findings of how users typically navigate through a video with a common video player [3,29].

Through an evaluation with a user study with interactive KIS tasks used in the VBS competition 2014 [24], we show that the proposed video browsing tool allows to solve known-item search tasks significantly faster than with a common video player. Moreover, our results show that users find the tool much more appropriate for search tasks in videos, like to use it, and find that the mental load as well as the physical load during usage is significantly lower than with a common video player.

The rest of the paper is organized as follows. First we discuss related work in Section 2. The idea of the Keyframe Navigation Tree (KNT) is presented in Section 3. Section 4 describes the keyframe selection method used to extract the keyframes from the videos. The two video browsing interface used for the evaluation in this work are described in Section 5 and the evaluation results are presented in Section 6.

2   Related Work

Many video browsing approaches have been presented in the literature, including methods for improved video navigation (a review is provided in [30]). Due to space limitations, here we focus only on video navigation proposed for mobile devices, a topic addressed by only a few papers.

Hürst et al. proposed the *Mobile ZoomSlider* interface [13] for stylus-based navigation on mobile handheld devices. The basic idea is a virtual seeker-bar that can be used on any position on the screen. The vertical click position is utilized as a parameter for navigation granularity. Moving the stylus left or right results in backward or forward navigation and the vertical position of the drag operation defines how fast the navigation is performed. The same concept has been used in later works for PDAs and smartphones [14].

The *PocketDRAGON* interface proposed by Karrer et al. [17] tries to avoid occlusion of video content during navigating. Instead of an overlaid seeker-bar, as typically used with video players on mobile devices, they propose direct manipulation of objects in the scene for navigation. A similar concept was also proposed by Dragicevic [6] earlier for desktop PCs. In their work, motion tracking is performed and object motion is used as basis for the dragging operation along a motion trajectory. Additionally to this object-based navigation mode, which rather improves the accuracy of navigation and typically does not allow quicker navigation over longer segments in the video, they also support two finger gestures. Here, a horizontal wipe gesture allows to jump to the previous or next scene in the video. However, as no evaluation has been performed by the authors, it remains unclear how well this kind of navigation supports interactive search in videos.

Huber et al. [12] focused on improving navigation in e-learning videos rather than entertaining videos and proposed the *Wipe'n'Watch* interface. Instead of using a timeline navigation with a seeker-bar they provide navigation between keyframes by wipe gestures. For the intended use case of e-lectures this works very well, since each keyframe represents a position of different slide appearance in the video. Their interface that operates in portrait mode is subdivided into two areas: (i) the upper area shows the actual video content and (ii) the lower area shows an overview of all available keyframes, i.e., available slides that act as direct access points. Their work also targets inter-video navigation, similarly to the idea of the *RotorBrowser* proposed by De Rooij et al. for desktop use [5]. Hence, vertical wiping allows to jump between semantically similar segments among videos; e.g., topically related segments. The availability of such related segments is indicated with an arrow in the upper right corner of the interface.

3   Keyframe Navigation Tree

The design of the *Keyframe Navigation Tree* (KNT) is inspired by the *frame stripes* visualization [31] (also known as *MO-images* [21]), where the (downscaled) center column of every frame is extracted and adjacently visualized in horizontal manner for video navigation support. While frame stripes can give clues on the video content structure in a very compact way, they cannot provide clear content information. Therefore, in difference to visualizing one vertical line for every frame in the video, we propose to visualize small excerpts from keyframes of the video. In order to make sure that the keyframes give a good representation of the actual video content, we use a special keyframe selection method that operates on the sub-shot level [19] (see Section 4 for details) and hence also works for video with long shots (e.g., camera pans).

Figure 1 shows an example of the KNT with three levels of detail. In the first level (1a) very small excerpts taken from the center of the keyframes are visualized. In the second level (1b) wider excerpts of the keyframes in the current browsing context are presented. Similarly, in the last level (1c) the entire keyframes are presented. All three levels are temporally synchronized (the red line represents the current position in the video in all three areas). A user can interact with each level independently by horizontal drag gestures for browsing and navigation, and by tap gestures for starting playback at a specific segment corresponding to the selected keyframe.

In our previous work on users' navigation behavior with common video players [3,29], we found that for known-item search tasks users typically navigate in a *coarse-to-fine* grained manner. They first try to quickly narrow down the search to a specific temporal area and then switch to closer inspection of that area. The idea of using three different granularity levels in the Keyframe Navigation Tree supports such a navigation behavior. It allows a user quickly narrowing down an interactive content search to a specific content area at the first level, and performing a fine-grained search with consideration of more details on the lower levels.

Fig. 1. The *Keyframe Navigation Tree* supports efficient navigation in video through a compact visualization of keyframes at three different levels of detail in a synchronized way: (a) In the top area only small excerpts of the keyframes are shown for coarse navigation. (b) In the area below the wider excerpts of keyframes are shown to provide more details. (c) Finally, in the last area the entire keyframe is shown to provide maximum detail but only for a fiew keyframes (i.e., a short range in the video). All three areas can be used independently and are kept temporally synchronized (red vertical line).

4   Keyframe Selection

Most keyframe selection approaches, proposed in the literature, are based on shot division. A shot is generally a group of continuous video frames with consistent visual characteristics such as color, texture, and motion, captured from a single camera at a time. A video sequence normally contains a (large) number of shots, and a shot boundary is the "gap" between two neighboring shots. The transitions between shots can be roughly classified into two groups: *hard cuts* and *gradual transitions*. A hard cut is an abrupt shot change that occurs between two continuous frames. Gradual transitions occur over multiple frames and the most typical are fade-in/out, dissolve and wipe, and many other types of gradual transition are also possible [10,18].

In order to handle video sequences of any kind, in this paper we make use of the very general information theoretic measure, the *Jensen-Shannon Divergence* (*JSD*) [34], for computing the difference between video frames. As a matter of fact, any general metrics (such as the widely used f-divergences [19]) that are computed efficiently can be used for this purpose. As for the computational mechanism of the keyframe selection, a simple and effective shot-based approach is utilized. In our approach, a video sequence is divided into non-overlapping shots. Considering the possibilities of gradual transitions between neighboring shots, a shot is then possibly divided into sub-shots. Finally one frame is chosen for each (sub-)shot. Here we use $D(f_i \| f_{i+1})$ to represent the difference between the i and $i+1$ video frames, calculated based on their corresponding normalized intensity histogram distributions, f_i and f_{i+1}. In practice, the distance between video frames is the sum of correspondences for the three RGB channels.

The computing procedure for the keyframe selection is concisely described as follows, and the details can be found in [19,32,33,34]. The Ds between each pair of two consecutive video frames are at first obtained. In order to locate D spikes indicating the existence of shot boundaries, a ratio $\delta = \frac{D}{D_w}$, where D_w is a local average of D on a w size temporal window, is used. A shot boundary is located where δ is greater than a pre-defined number δ^*. Next, a shot is grouped into several sub-shots if the content change of this shot is significant enough. The gradient of D_w, calculated as $\Delta(j) = D_w(f_j \| f_{j+1}) - D_w(f_{j-1} \| f_j)$, is employed to detect a significant content change. To prevent the use of small outliers of Δ, the local average of Δ on a temporal window, denoted as Δ_w, is actually used for this sake. Within a shot, if $|\Delta_w(j)|$ is greater than a pre-determined number Δ_w^*, then a significant content change inside this shot appears around the frame f_j. The left and right closest frames to f_j, f_m and f_n, which satisfy $\Delta_w(f_m) \simeq 0$ and $\Delta_w(f_n) \simeq 0$, are respectively located as the beginning and end of the significant content change. In fact, f_m and f_n are respectively the left and right boundaries of a sub-shot $[f_m, f_n]$ with the significant content change for this shot. In this way, a shot is segmented into several sub-shots according to the boundaries of all the sub-shots with significant content changes. For a sub-shot with a significant content change, the frame being most similar to all the others is selected as a keyframe. For a sub-shot without significant content change, the center frame of it is used as a keyframe.

5   Tested Interfaces

In order to evaluate the efficiency of the Keyframe Navigation Tree we implemented a video browsing tool with KNT navigation on an Apple iPad device (9.7-inch screen). Since no video browsing tools or video retrieval tools are available for such tablets (at least not for iOS devices), we used the default video player with a common seeker-bar as a comparative interface in our evaluation. Another good reason for using a common video player is the fact, that similar studies with known-item search tasks, to be solved with common video players on desktop computers, were performed in earlier work [3,25,29]. This allows for comparative considerations between different platforms.

Fig. 2. Left: The *Video Player* interface, which is the default video player widget available on iOS 7. **Right:** The *KNT Browser* interface, which is basically a common video player enhanced with an additional navigation area using the KNT approach.

Figure 2 shows the two different interfaces used for our evaluation. The interface on the left side is a common video player with the default configuration available on iOS 7. The right side of the figure shows the *KNT Browser* interface, which is basically also a video player (with a self-implemented seeker-bar/control set to avoid a fullscreen usage option, provided by the default video player) but with an additional navigation area using the KNT approach, as described in Section 3.

6   Evaluation

6.1   User Study

To evaluate the interfaces described in the last section we performed a user study with known-item search (KIS) tasks in videos. As dataset and search tasks we used the data from the *Single Run* session of the last Video Browser Showdown competition (VBS 2014) [27], which consists of ten different tasks to be performed in ten videos, each video with a duration of about one hour.

In total 20 participants (5 females) took part in the user study, aged from 18 to 40 (mean 28.15 years, SD 6.08). Every participant performed 5 tasks with the Video Player interface and 5 tasks with the KNT Browser interface, using a latin-square principle with random order to avoid familiarization effects.

The study procedure was similar to the one of the VBS: for each task a 20 seconds target clip was presented on the screen (without any interaction) and then the corresponding search interface appeared on the screen, opened with the video containing the target clip. For each task the user had a 3 minutes time frame to solve the task, otherwise the trial was marked as *timed out*. In contrast to the VBS we used no scoring – based on task solve time and the number of wrong submissions for the task, as done in the VBS – but only measured task solve time. If a user submitted the wrong segment, the trial was marked as *erroneous* (in accordance with the VBS we also used a tolerance of ±5 seconds at the beginning and the end of the target segment). A test session lasted for about half an hour per participant.

After a user finished the five trials with one interface, we requested her to fill out a questionnaire with Likert-scale ratings about the subjectively perceived workload of the interface, according to the NASA Task-Load-Index (TLX) [11].

6.2   Task Solve Time

For the evaluation of task solve time we used only correct trials and removed outliers (± 2 SD). The task solve times (for each interface) were positively skewed. Therefore, we performed a logarithmic transformation (that resulted in a close to normal distribution) before the data was analyzed. Figure 3a shows the geometric means for the two tested interfaces (i.e., the antilog of the mean of the log-transformed data): 28.11 seconds for the KNT Browser and 44.18 seconds for the Video Player interface.

This result is statistically significant, according to a *dependent paired-samples t-test* ($t(19) = -3.937, p < 0.005$). As also visible in the figure, the variance of search time was much smaller for the KNT Browser than for the Video Player, which indicates that the proposed interface much better supports KIS tasks.

When comparing these results to the average task solve time achieved by sophisticated video browsing tools in the VBS 2014 competition (which was 20.58 seconds for the *Visual KIS/Experts* run [24]), we can see that the achieved performance in our user study (mainly with novices) is quite remarkable. With the KNT Browser the users were even faster than the experts in the VBS 2013 competition (with sophisticated tools) [25], who required 40.5 seconds in average to solve KIS tasks in different but similarly long videos. For the same data set novices in a baseline study with a common video player on a desktop PC required 57.9 seconds in average [25].

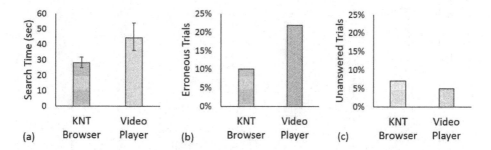

Fig. 3. (a) Geometric mean trial times per interface (error bars: 95% confidence interval), (b) Percentage of erroneous trials performed with each interface, (c) Percentage of timed out trials performed with each interface.

6.3   Errors and Timeouts

We also investigated the number of errors, which were performed with each interface. From a total of 100 search tasks performed with each interface (each of the 20 users solved 5 tasks with one interface), 22 were not correctly solved with the Video Player interface, whereas with the KNT Browser interface only 10 tasks were erroneous (see Figure 3b). A closer inspection revealed that almost half of the erroneous trials were caused by one specific task, which required to find a segment that reappeared in similar form in other locations of the corresponding video.

The number of timeouts (i.e., where a user needed more than 3 minutes for a task) was quite balanced between both interface. From a total of 200 trials with both interfaces, only 5 could not be solved with the Video Player interface, whereas 7 could not be solved with the KNT Browser interface. The timeouts were clearly task-dependent; all 12 timeouts were almost uniformly caused by three specific tasks.

6.4   Subjective Rating

Statistical analysis (Wilcoxon signed-rank tests) of the subjective evaluations of the interfaces showed that the KNT Browser performed significantly better in all seven categories of the NASA Tast-Load-Index [11], as shown in Figure 4. More specifically, all test participants found that the KNT Browser interface better supports known-item search tasks in videos ("Performance" in the figure, $Z = -3.920$, $p < 0.0005$), that is is less frustrating than the Video Player ($Z = -3.921$, $p < 0.0005$), and that it is more fun to use than the Video Player

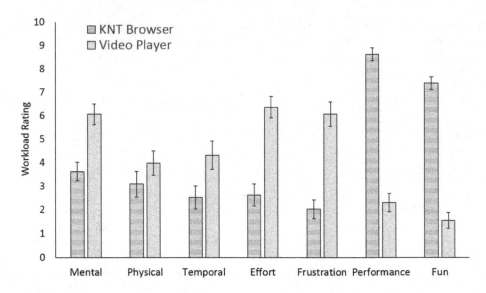

Fig. 4. Perceived workload rating (error bars:±s.e. of the mean)

($Z = -3.920$, $p < 0.0005$). Further, the vast majority found that the KNT Browser requires less effort ($Z = -3.659$, $p < 0.0005$), less mental demand ($Z = -2.576$, $p = 0.01$), less physical demand ($Z = -2.535$, $p = 0.011$), and produces less temporal pressure ($Z = -2.240$, $p = 0.025$) than the Video Player.

6.5 Preferred Interface

In the questionnaires at the end of a test session we also asked the participants about the preferred interface. The vast majority, more precisely 17 out of 20 users (85%), voted for the KNT Browsing interface, whereas only 3 preferred the Video Player.

7 Conclusions

A new approach for video navigation was presented in this paper. The navigation concept of the Keyframe Navigation Tree, inspired by frame stripes approach [31], gives much more insight to the actual content, while still providing a good overview of the content structure in a broad range. It allows for both coarse-grained navigation and fine-grained navigation, and therefor optimally supports the observed navigation behavior of video players, when used for KIS tasks, as reported in [3,29]. It provides both good overview of the video content structure (i.e., browsing context) and detailed information about keyframes. To allow for usage in several domains we used a novel method of keyframe selection, which operates on the sub-shot level and therefore also provides good content summarization not only for common videos with many short shots but also for videos with very long shots, or single-shot recordings (e.g., video recordings of intervention in the medical domain [22]).

We have evaluated the interface in a user study with 20 participants and the same known-item search tasks that were used in the most recent Video Browser Showdown competition (VBS 2014) at MMM 2014. The evaluation results show the proposed KNT Browser allows for significantly faster search tasks in videos than a common video player, which is typically used as a baseline in the evaluation of the VBS. Moreover, we could show that the proposed KNT Browser causes a very low workload (significantly lower than the Video Player) and is highly enjoyable to use. Furthermore, it is the preferred interface by the vast majority (85%) of tested users.

Acknowledgement. This work was funded by the Federal Ministry for Transport, Innovation and Technology (bmvit) and Austrian Science Fund (FWF): TRP 273-N15 and the European Regional Development Fund and the Carinthian Economic Promotion Fund (KWF), supported by Lakeside Labs GmbH, Klagenfurt, Austria. The work by Qing Xu has been funded by Natural Science Foundation of China (61471261, 61179067, U1333110).

References

1. Ahlström, D., Schoeffmann, K.: A visual search user study on the influences of aspect ratio distortion of preview thumbnails. In: Zhang, J., Schonfeld, D., Feng, D.D., Nanyang, J.C., Hanjalic, A., Magli, E., Pickering, M., Friedland, G., Hua, X.-S. (eds.) Proceedings of the 2012 IEEE International Conference on Multimedia and Expo Workshops, Los Alamitos, CA, USA, pp. 546–551. IEEE Computing Society (July 2012)
2. Bailer, W., Schoeffmann, K., Ahlström, D., Weiss, W., Del Fabro, M.: Interactive evaluation of video browsing tools. In: Li, S., El Saddik, A., Wang, M., Mei, T., Sebe, N., Yan, S., Hong, R., Gurrin, C. (eds.) MMM 2013, Part I. LNCS, vol. 7732, pp. 81–91. Springer, Heidelberg (2013)
3. Cobârzan, C., Hudelist, M.A., Del Fabro, M.: Content-based video browsing with collaborating mobile clients. In: Gurrin, C., Hopfgartner, F., Hurst, W., Johansen, H., Lee, H., O'Connor, N. (eds.) MMM 2014, Part II. LNCS, vol. 8326, pp. 402–406. Springer, Heidelberg (2014),
http://dx.doi.org/10.1007/978-3-319-04117-9_46
4. Rooij, O.d., Snoek, C.G., Worring, M.: Query on demand video browsing. In: Proceedings of the 15th International Conference on Multimedia, pp. 811–814. ACM (2007)
5. de Rooij, O., Snoek, C.G.M., Worring, M.: Mediamill: semantic video search using the rotorbrowser. In: Proceedings of the 6th ACM International Conference on Image and Video Retrieval, pp. 649–649. ACM Press (2007)
6. Dragicevic, P., Ramos, G., Bibliowitcz, J., Nowrouzezahrai, D., Balakrishnan, R., Singh, K.: Video browsing by direct manipulation. In: Proceedings of the SIGCHI Conference on Human Factors in Computing Systems, CHI 2008, pp. 237–246. ACM, New York (2008)

7. Del Fabro, M., Böszörmenyi, L.: AAU video browser: Non-sequential hierarchical video browsing without content analysis. In: Schoeffmann, K., Merialdo, B., Hauptmann, A.G., Ngo, C.-W., Andreopoulos, Y., Breiteneder, C. (eds.) MMM 2012. LNCS, vol. 7131, pp. 639–641. Springer, Heidelberg (2012)

8. Girgensohn, A., Shipman, F., Wilcox, L.: Adaptive clustering and interactive visualizations to support the selection of video clips. In: Proceedings of the 1st ACM International Conference on Multimedia Retrieval, p. 34. ACM (2011)

9. Goeau, H., Thièvre, J., Viaud, M.-L., Pellerin, D.: Interactive visualization tool with graphic table of video contents. In: 2007 IEEE International Conference on Multimedia and Expo, pp. 807–810. IEEE (2007)

10. Hanjalic, A.: Shot-boundary detection: unraveled and resolved? IEEE Transactions on Circuits, Systems, and Video Technology 12(2), 90–105 (2002)

11. Hart, S.G., Staveland, L.: Development of NASA-TLX (Task Load Index): Results of empirical and theoretical research. In: Hancock, P.A., Meshkati, N. (eds.) Human Mental Workload, pp. 139–183. Elsevier, Amsterdam (1988)

12. Huber, J., Steimle, J., Lissermann, R., Olberding, S., Mühlhäuser, M.: Wipe'n'watch: spatial interaction techniques for interrelated video collections on mobile devices. In: Proceedings of the 24th BCS Interaction Specialist Group Conference, BCS 2010, pp. 423–427. British Computer Society, Swinton (2010)

13. Hürst, W., Götz, G., Welte, M.: Interactive video browsing on mobile devices. In: Proceedings of the 15th International Conference on Multimedia, MULTIMEDIA 2007, pp. 247–256. ACM, New York (2007)

14. Hürst, W., Meier, K.: Interfaces for timeline-based mobile video browsing. In: Proceedings of the 16th ACM International Conference on Multimedia, pp. 469–478. ACM (2008)

15. Hürst, W., Snoek, C.G.M., Spoel, W.-J., Tomin, M.: Size matters! how thumbnail number, size, and motion influence mobile video retrieval. In: Lee, K.-T., Tsai, W.-H., Liao, H.-Y.M., Chen, T., Hsieh, J.-W., Tseng, C.-C. (eds.) MMM 2011 Part II. LNCS, vol. 6524, pp. 230–240. Springer, Heidelberg (2011)

16. Jansen, M., Heeren, W., van Dijk, B.: Videotrees: Improving video surrogate presentation using hierarchy. In: International Workshop on Content-Based Multimedia Indexing, CBMI 2008, pp. 560–567. IEEE (2008)

17. Karrer, T., Wittenhagen, M., Borchers, J.: Pocketdragon: a direct manipulation video navigation interface for mobile devices. In: Proceedings of the 11th International Conference on Human-Computer Interaction with Mobile Devices and Services, MobileHCI 2009, pp. 47:1–47:3. ACM, New York (2009)

18. Lienhart, R., Pfeiffer, S., Effelsberg, W.: Video abstracting. Commun. ACM 40(12), 54–62 (1997)

19. Luo, X., Xu, Q., Sbert, M., Schoeffmann, K.: F-divergences driven video key frame extraction. In: 2014 IEEE International Conference on Multimedia & Expo (ICME 2014). IEEE (2014)

20. Matejka, J., Grossman, T., Fitzmaurice, G.: Swifter: Improved online video scrubbing. In: Proceedings of the SIGCHI Conference on Human Factors in Computing Systems, CHI 2013, pp. 1159–1168. ACM, New York (2013)

21. Mueller-Seelich, H., Tan, E.: Visualizing the semantic structure of film and video (2000)

22. Münzer, B., Schoeffmann, K., Böszörmenyi, L.: Relevance segmentation of laparoscopic videos. In: Proceedings of the 2013 IEEE International Symposium on Multimedia (ISM 2013), pp. 1–8 (2013)

23. Over, P., Awad, G., Michel, M., Fiscus, J., Sanders, G., Shaw, B., Kraaij, W., Smeaton, A.F., Quénot, G.: Trecvid 2012 – an overview of the goals, tasks, data, evaluation mechanisms and metrics. In: Proceedings of TRECVID 2012 (2012)
24. Schoeffmann, K.: A user-centric media retrieval competition: The video browser showdown 2012-2014. IEEE Multimedia Magazine, 1–5 (to appear, 2014)
25. Schoeffmann, K., Ahlstrom, D., Bailer, W., Cobarzan, C., Hopfgartner, F., McGuinness, K., Gurrin, C., Frisson, C., Le, D.-D., Del Fabro, M., Bai, H., Weiss, W.: The video browser showdown: a live evaluation of interactive video search tools. International Journal of Multimedia Information Retrieval 3, 113–127 (2014)
26. Schoeffmann, K., Ahlström, D., Böszörmenyi, L.: A user study of visual search performance with interactive 2D and 3D storyboards. In: Detyniecki, M., García-Serrano, A., Nürnberger, A., Stober, S. (eds.) AMR 2011. LNCS, vol. 7836, pp. 18–32. Springer, Heidelberg (2013)
27. Schoeffmann, K., Bailer, W.: Video browser showdown. ACM SIGMultimedia Records 4(2), 1–2 (2012)
28. Schoeffmann, K., Boeszoermenyi, L.: Video browsing using interactive navigation summaries. In: Proceedings of the 7th International Workshop on Content-Based Multimedia Indexing, pp. 243–248. IEEE, Chania (2009)
29. Schoeffmann, K., Cobârzan, C.: An evaluation of interactive search with modern video players. In: IEEE International Conference on Multimedia and Expo Workshops (ICMEW), pp. 1–4 (July 2013)
30. Schoeffmann, K., Hopfgartner, F., Marques, O., Boeszoermenyi, L., Jose, J.M.: Video browsing interfaces and applications: a review. SPIE Reviews 1(1), 018004 (2010)
31. Schoeffmann, K., Taschwer, M., Boeszoermenyi, L.: The video explorer: A tool for navigation and searching within a single video based on fast content analysis. In: Proceedings of the First Annual ACM SIGMM Conference on Multimedia Systems, MMSys 2010, pp. 247–258. ACM, New York (2010)
32. Xu, Q., Li, X., Yang, Z., Wang, J., Sbert, M., Li, J.: Key frame selection based on jensen-rényi divergence. In: 2012 21st International Conference on Pattern Recognition (ICPR), pp. 1892–1895. IEEE (2012)
33. Xu, Q., Liu, Y., Li, X., Yang, Z., Wang, J., Sbert, M., Scopigno, R.: Browsing and exploration of video sequences: A new scheme for key frame extraction and 3d visualization using entropy based jensen divergence. Information Sciences 278, 736–756 (2014)
34. Xu, Q., Wang, P.-C., Long, B., Sbert, M., Feixas, M., Scopigno, R.: Selection and 3d visualization of video key frames. In: Proceedings of IEEE International Conference on Systems Man and Cybernetics (SMC), pp. 52–59 (2010)

Large-Scale Image Mining with Flickr Groups

Alexandru Lucian Ginsca[1,2], Adrian Popescu[1], Hervé Le Borgne[1],
Nicolas Ballas[3], Phong Vo[1], and Ioannis Kanellos[2]

[1] CEA, LIST, Vision and Content Engineering Laboratory, Gif sur Ivette, France
[2] TELECOM Bretagne, France
[3] Université de Montréal
{alexandru.ginsca,adrian.popescu,herve.le-borgne,phong.vo}@cea.fr,
nicolas.ballas@umontreal.ca, ioannis.kanellos@telecom-bretagne.eu

Abstract. The availability of large annotated visual resources, such as
ImageNet, recently led to important advances in image mining tasks.
However, the manual annotation of such resources is cumbersome. Exploiting Web datasets as a substitute or complement is an interesting but
challenging alternative. The main problems to solve are the choice of the
initial dataset and the noisy character of Web text-image associations.
This article presents an approach which first leverages Flickr groups to
automatically build a comprehensive visual resource and then exploits
it for image retrieval. Flickr groups are an interesting candidate dataset
because they cover a wide range of user interests. To reduce initial noise,
we introduce innovative and scalable image reranking methods. Then,
we learn individual visual models for 38,500 groups using a low-level
image representation. We exploit off-the-shelf linear models to ensure
scalability of the learning and prediction steps. Finally, *Semfeat* image
descriptions are obtained by concatenating prediction scores of individual models and by retaining only the most salient responses. To provide a
comparison with a manually created resource, a similar pipeline is applied
to ImageNet. Experimental validation is conducted on the ImageCLEF
Wikipedia Retrieval 2010 benchmark, showing competitive results that
demonstrate the validity of our approach.

1 Introduction

As predicted a few years ago [20], research in visual and multimedia recognition
has strongly benefited from the availability of manually labeled large-scale image
and video collections. The ImageNet representation [7] of nearly 22,000 concepts
with approximately 14 million images according to a hierarchy of concepts was
thoroughly exploited. For instance, a subset of ImageNet was used to train large
convolutional neural networks (CNNs), leading to the design of powerful visual
features such as Overfeat [21]. In the domain of large-scale recognition, another
important advancement came from the use of classifier outputs as descriptors,
with classemes [22] being one of the early methods of this type. The central
novelty of this approach was not the use of classifier outputs as feature in itself,
but that the signatures were built from many category-specific classifiers. It
therefore needs a sufficiently large-scale set of annotated data.

X. He et al. (Eds.): MMM 2015, Part I, LNCS 8935, pp. 318–334, 2015.

These approaches are very promising but raise new problems, concerning, in particular, the availability of the underlying resources. Manually labeled datasets are the result of sustained effort provided by motivated communities of researchers [21], eventually supplemented with crowdsourcing [20,7]. An important limitation of this approach is that manual annotation is a repetitive task and annotators tend to become demotivated. In addition, when conducted on a large scale, crowdsourcing has a non-negligible financial cost and dedicated funding is difficult to obtain. A promising way to circumvent the lack of annotated data is to use images shared on multimedia social networks (OSNs), such as Flickr. An advantage of this type of resource compared to formal "annotation tasks" is that data are annotated by a community of users motivated to make their content accessible.

Here, we firstly propose an approach to design semantic image features which are built on top of an automatically processed large-scale collection. Then we test these features in image retrieval; a particular attention is paid to the reduction of the noise inherent to Web collections. We address the following open and recurring research questions:

Q1 - Do features, established using automatically and manually built resources, have similar performances?

Q2 - Is it possible to build image representations whose parts efficiently convey semantic meaning?

Q3 - Can we design features which ensure a good coverage of the semantic space?

Q4 - How to build semantic representations which are at the same time compact and efficient?

Q5 - Can we learn semantic features with a resource and exploit them to mine other datasets?

Q1 is the central question addressed here. Our approach is validated only if performances obtained with the two types of resources are comparable. To our knowledge, the direct comparison of automatically and manually built large-scale resources was not properly addressed in literature. *Q2* relates to the use of semantic features as an alternative to low level features, such as bags-of-visual-words, Fisher Kernels [11] or CNN features [21]. Unlike low-level feature vectors, semantic features directly convey humanly understandable information. Consequently, they would be a promising candidate for bridging the semantic gap if they had been both precise and comprehensive, two conditions which are not yet met. For instance, the performances of meta-classes [1], which exploit a large part of ImageNet, lag behind those of Fisher Kernels, whose performances are in turn lower than those of CNN features [13]. We assume that Flickr groups are a good candidate to answer *Q3*, provided that they are properly selected and "cleaned" via image reranking. Regarding *Q4*, we note that textual documents can be searched efficiently because they are sparse and a similar property is desirable for semantic image features. Surprisingly, sparsification is not directly addressed in existing work, with the closest proxy being quantization done for more efficient signature storage [1]. *Q5* is a hot topic in computer vision and relates to transferring knowledge gained from a dataset to other datasets. Among others, it was recently tackled by [17] but not in the context of semantic features, as it is in this paper.

2   Related Work

Flickr Groups. Flickr groups are often used as a playground for investigating photo sharing communities. In a pioneering work, Negoescu *et al.* [16] looked at the involvement of users in groups and found, among others, that user group loyalty is generally low and most users share the same photos in different groups. Grabowicz *et al.* [10] make a distinction between topical and social groups and proposed several methods to classify a group into one of these two categories.

While informative for our purposes, such works do not tackle the exploitation of group content for image mining. Chen *et al.* [4] are among the first to exploit the visual content of groups. They use Flick group search for a set of 62 concepts and rank the returned groups based on 4 factors related to group popularity. They train dedicated SVM models for concepts and use them independently to recommend tags and groups. More closely related to our work is the idea presented in [24], where the similarity between two images is computed by leveraging the prediction scores of a set of 103 hand picked Flickr groups. Each probability is estimated using a SVM classifier trained over low-level visual features. The resulting vectors are also briefly tested in clustering and classification tasks, for which comparable results with visual features are reported.

Key differences with our work arise from the way groups are modeled. We propose several image ranking methods that improve individual classifier performance when using an initial training set of only 300 images, whereas in [24] the learning is performed on a large training set (15,000 to 30,000 images). In addition, we sparsify the features and thus enable fast retrieval over large datasets.
Semantic Image Representation. The availability of large image collections and of scalable machine learning techniques has led to a resurgence of semantic representation for image classification [14,22]. Li et al. [14] introduced Object Bank, where an image is represented as a scale-invariant response map of 200 pre-trained object detectors. Torresani et al. [22] also introduced a semantic representation using a fix number of hand selected binary classifiers. Each classifier is applied on the whole image input. Due to the relatively small number of visual concepts considered, early semantic representations ensured only a limited coverage of the semantic space. To tackle this issue, Bergamo and Torresani [1] learned the visual concepts of the semantic representation directly from the data. They however use 13 different features and "lift-up" each one to approximate a non-linear kernel, that is a much more costly approach than ours.
Convolutional Neural Networks (CNN) for Large-Scale Image Classification. CNNs have recently shown impressive image classification performances in the large-scale visual recognition challenge ILSVRC [13]. Compared to traditional low-level features such as Fisher Vector [18], the use of CNN brought down the ILSVRC error rate from 0.26 to 0.15 in 2012 and 0.11 in 2013 [13,21].

Moreover, CNN-based feature extractors, such as *Overfeat* [21], were publicly released. These extractors provide pre-trained weights files and facilitate the extraction of features for new image collections. The outputs of their final layer are semantic image representations but they are limited to the 1,000 ILSVRC concepts, due to computational complexity of the algorithm and their need of labeled training data. Very recently, the authors of [17] and [8] exploit CNN to build mid-level features and report impressive results on various image classification datasets. The focus here is not on building new CNN representations but rather on exploiting them as basic features in order to build powerful semantic representations.

3   Datasets

We collected Flickr groups starting with an initial list of 100 million metadata pieces from which we extracted the most frequently occurring groups. Then we downloaded group metadata for the most frequent 50,000 of them and retained the 38,500 groups which include at least 300 images. Given that some images were withdrawn by users before crawling, the initial dataset contains approximately 11 million images. This selection is done in order to ensure that a reasonable amount of data is fed into the visual classifiers which are build from Flickr groups.

ImageNet [7] is a visual resource built on top of WordNet. It contains manually labeled examples for nearly 22,000 concepts. From this dataset, we selected the 17,462 concepts which have at least 100 associated images so that the resulting subset includes around 13 million images.

3.1   Dataset Preprocessing

Visual preprocessing focuses on the extraction of Overfeat features [21] of Flickr groups and of ImageNet concepts. Groups and concept Overfeat features are written as $FG_o = \{I_i\}_{i=1...N}$ and respectively $IN_o = \{I_i\}_{i=1...N}$ where I_i is Overfeat feature associated to an image. The default configuration, i.e. layer 19 of the small network provided by Overfeat, is used for representing the datasets and for experiments. All images are represented by a vector of 4096 dimensions which is further normalized using L2. A similar extraction process is applied to a set of 4,000 diversified images from which we select negative examples during group/concept modeling.

Text preprocessing consists in extracting the most salient tags of each group. Groups are structured thematically but a single tag might not be sufficient to describe them. Tags are ranked by the number of unique users which annotate images of a group with them. This measure is chosen instead of tag frequency, which is sensitive to bulk uploads, in order to maximize the social relevance of tags. After an initial examination, we empirically retain the top three tags as a textual representation of groups and write this representation as $FG_t = \{T_1, T_2, T_3\}$.

4 Flickr Group Modeling

4.1 Group Analysis

Flickr groups may be formed around specific concepts (brands of cars, animals etc.), abstract concepts (beauty, frightening imagery) or they may gather images taken with a specific brand of camera or camera setting (black and white, light setting). We investigate how much visual variability there is among the images of a group. We measure the visual coherence of a group by the 5-fold cross-validation score reported by the model trained on that group, as described in Section 5. Figure 1 confirms that the ranking based on the cross-validation score manages to separate groups depicting concepts with a clear visual representation from those focusing on photography or generic concepts. Only the first 30,000 groups ranked by cross-validation are retained for the experiments described in Section 6. Most other groups are not conceptually oriented and have little added value as dimensions of a semantic image representation.

Fig. 1. Word clouds of the most frequent tags found in the last 10% groups (left word cloud) and the first 10% groups (right word cloud) in a ranking induced by the cross-validation accuracy score

Performances of semantic descriptors obtained from ImageNet and Flickr groups are compared throughout the paper. For a better understanding of the results, we are interested in the semantic overlap between ImageNet concepts and groups and the particularities of each data source. We consider that an ImageNet concept and a group match if at least one term describing the concept has an exact match in the FG_t representation of the group. From the total of 17,462 ImageNet concept names, only 2,567 are found in groups. The concepts from ImageNet that do not appear in groups include species of animals (*African elephant, eastern grey squirrel*) or technical equipment (*computer keyboard, microphone*). When first looking at Flickr groups, we find 28,243 groups that have at least one tag matching an ImageNet concept. Among the first groups ranked by the number of contributing users that do not have an ImageNet correspondent, we notice a high frequency of geographical locations ({*paris, france, eiffel*}, {*croatia, sea, dubrovnik*}) and car brands ({*bugatti, veyron, supercar*}, {*lamborghini, gallardo, murcielago*}). The main advantage over ImageNet comes from the nature of the concepts found in groups that are formed through social consensus, as opposite to ImageNet, where the specific concepts come from the leafs of the WordNet hierarchy and may be less encountered in images shared through online platforms.

4.2 Group Image Reranking

A part of the images associated to Flickr groups are irrelevant and direct learning of visual models with all group images is sub-optimal. We introduce image reranking techniques in order to automatically reduce the amount of noise present in groups. Existing approaches either exploit tags or rely both on tagging and/or visual content [3]. We test two classical methods and introduce moreover one which highlights a social cue [12], i.e. the identity of the uploader. Focus is put on scalability in order to be able to process Flickr groups efficiently. All methods use the Overfeat representation of group images described in 3.1. Furthermore, we implemented:

- avg_{sim} - this baseline method computes I_{avg}, the average Overfeat representation of each group, and ranks the images of the group by considering $sim(I_i, I_{avg}) = \frac{1}{||I_i - I_{avg}||^2}$, the inverse of the L2 distance from the average representation. The intuition supporting this method is that the similarity with group average is a good indicator of image relevance.
- kNN - classical method which compares group images with a set of diversified negatives in order to favor images which are best linked to other images of the same group. We keep computation cost low by choosing as many negative as there are images in the target group. The reranking score of each target image is given by the position of the 10th group image in the list of similar images which includes both positives and negatives. The higher this position is, the better the image rank will be. This reranking approach is motivated by the assumption that relevant images are more similar to other images of the group than to images from other groups.
- $skNN$ - is a "social" version of kNN in which all images which come from the same user as the target images are excluded from the list of similar images. Here we assume that an image is more likely to be relevant if it is visually similar to images uploaded by other users. $skNN$ is more robust to bulk upload behaviors than the simple kNN algorithm.

Training images are sorted according to one of the methods described above. Then only the top of the reranked list, referred as *cut*, is retained for group modeling.

5   Semantic Features Building

We combine a visual resource, initial low-level features, such as FG_o or IN_o, and an array of individual concept classifiers in order to compute a semantic representation of images. Simply put, our approach can be summed-up as: "use an appropriate large-scale visual resource, represent concepts with linear models, exploit a good low-level feature, and sparsify to retain the most salient dimensions". The focus is on the exploitation of automatically built resources, using for instance Flickr groups, but the pipeline is generic and is also applied to ImageNet. The two semantic features are named $Semfeat^{FG}$ and $Semfeat^{IN}$ respectively.

5.1   Model Learning and Aggregation

The individual dimensions of semantic features can be modeled with different types of classifiers. Considering the size of the problem we tackle, binary classifiers are used. In comparison with a multiclass classifier, they present the advantages of (i) remaining computable for any number of classes and having lower constraints on the training dataset size (ii) being easily "extensible", i.e. a given concept can be added/removed independently from other dimensions. For scalability, we model each concept with linear models, which are very fast to compute and exhibit good performance in practice [18,11]. Hence, each individual model is learned from a set $\{(I_i, y_i)\}_{i=1...N}$ of training images and their corresponding binary label ($y_i \in \{-1, +1\}$). Models are learned with L2-regularized logistic regression, which solves the following unconstrained optimization problem:

$$W^c = \arg\min_{\mathbf{w}} \frac{1}{2}\mathbf{w}^T\mathbf{w} + C \sum_{i=1}^{N} log(1 + e^{-y_i\mathbf{w}^T x_i}) \tag{1}$$

Where $x_i \in \mathbb{R}^{S_f}$ is an image low-level feature reflecting the visual content of image I_i. This feature is later augmented with a last dimension fixed to 1 to take into account the model bias: $\mathbf{f}_i^T \leftarrow [x_i^T\ 1] \in \mathbb{R}^{(S_f+1)}$. In practice, (1) is solved in the primal using a trust region Newton method, relying on the liblinear implementation [9].

Individual visual models map image features into a semantic space of size sup defined by Flickr groups or ImageNet concepts. We denote by $\mathbf{W} = \{\mathbf{W}^1, ..., \mathbf{W}^{sup}\} \in \mathbb{R}^{(S_f+1)\times sup}$ as the matrix concatenating all individual visual models learned by (1). Using the $\mathbf{W}$, an initial image feature $\mathbf{f} \in \mathbb{R}^{(S_f+1)}$ is mapped to its semantic representation $\mathbf{x} \in \mathbb{R}^{S_s}$ through

$$\mathbf{x} = \mathbf{W}^T\mathbf{f}. \tag{2}$$

$\mathbf{x}$ (a short notation for $Semfeat$) contains semantic information as it aggregates the classification scores of $\mathbf{f}$ given all available Flickr groups or ImageNet concepts. (2) is therefore comparable to a soft assignment encoding since all binary classifiers contribute to the semantic representation. This feature is dense since all classifier outputs are taken into account. One drawback of such representation is that the effects of a relevant classifier can be smoothed by the accumulation of weights of an array of poorer classifiers. For instance, a concept whose output is 0.95 has lower importance than a combination of 5 concepts with 0.2 outputs.

5.2 Semantic Feature Sparsification

Liu et al. [15] showed that soft assignment encoding is not optimal as it discards the manifold geometric structure of the mapped space. Beyer et al. [2] demonstrated that distances between points often become less meaningful in high-dimensional space. Consequently, the expressive power of high-dimensional features is limited. Moreover, empirical analysis of high-dimensional features [26,25] shows that they often lie on a manifold which has a much smaller intrinsic dimensionality. Such a manifold structure implies that the neighborhood of a feature point is homeomorphic to the Euclidean space into a local region only. Thus, computing distance (or proximity) between features is meaningful within a local region only. Outside of this region, two local points considered similar using a distance measure might actually be far from each other.

Given this manifold assumption, a classification score indicating the proximity of a feature $\mathbf{f}$ to a concept c is reliable only when $\mathbf{W}^c\mathbf{f}$ is large. The use of all classifiers degrades the semantic representation as concepts distant from $\mathbf{f}$ do not bring information. To obtain a more reliable representation, we leverage the manifold geometry by adding a locality constraint [25] to $\mathbf{x}$,

$$\mathbf{x} = \arg\min_{\bar{\mathbf{x}}} \|\mathbf{W}^T\mathbf{f} - \bar{\mathbf{x}}\|_2^2 + \lambda\|\mathbf{D}\bar{\mathbf{x}}\|_2^2, \text{ s.t. } \mathbf{1}^T\bar{\mathbf{x}} = 1, \tag{3}$$

where $\mathbf{D} \in \mathbb{R}^{S_s \times S_s}$ is a diagonal matrix such that each $(\mathbf{D})_{cc}$ penalizes the c-th basis of $\mathbf{x}$ when the c-th concept is unlikely to be observed in $\mathbf{f}$. More specifically,

$$(\mathbf{D})_{cc} = e^{-\gamma\mathbf{W}^{cT}\mathbf{f}} \,\forall c \in [1..S_s]. \tag{4}$$

In (4), $\mathbf{W}^c$ is the c-th linear model and γ a parameter which controls locality. In (3), the first term verifies that the semantic representation $\mathbf{x}$ is close to $\mathbf{W}^T\mathbf{f}$, i.e. the projection of $\mathbf{f}$ in the semantic space. The second term penalizes the semantic representation $\mathbf{x}$ that are non-local. It ensures that only the semantic dimensions of $\mathbf{x}$ which yield high-likelihoods will be selected. λ is a trade-off parameter between both terms.

The locality constraint in (3) chooses dimensions $\mathbf{x}$ based on the likelihood of the corresponding concepts with $\mathbf{f}$. Given this observation, an efficient sparse approximation of (3) can be derived. We define the matrix $\mathbf{W}' = \{\mathbf{W}'^1, ..., \mathbf{W}'^{S_s}\} \in \mathbb{R}^{(S_f+1)\times S_s}$ which considers only the top K individual visual models ($K << S_s$) that yield the highest scores on $\mathbf{f}$: $\mathbf{W}'^c = \mathbf{W}^c$ if the c-th concept achieves one of the K highest scores, 0 otherwise. We obtain an semantic representation which approximates locality through

$$\mathbf{x} = \mathbf{W}'^T\mathbf{f}. \tag{5}$$

Compared to (3), the only parameter is K which directly controls the locality of the solution.

6 Evaluation

Our objective is to evaluate the effectiveness of our semantic features in image retrieval. The central question is whether the performances obtained using an automatically constituted resource, based for instance on Flickr groups, can be similar to those of a manually built resource, such as ImageNet. We also evaluate the effects of semantic feature coverage and sparsification. Popular wisdom considers that an image is worth a thousand words. We test how many of these words should actually be used for image retrieval. In a preliminary experiment, we tune the learning algorithms and select the best Flickr group image reranking method. Then we perform the main experiment in a CBIR scenario using the ImageCLEF Wikipedia Retrieval 2010 collection [23]. Additionally, we present a simple fusion experiment to assess the effect of combining text and content based IR.

6.1 Preliminary Experiment

The purposes of this preliminary experiment are to tune individual model learning algorithm and to evaluate results obtained with the different image reranking methods. The validation dataset used here is created by matching Flickr groups, used for training, with ImageNet concepts whose images are used for testing. We first pre-select a list of groups which have the first tag of their textual representation FG_t, present in ImageNet. For instance, we match the Flickr group 1000405@N24 (top tags *memorial, war, warmemorial*) with the ImageNet concept *memorial*, defined as *a structure erected to commemorate persons or events*. To ensure diversity, each tag is used only once. We manually validate the alignment between groups and ImageNet concepts to obtain a final list of 367 pairs. Training is done using Flickr group images as positive examples and a diversified negative set extracted from Flickr. Test is performed with the images of the corresponding ImageNet concept as positives and a fixed list of over 4000 images of other concepts as negatives. Logistic regression tuning is done via a grid search on the parameter C. The best average classification scores are obtained with $C = 10$, which is used to learn all binary models included in *Semfeat*.

The reranking methods are compared by using them in an image retrieval scenario. For instance, assuming that *palm tree* is part of the 367 pairs, the purpose is to use the models trained with Flickr groups in order to classify ImageNet test images and ImageNet negatives. Using classification scores, we produce a ranking and assume that the best reranking method is the one which places the most *palm tree* among the top images. The precision at 100 (P@100), i.e. the number of positives among the first 100 results, is a used for assessment. This measure accounts for the capacity of the reranking methods to favor positive test examples over negatives and, indirectly, for the quality of the reranked training set. Also, compared to classical cross validation, all reranking cut-offs are evaluated using the same test set; thus, results are easier to compare.

The results obtained with the three reranking methods and different cut-off points are presented in Table 1. While the P@100 differences with the baseline are small, some improvement is obtained with all reranking methods. More interestingly, the use of *skNN* provides slightly better results compared to *kNN*, indicating that the use of social cues for reranking is beneficial. Given the

Table 1. P@100 results for different reranking methods and different cut-off percentages (*cut*) for the selection of reranked images. The baseline corresponds to a no cut-off, i.e. the rightmost column.

	cut[%]			
	70	80	90	100
avg_{sim}	0.915	0.917	0.92	0.917
kNN	0.918	0.92	0.92	0.917
$skNN$	**0.922**	**0.921**	**0.922**	0.917

results presented in Table 1, we will use *skNN* at different cut-off points in retrieval experiments.

6.2   CBIR and Fusion Evaluation

The main objective here is to assess the usefulness of *Semfeat* in a CBIR task performed over a diversified dataset. Wikipedia Retrieval 2010 was created as part of the ImageCLEF evaluation campaign[1] and is publicly available. It includes 237,434 Wikimedia images which were extracted from a large set of Wikipedia articles and includes a wide range of content. It it thus fitted for ad-hoc image retrieval experiments, in which any query can be submitted to the process. To ensure comparability with other methods tried on this dataset, we report mean average precision (MAP) performances. The 2010 query set contains 70 diversified queries, with 118 associated image examples. Retrieval over the Wikipedia collection is challenging because the image content is highly diversified. The Wikipedia Retrieval ground truth has been built using a pooling approach and is therefore incomplete [23]. To improve comparability, we extend the original ground truth (noted *origGT*) by pooling the new runs proposed here. This extension (noted *extGT*) is realized using similar topic narratives and a majority voting with three relevance judgments per image. We had limited resources and assessed only the new images appearing in the top 20 results of a selection of runs are annotated, compared to a pooling depth of 100 used for establishing the initial ground truth[2]. If a collection image is similar to two examples, the highest similarity score is retained. Similarities between images are computed using the cosine measure. The following runs are used in our main experiment:

- *Fisher* - existing baseline which exploits a version of Fisher Vectors adapted for CBIR [5].
- *Overfeat* - baseline which exploits the default outputs of the CNN-based feature extractor presented in [21], using the small network to speed-up feature computation. These features are also exploited to build individual models included in *Semfeat*.

---

[1] http://www.imageclef.org/

[2] The extended ground truth will be made available to other researchers.

- $SemfFeat^{IN}$ - semantic feature based on the 17462 ImageNet concepts which were modeled here. Results are reported for sparsification $K = 10$, corresponding to the best $Semfeat_{full}^{IN}$ MAP in Figure 2.
- $Semfeat_{cut}^{FG}$ - semantic feature based on the 30,000 groups with the highest cross-validation scores from the initial 38,500 group dataset.
- Txt [19] - to our knowledge, text run with best performances on the collection.
- $Fusion$ - fusion of image ranks from Txt and $Semfeat_{80}^{FG}$ with linear weighting.

Fisher Vectors and Overfeat are strong baselines, representing both pre-CNN and CNN features. Direct comparison to meta-classes and classemes was not performed since they show performances inferior to Fisher Vectors [1]. $Semfeat_{80}^{FG}$ and $Semfeat^{IN}$ are introduced here and, while based on $Overfeat$, they represent images semantically and cover significantly more concepts.

Sparsification Evaluation One of our central objectives is to create features which are in the same time efficient and compact. To this end, sparsity is a desirable property of document representations since it allows one to use inverted indexes in order to search massive datasets in real time. Sparsification is applied to semantic features built on top of 30,000 Flickr Groups and of ImageNet. These datasets have respectively 30,000 and 17,462 concepts available in each case. In Figure 2, we vary the sparsification factor $K = \{5, 10, ..., 100\}$ in order to thoroughly evaluate the effect of sparsification. Very interestingly, all group based features compare favorably with $Semfeat$ for $K > 10$. This finding confirms that Flick groups can be successfully used in CBIR as a substitute for manually built resources. The results presented in Figure 2 also show that the most interesting MAPs are obtained when 10 to 50 most salient concepts detected in them are used in the representation, with small peaks around $K = 30$. This finding has an important practical implication since $Semfeat$ features can be efficiently represented using inverted indexes. Inverted search is much faster than a brute force search, which is needed for dense low-level features, including Overfeat and Fisher Kernels.

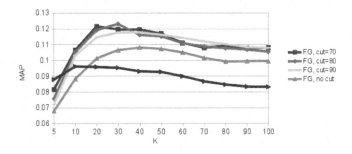

Fig. 2. Sparsification analysis in function of K, the number of most salient concepts retained in the semantic features built on top of Flickr groups and of ImageNet

Confirming the results presented in Table 1, the performances obtained with cut-offs $cut = \{70, 80, 90\}$ are close to each other. The constant gap between reranked versions of $Semfeat_{cut}^{FG}$ and $Semfeat_{full}^{FG}$ indicates that removing poorly reranked images has a clear beneficial effect regardless of the sparsification factor. Results obtained with the three cut values are similar, indicating that the positive effect of noise reduction and the negative effect due to training set shrinking compensate each other. Beyond $K = 100$, performances drop continuously, confirming the uselessness of concepts with low scores.

Conceptual Coverage Evaluation. Here we evaluate the influence of the semantic features support (sup), which is tightly linked with their capacity to cope heterogeneous datasets. We fix the $skNN$ reranking percentage $cut = 80$ and sparsification at $K = 30$ and vary the sup between 1000 and 30,000 for $Semfeat_{80}^{FG}$ and from 1000 to 17,462 for $Semfeat^{IN}$. The concepts/groups included in each support are selected randomly and we give the percentage of the performance obtained with the full features. Performances of $Semfeat^{IN}$ for $sup = \{1000, 5000, 10000, 15000\}$ are 70.9%, 87.7%, 89.7%, 96.4%. Performances of $Semfeat_{80}^{FG}$ for $sup = \{1000, 5000, 10000, 15000, 20000, 25000\}$ are 53.5%, 70.9%, 80.1%, 84.9%, 89.9%, 93.8%. sup effect is more important on $Semfeat^{FG}$, a behavior which is probably explained by the fact that automatically selected groups are more redundant than ImageNet concepts. Saturation is not reached in either case and this indicates that adding supplementary concepts or groups would probably further increase performances.

CBIR Results Analysis. In Table 2, we compare the MAP scores obtained with *Fisher* and *Overfeat* baselines to those with *Semfeat* versions proposed in this paper. The Fisher vector is dense and contains over 100,000 dimensions [5]. The Overfeat vector is also dense and is obtained with the small network configuration [21]. For $Semfeat^{IN}$, we use a sparsification factor $K = 30$, which corresponds to the best MAP reported in 2. For $Semfeat_{cut}^{FG}$, results are reported for sparsification $K = 30$ and $cut = 80$, corresponding to the best MAP in Figure 2.

The results presented in Table 2 show that $Semfeat^{IN}$ has roughly the same performances as *Overfeat* while both $Semfeat^{FG}$ are better than this strong baseline. The best overall results are obtained with $Semfeat_{80}^{FG}$. This feature

Table 2. Results for CBIR runs with the ImageCLEF Wikipedia Retrieval 2010. Both the original and the extended ground truths (*origGT* and *extGT*) are used. $Semfeat^{FG}$ results are reported for sparsification $K = 30$. *Fisher* performances do not change since this run was already pooled for *origGT*. For *Fusion*, contributions from image and text are weighted with 0.8 and respectively 0.2 but values are relatively stable for neighboring weights.

	Fisher	Overfeat	$Semfeat^{IN}$	$Semfeat_{full}^{FG}$	$Semfeat_{80}^{FG}$	Txt	Fusion
MAP *origGT*	0.0553	0.0986	0.0962	0.1065	0.1231	0.2786	0.2812
MAP *extGT*	0.0553	0.1149	0.1167	0.1267	0.1487	0.3358	0.3378

built on top of Flickr groups which exploits *skNN* reranking (24.8% and 29.4% relative improvement compared to *Overfeat* with *origGT* and *extGT*). A t-test shows that $Semfeat_{80}^{FG}$ is significantly different from all other CBIR runs with p at least 0.001. Compared to *Fisher*, the previous state-of-the-art method tested on this dataset, improvements are very consequent, 122.6% and 169% relative improvements for $Semfeat_{80}^{FG}$ for the two ground truths. As a complement to CBIR, we ran a simple fusion experiment. Even with a simple approach, results are however improved by over 20% compared to the text run and this difference is statistically significant with $p \leq 0.001$. This result confirms that text-image fusion is beneficial. The improvement could probably be further improved using more advanced fusion methods [6].

Table 3. Best and worst 10 topics ranked by MAP score using $Semfeat_{80}^{FG}$ with *origGT*

	MAP range	Textual topics
Best 10	0.52 - 028	stars and galaxies, tennis player on court, close up of bottles, polar bear, cyclist, race car, launching space shuttle, lightning in the sky, civil airplane, sailboat
Worst 10	0.003 - 0	paintings related to cubism, fractals, musician on stage, DNA helix, shiva painting or sculpture, solar panels, Oktoberfest beer tent, Rorschach black and white, videogame screenshot, Chernobyl disaster ruins.

The query set includes 70 topics and it would be therefore impractical to plot individual bars in order to visualize results. Instead, we present best and worst 10 topics ranked by MAP scores in Table 3. Confirming intuition, topics with high MAPs correspond to common Flickr topics, well represented in $Semfeat_{80}^{FG}$. Topics with low MAPs often depict non-natural scenes and the bad behavior of $Semfeat_{80}^{FG}$ is explained by two factors: *Overfeat* was trained mostly with natural images and many of these topics are poorly represented in Flickr groups. CNN retraining would be needed to deal with these cases but it falls outside the scope of the paper. Other examples of bad behavior include topics which are visually hard. For instance, *Oktoberfest beer tent* images often depict crowds which are difficult to distinguish, *solar panels* are visually similar to the surface of *skyscrapers* while *Chernobyl disaster ruins* can easily be mistaken for other ruins.

To give insight about $Semfeat_{80}^{FG}$ robustness, we compare its individual topic MAPs with those of two baselines. First $Semfeat_{80}^{FG}$ is better in 45 cases (average MAP gain of 0.068), *Fisher* in 22 cases (average MAP loss of -0.031) and there are 3 ties. The largest 3 gains are obtained for *tennis player on court* (0.469), *cyclist* (0.41) and *polar bear* (0.391). Inversely, the largest performance losses occur for *postage stamps* (-0.177), *brain scan* (-0.099) and *earth from space* (-0.095). Completing Table 3, these examples indicate that $Semfeat_{80}^{FG}$ is better for natural images whereas *Fisher* behaves better for other types of images.

We further illustrate the results obtained with $Semfeat_{80}^{FG}$ in Figure 3. The first two rows have high MAP scores in the original ground truth, the following two are in the middle of the topic ranking and the last two correspond to queries with poor results. Although not in focus in this paper, automatic image annotation with large vocabularies is a part of *Semfeat* pipeline and we present a list of 5 Flickr group tags which are automatically associated to query images. Interestingly, even though annotations are only partially relevant, their combination in *Semfeat* often favors the retrieval of relevant images, as this is the case for *tennis player*. The only image whose annotations are all conceptually unrelated to the image is *DNA helix*. Confirming the results in Table 3, the last two rows indicate that the conceptual support of *Semfeat* should be further extended.

7   Discussion

Advantages. CBIR results show that the *Semfeat* version based on reranked Flickr groups significantly outperforms other existing methods which were tried on the Wikipedia Retrieval dataset. Moreover, text-image fusion further improves the quality of results. Beside competitive performances and contrary to widely used image features, such as bags-of-visual-words, Fisher Kernels [11] or CNN features [21], *Semfeat* directly conveys semantic meaning. Image similarities are based on the comparison of humanly understandable dimensions, a characteristic which enables result explainability and reduces the *semantic gap*.

Another advantage of *Semfeat* is its sparsity. The best performances are obtained when only a few dozens of concepts are kept for each image. In this configuration it is straightforward to efficiently represent images as inverted indexes in order to speed up retrieval. We tested inverted search with a simple in-memory C++ implementation and simulated datasets up to 100 million images with sparsity $K = 100$. Retrieval time grows linearly and is under 1 ms for 10 million images and under 10 ms for 100 millions. For comparison, we also tested forward search with Overfeat (4096 dimensions) and obtained a retrieval time in the range of 15 s for 10 million images. Even if one would use compressed versions of dense features, inverted search would still be faster.

Last but not the least, $Semfeat^{FG}$ is built with an automatically mined dataset, using simple reranking and learning. This pipeline facilitates resource extension to new concepts/groups.

Limitations. We have mentioned some *Semfeat* limitations in the experimental section and extend the analysis. The learning methods used here are scalable but can be improved. With the use of more sophisticated models, the predictions associated to *Semfeat* dimensions would probably be more robust and have a positive impact on the overall results. However, when choosing the learning models to use, one should keep in mind that the prediction process needs to be fast, a constraint which is particularly relevant when the semantic features include a large number of dimensions. Consequently, as the authors of [1], we advocate for the use of linear models and will test different such models in the future. Another important limitation is the choice of positive examples which model individual groups/ concepts. We implemented a first version of *Semfeat* with a maximum of 300 images per group. While this volume is sufficient for simple visual concepts, it is probably insufficient to model complex concepts

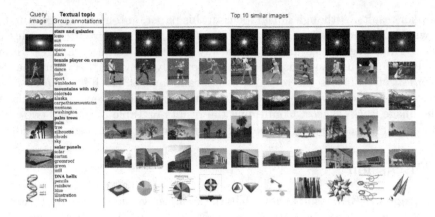

Fig. 3. Illustration of the CBIR process based on $Semfeat_{80}^{FG}$. We present the query image, the associated textual topic (bold face), 5 automatic annotations from Flickr groups and the most similar images from the Wikipedia collection. We present two highly ranked topics, two from the middle of the ranking and two from the bottom according to *origGT*. The *mountain* example illustrates well the incompleteness of the *origGT* because, while relevant, many of its neighbors were not found by official campaign runs.

and future experiments should focus on enlarging the positive examples set. In such a setting, a larger amount of potentially noisy images could be removed while still having a sufficiently rich and diversified representation of the concept. Flickr groups mirror users' interests but they are often redundant. For instance, there are tens of different groups which focus on *classic cars* and several of them can be jointly activated in the *Semfeat* representation. These groups could probably be merged into larger meta-groups to reduce redundancy and propose more informative features.

8   Conclusions

We proposed a technique for the automatic mining of large-scale visual resources from Web data. With the support of this, we proposed a new semantic image representation for image retrieval. Returning to our initial research questions we can conclude that:

Q1 The direct use of Web corpora, as proposed in [22] or [1], yields lower performances compared to manually curated datasets. However, with an appropriate choice of the initial collection and with the introduction of efficient image reranking techniques, the results obtained with the automatically built resource can rival with those of the manual resource.

Q2 Efficient semantic representations can be built through the combined use of powerful initial features, such as ImageNet, and of an appropriate visual representation of feature components. Further investigation is needed concerning the choice of machine learning models and the number of images, both positive and negative, used for learning individual models. It is probable that more sophisticated models combined with a larger number of training images will improve results.

Q3 A good coverage of the conceptual space is obtained with a good choice of the Web dataset. We explored the use of Flickr groups by other concept sets; but the pipeline presented here is easily applicable to larger datasets. The only potential constraints are the availability of data and the processing power needed to build individual models.

Q4 In image retrieval, compactness is achieved by sparsifying semantic features and by using inverted indexes. With this scheme, very large volumes of data can be search without precision loss, as it is the case for existing dense features [11].

Q5 The obtained results indicate that semantic features are useful for retrieval and, compared to [22,1], we propose an efficient way to clean and exploit large-scale noisy Web corpora.

Acknowledgments. This work is supported by the MUCKE FP7 CHIST-ERA project, partly funded by ANR, France, and by the USEMP FP7 project, partly funded by the EC under contract number 611596.

References

1. Bergamo, A., Torresani, L.: Meta-class features for large-scale object categorization on a budget. In: CVPR (2012)
2. Beyer, K., Goldstein, J., Ramakrishnan, R., Shaft, U.: When is nearest neighbor meaningful? In: Beeri, C., Bruneman, P. (eds.) ICDT 1999. LNCS, vol. 1540, pp. 217–235. Springer, Heidelberg (1998)
3. Chatzilari, E.: Using tagged images of low visual ambiguity to boost the learning efficiency of object detectors. ACM Multimedia (2013)
4. Chang, M.-H., et al.: Sheepdog: group and tag recommendation for flickr photos by automatic search-based learning. ACM Multimedia 2008 (2008)
5. Clinchant, S., et al.: Xrce's participation in wikipedia retrieval, medical image modality classification and ad-hoc retrieval tasks of imageclef 2010. In: CLEF 2010 (2010)
6. Clinchant, S., et al.: Semantic combination of textual and visual information in multimedia retrieval. In: ICMR (2011)
7. Deng, J., et al.: ImageNet: A Large-Scale Hierarchical Image Database. In: CVPR 2009 (2009)
8. Donahue, J., et al.: Decaf: A deep convolutional activation feature for generic visual recognition. CoRR (2013)
9. Fan, R.-E., et al.: Liblinear: A library for large linear classification. JMLR 9 (2008)
10. Grabowicz, P., et al.: Distinguishing topical and social groups based on common identity and bond theory. In: ACM WSDM 2013 (2013)
11. Jégou, H., et al.: Aggregating local image descriptors into compact codes. PAMI (2012)

12. Kennedy, L.S., Naaman, M.: Generating diverse and representative image search results for landmarks. In: WWW 2008 (2008)
13. Krizhevsky, A., et al.: Imagenet classification with deep convolutional neural networks. In: NIPS (2012)
14. Li, L.-J., et al.: Object bank: A high-level image representation for scene classification & semantic feature sparsification. In: NIPS (2010)
15. Liu, L., Wang, L., Liu, X.: In defense of soft-assignment coding. In: ICCV (2011)
16. Negoescu, R.A., Gatica-Perez, D.: Analyzing flickr groups. In: ACM CIVR 2008 (2008)
17. Oquab, M., et al.: Learning and Transferring Mid-Level Image Representations using Convolutional Neural Networks. In: CVPR (2014)
18. Perronnin, F., et al.: Large-scale image retrieval with compressed fisher vectors. In: CVPR 2010 (2010)
19. Popescu, A., Grefenstette, G.: Social media driven image retrieval. In: ACM ICMR 2011 (2011)
20. Russell, B., et al.: Labelme: a database and web-based tool for image annotation. IJCV 77 (2007)
21. Sermanet, P., et al.: Overfeat: Integrated recognition, localization and detection using convolutional networks, CoRR (2013)
22. Torresani, L., Szummer, M., Fitzgibbon, A.: Efficient object category recognition using classemes. In: Daniilidis, K., Maragos, P., Paragios, N. (eds.) ECCV 2010, Part I. LNCS, vol. 6311, pp. 776–789. Springer, Heidelberg (2010)
23. Tsikrika, T., et al.: Building reliable and reusable test collections for image retrieval: The wikipedia task at imageclef. IEEE MultiMedia (2012)
24. Wang, G., et al.: Learning image similarity from flickr groups using stochastic intersection kernel machines. In: CVPR (2009)
25. Wang, J., et al.: Locality-constrained linear coding for image classification. In: CVPR (2010)
26. Yu, K., Zhang, T., Gong, Y.: Nonlinear learning using local coordinate coding. In: NIPS (2009)

FISIR: A Flexible Framework for Interactive Search in Image Retrieval Systems

Sheila M. Pinto-Cáceres[1,2], Jurandy Almeida[1,3],
M. Cecília C. Baranauskas[1], and Ricardo da S. Torres[1]

[1] Institute of Computing, University of Campinas – UNICAMP, Campinas, SP, Brazil
{cecilia,rtorres}@ic.unicamp.br
[2] Institute of Science and Technology, Paulista University – UNIP, SP, Brazil
sheila.caceres@unip.br
[3] Institute of Science and Technology, Federal University of São Paulo – UNIFESP,
São José dos Campos, SP, Brazil
jurandy.almeida@unifesp.br

Abstract. This paper presents a flexible framework for interactive search in image retrieval systems. Our approach allows for the visual structure change in order to produce coherent layouts, which highlight the most relevant results according to user needs. This innovative framework is flexible in the sense it supports the dynamic creation of several hybrid visual designs, based on the combination of different visualization strategies. Results from a subjective evaluation demonstrate that the dynamic hybrid layouts created by the proposed framework provide the end-users with an effective user interface for intuitive browsing and searching experience.

Keywords: image retrieval, interactive search, result visualization, visual structure.

1 Introduction

Recent advances in technology have increased the availability of image data, creating a strong requirement for efficient systems to support image retrieval tasks. Making efficient use of image information requires that data be accessed in a user-friendly way. For this, it is important to provide users with a browsing tool to interactively search for (or query) an image in large collections, without having to look through many possible results at the same time, and therefore helping users to easily find images in which they are interested [1].

One of the main challenges of designing an interactive retrieval system is to minimize the fatigue and frustration that a user might experience when not obtaining useful information. Ideally, the layout of those systems should allow the inspection of large image collections in a coherent manner, by displaying appropriately the most relevant images according to user needs. To fulfill such requirements, a competitive layout must provide more than just images, rather, it should take advantage of different visualization strategies to improve and accelerate the comprehension of how images are correlated to each other.

A lot of research has been conducted on the proposal of browsing techniques for the interactive search on image collections [2]. In spite of all the advances,

X. He et al. (Eds.): MMM 2015, Part I, LNCS 8935, pp. 335–347, 2015.

many of those research approaches have displayed a set of images in a pre-computed rigid structure, which cannot change (or dynamically be changed) to visually emphasize particular user interests. Usually, they support the user interaction just by exploring and showing different parts of a static structure.

In this paper, we propose a novel framework for searching and browsing in image retrieval systems. Our approach is totally flexible, supporting the dynamic creation of several hybrid and coherent layouts, based on the combination of different visualization strategies, as shown in Figure 1. In this way, this innovative framework allows the visual structure change according to user needs and, hence, it can be easily adapted to different search scenarios. Experiments were conducted both for evaluating the dynamic hybrid layouts created by the proposed framework and for comparing it with several static visualization techniques. Results from a rigorous empirical comparison with a subjective evaluation show that the hybrid visual designs produced by our framework are effective in supporting intuitive browsing and searching experience.

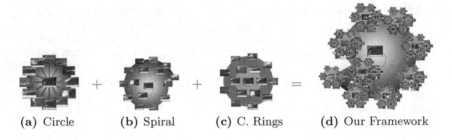

    (a) Circle   (b) Spiral   (c) C. Rings   (d) Our Framework

Fig. 1. An illustration of our framework. Users can dynamically create a hybrid layout (d) by combining different visual designs (a, b, and c). Our framework leads to a better searching experience as it is adaptive to users' needs.

2   Related Work

A comprehensive review of searching and browsing methods for image retrieval can be found in [2]. Some of the main ideas and results among the previously published methods are briefly discussed next.

Most of image retrieval systems have been designed to support a targeted search. In this kind of navigation, users provide a query pattern to the system and then receive a single ranked list containing the most similar results. The most popular approach for arranging the result set is to use list- or grid-based layouts, where images are disposed in a linear or grid manner according to their relevance to the query pattern [3]. Several techniques display the result set in a circular or elliptical form [5]. These radial methods try to centralize the user vision by setting the query at the center of the available space and then place similar images around it, allowing easy access and exploration of the result set [6]. Despite the simplicity of those methods, they fail on visually representing the intrinsic relationships among similar images.

As an alternative, extensive research has focused on the development of image retrieval systems for exploratory search. This kind of navigation is well suited when the user is not able to formulate a specific query. In this case, the similarities between image data are exploited in order to display similar images close to each other. The most popular approach in this category is *mapping*. The main idea behind the mapping methods is to project the feature space in which the images are represented into a two-dimensional space to be displayed in a computer or other device screen. Other methods employ clustering algorithms to create groups of images, which are displayed in a grid layout or in such a way prone to overlapping. Overlapping occurs because similar clusters can be very close to each other in the feature space. For this reason, sometimes only a single image (representative) or a subset for each group is displayed in order to free up visualization space and reduce overlapping as in [7]. In [8], mapping and clustering techniques are combined to improve the provided layout.

In general, all the previous research work provide the end-users with a static structure, penalizing any customization to fit their specific needs. Therefore, they can interact just with parts of a same structure, without changing the precomputed rigid structure. This paper aims to fill such a gap. Here, we introduce a flexible framework, which is able to dynamically create several hybrid and coherent layouts by combining different visualization strategies, making our approach very adaptive to user interests.

3   Proposed Approach

We present an innovative framework for supporting the dynamic creation of several visual designs. Thus, we provide a more coherent space for searching and browsing images by organizing them in a well-defined visual structure, which can be adapted according to user interests. In this way, we introduce several original layouts, which are produced by our framework in a straightforward manner.

3.1   Features and Similarity

In our framework, we implement a traditional image retrieval system. Thus, an image descriptor is used to encode visual properties of an image into a feature vector and to measure the similarity between images by means of a distance function used for comparing their feature vectors. For that, we have adopted the Border/Interior pixel Classification (BIC) [9]. However, our framework does not impose any restriction on the descriptor used. It only uses the set of elements to be analyzed, which are not the images themselves but rather their extracted features, along with a suitable distance function. Therefore, it can be used with any searching and browsing scheme.

3.2   Intuitive Layout

An intuitive layout should ideally assist users to quickly comprehend search results preventing them from wasting time. For that, we need to coherently distribute the images over the available space. With this goal, we use clustering techniques to place search results over well-defined radial layouts. The advantage of using a hybrid scheme is to combine positive aspects of different visualization strategies. One first step on that direction was the proposal of the Clustering Set strategy [10]. Figure 2c illustrates such strategy.

In this work, we are extending the Clustering Set principles to create novel visual structures. First, the query image is placed in the center of the visualization display. Then, the clusters are circularly distributed around the query in a clockwise order of similarity regarding the query, which is represented by the size of each cluster. This strategy is also applied inside each of the clusters. At the center, we place its most representative element. The remaining elements are sorted in a circular manner according to a clockwise order of similarity with respect to the center element, which is denoted by their size and border color.

In this scenario, we can observe the presence of two levels of radial visual structures. Notice that several radial visual structures are possible to be used for each level. It would also be possible to increment the number of levels. If we keep the same radial structure for $2 \ldots n$ levels, we can recursively create fractal-based structures, as discussed later.

3.3   Flexible Design

By using our framework, users can dynamically create new visual designs and modify the previous ones independently of the core and additional functionalities of the system. In order to provide a flexible framework, we need to support the easy creation and change of layouts. For that, we assume that visual structure developers may be interested in constructing complex diagrams from simple components. Thus, they may group components to form more complex structures.

A simple implementation could define classes for basic elements, such as a simple node containing an image, and other classes that act as containers of these basic elements, forming a structure. The problem with this approach is that it needs to handle the basic elements and the containers in a different way, which makes the application development more complex.

We can avoid this distinction using a recursive composition provided by the composite design pattern [11]. It establishes three basic elements: a component, a composite, and other primitive objects. The composite and the other primitive objects inherit from component. A component is a common interface that represents the object to be manipulated, it can be a primitive object or a complex set of other components (composite). It is possible to group together simple components to form bigger components that, at the same time, can be combined in order to form even more complex components.

In this way, we have separated elements and use them alone or in conjunction according to our needs. For example, when resizing the structure, we can recursively call the same function of every individual element so that all of them are resized as well, thus acting as a single unity. Thus, we can use a single interface to represent both the basic elements and the containers. This strategy enables us to treat an image or a structure of images in the same way as a simple element.

In Figure 2a, we define a primitive object that represents the basic cell (*ImageView*) containing an image, a composite object (*ComposedView*) that encompasses a list of objects (primitive of not) forming a visual structure, and a component (*View*) that can be either a primitive or a composite object. Note that a *ComposedView* is composed of a set of *View* elements. Thus, complex visual structures can be produced by the successive composition of simpler structures. If we use a hierarchical clustering method, several levels can be included in our structure. This extension is absolutely simple, since it does not require that an element has a specific nature: a single element (image) or a complex one (set of elements).

To illustrate the potential of the composite design pattern, consider that we initially have a *ComposedView* element, which is formed by a set of basic *ImageView* elements, as shown in Figure 2b. Once they share the same interface *View*, we can equally treat a basic or a complex element. Then, suppose that we replace each basic *ImageView* element in this structure by a *ComposedView* element. In this way, we can easily create a more complex structure, as presented in Figure 2c.

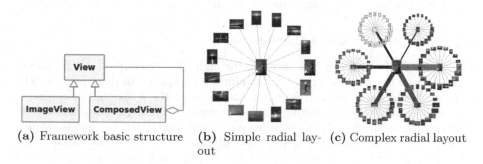

(a) Framework basic structure   (b) Simple radial layout   (c) Complex radial layout

Fig. 2. The potential of the proposed framework in creating complex layouts from simpler ones. A layout is composed by Views, which can be either a single ImageView or a ComposedView made of a set of Views, as shown in (a). Thus, a simple radial layout (b) can become a more complex layout (c).

This approach can be extended to any radial-based visual design, such as illustrated in Figures 1a, 1b, 1c. In other words, any *ImageView* element can be replaced by a *ComposedView* element. Therefore, many complex radial layouts can be created from those simpler ones. Moreover, we can repeat this process many and many times so to produce fractal-based visual designs.

3.4   Visual Marks

We named as visual mark to any visual property used in a graphical display (e.g., size, color, and border). A visual mark is extremely useful tool to emphasize user needs. For instance, the background color applied in other approaches (e.g., Concentric Rings) has assisted the users to understand the distribution of the results according to their opinions. Such a characteristic is useful to indicate relevance. Clearer tone is applied to more relevant information and displayed elements get darker as they are placed far away from the query image (most relevant information tends to be nearest to the center where we locate the query image). Figure 3a shows an example of improved Clustering Set layout that considers a different background color.

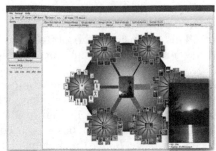

(a) Improved Clustering Set

(b) The pop-up window where a specific image is handled

Fig. 3. Some visual marks and browsing capabilities provided by our framework

3.5 Advanced Browsing Capabilities

The addition of a dynamic user control through direct manipulation principles enables a user to navigate through large information spaces. In this way, given the increasing volumes of data, it is important to allow the dynamic manipulation through the image set by making available several browsing capabilities. When trying to offer a comfortable navigation, a useful interaction technique is output-as-input. It consists of providing the possibility of performing a new input into a system based on the previous output. In our proposal, anytime a user right-clicks on any already presented image (output) on the screen, it is possible to make a new query (input). The display will be automatically rearranged in order to show a new result of the most relevant clusters in relation to the last selected image. Thus, they can interleave between targeted search and exploratory search. Using different searching and browsing methods into a single environment enhance the user's inspection ability.

Our approach also offers zoom and pan controls. Furthermore, our framework takes advantage of the inspection ability of the users by adding a pop-up functionality to any clicked image. Specifically, when a user clicks on an image, a pop-up window is opened on the screen providing more information about the selected element, as illustrated in Figure 3b. On the top and center of the displayed pop-up window, we make available a image visualizer. On the bottom part, we display a set of image characteristics. This technique is known as details-on-demand. The details-on-demand technique allows to interactively select parts of data to be visualized in more detail while providing, at the same time, an overview of the whole context.

3.6 Hybrid Combinations

Our framework allows the straightforward creation of new structures from the combination of basic ones. Thus, we present several hybrid visualization structures as an original contribution from this work. This set of structures is guided in the Clustering Set principles. Figure 4a presents an hybrid structure that combines the successful Clustering Set approach with the Spiral method. In this case, we apply the clustering strategy and an approach for placing results sequentially inside each cluster. Thus, we maintain the intrinsic relationship among images arranged in a radial well-organized structure that offers a better utilization of the space. Figure 4b presents a similar approach that uses concentric rings to display an ordered list inside each cluster. As we can see in Figures 4a and 4b, we have dealt with some strategies for sequential results in order to increase the provided information inside each cluster. Then, we can follow a similar criterion to display the sequential list of clusters. Thus, we propose other methods (Figure 4c and 4d) to offer more information using the same space. In these layouts, more clusters are displayed in a coherent manner.

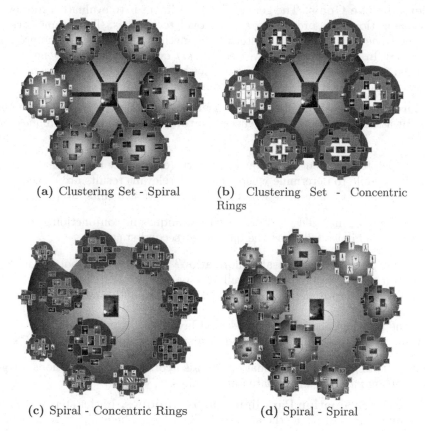

(a) Clustering Set - Spiral   (b) Clustering Set - Concentric Rings

(c) Spiral - Concentric Rings   (d) Spiral - Spiral

Fig. 4. Several layouts created in a straightforward manner by our framework

4 Evaluation

Unlike other research areas, evaluating a visualization strategy is not a straightforward task due to the lack of an objective ground truth. A consistent evaluation framework is seriously missing for visualization research. Presently, every work has its own evaluation methodology, often presented without any performance comparison with previously existing techniques.

4.1   Experimental Protocol

In this work, we adopted a subjective evaluation framework to assess the quality of visualization structures, known as *DECIDE* [12]. This framework incorporates the judgement of the user in evaluating the quality of a visualization structure. It guides the experimental evaluation through six well-defined steps: **D**etermine the goals, **E**xplore the questions, **C**hoose the evaluation paradigm and techniques, **I**dentify the practical issues, **D**ecide how to deal with the ethical issues, **E**valuate, interpret, and present the data. In the following subsections, each of those steps is explained in more detail.

Determine the Goals. The goal of the experiment is to evaluate a framework for visualization that provides a set of novel techniques. This framework follows the same strategy of the clustering set method, thus likewise the proposed techniques. In order to do that we need to *(i)* evaluate the effectiveness of the proposed techniques and *(ii)* compare it with current state-of-the-art methods. Therefore, we defined two stages to accomplish both aims.

Explore the Questions. In order to achieve our goals, we defined some questions for assessing the user preferences regarding the proposed techniques (first stage):

- Is it possible to identify the query pattern (e.g., query image)?
- Is it possible to determine where the most relevant results are placed?
- Is it possible to determine where the least relevant results are placed?

We have also analyzed the introduced techniques in conjunction with state-of-the-art techniques following the questions (second stage):

- Does it provide the necessary information? (not too much, not too little)
- Does it appropriately use the available space?
- Is it possible to understand the visual structure and how it distributes the result set?
- Is the user satisfied with the presented layout?
- What is the most suitable method to visualize the query results?

For obtaining unaware information, we also consider open questions for general comments about the evaluated approaches.

Choose the Evaluation Paradigm and Techniques. In this step, we adopted the Usability Test method (the standard ISO 9241). The evaluation instrument used in our experiments was a questionnaire. In this way, we are able to analyze a bigger sample in a faster manner. A bigger sample can give us a better statistical significance.

Identify the Practical Issues.

- **Users.** In order to obtain significant results, we invited a set of undergraduate students from the Institute of Computing, University of Campinas, totalizing 28 volunteers. It is important to point out that nobody had previous knowledge of the proposed methods and, hence, they had never seen our strategy before the experiment.

- **Equipment.** Each participant had access to one computer and was free to use the operating system he/she prefers.
- **Material.** Each participant accessed a set of files. This set included:
  - User Instructions. It details what the user needs to know about the evaluation process.
  - Free and Clear Consent Terms. It clearly describes the conditions in which the user will participate in the experiment detailing risks, benefits, confidentiality, etc.
  - User Profile Form. It allows to register relevant data of an user, such as his/her familiarity with the system and frequency of using the computer.
  - Evaluation Forms. We use two evaluation forms. By using the first one, we were able to objectively capture the users' ability to find some elements in the proposed structures (first stage), and by using the second one, we were able to capture the users' judgment about the proposed methods along with state-of-the-art methods (second stage).

Decide How to Deal with the Ethical Issues. Ethical questions were clearly explained to each user verbally and in the user instructions.

Evaluate, Interpret, and Present the Data. Initially, we gave the users the "Users Instruction" and the "Free and Clear Consent Terms." Then, users read both documents and signed them. After that, they filled the user profile form online. Finally, they analyzed each visual structure prototype and evaluate the visual structures by answering the two evaluation forms.

4.2   Results

Each form allows us to obtain specific information from the users and their perceptions regarding each evaluated method.

The profile form served us to obtain personal information such as age, gender, studies, familiarity, and frequency of using computers and search systems. From the 28 participants at the moment of the experiment, nearly all of them (96.42%) were from 19 to 26 years old and only one user (representing the 3.6%) was 33 years old. The 78.57% of the users were male and the remaining 21.43% were female. All the users were undergraduate students in a computing course with a high-level familiarity and frequency use of computers. Most of them had a lot of experience with searching systems.

The first stage of the evaluation helped us to objectively analyze the real effectiveness of the used approach over the several novel methods proposed in this work and the previously proposed Clustering Set method. These methods follow the same basis: they use clustering algorithms to offer a coherent space in conjunction with well-defined radial layouts to coherently organize that information.

Figure 5 exhibits the percentage of users that correctly found the requested element or region inside each structure (% of Correct Responses). The results for the different questions are grouped by the visual designs. This graph shows that our strategies have achieved a good performance. In almost all questions involving all the proposed structures, more than 80% of the users were able to correctly identify the requested elements, demonstrating a high level of usability. No remarkable difference between the methods statistics was found, probably because

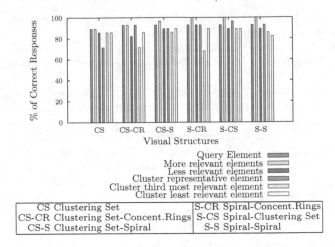

CS Clustering Set	S-CR Spiral-Concent.Rings
CS-CR Clustering Set-Concent.Rings	S-CS Spiral-Clustering Set
CS-S Clustering Set-Spiral	S-S Spiral-Spiral

Fig. 5. Usability of each method according correct answers

they all follow the same philosophy: coherent clustered information organized in well-defined radial structures.

A summarized plot of the answers of this stage is displayed in Figure 6a which represents a Tukey-style boxplot [13] of the number of correct answers by structure for all users. The x axis varies from 1 to 6 as the evaluation form for this stage requested six questions for each method. The y axis represents each evaluated method. We can observe that the hybrid methods involving a combination of Clustering Set with Spiral (S-CS and CS-S) obtained very good results with a low dispersion of data and a highly compact boxplot, representing that most users was able to correctly identify all the elements in such structures. Thus, spiral layouts seem to be more intuitive for users.

The second stage of the evaluation consisted of a rigorous subjective evaluation to capture the users judgments of several methods from literature and from the proposed framework. After collecting the data from this second evaluation form, we obtained the results shown in Figures 6b, 6c, 6d, 6e. These graphs present an overall analysis of each criterion also using the Tukey-style boxplots. Such figures indicate that there is no statistical difference between the proposed method and the already familiar Grid method. Therefore, the strategies created by our framework can be seen as a truly alternative to the popular Grid method. This is a significant achievement since it is difficult to compete with a widely known method (Grid) considering that users are already familiarized with it.

Users were also asked to rank the structures according to their suitability for visualizing query results. Figure 6f exhibits the distribution of ranks attributed by the participants for different approaches. The comparison of the average rank of those methods is showed in Table 1.

The non-parametric Friedman test [14] was performed to verify the statistical significance of those results. It checks whether the measured average ranks are significantly different from the mean rank. The analysis of the experiment confirms that there is no statistical difference among those methods (p-value

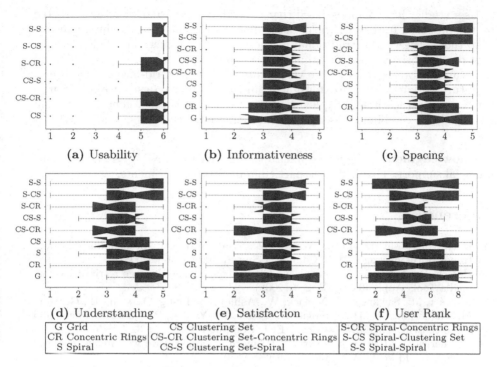

Fig. 6. Results from our experiments

Table 1. Average rank achieved by several approaches

Method	Average Rank
Clustering Set	5.65
Spiral - Clustering Set	5.48
Grid	5.19
Spiral - Spiral	5.11
Concentric Rings	4.98
Spiral	4.80
Clustering Set - Spiral	4.80
Clustering Set - Concentric Rings	4.65
Spiral - Concentric Rings	4.35

= 0.788). According to a Nemenyi test, the critical distance (CD) for pairwise comparisons between different strategies at a confidence of 95% is 2.31.

5   Conclusion

This paper presents a novel framework for creating visual structures to support search and browse services in image retrieval systems. The proposed framework allows the construction of coherent layouts that emphasize the more relevant results and, at the same time, the comprehension of the relations among them.

User studies have demonstrated that several hybrid visual structures created by the proposed framework yield, according to different criteria, similar results than those observed for traditional methods (e.g., Grid method). This is a relevant result considering the fact that the users have never saw our proposed visual structures before, and they were very familiar with the Grid-based structures.

Future work includes the extension of our framework to consider other descriptors (e.g., local features [15]), learning-to-rank methods (e.g., genetic programming [16]) for combining descriptors and hierarchical clustering methods [17].

Acknowledgments. We would like to thank CAPES, CNPq, FAPESP for the financial support. Thanks to the participants of the experiments.

References

1. Datta, R., Joshi, D., Li, J., Wang, J.Z.: Image retrieval: Ideas, influences, and trends of the new age. ACM Computer Survey 40(2), 1–60 (2008)
2. Heesch, D.: A survey of browsing models for content based image retrieval. Multimedia Tools Appl. 40(2), 261–284 (2008)
3. Flickner, M., Sawhney, H., Niblack, W., Ashley, J., Huang, Q., Dom, B., Gorkani, M., Hafner, J., Lee, D., Petkovic, D., Steele, D., Yanker, P.: Query by image and video content: The QBIC system. IEEE Computer 28(9), 23–32 (1995)
4. Mukherjea, S., Hirata, K., Hara, Y.: Amore: A world wide web image retrieval engine. World Wide Web 2(3), 115–132 (1999)
5. Draper, G.M., Livnat, Y., Riesenfeld, R.F.: A survey of radial methods for information visualization. IEEE Trans. Visualization and Computer Graphics 15, 759–776 (2009)
6. Torres, R.d.S., Silva, C.G., Medeiros, C.B., Rocha, H.V.: Visual structures for image browsing. In: ACM CIKM, pp. 49–55 (2003)
7. Schaefer, G., Ruszala, S.: Hierarchical image database navigation on a hue sphere. In: Bebis, G., et al. (eds.) ISVC 2006. LNCS, vol. 4292, pp. 814–823. Springer, Heidelberg (2006)
8. Nguyen, G.P., Worring, M.: Interactive access to large image collections using similarity-based visualization. J. Visual Languages and Computing 19(2), 203–224 (2008)
9. Stehling, R.O., Nascimento, M.A., Falcão, A.X.: A compact and efficient image retrieval approach based on border/interior pixel classification. In: ACM Intl. Conf. Information and Knowledge Management, pp. 102–109 (2002)
10. Pinto-Caceres, S.M., Almeida, J., de Almeida Néris, V.P., Baranauskas, M.C.C., Leite, N.J., da Silva Torres, R.: Navigating through video stories using clustering sets. Intl. J. Multimedia Data Engineering and Management 2(3), 1–20 (2011)
11. Erich Gamma, R.J., Helm, R., Vlissides, J.: Design Patterns: Elements of Reusable Object-Oriented Software. Addison-Wesley (1994)
12. Preece, J., Rogers, Y., Sharp, H.: Interaction Design. John Wiley & Sons, Inc., Chichester (2002)
13. Tukey, J.W.: Exploratory Data Analysis. Addison Wesley (1977)
14. Friedman, M.: The use of ranks to avoid the assumption of normality implicit in the analysis of variance. J. American Statistical Association 32(200), 675–701 (1937)
15. Almeida, J., Rocha, A., da, R., Torres, S., Goldenstein, S.: Making colors worth more than a thousand words. In: ACM Intl. Symp. Applied Computing (ACM-SAC 2008), pp. 1180–1186 (2008)

16. Andrade, F.S.P., Almeida, J., Pedrini, H., da, R., Torres, S.: Fusion of local and global descriptors for content-based image and video retrieval. In: Iberoamerican Congress on Pattern Recognition (CIARP 2012), pp. 845–853 (2012)
17. Rocha, A., Almeida, J., Nascimento, M.A., Torres, R., Goldenstein, S.K.: Efficient and flexible cluster-and-search for CBIR. In: Blanc-Talon, J., Bourennane, S., Philips, W., Popescu, D., Scheunders, P. (eds.) ACIVS 2008. LNCS, vol. 5259, pp. 77–88. Springer, Heidelberg (2008)

Auditory Scene Classification with Deep Belief Network

Like Xue and Feng Su*

State Key Laboratory for Novel Software Technology,
Nanjing University,
Nanjing 210023, China
suf@nju.edu.cn

Abstract. Effective modeling and analyzing of an auditory scene is crucial to many context-aware and content-based multimedia applications. In this paper, we explore the effectiveness of the multiple-layer generative deep neural network model in discovering the underlying higher level and highly non-linear probabilistic representations from acoustic data of the unstructured auditory scenes. We first create a more compact and representative description of the input audio clip by focusing on the salient regions of data and modeling their contextual correlations. Next, we exploit deep belief network (DBN) to unsupervisedly discover and generate the high-level descriptions of scene audio as the activations of units on higher hidden layers of the trained DBN model, which are finally classified to certain category of scene by either the discriminative output layer of DBN or a separate classifier like support vector machine (SVM). The experiment reveals the effectiveness of the proposed DBN-based classification approach for auditory scenes.

Keywords: auditory scene, scene classification, deep belief network, saliency detection, context description.

1 Introduction

Automatic classification of auditory scenes simulates the human's ability to analyze and classify accurately the surrounding scene or context based on acoustic signals received, and plays an important role in various context-aware applications, which automatically change the behavior according to the category of the scene or provide the user with customized information as the circumstance changes. Information about the category and high-level semantics of the scene can also be effectively exploited in many other aspects of content-based multimedia processings.

By auditory scene, we refer to a specific location or site with distinctive acoustic characteristics such as a restaurant or beach. As illustrated in Fig. 1, one auditory scene typically contains and is characterized by various irregular occurring auditory events caused by specific objects or events in the scene such as laughing

---

* Corresponding author.

X. He et al. (Eds.): MMM 2015, Part I, LNCS 8935, pp. 348–359, 2015.

and whistling. Different from speech or music, however, which can be categorized to structured sounds due to their formantic or harmonic structure and receive much more research interests in past years, much of the composition of auditory scenes can be categorized to the *environmental sounds* [4] that are typically unstructured with a broad noise-like flat spectrum and inherently diverse variety of signal composition, and considerably less of existing work addresses such type of audio.

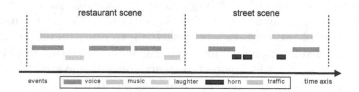

Fig. 1. Illustration of auditory scenes

Research concerning auditory scenes can be roughly categorized into two schemes. The first category employs the explicit modeling of specific events and characterize the scene on basis of them [3,5,15]. For example, [3] employed a Bayesian network-based model to discover the high-level semantics of an auditory scene embedded in key effect sequences, while in [5], both hidden Markov model (HMM) and support vector machine (SVM) are used to fuse the characteristics and correlations among various audio events at the semantic context level. Both methods exploited the HMM for the model of key audio effects/events in the auditory scene. One main difficulty of these methods is the insufficiency of the fixed and limited set of audio events explicitly defined, compared to the diversity of constituent events of a complicated or unknown scene. The other category of methods focuses on searching or developing efficient hand-designed features characterizing low-order statistics of signals of auditory scenes and treats the scene as a discrete or temporal combination of such features [7,16,4,9,17], while omitting the modelling of higher level structures inside the audio data of the scene, which could be discriminative for particular auditory scenes. In spite of being more generalized, this category of algorithms highly depends on the effectiveness of the feature description, therefore many of them employ certain forms of feature learning from the data.

In this paper, we explore the effectiveness of the higher level representations of acoustic data discovered by one category of sophisticated multiple-layer generative models - the deep neural networks (DNNs) [10] and particularly the deep belief network (DBN) [11] in the context of auditory scene classification. Different from the traditional relatively shallow one-layer representations like Gaussian mixture model (GMM) and hidden Markov model (HMM), deep neural network related models have many hidden layers of different abstract representations for auditory signals, which fit in with human acoustic perception that appears to use many layers of feature extractors and event detectors, and have been shown to outperform state-of-the-art GMM/HMM systems on a variety of speech recognition benchmarks [10,6]. As the first work that trains a very deep neural network successfully and also a particularly popular model of DNN, the goal of deep belief network is to progressively and unsupervisedly learn more abstract representations of the input data at the higher layers of an hierarchical model. Since first proposed in 2006, DBN has received lots of research interests and has been widely used to solve problems in signal processing, speech recognition and many other fields [13,14,18,8]. As far as we know, however, the DBN model has been seldom explored for recognizing auditory scenes that are relatively unstructured.

Our basic idea is to avoid the explicit modeling of audio events and rely on the higher layers of the *deep acoustic model* [13] to discover and better model the intrinsic higher-order and highly non-linear statistical structures embedded in auditory scene audio, which potentially correspond to the structured components (e.g. the events) of the scene and may play a critical role in characterizing specific scene categories. Furthermore, for more efficient learning of such higher-level auditory structures from the audio stream, we propose to concentrate on the signals of the *salient* parts of the stream instead of most stationary background sounds in the unstructured scene, by detecting salient spectral points in signals and selecting representative subset of the original data around them as the input to the DBN model. We also introduce an effective intermediate *contextual description* for audio signals based on the saliency analysis, to capture the relatively longer-term correlations of signals, in complement of the piecewise short-term window of signals used for low-level description extraction. Finally, on the basis of representing an audio clip as a reduced set of low-level and contextual descriptions extracted at salient points, we exploit the DBN model to generatively discover latent structures from audio of scenes, which are represented by the high-level features formed by the activations of units on the hidden layers of DBN. The audio clip is then classified to certain scene category on basis of the high-level features. Fig. 2 shows a summary of our method.

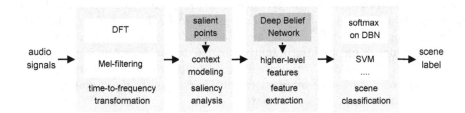

Fig. 2. Flow chart of the proposed method

The remainder of the paper is organized as follows. We introduce the saliency based representation of audio data in Section 2. Section 3 describes the DBN-based learning and generation of high-level features for auditory scenes, while the classification of audio clip based on high-level features is depicted in Section 4. We present the experiment results and discussions in Section 5.

2   Data Representation with Saliency Analysis

In one clip of auditory scenes, most parts of the audio data are the stationary background sounds between the audio events located isolatedly along the temporal axis. For their property of unstructuredness, these background sounds usually contribute less to the characterization of the scene. Thus, in this paper, unlike most other works exploiting DBN such as [18], which sends all the windows of spectral coefficients of audio signals to DBN, we propose to create a representative subset of the original audio data for feature learning and generation by the DBN model based on the saliency analysis. This scheme brings two advantages: 1) Potentially more characteristic parts of the audio data are used to train the DBN model and perform prediction, which may yield more representative features for classification. 2) The number of the total features input to DBN is reduced to include only those 'salient' ones, which helps to speed up the training and prediction of DBN, while on the other hand, the number of these training features can be adjusted by threshold parameters to meet scalability needs.

2.1   Time-to-Frequency Transformation

For an input audio clip, the short term DFT (discrete Fourier transform) is performed on the signals within a short-term window. A Human Auditory System (HAS) filtering, i.e. the Mel filter-bank in our work, is then taken on the power spectrum output of DFT to equalize the frequency bins to values that closely approximate the human auditory perception system's response to signals, as well as to reduce the number of total frequency bins.

2.2   Salient Spectral Point Detection

Since audio events play an critical role in characterizing auditory scenes, ideally we intend to place salient marks on the signals within the event duration, instead of those corresponding to background sounds. Without detection of events in advance, however, precisely locating temporal points of events in the signal sequence is impossible. Thus, as an approximate alternative, we propose to take the local spectral peak in the spectrogram as the *salient point*, whose effectiveness was demonstrated in the experiments. Fig. 3 is a example of the distribution of the salient point in the Mel-scale spectrogram.

Specifically, the salient points are selected as the time-frequency positions in the spectrogram, whose energy is strictly higher than all energies in immediately adjacent frequency (band) and time locations, i.e. the local maxima (peak). Such peaks are found to be resilient to usual audio transformations [2]. However, by definition, the local temporal-spectral peaks could be dense in the spectrogram. To reduce the total number of salient points by selecting the most representative ones, a temporal filtering is performed on the set of candidate peaks to preserve only peaks whose energy stays above a temporal masking threshold as defined by Eq. (1), and remove the peaks temporally close to its neighbors and having similar intensities.

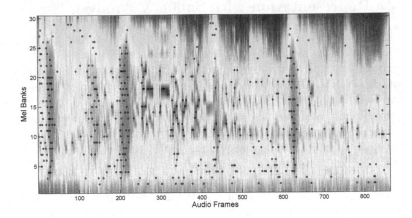

Fig. 3. Illustration of salient points (denoted by circles) in the Mel-scale spectrogram

$$Thr[n] = \alpha^{\Delta t} E[n-1] \exp - \frac{(\Delta t)^2}{2 * \sigma^2} \tag{1}$$

where, Δt is the temporal distance (in frames) between the considered peak n and the previous peak chosen in the same Mel band whose energy is denoted by $E[n-1]$, and α and σ are two parameters controlling the temporal falling speed of the threshold, which can be experimentally adjusted according to the required density of the final salient points.

2.3   Low-Level Audio Description Based on Salient Points

On the basis of the salient points detected, we represent one clip as a reduced (i.e. subsampled) set of low-level descriptions, which are constructed only at the salient points while discarding short-term windows (frames) of signals that contain no salient point. To be precise, we name a frame i containing at least one salient point the *salient frame*, for which we generate one local low-level description, and associate a *saliency weight* $\mathbf{w}_i = g(n_i)$ with it, in which n_i denotes the number of salient points in frame i and $g()$ is an increasing function, for which we use a linear function $g(n_i) = \beta n_i$, $\beta = \frac{1}{\#of Mel-bands}$ in this work.

We consider two alternative local spectral descriptions for one salient frame. The first is composed of the set of the Mel-frequency filtering output coefficients. Since a total of 30 Mel bands is used in the work, which results in a 30-dim coefficient vector for one short-term window, to acquire the reasonable large dimension of the description as the input to DBN for effective feature learning, we define a larger *analysis window* centered at the salient frame and covering several successive short-term windows, and then stack the Mel-filtering coefficients in these windows to form the description vector, which is denoted by $\boldsymbol{f}_{MEL}$.

On the other hand, as noted in [4], the Mel-frequency filtering on the basis of DFT may cause loss of certain representative information of audio signals. Therefore, we can use the DFT coefficients within the short-term window directly as the low-level spectral description f_{DFT}. We evaluate the two descriptions f_{MEL} and f_{DFT} in experiments.

2.4   Saliency-Based Context Description

Given the time-frequency distribution of salient points as shown in Fig. 3, we define a *context description* f_{CTX} for each salient frame, or equivalently, for each low-level audio description (f_{DFT} or f_{MEL}), describing its correlations with other local descriptions along the temporal axis.

Specifically, given a set of N normalized low-level audio descriptions $\{f_{i=1..N}\}$ and their temporal locations $\{t_{i=1..N}\}$ in the clip, the context feature of f_i is computed as a coarse 2-D histogram:

$$f^i_{CTX} = \mathbf{h}_i(\theta, l) \tag{2}$$

over the temporal distance $l = \log(t_i - t_j)$ in log scale between f_i and any other local description f_j and their angle $\theta = \arccos(f_i \cdot f_j)$ in the vector space. Due to its sparsity, this context feature can be further transformed by PCA to the lower-dimensional and denser representation space, and then be incorporated into the audio description by concatenating with low-level features like $[f_{MEL}, f_{CTX}]$.

The context description provides a complementary characterizing capability to the piece-wise low-level audio descriptions by incorporating long-range correlation between signals exceeding the boundary of the analysis window. It's effectiveness is revealed by the experiments.

3   Feature Learning and Extraction by Deep Belief Network

As a powerful hierarchical generative model for feature extraction, the deep belief network is a neural network constructed by stacking many layers of restricted Boltzmann machines (RBMs), which in turn is structured as two layers of neurons - a visible layer representing observations and a hidden layer learned to represent features. Each neuron is fully connected to the neurons of the other layer, but there is no connection between neurons of the same layer. One RBM is stacked on top of another RBM by linking the hidden layer of the latter to the visible layer of the former. A schematic representation of DBN is shown in Fig. 4. The intermediate hidden layers of DBN form a top-down directed acyclic graph of many stochastic latent variables, while the units in the lowest layer are called visible units, which represent an input feature vector. The highest layer of DBN could be a softmax output layer of label units for discriminative prediction.

In this work, with the low-level feature vector used to set the states of the visible units on the lowest (input) layer, the DBN is first pre-trained as a multi-layer

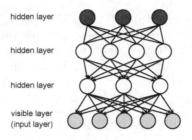

Fig. 4. Schematic representation of the DBN model

Table 1. Typical parameter setting of the DBN structure and learning algorithm

Number of hidden layers	3
Units per layer	300, 150, 60
Unsupervised learning rate	0.1
Number of unsupervised epochs	50
Supervised learning rate / Scaling factor	5 / 0.99
Number of supervised epochs	500

generative model, which can be learned efficiently, one layer at a time, by using an unsupervised learning procedure that maximizes a variational lower bound on the log probability of the input and also prevents over-fitting to the training set, without making use of any discriminative information. Once it's completed, the resulting feedforward neural network is discriminatively fine-tuned using backpropagation to slightly adjusts the weights in every layer to make them better at the predicting task.

Once trained, the low-level acoustic descriptions in the previous section are set as the input of the DBN and further transferred through the multiple nonlinear hidden layers of DBN, and we use the activations of the DBN hidden units as a higher level feature representation f_{DBN} of the input audio, which better characterizes the audio with the learned intrinsic high-order structures of data on the deeper layers of DBN, compared to the low level 'shallow' features at the input layer. Table 1 shows a typical setting of DBN parameters among more than 50 configurations of parameter combination tested in our experiments.

4   Scene Classification

Given the features $\{f^i_{DBN}\}$ learned and extracted by the DBN for one audio clip as well as their corresponding saliency weights $\{\mathbf{w}_i\}$, we rely on the DBN to classify the feature to certain scene category by using the linear classifier (i.e., softmax) as the final output layer of label units. Alternatively, we can use a separate shallow classifier model like support vector machine (SVM) to yield a class label for each feature.

With the scene label c_i predicted for each DBN feature $\boldsymbol{f}^i_{DBN}$ of an input clip, we exploit a simple voting scheme over all predictions, which aggregates the weighted vote by $\boldsymbol{f}^i_{DBN}$ for scene class c_i as $\mathcal{V}(c_i) = \mathcal{V}(c_i) + \mathbf{w}_i$, where $\mathcal{V}()$ is the vote accumulator for all classes, and choose the highest score of the potential scene category as the final output.

5 Results

We evaluate the proposed auditory scene classification method on the public auditory scene dataset [1] created by us, whose samples are collected from the Freesound.org website and some movie/TV clips. The dataset contains 1,281 clips of 10 categories of auditory scenes (6 outdoor and 4 indoor) - inside vehicle, beach, train station, street, restaurant, auditorium, forest, raining, playground and war field. Table 2 shows the distribution of sample durations for each scene category in the dataset. The sampling rate ranges from 22.05kHz to 48kHz.

Table 2. Dataset distribution on sample category and duration

Dur. (sec.)	Auditory Scene Category (# of samples)									
	vehicle	beach	station	street	restau.	audito.	forest	rain	playgnd	war
< 5	0	2	0	1	0	0	0	1	0	1
5 ∼ 15	5	6	8	21	4	31	10	7	6	42
15 ∼ 30	48	101	67	84	99	77	96	104	98	58
30 ∼ 60	67	20	32	17	26	14	14	9	18	13
60 ∼ 120	10	0	8	3	5	4	6	13	11	3
> 120	1	1	1	2	1	1	1	1	1	1
Total	131	130	116	128	135	127	127	135	134	118

In the experiments, each clip is first resampled to the uniform sampling rate of 22.05kHz with 16 bits per data point, then a short-term window of 1024 data points with 50% overlapping between successive windows is applied on the signal sequence to extract DFT and Mel-scale (30 bands) low-level spectral features.

To detect salient peak points in the spectrogram and further generate different sizes of the resulting low-level description subset, we set α and σ in Eq. (1) to 0.98, 0.94, 0.86 and 40, 40, 40, respectively, which gives feature subsets of 23%, 42% and 62% of the original data set. Additionally, we use an analysis window covering 11 short-term windows (5 before and 5 after each salient point) in Section 2.3 to concatenate the 30-dim Mel-scale coefficients, and a 200-dim context description (with 20 angular and 10 temporal bins) is used in experiment.

For the DBN model used in the experiment, we tested different number of hidden levels of DBN, among other combinations of hyper-parameter values, ranging from 2 to 5, and found the Layer 3 of the DBN with 3 hidden layers yields the average highest accuracy, in which the number of units on each hidden layer is set to 300, 150 and 60, respectively.

To evaluate the performance of different classification models, we use 20% of sample clips in the dataset to train the model and use the left 80% clips for testing. We use the Gaussian mixture model (GMM) on Mel-frequency cepstral coefficients (MFCCs) of scene as the baseline algorithm for comparison. Fig. 5 shows the confusion matrix of the proposed method using softmax on DBN-based features learned from the saliency-based reduced low-level description, which gave an average prediction accuracy of 82% and, 79.7% if using SVM as the final classifier. The baseline GMM-MFCC method yielded an average accuracy of 75.2%.

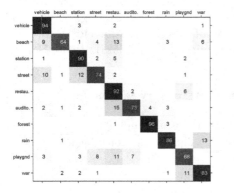 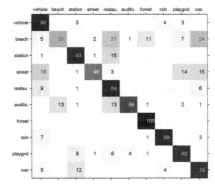

Fig. 5. Confusion matrix of the proposed DBN-based method (left) and the baseline GMM-MFCC method (right) for 10 auditory scenes

In Fig. 6, we compare the overall recognition accuracy of algorithm on the 10 scene categories by: the proposed DBN-based model on low-level description of 1) DFT, 2) DFT concatenated with the context description, 3) Mel-scale coefficients with the context description, respectively, and 4) a HMM-based scene model with MFCC feature used in [15], and 5) the baseline GMM-based scene model with MFCC feature similar to that used in [4]. The results show that DBN-based classification model yields better results than the models based on MFCC - the most common low-level spectral feature, revealing the effectiveness of the DBN model in discovering underlying higher-order statistical structures from the unstructured audio data of auditory scenes. Moreover, the proposed context feature is shown contributing to a further enhancement of accuracy, owing to its complementary descriptive ability on correlations of the signals beyond the analysis window of DBN. On the other hand, compared to the more primitive DFT, as shown by results especially on the beach and street scenes, the Mel-scale feature loses certain discriminative information that is important for some categories of scenes, while being more efficient for the other scenes.

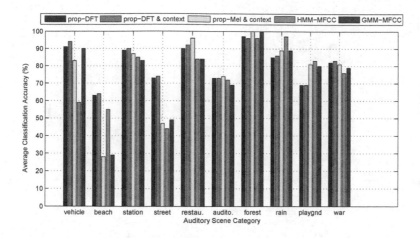

Fig. 6. Comparisons of accuracies over 10 scene classes by the proposed DBN-based model on low-level description of DFT (81.2%), DFT & context (82%), Mel-coef. & context (77.6%) respectively, and the HMM-MFCC (74.1%), GMM-MFCC scene model (75.2%)

In Fig. 7, the scatter of the original DFT coefficients, the high-level features learned by DBN on top of DFT and the MFCC features wihtin a random subset of testing data are projected into 2-dimensional by the t-SNE algorithm [12] and plotted. Notice the clustering of the feature discovered by DBN is much more definite than DFTs and also has more compact intra-class distribution than MFCC (e.g. class 4 and 7), which helps improving the classification accuracy.

Fig. 8 shows the effectiveness of the proposed saliency-based low-level feature subsampling scheme by comparing average accuracies of the DBN model when fed with the feature subsets of different sizes. Classification on salient points of audio gives the averagely higher accuracy than that on the complete feature set, showing the signals around salient points capture most of discriminative information of the audio stream of the scene.

6 Conclusions

We propose an auditory scene classification method based on the deep belief network model. The main contribution of the paper is twofold: First, we explored and verified the effectiveness of exploiting the DBN model to learn the intrinsic and characteristic higher level representations from the unstructured auditory scenes. Second, we proposed a saliency-based feature subsampling and context modeling scheme, which was demonstrated effective in experiments to improve the performance and scalability of the DBN-based scene classification model.

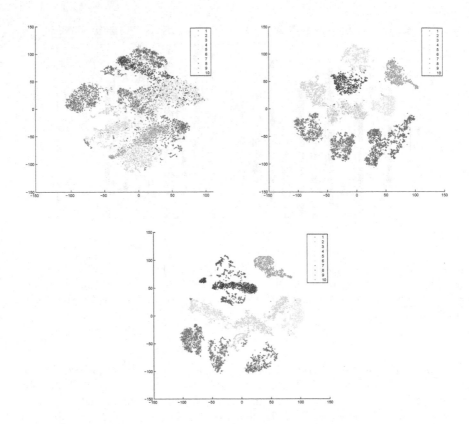

Fig. 7. 2-Dimensional projections of DFT coefficients (left), DBN Layer-3 feature learned on DFT (right) and MFCC feature (bottom) of the audio with respect to 10 scene categories as differently colored

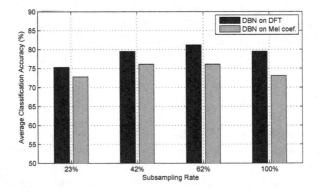

Fig. 8. Comparisons of accuracies of the proposed DBN-based model on different sizes of the subsampled low-level feature set based on saliency analysis

Acknowledgments. Research supported by the National Science Foundation of China under Grant Nos. 61003113, 61272218 and 61321491.

References

1. Auditory scene analysis & classification dataset NJU-AS-10, http://cs.nju.edu.cn/sufeng/data/audioscene/
2. Baluja, S., Covell, M.: Audio fingerprinting: Combining computer vision & data stream processing. In: ICASSP 2007, pp. 213–216 (April 2007)
3. Cai, R., Lu, L., Hanjalic, A., Zhang, H., Cai, L.: A flexible framework for key audio effects detection and auditory context inference. IEEE TASLP 14(3), 1026–1039 (2006)
4. Chu, S., Narayanan, S., Kuo, C.C.J.: Environmental sound recognition with time-frequency audio features. IEEE TASLP 17(6), 1142–1158 (2009)
5. Chu, W., Cheng, W., Wu, J.: Generative and discriminative modeling toward semantic context detection in audio tracks. In: MMM 2005, pp. 38–45 (2005)
6. Deng, L., Hinton, G., Kingsbury, B.: New types of deep neural network learning for speech recognition and related applications: An overview. In: ICASSP 2013, pp. 8599–8603 (2013)
7. Eronen, A.J., Peltonen, V.T., Tuomi, J.T., Klapuri, A.P., Fagerlund, S., Sorsa, T., Lorho, G., Huopaniemi, J.: Audio-based context recognition. IEEE TASLP 14(1), 321–329 (2006)
8. Hamel, P., Eck, D.: Learning features from music audio with deep belief networks. In: ISMIR 2010, pp. 339–344 (2010)
9. Han, B., Hwang, E.: Environmental sound classification based on feature collaboration. In: ICME 2009, pp. 542–545 (2009)
10. Hinton, G., Deng, L., Yu, D., Dahl, G.E., Rahman Mohamed, A., Jaitly, N., Senior, A., Vanhoucke, V., Nguyen, P., Sainath, T.N., Kingsbury, B.: Deep neural networks for acoustic modeling in speech recognition: The shared views of four research groups. IEEE Signal Processing Magazine 29(6), 82–97 (2012)
11. Hinton, G.E., Osindero, S., Teh, Y.W.: A fast learning algorithm for deep belief nets. Neural Computation 18(7), 1527–1554 (2006)
12. van der Maaten, L., Hinton, G.: Visualizing high-dimensional data using t-sne. JMLR 9, 2579–2605 (2008)
13. Rahman Mohamed, A., Dahl, G.E., Hinton, G.: Acoustic modeling using deep belief networks. IEEE TASLP 20(1), 745–770 (2012)
14. Rahman Mohamed, A., Sainath, T.N., Dahl, G., Ramabhadran, B., Hinton, G.E., Picheny, M.A.: Deep belief networks using discriminative features for phone recognition. In: ICASSP 2011, pp. 5060–5063 (2011)
15. Su, F., Yang, L., Lu, T., Wang, G.: Environmental sound classification for scene recognition using local discriminant bases and HMM. In: ACM Multimedia 2011, pp. 1389–1392 (2011)
16. Wang, J., Wang, J., He, K., Hsu, C.: Environmental sound classification using hybrid SVM/KNN classifier and mpeg-7 audio low-level descriptor. In: IJCNN 2006, pp. 1731–1735 (2006)
17. Yang, L., Su, F.: Auditory context classification using random forests. In: ICASSP 2012, pp. 2349–2352 (2012)
18. Zhang, X.-L., Wu, J.: Deep belief networks based voice activity detection. IEEE TASLP 21(4), 697–710 (2013)

An Improved Content-Based Music Recommending Method with Weighted Tags

Lu Ding, Ning Zheng, Jiang Xu, and Ming Xu

College of Computer, Hangzhou Dianzi University 310018 Hangzhou, China
ddbb-love@163.com, {nzheng,jian.xu,mxu}@hdu.edu.cn

Abstract. Content-based filtering is widely used in music recommendation field. However, the performance of existing content-based methods is dissatisfactory, because those methods simply divided the listened songs into like or unlike set, and ignored user's preference degree. In this paper, an enhanced content-based music recommending method was proposed by quantifying the user preference degree to songs with weighted tags. Firstly, each listened song was classified into like or unlike set according to user's playing behaviors, such as skipping and repeating. Secondly, the songs' social tags were collected from LastFm website and weighted according to their frequency in the collected tags.Finally, the user's preference degree for each song was quantified with the weighted tags, and the candidate songs with high preference degrees would be recommended to him. On the LastFm dataset, the experimental results demonstrate that the proposed method outperforms those traditional content-based methods in both rating and ranking prediction.

Keywords: Recommend Systems, Social Tagging, Content Filtering.

1 Introduction

Along with the development of World Wide Web, millions of free online music makes it hard for people to find out what they like manually. The recommender systems provide a widely adopted solution to the information overload problem, and can automatically help people to decide what to listen.

The current music recommendation technologies mainly fall into two categories: content-based filtering [1] and collaborative filtering [2]. Content-based filtering technology analyzes the similarity between users or items by the metadata, such as user profiles and music acoustic features. In contrast, collaborative filtering technology analyzes the similarity between users or items by users' past behaviors, no requiring domain knowledge. The Latent factor models like matrix factorization (MF) and the neighborhood model are typical approaches in collaborative filtering. They found the relationships between users and items by analyzing users' listening histories. Besides, there are other new techniques, such as LDA methods [3] and graph-based models [4].

Though the traditional recommenders can effectively predict which song a user likes, they can hardly make an understandable explanation why they do

X. He et al. (Eds.): MMM 2015, Part I, LNCS 8935, pp. 360–371, 2015.

these recommends, and difficultly answer the question of what kind music a user likes or unlike

In this paper, a personalized tag model was proposed for music recommendation. The music's metadata was replaced with the social tags to represent songs. Here, the social tags are keywords generated by internet users on a platform and they are used to describe and categorize an object, concept or idea. As they are originally created by users' own way, they contain meaningful concepts to users. Furthermore, a song's top tags are the most popular ones which are attached by the users on this platform, so that they can stand for the social opinion to songs. Comparing with the audio features, such as pitch and tempo, social tags can better classify and label resource, besides they can make the recommender more understandable. Moreover, the tags were weighted on the perspective of statistic analysis of each user's implicit feedbacks to build each user's personalized tag model. On the LastFm dataset, experiments demonstrate that the proposed method can outperform the traditional content-based method in both rating and ranking prediction.

The remainder of this paper is organized as follows. Section 2 introduces prior work related to this paper. Then the details of the proposed method are presented in Section 3. In Section 4, experiments are discussed, followed by conclusion and future work in Section 5.

2   Related Work

In this section, some of research works related to tag-based music recommender systems are presented.

The methods which recent research papers apply in tag-based music recommendation can be classified into two categories. One way is to use tag data as content information to compute users or items similarity. For examples, Bosteels et al. [5] used social top tags to calculate similarity values between listened songs and candidate songs, and then used those values as the fuzzy relationship degrees to compare performance of some different heuristics. It can reduce the predicting failure rate comparing with the method in [6], which used the audio-based similarity. The main difference between the proposed method and [5] is that the tags' weights were equal in [5] but in the proposed method tags were assigned with different weights according to their frequency in the collected tags. Kim [7] assigned weight to tags according to the intensity of tag's emotion which was judged by the SentiWordNet [8], after then, user profiles using the weighted tags were generated and a user-based collaborative filtering algorithm was executed. The main barrier in [7] is the sparsity problem as the tags each user assigned to a song are very poor.

The other way is to make use of the tag data to build recommend models. For example, Zhang et al. [9] proposed a random walk model, which was based on Page Rank liked random walks among the user-item, user-tag and item-tag bipartite graphs. Besides, it also made use of the tag information to build item graphs and user graphs based on a probabilistic method. As it applied random walks on ternary interaction graphs to capture transitive associations between users and items, the sparsity problem had been alleviated but it brought the huge time consumption on building graphs and searching. Hariri et al. [10] proposed a LDA model to predict what topics next song contains. In that paper, the tag data was used to establish the topic modeling module. Each song was represented as a set of topics. By matching the current topic sequence with history frequent topic sequential pattern, the next topics were found. Then songs which contain these topics were recommended. The main problem is that the recommender's performance relies on the number of LDA topics, so that different settings result in different frequent sequential patterns and affect the recommendation performance. Taramigkou et al. [3] also used the LDA model, the difference is the tags are artists' tags which are assigned by users. Then a users graph was build and the weights of edges were assigned using the cosine similarity between topic vectors generated for each user by LDA model. Finally, the Dijkstra algorithm [11] was used as a graph search approach to recommend a list of artists. But all these mentioned models didn't do a good job of explaining why the song is recommended, and answering the question what users like or unlike. In this paper, those weighted tags are straightly used to form users' profiles. The greater weight the tag has, the more a user like the tag, vice versa.

3   The Proposed Method

Give the history logs which contain songs that have been listened previously by users, each user's music taste can be established by analyzing these logs. This paper uses social tagging web sites to retrieve each song's top tags and use these tags as features to build up users' personalized tag model. Then for a given user, a list of songs which are close match with his/her tag model will be recommended.

This section first presents the format of users' listening logs from LastFm website, and the method of capturing extra information for each entry. Then these complementary logs are used to analysis users' behaviors. At last, the personalized tag model for music recommendation is presented.

3.1   Acquiring the Extra Information

The logs is formatted one entry per line as follows: userid \ timestamp \ music-brainz-artist-id\ artist-name\ musicbrainz-track-id \ track-name. The timestamp was the moment when a user started this track. The musicbrainz-artist-id and musicbrainz-track-id are MusciBrainz Identifiers (MBIDs),[1] an MBID is a 36 character Universally Unique Identifier that is permanently assigned to each entity in the MusicBrainz database. For example, the artist Adele has an artist MBID of cc2c9c3c-b7bc-4b8b-84d8-4fbd8779e493, and her song Best for Last has a recording MBID of 84c00aff-b2cf-4bf7-a2e3-9460820efb03. The aim of MusicBrainz is to be the universal lingua franca for music by providing a reliable

---

[1] `http://musicbrainz.org/doc/MusicBrainz_Identifier`

and unambiguous form of music identification. So if the logs'multi-entries have the same musicbrainz-track-id, then it means they are the same track objects, and different musicbrainz-track-ids mean different track objects.

Having these musicbrainz-track-ids, the duration and top5 tags of each track in the users' history logs can be got by using the method named track.getinfo, which is provided by the LastFm API. This extra information is very important, and it will be combined with the original logs to analysis users' behaviors and establish each user's personalized tag model in the following sections.

3.2   Analysis of Users' Behaviors

To capture users' music bias, it is necessary to find the implicit feedbacks through looking inside the logs. The proposed method is based on a simplistic assumption that a user like a song if he continuously repeated it, and a user dislike a song if he skipped it several times. Despite once skip didn't infer dislike, you skipped a special song many times may infer you dislike it. The main goal of this subsection is to present the key method of finding out such behaviors.

Supposing you listen to the songs like this: there are some buttons which can be used to control the music player to "play next", "play back", and when you started a song the timestamp and its' mbid was recorded in the log. At the beginning, you listened to two songs and during this time you didn't press any button. Then, you listened to the third one but in the half way you pressed the "play next" button because you felt it's not good. After that, you completely finish listening to the fourth one and found this song is very fit for you, so you pressed the "play back" button in order to re-listen it, after listening it twice you closed the player. During this procedure five entries were wrote into the log, which can be simply formatted as follows:

Session 1=$\{\langle$ t1,m1$\rangle,\langle$t2,m2$\rangle,\langle$t3,m3$\rangle,\langle$t4,m4$\rangle,\langle$t5,m4$\rangle\}$. One tuple stands for one entry, and the notations are defined in Table 1. All tuples are in chronological order. What we need to do is to find out the "play next", "play back" actions in this session. An algorithm was proposed:

1. Step 1, calculate the length of each song played. As the timestamp is the start moment, so p1=t2-t1;
2. Step 2, compare the length with song's duration. If you pressed "play next", then length<duration, in this session p3<d3;
3. Step 3, compare adjacent songs' mbids, if you re-listened a song, then adjacent songs' mbids were the same. Algorithm 1 shows the details for analyzing the whole logs.

More formalized description of users' behaviors is displayed below:

1. Shuffling, it means in one session the user u just played the track i once. It corresponds to "nothing", and is marked as "Normal"
2. Skipping, it means in one session the user u skip the track i. It corresponds to "play next", and is marked as "Skip"

3. Repeating, it means in one session the user u continuously plays the track i. It corresponds to "play back", and is marked as "Repeat"

The commonly used notations for this work are displayed in Table 1. Two main functions are the length's calculation of each entry in the logs heard, and the method to find repeating behavior. They are defined as follows:

Table 1. Commonly Used Notations

Representations	Descriptions
L	the set of entries in the completed users' logs
U	user set
l,u	entry l,user u
p_l	the length of track in the l be heard
S_l,R_l,N_l	the Skip/Repeat/Normal Flag of l
$L(u)$	the sequence of entries produced by user u
d_l	the duration of track in l
t_l	the timestamp of l
m_l	the mbid of track in l

$$\Delta(i,j) = t_i - t_j (i,j \in L) . \tag{1}$$

$$com(i,j) = \begin{cases} 1 & \text{if } m_i \text{ equals to } m_j \\ 0 & \text{otherwise} \end{cases} (i,j \in L) . \tag{2}$$

In Algorithm.1 the purpose of line 6-9 is to calculate the length of each song had been listened. The purpose of line 13-14 is to find the Repeating behavior, and mark it "Repeat". The purpose of line 15-16 is to find the Skipping behavior, and mark it "Skip". The purpose of line 17-18 is to find the Shuffling behavior, and mark it "Normal". The purpose of line 23 is to add these marks into user's original datasets.

3.3   The Proposed Model

As it was mentioned in section 3.1, the extra information of top tags will be used in the proposed model. The reason for choosing these tags is that they not only describe various features of the songs including genre, artist name and era, but also describe users' attitudes toward the songs, such as sadness, mellowness, and so on. Although people may have different opinions about songs, top tags with frequency above a minimum threshold capture the social opinion about each song. These features can often be very helpful in explaining the commonalities in a set of songs selected by a user. Table 2 shows a set of songs' top tags as example.

Algorithm 1. Analyzing users' behaviors

Input: L, U, Session Threshold value λ, Skip Threshold value θ
Output: L^+
1: Initialize the Skip, Repeat, Normal Flags with 0 for each l in L;
2: **for** each user u in U **do**
3:     **for** $i = 0 \rightarrow length.L(u)$ **do**
4:         calculate $\Delta(l_i, l_{i+1})$ use function(1);
5:         // l_{i+1} denotes the $(i+1)$th entry in the $L(u)$
6:         **if** $\Delta(l_i, l_{i+1}) < \lambda$ **then**
7:             $p_{l_{i+1}} \leftarrow \Delta(l_i, l_{i+1})$
8:         **else**
9:             $p_{l_{i+1}} \leftarrow -1$
10:         **end if**
11:         **if** $p_{l_{i+1}} \neq -1$ **then**
12:             calculate $com(l_i, l_{i+1})$ use function(2);
13:             **if** $com(l_i, l_{i+1})$ **then**
14:                 $R_{l_i} \leftarrow 1$ and $R_{l_{i+1}} \leftarrow 1$
15:             **else if** $p_{l_{i+1}} < \theta$ **then**
16:                 $S_{l_{i+1}} \leftarrow 1$
17:             **else**
18:                 $N_{l_{i+1}} \leftarrow 1$
19:             **end if**
20:         **else**
21:             $R_{l_{i+1}} \leftarrow 0, S_{l_{i+1}} \leftarrow 0, N_{l_{i+1}} \leftarrow 0$
22:         **end if**
23:         update the Skip, Repeat, Normal Flags with $S_{l_i}, R_{l_i}, N_{l_i}$ for each l in $L(u)$;
24:     **end for**
25:     $L^+ \leftarrow L^+ \cup updated\_L(u)$
26: **end for**
27: **return** L^+

Table 2. Top tags of songs

Artist name	Track name	Top tags
Valerie Etienne	The Birds Sing	jazz, chillout, lounge, groovy, female vocalist
U2	If God Will Send His Angels	rock, u2, soundtrack,90s, irish
Beck	Lost Cause	mellow, alternative, indie, sad, singer-songwriter
Bjork	Show Me Forgiveness	alternative, icelandic, bjork, female vocalists, electronic

Genres like "jazz", eras like "90s" ,artists likes "u2", mood likes "mellow" can be found in the Table 2. Using these top tags, more diverse and deeper factors can be found, and the recommender can be more precise and understandable.

To build the proposed model, all the songs' top tags were collected in skip set, repeat set, and normal set respectively. After getting rid of the same tags, weights were assigned to them according to frequency.

Marking a user u's skip tags set, repeat tags set and normal tags set as ULT, LT and NT respectively, the function to compute weight is:

$$score(t_i|u, ULT) = fr(t_i|ULT) \,. \tag{3}$$

$$score(t_i|u, LT) = fr(t_i|LT) \,. \tag{4}$$

$$score(t_i|u, NT) = fr(t_i|NT) \,. \tag{5}$$

Where $fr(t_i|ULT)$ is the overall frequency of tag t_i in user u's skip tags set. The personalized tag model for music recommendation is formalized as follows:

$$TModel(u) = \alpha \cdot score(u, ULT) + \beta \cdot score(u, LT) + \gamma \cdot score(u, NT) \,. \tag{6}$$

Where α, β, γ are independent params to control the power of different tags in ULT, LT and NT , $-1 \leq \alpha < 0$; $0 < \gamma \leq \beta \leq 1$; denoting a candidate song i's top tag set is iT , then its predict score is :

$$
\begin{aligned}
\widehat{r}_{ui} = \alpha \cdot &\sum_{t_i \in \{iT \cap ULT\}} score(t_i|u, ULT) \\
+ \beta \cdot &\sum_{t_i \in \{iT \cap LT\}} score(t_i|u, LT) \\
+ \gamma \cdot &\sum_{t_i \in \{iT \cap NT\}} score(t_i|u, NT)
\end{aligned}
\tag{7}
$$

Where $\sum_{t_i \in \{iT \cap ULT\}} score(t_i|u, ULT)$ is the accumulated weight of tags in both of iT and ULT , we treat this value as the possible degree of user u hate this candidate song i. Where $\sum_{t_i \in \{iT \cap LT\}} score(t_i|u, LT)$ is the possible degree of user u like this song i, and $\sum_{t_i \in \{iT \cap NT\}} score(t_i|u, NT)$ is the possible degree of user u neither hate nor like this song i.

In Algorithm.2 the purpose of line 3-5 is to remove the same tags in intersections. The purpose of line 7 is to assign weight to tags. The purpose of line 8 is to establish user's personalized tag model. Based on the user's personalized tag model, the predicted score of a candidate song can be calculated using the function (7). The higher the song's score is, the more the user like it, vice versa.

Algorithm 2. Building the proposed model

Input: L^+, U ,Params α, β, γ
Output: A list of personalized tag models $TModel$
1: **for** each user u in U **do**
2:     split $L^+(u)$ into $LT^+(u), ULT^+(u), NT^+(u)$
3:     $\overline{LT}(u) \leftarrow LT^+(u) - (LT^+(u) \cap ULT^+(u)) - (LT^+(u) \cap NT^+(u))$
4:     $\overline{ULT}(u) \leftarrow ULT^+(u) - (LT^+(u) \cap ULT^+(u)) - (ULT^+(u) \cap NT^+(u))$
5:     $\overline{NT}(u) \leftarrow NT^+(u) - (LT^+(u) \cap NT^+(u)) - (ULT^+(u) \cap NT^+(u))$
6:     **for** each tag t in $\overline{LT}(u)/\overline{ULT}(u)/\overline{NT}(u)$ **do**
7:         calculate $f_{\overline{LT}(u)}(t), f_{\overline{NLT}(u)}(t), f_{\overline{NT}(u)}(t)$ use function (3),(4),(5);
8:         $TModel(u) \leftarrow \alpha \cdot f_{\overline{LT}} + \beta \cdot f_{\overline{NLT}} + \gamma \cdot f_{\overline{NT}}$
9:     **end for**
10:    append $TModel(u)$ to $TModel$;
11: **end for**
12: **return** $TModel$

---

4 Experimentations

4.1 Dataset

The public dataset (Last.fm Dataset-1K users) [2] from the website of Music Recommendation Datasets for Research was chosen to evaluate the proposed method in this paper. This dataset contains the whole listening habits during May, 5th 2009 to May 2010 for 992 users. The total number of records is 19,150,868. Each entry includes "userid", "timestamp", "artid", "artname", "trackid" and "trackname", and there have 961,416 unique trackids.

4.2 Evaluation Measures

To evaluate the proposed model, two different kinds of evaluation metric are chosen. One is the rating prediction, and the other is the traditional top-K recommendation evaluation metric. The recent $1/11$ entries of each user's records are chosen as test set, and the remaining as training set.

The first task of this recommender is to predict whether the user u will repeat/skip/normal play the track i in test set according to its' predict score which is produced by u's personalized tag model. Then counted how many times it does the right decision. More formally, each user corresponds to a (R, S, N, rr, ss, nn) tuple, where R, S and N are the sets of repeated, skipped and normal played songs by the user, respectively. And "rr" means this repeated song is judged to be repeated, and it's a correct judgment, and the same to "ss" and "nn". Then the hit rate of repeated songs is $Hit(R) = \frac{\sum rr}{\sum |R|}$, the hit rate of skipped songs is $Hit(S) = \frac{\sum ss}{\sum |S|}$, and the hit rate of normal played songs is $Hit(N) = \frac{\sum nn}{\sum |N|}$. The overall hit rate is $Hit = \frac{\sum (rr+ss+nn)}{\sum (|R|+|S|+|N|)}$. The second task of our recommender

---

[2] http://www.dtic.upf.edu/ ocelma/MusicRecommendationDataset/lastfm-1K. html

is to rank the songs in test set according to the predicted scores, then recommend top-K songs to the user. We compute the precision and recall of top-K recommendations as follows:

$$Precision@K = \frac{\sum\limits_{u \in U} |R_k(u) \cap T(u)|}{\sum\limits_{u \in U} |R_k(u)|} . \tag{8}$$

$$Recall@K = \frac{\sum\limits_{u \in U} |R_k(u) \cap T(u)|}{\sum\limits_{u \in U} |T(u)|} . \tag{9}$$

Where $R_k(u)$ is the recommended songs set for user u, $T(u)$ is the accepted songs set of user u.

And compute the F1-meature value as follows:

$$F1@K = \frac{2 \times Precision@K \times Recall@K}{Precision@K + Recall@K} . \tag{10}$$

4.3   Performance Comparison

In this section, the proposed model and the Fuzzy theory method [5] are compared to evaluate whether the weighted tags can improve the recommender's performance when compared with the unweighted tags. Tracks were considered skipped when the user listened to less than 50% of them.For the listening session segmentation, we let the songs to be in the same session, if the user's inactive intervals between adjacent songs are less than 1 hour, which refer to Xiang et al. [12]. And the parameters such as α, β, γ are determined experimentally. In this experiment we set $\alpha = 1, \beta = -1, \gamma = 0.5$.

First Task Result. After getting song i's predicted score , do the judgments as follow: 1) song i will be skipped, if $\hat{r}_{ui} < -|threshold|$. 2) song i will be repeated, if $\hat{r}_{ui} > |threshold|$. 3) song i will be normal played, if $-|threshold| \leq \hat{r}_{ui} \leq |threshold|$. The threshold value is changed from 0 to 1 by step 0.05. In fuzzy theory method the skip hit rate is 8.5% and the overall hit rate is 94.78%.

Fig. 1 shows the hit ratios of the proposed model for different threshold values. It can be found that along with the bigger value of threshold, both of the $Hit(R)$ and $Hit(S)$ decrease, and both of the $Hit(N)$ and overall Hit increase. The reason of these phenomena is that the threshold divides user bias space into three areas. The bigger of threshold the larger of normal area, the smaller of repeat area and skip area. The best performance of overall hit rate is 91.73% when the threshold is 0.25, meanwhile the normal hit rate is 99.9%, the skipped hit rate is 9.02%, and the repeated hit rate is 1.91%.

Comparing to the fuzzy theory method, the proposed approach do have significant improvement in the skip hit rate. It has a gain of 6.12% over the fuzzy model (8.5%), although the overall hit rate slightly reduces 3.22%. It indicates the proposed model can better identify what users unlike.

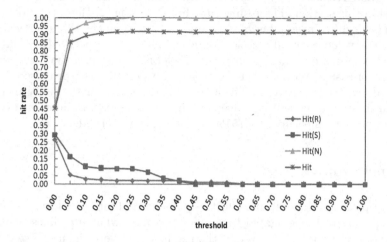

Fig. 1. Hit ratio of personalized Tag Model for different threshold

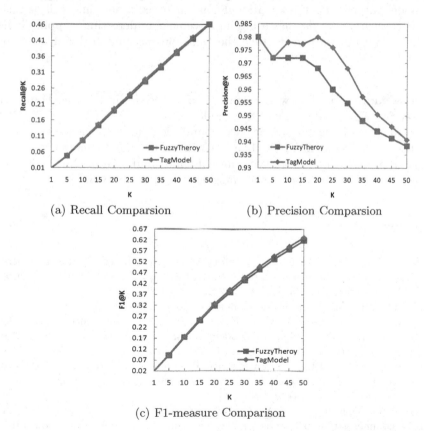

(a) Recall Comparsion

(b) Precision Comparsion

(c) F1-measure Comparison

Fig. 2. Comparing Fuzzy theory method and personalized Tag Model with $k \in [1, 50]$

Second Task Result. The proposed model and the fuzzy theory method are compared in this section. The results are showed in Fig. 2. Here, the horizontal axis stands for different number of recommendations, and the three vertical axes stand for the value of recall, precision and F1-measure respectively. From Fig. 2, it could be observed that the proposed model is better than the fuzzy theory method in precision, recall and F1-measure. Specifically, the proposed Model achieves an average improvement in precision of 1.68% over the fuzzy method. And it achieves an average improvement in recall of 0.748% and an average improvement in F1-measure of 0.751% over the fuzzy method.

5   Conclusion

This paper proposes a simple and effective personalized tag model to mine users' music bias. Due to the weighted tags, it can explain what user likes or unlike understandably as the greater weight the tag has, the more the user like it, vice versa. Experiments conducted on the LastFm dataset show that the personalized tag model outperforms the traditional content-based model in recall, precision and F1-measure, and also has a big improvement in detecting skipping behavior. In the future, we plan to take the time into account to build a dynamic personalized tag model.

Acknowledgments. This work is supported by the Natural Science Foundation Natural Science Foundation of China under Grant No.61070212 and 61003195, the Zhejiang Province Natural Science Foundation Natural Science Foundation of China under Grant No.Y1090114 and LY12F02006, the Zhejiang Province key industrial projects in the priority themes of China under Grant No 2010C11050, the soft science research project of Hangzhou (No. 20130834M15).

References

1. Ferman, A., Errico, J., Beek, P., Sezan, M.: Content-based filtering and personalization using structured metadata. In: 2nd ACM/IEEE-CS Joint Conference on Digital Libraries, pp. 393–393. ACM, Portland (2002)
2. Koren, Y.: Factorization meets the neighborhood: a multifaceted collaborative Filtering model. In: 14th ACM SIGKDD International Conference on Knowledge Discovery and Data Mining, KDD 2008, pp. 426–434. ACM, Las Vegas (2008)
3. Taramigkou, M., Bothos, E., Christidis, K., Apostolou, D., Mentzas, G.: Escape the Bubble: Guided Exploration of Music Preferences for Serendipity and Novelty. In: 7th ACM Conference on Recommender Systems, RecSys 2013, pp. 335–338. ACM, Hong Kong (2013)
4. Xiang, L., Yuan, Q., Zhao, S., Chen, L., Zhang, X., Yang, Q., Sun, J.: Temporal recommendation on graphs via long-and short-term preference fusion. In: 16th ACM SIGKDD International Conference on Knowledge Discovery and Data Mining, KDD 2010, pp. 723–732. ACM, Washington, DC (2010)

5. Bosteels, K., Pampalk, E., Kerre, E.E.: Evaluating and Analysing Dynamic Playlist Generation Heuristics Using Radio Logs and Fuzzy Set Theory. In: Proceedings of the 10th International Society for Music Information Retrieval Conference, ISMIR 2009, Kobe, pp. 351–356 (2009)
6. Pampalk, E., Pohle, T., Widmer, G.: Dynamic Playlist Generation based on Skipping Behavior. In: Proceedings of the 6th International Society for Music Information Retrieval Conference, ISMIR 2005, London, pp. 634–637 (2005)
7. Kim, H.H.: A Semantically Enhanced Tag-Based Music Recommendation Using Emotion Ontology. In: Selamat, A., Nguyen, N.T., Haron, H. (eds.) ACIIDS 2013, Part II. LNCS, vol. 7803, pp. 119–128. Springer, Heidelberg (2013)
8. Esuli, A., Sebastiani, F.: SentiWordNet: A Publicly Available Lexical Resource for Opinion Mining. In: Proceedings of the 5th Conference on Language Resources and Evaluation, LREC 2006, Genova, Italy, pp. 417–422 (2006)
9. Zhang, Z., Daniel, D.Z., Ahmed, A., Jing, P., Zheng, X.L.: A Random Walk Model for Item Recommendation in Social Tagging Systems. ACM Transactions on Management Information Systems 4(2) 8, 1–24 (2013)
10. Hariri, N., Mobasher, B., Burke, R.: Context-aware music recommendation based on latent topic sequential patterns. In: 6th ACM Conference on Recommender Systems. RecSys 2012, pp. 131–138. ACM, Dublin (2012)
11. Dijkstra, E.W.: A note on two problems in connexion with graphs. Numerische Mathematik 1(1), 269–271 (1959)
12. Xiang, W., Qi, L., Enhong, C., Liang, H., Jingsong, L., Can, C., Guoping, H.: Personalized Next-song Recommendation in Online Karaokes. In: 7th ACM Conference on Recommender Systems, RecSys 2013, pp. 137–140. ACM, Hong Kong (2013)

A Unified Model for Socially Interconnected Multimedia-Enriched Objects

Theodora Tsikrika, Katerina Andreadou, Anastasia Moumtzidou,
Emmanouil Schinas, Symeon Papadopoulos,
Stefanos Vrochidis, and Ioannis Kompatsiaris

Information Technologies Institute, CERTH, Thessaloniki, Greece

Abstract. Enabling effective multimedia information processing, analysis, and access applications in online social multimedia settings requires data representation models that capture a broad range of the characteristics of such environments and ensure interoperability. We propose a flexible model for describing Socially Interconnected MultiMedia-enriched Objects (SIMMO) that integrates in a unified manner the representation of multimedia and social features in online environments. Its specification is based on a set of identified requirements and its expressive power is illustrated using several diverse examples. Finally, a comparison of SIMMO with existing approaches demonstrates its unique features.

Keywords: multimedia description, social interactions, interoperability.

1 Introduction

The massification of web publishing, together with the proliferation of social media hosting and sharing platforms, and the widespread use of mobile devices have led to an unceasing generation of large volumes of online content objects, such as Web pages and social media posts. Such objects are enriched with multimedia and are interconnected in various ways, including through social interactions. To satisfy the information needs of the diverse users in such online social multimedia environments (ranging from laypeople simply searching for information to media monitoring professionals) requires effective multimedia information processing, analysis, and access applications that support tasks such as clustering, classification, summarisation, search, recommendation, and retrieval. To enable such applications, there is a need to employ data representation models that capture a broad range of the characteristics of such settings and ensure interoperability across diverse multimedia objects, hosting services, and tasks.

Existing models for the description of multimedia content and structure, such as MPEG-7 [5], RUCoD [6], and WebLab [7], do not take into account the social characteristics and interconnections in current web settings, since most were developed prior to the advent of social Web or/and for different purposes. On the other hand, more recent approaches that capture such social aspects,

X. He et al. (Eds.): MMM 2015, Part I, LNCS 8935, pp. 372–384, 2015.
© Springer International Publishing Switzerland 2015

such as the SIOC [1] and FOAF [2] ontologies, do not consider the potential multimodality of online content and the variety of its extracted annotations.

This work aims to bridge this gap by proposing a flexible and expressive framework for the represention of Socially Interconnected MultiMedia-enriched Objects (SIMMO), that avoids the complexity of previous models (e.g., [5]). Our main contribution is a model that describes in a unified manner interconnected multimedia content on the Web, generated by a variety of sources and hosted in diverse platforms, by also considering social features, such as users' interactions with such multimedia objects and with each other. The data representation requirements in such online social multimedia environments are identified (Section 2) in order to support the specification of the model and its implementation (Section 3). The expressive power of SIMMO is illustrated using a number of diverse examples of online multimedia objects with social interconnections, such as multimodal Web pages and YouTube videos (Section 4), while a comparison with existing approaches demonstrates its unique features (Section 5).

2   Data Representation Requirements

The proposed model needs to capture a number of characteristics stemming from the nature of online social multimedia with particular focus on those that are typically taken into account in the context of multimedia information processing, analysis, and access applications. The most salient such characteristics [8,9] are:

C1 Host heterogeneity and fragmentation: There is a plethora of diverse types of online services hosting and sharing media content, ranging from Web sites hosted on dedicated servers to social media sharing platforms, such as Flickr, Instagram, YouTube, etc. This heterogeneity is further exacerbated by the fact that each of the latter is typically associated with a number of attributes that are particular to the platform. It is possible though to identify cross-platform mappings for several of these attributes, such as among those conveying endorsement, e.g., likes in Facebook and favourites in Twitter.

C2 Media objects diversity: Online media content is expressed in a variety of modalities (such as text, images, video, and audio) and contained within diverse media objects, ranging from simple media items (e.g., an online image or video file) to complex multimedia documents (e.g., Web pages and social media posts) consisting of heterogenous media items.

C3 Online links and relations: Online media content does not live in isolation; there are in fact various relations that can be established among media objects. As mentioned above, multimedia documents can contain media items (e.g., YouTube videos can be embedded in Web pages or be shared through tweets), while they can also be interconnected with other media objects (e.g., Web pages or social media posts can contain links to other Web pages).

C4 Social links and interactions: The users of social media sharing and networking platforms are connected with each other through explicit links (e.g., followship, friendship) and interact with the posted content and with each other (often using content), e.g., they like Facebook posts, comment on YouTube videos, add Vines to replies in Twitter, etc. Such social user behaviour is also supported outside the context of such platforms by several Web sites that allow social interactions with their Web pages through, e.g., posting comments on them or sharing them on Twitter or Facebook.

C5 Dynamic content: Multimedia documents can also be classified based on their relationship with time. *Static* multimedia documents do not have a temporal dimension, whereas *dynamic* Web pages change over time, e.g., through comments being continuously posted on them.

C6 Automatically generated metadata: The digital devices currently used for generating media items (e.g., images) have the capability of automatically creating a wealth of metadata to annotate them, such as the geographical identification metadata tagging the photographs taken by smartphones and digital cameras. Such automatically generated metadata typically accompany the raw content, but social media sharing platforms may replace them with explicit metadata fields or even completely remove them in some cases.

The proposed model also needs to support a number of tasks commonly performed in multimedia information processing, analysis, and access applications [8,9], such as search, clustering, and summarisation. Typical such tasks in online social multimedia settings include those listed below; their characteristics are influenced, to a large extent, by the properties of such settings outlined above:

T1 Cross-host search: In a variety of settings, end users are interested in retrieving media content in response to their information needs irrespective of the environment hosting the relevant media objects (see also C1), e.g., both Web pages and tweets relevant to a submitted query. Establishing the relevance and importance of media objects hosted in multiple and widely different environments is particularly challenging given their heterogeneity.

T2 Multimodal search: End users are interested in retrieving relevant information irrespective of the media in which it is encoded, while also having the freedom to express their queries in whichever media they find intuitive, e.g., using similar images when searching for an image or keywords to find their favourite song, and combinations thereof. Enabling unified retrieval that is transparent to users given queries expressed in any number of modalities is a difficult task given also the heterogeneity of available media objects (see also C2) and annotations (as discussed next in T3).

T3 Layered annotation: Multimedia content can be currently described in a multitude of ways and at different levels of abstraction, including descriptive metadata (e.g., creation date) (see also C6), textual annotations (e.g., keywords), low-level features (e.g., visual features such as SIFT), high-level features (e.g., concepts), and events. Many such annotations are interdependent, e.g., high-level features are generated based on the extracted low-level features, while events may be determined using the identified concepts. Establishing relations among annotations (e.g., determining which visual features were used for the concept annotation process) is important in many settings, particularly when end users are search professionals or researchers.

T4 Varied granularity access: In many cases, end users are interested in accessing media content at a granularity level different to that of a multimedia object. When searching for information, for instance, retrieval of only the specific media segments that contain relevant information, instead of entire multimedia objects, reduces users' cognitive load and increases their satisfaction. Such focussed retrieval applications include finding only the shots in a news video relevant to a story or only the image segments where e.g., a specific logo appears. Furthermore, representation at higher levels of granularity, e.g., multimedia collections, is also useful in many contexts. For instance, an aggegated view created by summarising a set of social media posts on the same subject or story provides a snapshot of public opinion on that topic.

T5 Content provenance: In several applications, it is important to track the original source of a content item posted online, e.g., to establish whether an image has been previously published in a different context. The ease with which media content is embedded within multimedia documents and shared across diverse platforms (see also C3 and C4) indicates the significance, but also the difficulty of this task. This is further the case when online content undergoes manipulations and alterations, and is subsequently reposted for entertainment (e.g., memes) or malicious (e.g., propaganda) purposes.

T6 Influentials identification: When researching a particular story or topic, several types of users (e.g., journalists, analysts, etc.) are interested in identifying influential and relevant content and also, particularly in the case of social media, the content contributors who publish such content. As this is typically achieved by analysing the Web and social link structures, it is paramount to model such relations between multimedia objects, between users and multimedia objects, and also between users (see also C3 and C4).

Based on the above characteristics and tasks, we extract a set of requirements for an effective data representation model that would enable multimedia information processing, analysis, and access applications in online socially interconnected environments. Without claiming that the above lists are exhaustive, they do cover both the principal aspects of online multimedia settings and their social features (see also Section 5). Therefore, our model should represent (in brackets the relevant items from the above lists giving rise to each requirement):

R1 media content of various modalities (C2, T2),

R2 diverse media objects, ranging from mono-modal media items to composite multimedia documents (C2, C3, T2),

R3 media objects across heterogeneous hosting environments in a unified manner (C1, T1),

R4 online relations between media objects (C2, C3, T5, T6),

R5 social interactions between users and media objects (C3, C4, C5, T5, T6),

R6 content contributors, their relations and interactions, as expressed through their accounts in social Web platforms (C4, T6),

R7 granularity at various levels, ranging from media segments to multimedia collections (T4),

R8 rich heterogeneous annotations describing media objects of any granularity, and the relationships between such annotations (T2, T4, T3), and

R9 descriptive metadata as attributes of media objects (C6, T3).

Next, we propose a data representation model based on the above requirements.

3 Data Representation Model

This section presents the proposed framework for the unified representation of Socially Interconnected MultiMedia-enriched Objects (SIMMO) available in web environments. SIMMO consists of a number of core entities and their sub-classes, attributes, and relations that have been determined based on the requirements (R1-R9) identified in Section 2. While similar entities are also encountered, at least in part, in other models (e.g., [5,6,1]) that have also formed part of our inspiration (see Section 5), it is the interconnections among SIMMO elements and the novel approach of bridging the gap between multimedia and social features that make SIMMO unique in its ability to support a wide range of applications.

Figure 1 presents a conceptual model of SIMMO with the following core entities and their sub-classes:

- **Object** is a generic entity representing media content ranging from mono-modal **Item**s to multimedia **Document**s. Each Item represents the actual media content consisting of a single modality, such as **Text, Image, Video**, or **Audio**, whereas Documents may be viewed as container objects consisting of potentially multiple such Items, and thus modalities. The most common instantiations of Web Documents are **Webpage**s (e.g., pages in news sites, in entertainment portals, etc.) or **Post**s in media sharing platforms with social characteristics (e.g., Facebook posts, tweets, etc.). There are also cases of Webpages consisting of Posts; a forum page, for instance, can be viewed as a container object consisting of posts on the same topic. The **Media** entity is introduced as an abstraction of Image, Video, and Audio so as to represent their common characteristics, such as the fact that they all may be associated with a Text item modelling the text associated with them (e.g., a caption) or extracted from them through e.g., ASR (Automatic Speech Recognition) for Video and Audio, and OCR (Optical Character Recognition) for Image and Video. Finally, further media (e.g., 3D objects) may be added as Item instantiations depending on the requirements of the particular application.
- **Source** is a generic entity representing media content contributors. This includes **UserAccount**s representing users generating content, mainly posts in social media sharing platforms where they hold accounts, and **WebDomain**s representing the Web sites hosting media content generated by their contributors. WebDomains are viewed as content contributors, even though they do not correspond to the actual person who contributed the content, given that in many cases the information regarding such people may not be available, or may be of much lesser importance in this specific context.

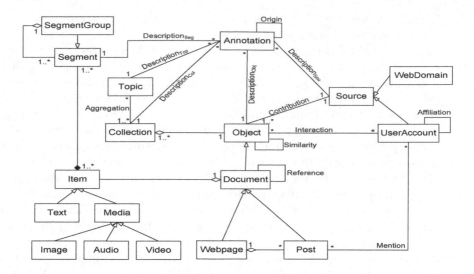

Fig. 1. SIMMO conceptual model presenting its elements and their relations. For simplicity, association relations that have attributes are depicted as simple associations.

- *Segment* locates the media content of Items at a finer level of granularity (e.g., a passage in text, a region in an image, or a portion of a video) by including positional information as attributes. Instantiations of Segments (not depicted in Figure 1) include *LinearSegment*s (e.g., with start/end positions as attributes for referring to text parts), *SpatialSegment*s (e.g., with (x, y) pairs as attributes for referring to image regions), *TemporalSegment*s (e.g., with start/end times as attributes for referring to video frames/shots/scenes), and *SpatioTemporalSegment*s. A *SegmentGroup* represents a collection of Segments; it is also modelled as a sub-class of Segment, thus allowing it to contain both Segments and other SegmentGroups.
- *Collection* models aggregates of Objects (i.e., higher levels of granularity), such as corpora of Web documents, sets of tweets, and image collections.
- *Annotation* is a generic entity representing together with its sub-classes (not depicted in Figure 1) a wide range of descriptions extracted from media content. These include annotations typically extracted from text (e.g., keywords, named entities, summaries, categories, etc.), media content features (e.g., low level descriptors, concepts and events), affective descriptions (e.g., sentiment and polarity), veracity scores reflecting the reliability of information and thus the trust that should be placed on it, and many others.
- *Topic* refers to any subject of interest in the context of an information processing, analysis, or access applications that users would like to keep track of. Its explicit representation allows to support a broad range of tasks, such as information filtering, topic tracking, and classification.

The main relations between these SIMMO elements, excluding the generalisation and aggregation/composition relations already discussed, are:

- The generation of media objects is modelled through a **Contribution** association between Source and Object.
- Explicit relations between Documents are modelled as **Reference** associations, with attributes such as the type of the relation. By considering that a Document may Reference another Document, we also consider (through inheritance) that a Webpage may Reference another Webpage (e.g., link to it) and a Post may Reference another Post (e.g., reply to it or comment on it). We consider that this association is also able to model the References to Webpages from Posts (e.g., the Web links embedded in tweets) and to Posts from Webpages (e.g., to the comments dynamically posted on a Webpage).
- Objects may also be implicitly related to other Objects, e.g., through a computation of their similarity. Such **Similarity** relations are modelled as recursive associations between Objects, with attributes such as the type of the relation and the similarity score. This is useful in several applications and tasks, including clustering and verification of content provenance.
- A UserAccount may be involved in several relations, e.g., (i) be mentioned in a Post, (ii) be affiliated with (be friends with, follow etc.) another UserAccount, or (iii) interact with an Object (through likes, shares, views, etc.); the latter is more common for Posts, but users also interact with (e.g., like) whole Webpages. These three relations are modelled through the **Mention**, **Affiliation**, and **Interaction** assosiations, respectively, with attributes, such as the type of relation and the date it was established. As mentioned above, commenting is not modelled as a relation between Documents and UserAccounts, but rather as a Reference between two Documents (e.g., two Posts).
- All types of entities (i.e., Objects, Segments, Collections, Sources, and Topics) and their sub-classes may be associated with Annotations that are used for describing them. Such **Description** relations represent, for instance, the annotation of an Image with the SIFT features extracted from it, a TemporalSegment of a Video (such as a shot) with Concepts, or a UserAccount with Keywords reflecting the users' profile. Furthermore, links between different annotations (e.g., low-level descriptors and the concepts obtained from them) are modelled through the reflexive relation **Origin** between Annotations to denote the provenance of one with respect to the other.
- Each Topic is associated with a Collection of Objects on the particular subject of interest and may also be annotated itself. For instance, the Topic "Tour de France 2014" bicycle race would be associated with a Collection of Documents, such as Webpages and tweets on the subject, and could be annotated with the concepts "cycling" and "yellow yersey", the entity "Union Cycliste Internationale", and extracted locations, such as "Grenoble, France".

SIMMO elements and their relations also have several attributes representing their properties. For example, each Object is associated with a *URI, creation date*, and *crawl date*. Text is described by its *format* (e.g., HTML), an Image by its *size, EXIF* data, and associated *thumbnail*, a Video by its *duration, number of frames*, and associated *thumbnail*, and an Audio by its *duration*. Documents also have attributes related to the statistics regarding their social interactions, e.g., numbers of likes, comments, views, etc. The properties of a UserAccount include a *stream ID* denoting the platform hosting the account, the user's *name*, and the *number of followers/following/friends*. A complete list of the available attributes can be found in our implementation of SIMMO, discussed next.

Implementation: We have implemented the SIMMO framework in Java 1.7. We used Maven for controlling the project's build process, unit testing, and documentation creation, and the Google GSON library for converting Java objects into their JSON representation. This does by no means constrain the user from choosing another JSON library or another serialisation method. The SIMMO framework is open-source, released under the Apache License v2, and available at: `https://github.com/MKLab-ITI/simmo`. Based on this implementation, examples of commonly encountered online multimedia objects are presented next.

4   SIMMO Examples

To illustrate the flexibility and expressive power of the proposed framework, this section presents (using JSON) the SIMMO specifications of three diverse examples of online multimedia objects with social interconnections: (i) a Web page with content in different modalities and various annotations, (ii) a YouTube video with comments by several users, and (iii) a tweet with social interactions.

Consider, for instance, a Web page from an online newspaper discussing the recent World Cup 2014 final (Figure 2 (left)). SIMMO models this as a Webpage corresponding to the following JSON (some URLs have been shortened):

```
{  /* Webpage.json */
   "id":"180444840287",
   "url":"http://goo.gl/5JRsHi",
   "title":"World Cup was biggest event yet for Twitter with 672m tweets",
   "description":"Germany's demolition of Brazil ... peak tweets-per-minute",
   "tags":["Twitter", "World Cup", "Social networking", ...,  "Digital media", "Internet" ],
   "creationDate":"Jul 15, 2014 8:01:20 AM",
   "items":[
       {
          "type": "TEXT",
          "textFormat":"HTML",
          "content":"<p>Germany may have beaten Argentina to win the World Cup, ... "
       },
       {
          "type":"IMAGE",
          "url":"http://goo.gl/Uh4ok0",
          "width":620,
          "height":372
          "describedBy":[ { "type":"LOWLEVELDESCRIPTOR", "annotationId":"A8423" }, ... ],
       }
   ],
   "references":[ { "type":"LINK", "referencedDocumentId":"180444840289" }, ... ],
   "describedBy":[ { "type":"SUMMARY", "annotationId":"A9765" }, .... ]
}
```

The Webpage has particular attributes, such as title and description, and contains HTML Text and an Image, each with its own properties. Both the Webpage and its constituent Items may be annotated (e.g., the Webpage with a summary and the Image with visual features, listed below as separate JSON entries). The Webpage also connects to other Webpages through References of type "LINK".

Fig. 2. Illustrative examples of multimedia documents (l-r): a Web page in the news domain, a YouTube video with comments, and a tweet with replies

```
{  /* Summary.json */
   "id":"A9765"
   "summaryMethod":"Manual",
   "content":"Germany may have beaten Argentina to win the World Cup, ..."
}
{  /* LowLevelDescriptor.json */
   "id":"A8423"
   "descriptorType":"SURF",
   "numberOfFeatures":128,
   "descriptorValue":"128 1035 <CIRCLE 470 276 1 0 0>; 0.000537204 0.000681045 ... 0.00020111"
}
```

The next example corresponds to a YouTube video (Figure 2 (middle)) contributed by a UserAccount and modelled as a Post consisting of the actual video content and References to its comments, each also modelled as a Post. Several social interaction features are also modelled as attributes, such as the number of subscriptions to the UserAccount and the number of views of the video.

```
{  /* Post.json */
   "id":"wtt2aSV8wdw",
   "url":"https://www.youtube.com/watch?v=wtt2aSV8wdw",
   "title":"Internet Citizens: Defend Net Neutrality",
   "description":"Tell the FCC to reclassify broadband internet ...",
   "creationDate":"May 5, 2014 4:07:17 PM",
   "createdBy":"acc98754",
   "items":[
       {
           "type":"VIDEO",
           "url":"https://www.youtube.com/v/wtt2aSV8wdw",
           "width":1280,
           "height":720,
           "duration":213,
       }
   ],
   "numComments":4538,
   "numViews":919353,
   "positiveVotes":43615,
   "negativeVotes":394,
   "references":[ { "type":"COMMENT", "referencedDocumentId":"409sfh" }, ...  ]
}
{  /* UserAccount.json */
   "id":"acc98754",
   "name":"CGP Grey",
   "numSubscriptions":1361024,
   "avatarSmall":"http://goo.gl/YJS4PG"
}
{  /* Post.json */
```

```
    "id":"409sfh",
    "createdBy":"acc74528"
    "items":[
        {
            "type":"TEXT"
            "textFormat":"HTML",
            "content":"<div class="Ct">Learn about this and pass it on! ... </div>",
        }
    ],
    "numComments":72,
    "positiveVotes":739,
    "negativeVotes":0
}
```

The final example corresponds to a tweet (Figure 2 (right)) modelled as a Post that contains both Text and an Image, together with Mentions to specific UserAccounts, while statistics of social interactions are represented by attributes. Replies to the tweet are also modelled as Posts (not listed here).

```
{  /* Post.json */
   "id": "491252639225901056",
   "createdBy": "digitalocean"
   "creationDate": "Jul 21, 2014 4:05:30 PM",
   "items": [
      {
         "type": "TEXT"
         "textFormat": "HTML",
         "content": "We sent @jedgar out to meet DigitalOcean customer @KrakenIO and all ..."
      }
      {
         "type": "IMAGE"
         "url": "http://pbs.twimg.com/media/BtFHq9ZCUAAhyho.jpg:large",
         "height": 768,
         "width": 1024,
      }
   ],
   "mentions":[ { "mentioned":"jedgar" }, ... ],
   "numShares": 4,
   "positiveVotes": 19,
   "negativeVotes": 0,
   "references":[ { "type":"REPLY", "referencedDocumentId":"491255375912370176" }, ... ]
}
```

5   Comparison to Existing Approaches

To assess the expressive power of SIMMO, we compare it to other multimedia data representation models. First, existing approaches are presented and then a comparison is performed on the basis of the requirements identified in Section 2.

Early attempts to describe the content and structure of multimedia data (e.g., [4]) were soon superseded by the MPEG-7 standard [5], a generic, but complex, framework that enables highly structural, detailed descriptions of multimedia content at different granularity levels. MPEG-7 relies on: (i) *Descriptors (D)* defining the syntax and semantics of diverse features, (ii) *Description Schemes (DS)* describing the structure and semantics of relations among D or DS, (iii) a *Description Definition Language* allowing the creation and modification of DS and D, and (iv) *Systems Tools*, supporting various tasks, e.g., synchronisation of descriptions with content.

MPEG-21 [3] followed soon as an open framework for multimedia delivery and consumption, focussing on how the elements of a multimedia application infrastructure should relate, integrate, and interact. To this end, it centres around the concept of *Digital Items*, i.e., structured objects with multimedia content and metadata, and *Users* interacting with them; it also puts particular emphasis on Intellectual Property issues and mechanisms for the management of rights.

More recently, the Rich Unified Content Description (RUCoD) [6] framework was introduced for representing intrinsic properties of multimedia *Content Objects*, enhanced with real-world information (e.g., geo-location) and affective descriptions (e.g., in the valence/arousal 2D space). Each RUCoD consists of: (i) a header containing descriptive metadata (e.g., id and creation date) together with information about the media it contains and their descriptors, (ii) low-level descriptors, (iii) real-world descriptors, and (iv) user-related descriptors.

In addition, the infrastructure developed by WebLab [7] for integrating multimedia information processing components also defined a common exchange format to support the communication between such components. This exchange format, in essence a multimedia data representation model, centres on the notion of *Resource* that models several types of entities, including content in various modalities, multimedia documents and their segments, and diverse annotations.

The models discussed above focus on the description of multimedia content and thus satisfy requirements R1, R2, R7, R8 and R9, listed in Section 2; see also Table 1 for an overview of the requirements satisfied by each model. Given though that most were developed prior to the explosion of social media, they do not take into account the social characteristics and interconnections in current web environments (requirement R5). Such aspects have been addressed by ontologies, such as SIOC [1] and FOAF [2]. SIOC (Socially-Interlinked Online Communities) captures the nature, structure, and content of online communities (such as forums) through the representation of *Users* creating *Posts* organised in *Forums* that are hosted on *Sites*; modelled as sub-classes of the generic concepts *Item*, *Container*, and *Space*, respectively. SIOC is commonly used in conjunction with the FOAF (Friend Of A Friend) vocabulary to express users' personal information and social networking interactions. These approaches are not concerned though with the potential multimodality of Posts/Items and the annotations

Table 1. Comparison of different models w.r.t. the requirements identified in Section 2 ($\checkmark$ = requirement is satisfied; $\sim$ = requirement is partly satisfied)

Requirement: brief description	MPEG-7	RUCoD	WebLab	SIOC+FOAF	SIMMO
R1: multiple modalities	$\checkmark$	$\checkmark$	$\checkmark$		$\checkmark$
R2: diverse media objects	$\checkmark$	$\checkmark$	$\checkmark$		$\checkmark$
R3: heterogenous hosts	$\checkmark$	$\checkmark$	$\checkmark$	$\checkmark$	$\checkmark$
R4: online links		$\checkmark$		$\checkmark$	$\checkmark$
R5: social interactions				$\checkmark$	$\checkmark$
R6: contributors(description+relations)	$\sim$			$\checkmark$	$\checkmark$
R7: granularity at different levels	$\checkmark$	$\checkmark$	$\checkmark$		$\checkmark$
R8: various annotations	$\checkmark$	$\checkmark$	$\checkmark$		$\checkmark$
R9: descriptive metadata	$\checkmark$	$\checkmark$	$\checkmark$	$\checkmark$	$\checkmark$

extracted from such multimedia content (requirements R2 and R8), that are of paramount importance in information processing, analysis, and access tasks.

SIMMO bridges the gap between these perspectives by modelling both multimedia content (and its descriptions) and also users' social interactions with such content and with each other; see Table 1. To this end, SIMMO has borrowed several elements from the aforementioned approaches, while it has also introduced new aspects to support the emerging needs and requirements. For instance, the SIMMO multimedia content description draws many ideas from MPEG standards, but eschews their complexity, while SIMMO Annotations instantiated as LowLevelDescriptors could be mapped to standardised MPEG-7 Descriptors. Modelling granularity at the Segment level has been inspired by WebLab, while RUCoD has motivated the incorporation of affective and real-world features. The concept of UserAccount has been borrowed by FOAF, while the Post and Forum SIOC elements could be mapped to the Post and Webpage SIMMO components. Finally, many of the attributes of media Objects, Documents, and Items are equivalent to those proposed by Dublin Core (http://www.dublincore.org).

6 Conclusions

We have proposed a model that integrates in a unified manner the representation of multimedia and social features in online environments. Its flexibility and expressive power allow it to embrace the heterogeneity of multimedia content and its interconnections, thus making it unique in its ability to support a wide range of multimedia information processing, analysis, and access applications. Our aim is for SIMMO to be a reusable data model across such applications; to facilitate its adoption, we plan to extend its documentation, add utility methods (such as the implementation of standard indexing and retrieval operations), and identify and implement mappings to established data models (such as SIOC).

Acknowledgements. This work was supported by MULTISENSOR (contract no. FP7-610411), SocialSensor (contract no. FP7-287975), and REVEAL (contract no. FP7-610928) projects, partially funded by the European Commission.

References

1. Bojars, U., Breslin, J.G., Peristeras, V., Tummarello, G., Decker, S.: Interlinking the social web with semantics. IEEE Intelligent Systems 23(3), 29–40 (2008)
2. Brickley, D., Miller, L.: FOAF vocabulary specification 0.98. Namespace Document 9 (2012)
3. Burnett, I.S., de Walle, R.V., Hill, K., Bormans, J., Pereira, F.: MPEG-21: Goals and achievements. IEEE MultiMedia 10(4), 60–70 (2003)
4. Caetano, A., Guimaraes, N.: A model for content representation of multimedia information. In: Proceedings of the 1st workshop on the Challenge of Image Retrieval, CIR 1998 (1998) organised by the British Computer Society

5. Chang, S.-F., Sikora, T., Purl, A.: Overview of the MPEG-7 standard. IEEE Transactions on Circuits and Systems for Video Technology 11(6), 688–695 (2001)
6. Daras, P., Axenopoulos, A., Darlagiannis, V., Tzovaras, D., Bourdon, X.L., Joyeux, L., Verroust-Blondet, A., Croce, V., Steiner, T., Massari, A., Camurri, A., Morin, S., Mezaour, A.-D., Sutton, L.F., Spiller, S.: Introducing a unified framework for content object description. International Journal of Multimedia Intelligence and Security 2(3), 351–375 (2011)
7. Giroux, P., Brunessaux, S., Brunessaux, S., Doucy, J., Dupont, G., Grilheres, B., Mombrun, Y., Saval, A., des Portes, P.d.: Weblab: An integration infrastructure to ease the development of multimedia processing applications. In: Proceedings of the International Conference on Software and System Engineering and their Applications, ICSSEA (2008)
8. Papadopoulos, S., Kompatsiaris, Y.: Social multimedia crawling for mining and search. IEEE Computer 47(5), 84–87 (2014)
9. Ramzan, N., van Zwol, R., Lee, J.-S., Clüver, K., Hua, X.-S.: Social Media Retrieval. Computer Communications and Networks. Springer (2013)

Concept-Based Multimodal Learning for Topic Generation

Cheng Wang, Haojin Yang, Xiaoyin Che, and Christoph Meinel

Hasso Plattner Institute, University of Potsdam,
Prof.-Dr.-Helmert-Str. 2-3 14482 Potsdam, Germany
{cheng.wang,haojin.yang,xiaoyin.che,christoph.meinel}@hpi.de

Abstract. In this paper, we propose a concept-based multimodal learning model (CMLM) for generating document topic through modeling textual and visual data. Our model considers cross-modal concept similarity and unlabeled image concept, it is capable of processing document which has modality missing. The model can extract semantic concepts from unlabeled image and combine with text modality to generate document topics. Our comparison experiments on news document topic generation shows, in multimodal scenario, CMLM can generate more representative topics than latent dirichet allocation (LDA) based topic for representing given document.

Keywords: multimodal learning, LDA, concept detection, textual and visual concept, topic generation.

1 Introduction

The rapid development of digital world has boosted the production of image, audio, and video content in the last decade, making the WWW a truly multimedia platform. As an important part in multimedia retrieval, modeling multimedia content attracted much attention in recent years. But some problems still not appropriately solved. First, most of existing retrieval methods focus on single information modality only, such as image retrieval [7] and video retrieval [5]. However, information we received in real world comes from various modalities. Semantically combining information from different modalities would definitely enhance our understanding about multimedia content. But the issue of how to bridge the semantic gap between modalities have yet to be explored. Secondly, current approaches for modeling multimodal data generally focuses on paired modalities, for example, text-image or audio-video pairs, which don't take loss of modality into account. Thirdly, image tags, labels or relative text description are provided in most of researches, but the fact is unlabeled images are much more than those labeled or associated with relative text on WWW. Therefore it is difficult to apply existed models to practical application.

In order to find semantic models that are able to deal with modality missing and consider the concept of unlabeled image, many methods have been explored

X. He et al. (Eds.): MMM 2015, Part I, LNCS 8935, pp. 385–395, 2015.

and used to deal with parts of difficulties we stated in last paragraph. In processing text modality, Latent Dirichlet Allocation (LDA) [2] is widely used to discover the underlying topics in text content. It attracted much attention from researchers and some extensions of the standard LDA were applied to image domain, i.e. image annotation [6]. Unfortunately, it cannot be applied to multimodal scenarios directly. In [14,10,1,13,9,11,3], multimodal topic models are proposed for multimedia retrieval, but a common weak of those work is that all of them assume each image (text) is associated with related text (image). Recently, Jia et al.[9] proposed Multi-modal Document Random Field Model (MDRFM) which promoted modeling multimodal data based on previous work and considered modality missing in document. Another important contribution in his paper is that MDFRM took document-level similarity into consideration for discovering the topics shared across documents and modalities. Nevertheless, some shortcomings still exist. At first, it cannot handle unlabeled image because training data used in that paper are text-image pairs (Wikipedia dataset), where a short paragraph of about 100 words is used to describe image. On the other hand, this model considered document-level similarity across documents but ignored concept-level similarity between image and text within one document. S. Clinchant et al. [3] proposed semantic combination approach, called Late Semantic Combination (LSC), which considered cross-media similarities, and focused on more natural image dataset rather than Wikipedia-liked dataset, but it is also weak to process unlabeled image.

In this paper,to address those problems for complementing previous research and discover document topic with image concept. We propose a concept based multimodal learning model, named as CMLM. This model mainly leverages unlabeled image and cross-modal similarity within single document to tackle multimodal data modeling problems. Our model can not only apply to multimodal scenarios but also have the capability in processing documents that loss of modality. To simplify our research, we primarily concentrate on modeling text and image by extending LDA to explore shared hidden concept across modalities and documents. Concretely, the contributions in this paper can be summarized as follows:

(1) This paper proposes multimodal learning model base on the concept of unlabeled image for document topic generation.
(2) Multimodal learning model is able to tackle documents with missing modality in learning procedure.
(3) This paper extends application scenarios in previous research and applies our model to news documents, rather than wikipedia dataset that mainly for research purpose.

The rest of this paper is structured as follows, Section 2 gives background knowledge of Latent Dirichlet Allocation (LDA), meanwhile, introduces multimodal learning model and illustrates its learning procedure. Section 3 is experiment and evaluation, in this part, the approaches for image concept detection and concept based multimodal learning for document topic generation described, in addition, the evaluation is conducted with collected news document to demonstrate the benefit of our approach. The conclusions we draw from experiment are given in Section 4.

2   Learning Multimodal Data

In this section, LDA model is briefly introduced at first, and we describe our concept based multimodal learning model, which extends standard LDA and considers the concept of unlabeled image. At the same time, the learning procedure is also given in this section.

2.1   Latent Dirichlet Allocation

Latent Dirichlet Allocation is a generative probabilistic model for discover latent topics from given corpus. The generative process can be decomposed into doc-topic and topic-word generative process. As shown in Figure 1. (a), for a given document D , the topic proportion θ follows a Dirichlet distribution with prior probability α . For given θ , the specific topic z_n is draw from a multinomial distribution. Similarly, in topic-word distribution, a word w_n follows multinomial distribution with φ that is drawn from a Dirichlet distribution with prior probability β .Thus joint probability distribution can be described as

$$p(w, z, \varphi, \theta \mid \alpha, \beta) = \prod_{n=1}^{N} p(w_n \mid \varphi)p(z_n \mid \theta)p(\varphi \mid \beta)p(\theta \mid \alpha) , \qquad (1)$$

where w_n means the n-th word in document, z_n means the topic of n-th word. N is the number of words in corpus.

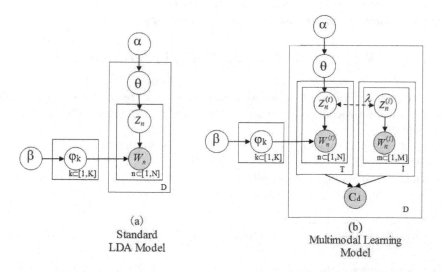

(a)
Standard
LDA Model

(b)
Multimodal Learning
Model

Fig. 1. LDA graph model and multimodal learning model

2.2   Multimodal Learning Model

Concept based multimodal learning model (CMLM) (Figure 1.(b)) extends the standard LDA by taking account of the concept of unlabeled image. Different to the models proposed in [10,1,13], which assume that each image is associated with caption or textual description. CMLM model firstly detects the concept of unlabeled image and then generates document topic with text concept accordingly. Similar to research [9], our model is also suitable for processing documents that contain single modality, rather than limit to text-image pair. Specifically, CMLM is able to deal with $< text >$, $< text, image >$, $< image >$ and $< text, images >$ modality or modalities. Besides considering modality missing case, another innovation is that CMLM through measuring the concept-level similarity between text concept z_n and image concept z_n' within document for generating document concept. With respect to similarity measurement, many methods can be employed [8]. In this study, Euclidean distance was applied as

$$S = E(d^{(t)}, d^{(i)}) = \sqrt{\sum_{i=1}^{m} |d^{(t)} - d^{(i)}|^2} .$$ (2)

In that term $T = t_1, t_2...t_m$, $d^{(t)}$ and $d^{(i)}$ represent text and image modality respectively, $d^{(t)}$ is produced by image concept detection, $d^{(t)}$ and $d^{(i)}$ is derived as

$$d_t^{(t)} = tfidf(d^{(i)}, t) , d_t^{(t)} = tfidf(d^{(i)}, t) .$$ (3)

Generally, the concept generative procedure for a given document goes through following steps:

Step 1: detect image concept for generating which consists of one or several keywords and descriptive text for image.

Step 2: sample mixture components $\theta_d \sim Dir(\theta|\alpha)$.

Step 3: sample topic proportion parameters $\varphi_1, \varphi_2...\varphi_k$

$$P(\varphi_1, \varphi_2...\varphi_k|\beta, S) = \lambda E(d^{(t)}, d^{(i)}) \prod_{k=1}^{K} Dir(\varphi_k|\beta) ,$$ (4)

λ is a constant parameter, which demonstrates text-image concept similarity strength within a specific document, and it varies according to different type of documents.

Step 4: sample a specific topic $z_n \sim Multi(z|\theta)$ and sample a specific word $w_n \sim Multi(w|\varphi_{z_n})$.

With such procedure, considering the cross-modal similarity within document, concept multimodal learning model is formulated as

$$P(c_d, \theta, z, \varphi_1, \varphi_2...\varphi_k|\alpha, \beta, S)$$
$$= \lambda E(d^{(t)}, d^{(i)}) \prod_{k=1}^{K} Dir(\varphi_k|\beta) Dir(\theta|\alpha) \prod_{n=1}^{N} Multi(w_n|\varphi_{z_n}) Multi(z_n|\theta)$$ (5)

For document d , the final output c_d is the concept that consisting of a set of key-words and each keyword has concept weight, which is used to express its contribution in representing a certain document. More specifically, the concept that generated from text and image can be represented as

$$c_d^{(t)} = \bigcup_{n=1}^{N} w_n^{(t)} \ and \ \ c_d^{(i)} = \bigcup_{m=1}^{M} w_m^{(i)} , \tag{6}$$

respectively, where N and M are numbers of text words and generated words from image concept detection. For each $w_n^{(t)} (n \in [1, N])$ and $w_m^{(i)} (m \in [1, M])$ in c_d, their concept weights are written as $v_n^{(t)}$ and $v_m^{(i)}$. By combining image concept we generate topics for document. And thus the concept of document d is

$$c_d = (c_d^{(t)}, c_d^{(i)}) = \bigcup_{n=1}^{N} \bigcup_{m=1}^{M} w_n^{(t)} w_m^{(i)} \tag{7}$$

CMLM aims to find the most representative topic for given document, that is, the topic has highest concept weight, therefore, the last step is to learn the most representative topic z. In this research, we only consider top-5 words $w = \{w_1, w_2, w_3, w_4, w_5\}$ for each topic, correspondingly, the weight vector $v = \{v_1, v_2, v_3, v_4, v_5\}$. The topic learning represented as

$$z = \max_{w \in z, z \in C_d} \sum_{n=1}^{5} v_n^{(z)} \tag{8}$$

Through Gibbs sampling and optimizing concept weight of topic, CMLM ultimately generate the topic which is more appropriate for representing given document than other topics. It is definitely beneficial to multimodal information retrieval.

3   Experiment And Evaluation

In order to prove the effectiveness of CMLM in this section. First, we describe the approach for image concept detection. And compare our model with LDA on collected news documents.

3.1   Experimental Configurations

The effectiveness of multimodal learning model we proposed in section 2.2 is evaluated with schemes in Figure 2. We conducted our experiments in unimodal and multimodal scenario respectively. The dataset contains 20 news articles that we collected through query image category labels on search engine.

In experiment scheme (a), we remove image within document so that concept is generated only from textual data. In (b), both text and image concept are considered with multimodal learning model. We firstly detect visual concept from image and then combine it with textual content for generating document topic. Our evaluation based on comparing the concepts from two schemes.

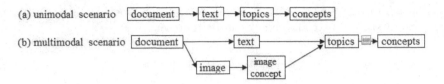

Fig. 2. Experiment scheme

3.2   Image Concept Generation

In image concept generation, we developed an image concept detection approach and used some image categories of ImageNet [4] dataset, which is a large scale image dataset, images in dataset are annotated with labels and descriptions therefore much more information about image content can be found. We randomly selected 20 categories (shows in Figure 3) as experimental dataset, each category contain 250 training images and 80 test images. Image concept detection relies on image classification technique and thus we implemented our model with Vlfeat library [12] for image classification. But from different perspective, we consider not only the label but also image description for generating more precise image concept.

Fig. 3. image training categories

Image features are extracted by dense scale-invariant feature transform (DSIFT) descriptor, K-means algorithm is used for clustering and building visual dictionary. In training stage, Stochastic Gradient Descent (SGD) is used as solver for training linear Support Vector Machine (SVM) classifiers. The optimization problem can be represented as:

$$E(\mathbf{w}) = \frac{1}{n} \sum_{i=1}^{n} E_i(\mathbf{w}), E(\mathbf{w}) = \frac{1}{2\lambda} \|\mathbf{w}\|^2 + l_i(< \mathbf{w}, \mathbf{v} >) \qquad (9)$$

SGD optimizes learning process by selecting term randomly in each iteration time. With this method, we trained 20 image categories and achieved average accuracy 65.75% shown in Figure 4. Table 1 show the result of image concept, we extracted unlabeled image from document and assigned it specific category with classification, the final output are labels and descriptions.

Table 1. image inputs and detected concept

input image	label&description
	Label: Life boat ***Description***: a strong sea boat designed to rescue people from a sinking ship
	Label: Ambulance ***Description***: a vehicle that takes people to and from hospitals
	Label: Minivan ***Description***: a small box-shaped passenger van; usually has removable seats, used as a family car
	Label: Fireboat ***Description***: a boat equipped to fight fires on ships or along a waterfront

Generated image concept provides useful representation for understanding image. In multimodal setting, through fusing the concepts from image and text modalities for learning document topics semantically enhance comprehensive understanding of document.

3.3   Multimodal Concept Generation

Concept based multimodal learning model (CMLM) we proposed in previous section is developed and implemented for generating multimodal concept, particularly, textual and visual concept. Although CMLM can be generalized and applied to other multimodal cases, for example, two or more modalities in text, image, audio and video. In this research we only consider text and image modalities. The influence of image concept for multimodal concept generation is measured through comparing the textual concept and multimodal concept separately. For illustration purpose, a news document is given to generate textual concept and multimodal concept. In this case, we set initialization model parameters as $\alpha = 16.67, \beta = 0.1, K = 3$, which means 3 topics would be generated from document.

The rescue took place last night after police saw a man in difficulty in the sea off Links Road in north Lowestoft at about 11pm. The RNLI Spirit of Lowestoft lifeboat was then called out along with the Gorleston and Great Yarmouth inshore lifeboat, Lowestoft and Southwold coastguard rescue teams and a RAF search and rescue helicopter from Wattisham airfield. John Fox, coxswain of Lowestoft lifeboat, said: It took us just a few minutes to reach the incident and as soon as we arrived second coxswain Karl Jackson jumped into the sea to support the man. He was beyond the end of the groynes, roughly 50 metres from the shore and was very cold and in an exhausted state. The crew threw a rope to the pair and then pulled them alongside before lifting the casualty on to the deck of the lifeboat where they administered first aid as the lifeboat sped back to port. An ambulance and paramedic were waiting for the lifeboat when it returned to Lowestoft harbour and the man was then taken to the James Paget University Hospital, in Gorleston, for further treatment. The man was said to have been in a depressed state.[1]

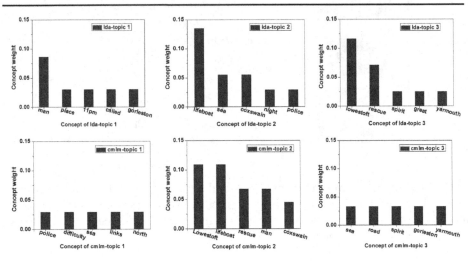

Fig. 4. Examples for textual and multimodal concept

Figure 5 shows the topics generated from textual information almost have equal concept weights. Through summing words concept weights within each topic, the topic concept weights are 0.206, 0.303 and 0.262 respectively. In multimedia retrieval we

---

[1] http://www.edp24.co.uk/news/lifeboat_rescues_man_in_sea_off_lowestoft_1_
3515860

generally use inquiry words for searching related information. The most important keywords would definitely very helpful for retrieving information. If the semantic contribution of topics are equal or similar, it would different to find the most representative topic for given document. In multimodal topics,when involving image concept, our model achieved 0.145, 0.397 and 0.165 for three topics respectively. The cmlm-topic 2 is highly distinguished from topic 1 and topic 3, because CMLM generated more representative topic. To make it clarify, we can see from following table.

Table 2. True concept, textual concept and multimodal concept

True concept	Textual concept	Multimodal Concept
Lifeboat, rescues, man, sea off, lowestoft	Lifeboat, sea, coswain, night, police	Lifeboat, lowestoft, rescue, man, coxswain

It is obvious that CMLM based concept can represent document in better degree when comparing to the true concept that we extracted from article link. To generalize over this, we use 20 news documents to measure the influence of image concept in document topic generation. Before that, we introduce a definition of Topic Weight Variance (TWV), which used as criterion for measuring the change of topic concepts that influenced by image concept.

Definition 1: For a given document D in which N topics are generated and each topic contains M words,the concept of $m - th$ word represented as v_m, the TWV of D can be formulated as

$$TWV = \frac{1}{N} \sum_{n=1}^{N} (T - \bar{\mu})^2, \bar{\mu} = \frac{1}{N} \sum_{n=1}^{N} \sum_{m=1}^{M} v_m \qquad (10)$$

T is the sum of concept weight of each word within one topic and $\bar{\mu}$ is the average weight of N topics. Figure 6 describes the topic weight variance of 20 news document in different categories. For reason of comparison, we have generated two versions of document concept by using and without using image concept respectively. For each document, the concept generation process performed 10 times, each TWV in document was calculated. The mean value of 10 TWVs is final TWV value for specific topic. In the first experiment scheme, we generated document concept using LDA without considering image concept. In second experiment scheme, we combined textual and visual concept and generated document concept with CMLM. From figure 6 the cmlm-twv of 70% of categories are higher than its corresponding lda-twv. It suggests that cmlm can highlight the main topic of document and boost document representation in textual form. The more precise understanding of documents can be achieved, this would be useful in improving multimedia retrieval performance.

3.4   Discussion

Image concept involved document topic generation can provide more integrated understanding of document content. Due to image concept detection relies on image classification technique thus image classification accuracy can influence the result of document topic generation. The image classification accuracy in our case is 65.75% and there are still error classification result considered in topic generation, for example, taxi, android and prairie chicken categories, those error classification results are

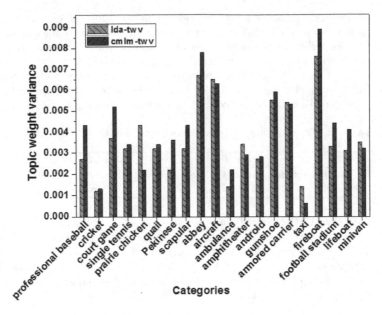

Fig. 5. Topic weight variance for news document

negative image concept. However, those negative image concept produce little influence in main topic generation, especially in long text content document, for example, in minivan and aircraft. To improve the performance of multimodal concept generation, on one hand, increasing image classification accuracy is important to generate more appropriate concept. On the other hand, extending training image categories is also needed so that images appear document can be covered by pre-trained categories and recognized by classification model.

4   Conclusion

We proposed a concept-based multimodal learning model which considers unlabeled image concept and cross-modal similarity between text and image within document for generating document topic. The model is able produce topics even through in the condition of missing modality. Our model has been proved its effectiveness in document topic generation and enhanced our understanding of document content.

Our future work will focus on semantically building more reasonable model for discover hidden concept from documents and extending this model to more application availability scenarios. Particularly, we will enlarge image categories to cover more image categories. In this framework, image concept will be detected by using multi-labels and multi-description for demonstrating one image, so that more hidden concept can be discovered. An important application of this model is to realize multi-modal media retrieval from concept level, because CMLM have the capability in generating more comprehensive and appropriate document concept by combining textual and visual concept.

References

1. Blei, D.M., Lafferty, J.D.: A correlated topic model of science. The Annals of Applied Statistics, 17–35 (2007)
2. Blei, D.M., Ng, A.Y., Jordan, M.I.: Latent dirichlet allocation. The Journal of Machine Learning Research 3, 993–1022 (2003)
3. Clinchant, S., Ah-Pine, J., Csurka, G.: Semantic combination of textual and visual information in multimedia retrieval. In: Proceedings of the 1st ACM International Conference on Multimedia Retrieval, p. 44. ACM (2011)
4. Deng, J., Dong, W., Socher, R., Li, L.-J., Li, K., Fei-Fei, L.: Imagenet: A large-scale hierarchical image database. In: IEEE Conference on Computer Vision and Pattern Recognition, CVPR 2009, pp. 248–255. IEEE (2009)
5. Fan, J., Elmagarmid, A.K., Zhu, X., Aref, W.G., Wu, L.: Classview: hierarchical video shot classification, indexing, and accessing. IEEE Transactions on Multimedia 6(1), 70–86 (2004)
6. Feng, Y., Lapata, M.: Topic models for image annotation and text illustration. In: Human Language Technologies: The 2010 Annual Conference of the North American Chapter of the Association for Computational Linguistics, pp. 831–839. Association for Computational Linguistics (2010)
7. He, X., Ma, W.-Y., Zhang, H.-J.: Learning an image manifold for retrieval. In: Proceedings of the 12th Annual ACM International Conference on Multimedia, pp. 17–23. ACM (2004)
8. Huang, A.: Similarity measures for text document clustering. In: Proceedings of the Sixth New Zealand Computer Science Research Student Conference (NZCSRSC 2008), Christchurch, New Zealand, pp. 49–56 (2008)
9. Jia, Y., Salzmann, M., Darrell, T.: Learning cross-modality similarity for multinomial data. In: 2011 IEEE International Conference on Computer Vision (ICCV), pp. 2407–2414. IEEE (2011)
10. Putthividhy, D., Attias, H.T., Nagarajan, S.S.: Topic regression multi-modal latent dirichlet allocation for image annotation. In: 2010 IEEE Conference on Computer Vision and Pattern Recognition (CVPR), pp. 3408–3415. IEEE (2010)
11. Rasiwasia, N., Pereira, J.C., Coviello, E., Doyle, G., Lanckriet, G.R., Levy, R., Vasconcelos, N.: A new approach to cross-modal multimedia retrieval. In: Proceedings of the International Conference on Multimedia, pp. 251–260. ACM (2010)
12. Vedaldi, A., Fulkerson, B.: Vlfeat: An open and portable library of computer vision algorithms. In: Proceedings of the International Conference on Multimedia, pp. 1469–1472. ACM (2010)
13. Yu, J., Cong, Y., Qin, Z., Wan, T.: Cross-modal topic correlations for multimedia retrieval. In: 2012 21st International Conference on Pattern Recognition (ICPR), pp. 246–249. IEEE (2012)
14. Zhai, X., Peng, Y., Xiao, J.: Cross-media retrieval by intra-media and inter-media correlation mining. Multimedia Systems 19(5), 395–406 (2013)

Audio Secret Management Scheme
Using Shamir's Secret Sharing

M. Abukari Yakubu[1], Namunu C. Maddage[2], and Pradeep K. Atrey[1,3]

[1] Department of Applied Computer Science, University of Winnipeg, Winnipeg, MB,
Canada
[2] NextGmultimedia, Melbourne, Australia
[3] Department of Computer Science, University at Albany - State University of New
York, Albany, NY, USA

Abstract. Audio Secret Sharing (ASS) is a technique used to protect
audio data from tampering and disclosure by dividing it into shares such
that qualified shares can reconstruct the original audio data. Existing
ASS schemes encrypt binary secret messages[1] and rely on the human
auditory system for decryption by simultaneously playing authorized
shares. This decryption approach tends to overburden the human auditory system when the number of shares used to reconstruct the secret
increases [3]. Furthermore, it does not create room for further analysis or
computation to be performed on the reconstructed secret since decryption ends at the human auditory system. Additionally, schemes in [2],
[3], [4], [6] do not extend to the general (k, n) threshold. In this paper we
propose an ASS scheme based on Shamir's secret sharing, which is (k, n)
threshold, ideal[2], and information theoretically secure and it provides
computationally efficient decryption.

Keywords: Shamir's secret sharing scheme (SSS), threshold schemes,
information theoretically secure.

1 Introduction

Audio is one of the key types of multimedia content which may contain confidential information such as names, addresses, social security numbers, credit
card numbers, evidence to be used in a court of law by a jury, and information
with national security implications. Such sensitive information might be mis-
used when it falls into the wrong hands. For instance, call centers record several
hours of customer calls, most of which contain confidential information. In order
to save cost, call centers often store data on Cloud Data Centers (CDCs). A
rogue or malicious employee within the call center or CDC may use this confidential information to their own benefit. Therefore the security of such sensitive
audio records is of utmost importance.

---

[1] A binary representation of a secret plaintext message.
[2] In a perfect secret sharing scheme, any unauthorized subset of participants cannot
obtain any information about the secret. We will say that a perfect sharing scheme
is ideal if all of the shares are from the same domain as the secret.

X. He et al. (Eds.): MMM 2015, Part I, LNCS 8935, pp. 396–407, 2015.

One way to secure the audio could be to encrypt it using Advanced Encryption Standard (AES). However, AES suffers from single point vulnerability meaning that the security of the method lies in securing the encryption key which is usually entrusted to the sender and receiver. This problem can be overcome by employing a secret sharing scheme to divide the audio secret into a number of shares and distribute them among a number of participants such that only more than a certain number of participants can reconstruct the secret by putting their shares together; individual shares are of no use on their own. Thus, a group of participants collectively protect and control access to the secret. In this case, the audio shares are distributed amongst multiple CDCs. Unless the required number of CDCs are compromised, an adversary cannot get the secret audio.

Some of the existing Audio Secret Sharing (ASS) schemes [2], [3] are designed to encrypt text secrets. In these schemes a binary representation of the text secret is embedded into an audio cover and shares of the cover signal are created. This approach combines cryptography to encrypt the plaintext and steganography to hide the existence of the ciphertext. Such schemes only had $(2, n)$ threshold and never extended to the general (k, n). The ciphertext was decrypted by the Human Auditory System (HAS) by simultaneously playing authorized shares which is analogous to Visual Cryptographic System (VCS) where the human visual system is used for decryption in image secret sharing. There is no computational cost to decrypt with HAS, however it has following limitations: 1) People with hearing impairments cannot participate in the decryption process 2) It requires manpower to decrypt the secret and also overburdens the human ear with increasing numbers of shares required to reconstruct the secret [3]. While schemes proposed in [2], [3] encrypt a binary secret message, schemes in [5], [6], [7] encrypt an audio secret. However, decryption still requires the human auditory system.

The scheme proposed in [5] is (k, n) threshold secret sharing scheme, where k out of n generated secret shares are required to reconstruct the secret audio. The security of this scheme is not proven from an information theoretical point of view and is highlighted in [6], [7]. Authors in [6], [7] propose schemes whose security is evaluated in terms of the mutual information between secret and shares from an information theoretical perspective. The scheme in [7] is an improvement to [6] where the encryption function uses normal distribution over a bounded domain in order to create bounded shares. However, both schemes do not extend to (k, n) threshold.

In practical applications of secret sharing schemes to an audio secret and to address the limitations of HAS decryption, there are instances where decryption is required to be performed on a computer. The scheme in [4] achieved decryption computationally, but is limited to binary audio and does not extend to the general (k, n) threshold scheme. Moreover, the security of this scheme is not proven from an information theoretical point of view. In summary, each one of previous schemes has at least one of these limitations: 1) It does not extend to (k, n) threshold scheme, 2) Information theoretical security is not proven and 3) It has the limitations of HAS decryption.

In this paper we propose a method to protect audio secrets using Shamir's secret sharing (SSS) scheme to address the above limitations. To the best of our knowledge, this is the first ASS scheme based on SSS which is (k, n) threshold and information theoretically secure and it offers a computationally efficient decryption. SSS in general does not have the above limitations described in points 1 and 2. Because of the proven security properties of the SSS scheme, many researchers have applied it to protect secret text, images, video, digital signatures and encryption/decryption keys [10]. Another work [11] uses SSS to protect an image and PDF secret by creating shares and applying steganography to hide each share in an MP3 cover. Such an approach is different from our method since we are protecting an audio secret. Table 1 compares the limitations of previous techniques and highlights that the proposed scheme does not have such limitations.

The rest of this paper is organized as follows. In Section 2, we discuss Shamir's secret sharing scheme. The proposed method for managing audio secrets is detailed in Section 3 and Section 4 discusses the experimental results. We conclude the paper in Section 5.

Table 1. A comparison of the proposed scheme with previous schemes

Scheme	Threshold	Information theoretically secure	Decryption
Desmedt et al. in 1998 [3]	$(2, n)$	Yes	Human auditory system
Lin et al. in 2003 [2]	$(2, n)$	Yes	Human auditory system
Nishimura et al. in 2005 [4]	(n, n)	Not proven	Computer
Ehdaie et al. in 2008 [5]	(k, n)	Not proven	Human auditory system
Yoshida and Watanabe in 2012 [6]	(n, n)	Yes	Human auditory system
Washio and Watanabe in 2014 [7]	(n, n)	Yes	Human auditory system
Proposed scheme	(k, n)	Yes	Computer

2   SSS Scheme

Shamir introduced his scheme in 1979 [1], which is based on polynomial interpolation. The goal of this scheme is to divide data into n shares such that:

1. Any k or more shares can reconstruct the secret.
2. $k - 1$ or fewer shares cannot reconstruct the secret.

Such a scheme is called a (k, n) threshold scheme where $2 \leq k \leq n$, n is the number of shares and k is the least number of shares required to reconstruct the secret.

To share a secret S among n participants, a polynomial function $f(x)$ is constructed of degree $k - 1$ using k random coefficients $a_1, a_2 \ldots a_{k-1}$ in a finite field $GF(q)$ where a_0 is S, and q is prime number $> a_0$.

$$f(x) = (a_0 + a_1 x + \ldots + a_{k-1} x^{k-1}) mod \ q \qquad (1)$$

Any k out of n shares can reconstruct the secret using Lagrange interpolation to reconstruct the polynomial $f(x)$; the secret can be obtained at $f(0)$ i.e. $f(0) = a_0 = S$

$$f(x) = \sum_{j=1}^{k} \left(y_j \prod_{i=1, i \neq j}^{k} \left(\frac{x - x_i}{x_j - x_i} \right) \right) \mod q \qquad (2)$$

3 Proposed Method

As described in Section 2, we apply the SSS scheme to create an audio secret sharing method as depicted in Fig. 1. In our method we create shares of amplitude samples since they contain information of an audio signal. The following section details share generation and reconstruction of the secret audio. Notations of variables, symbols and functions used throughout this section are summarized in Table 2.

Table 2. Notation of variables

Variable	Description
A	Original secret audio signal
A'	Preprocessed secret audio signal
a_o	Original secret audio sample ($a_0 \in A$)
a'_o	Preprocessed secret audio sample ($a'_0 \in A'$)
ϵ	Round-off error
d	Rounding precision
γ	DC shift of signal to first quadrant
$Pr(.)$	Probability function
$GF(.)$	Finite field
q	First prime number greater than the maximum original secret audio sample (a_o)
q'	First prime number greater than the maximum preprocessed secret audio sample (a'_o)
b	Number of bits to represent q
b'	Number of bits to represent q'

3.1 Preprocessing and Share Generation

Using the SSS (k, n) threshold, we generate n shares such that at least k shares can reconstruct the secret. Using real numbers in a cryptosystem means excluding the modular prime operation which in the case of SSS degrades security. Therefore, we have to preprocess amplitude samples of the secret audio from real to positive integer values. During preprocessing, we first round-off the real amplitude samples by multiplying by 10^d where d is some integer value. Roundoff error is bounded by:

$$-\frac{1}{2} \times 10^{1-d} \leq \epsilon \leq \frac{1}{2} \times 10^{1-d} \qquad (3)$$

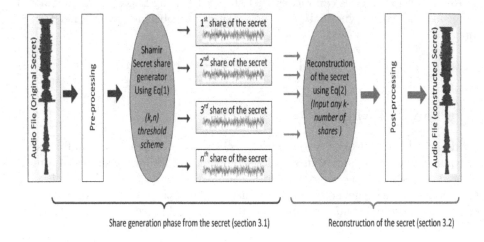

Share generation phase from the secret (section 3.1)   Reconstruction of the secret (section 3.2)

Fig. 1. Audio secret sharing framework

where ϵ is the rounding error and d is the rounding precision. Each amplitude secret a_0 is converted to an integer and shifted to the first quadrant by a threshold γ to obtain positive sample values within $\mathbb{Z}_p$. Shifting the signal to first quadrant does not distort the waveform as illustrated in Fig. 2.

$$a_0' = \big((a_0 + \epsilon) \times 10^d\big) + \gamma \tag{4}$$

Using Equation (1) from Section 2, n shares are created and distributed to n participants. The algorithm is shown below.

Algorithm 1: Share Generation
Input: Secret audio $A = \{A_1, A_2 \ldots A_m\}$; where A_m is the amplitude at the mth time interval
Output: Secret Shares $S_1, S_2 \ldots S_n$
Description:
1. Read wav file i.e. $[A, fs] = \text{wavread}('wavfile')$
2. $A = \text{round}((A + \epsilon) \times 10^d)$
3. $A' = A +$ absolute of the minimum value of A
4. Compute the first prime number q' greater the than maximum value of A'
5. **for** $i = 1$ to length of A' **do**
   amplitude value at the ith time interval is the secret i.e. $a_0' = A_i'$ and randomly choose coefficients $a_1, a_2 \ldots a_{k-1}$ from a set a positive integer field $\mathbb{Z}_p$
6.   **for** $j = 1$ to n; number of shares to create **do**
      Compute $share(i, j)$ from the polynomial obtained in 5. $share(i, j)$ is the jth share for the ith amplitude value
7.   **end for**

8. **end for**
9. **for** $j = 1$ to n **do**
10. S_j = combine all amplitude share values for each share index
11. **end for**
12. **return** $S_1, S_2 \ldots S_n$;

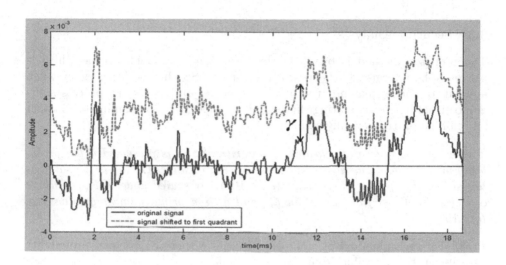

Fig. 2. Shifting signal to first quadrant

3.2   Secret Reconstruction and Post-processing

To reconstruct the secret audio we need at least k out of the n shares. Referring to Fig. 1, there are two blocks at the secret reconstruction phase: 1) reconstruct the secret by using Equation (2) to solve the polynomial function in Equation (1) and obtain the secret sample at evaluation point $x = 0$, (this is done for all samples) and 2) post-process to reverse engineer the preprocessing done during share generation. We first subtract the signal shift threshold from the obtained signal in step 1 and then divide by 10^d to get the secret audio signal. The algorithm is shown below.

Algorithm 2: Secret Reconstruction
Input: Any $k \leq n$ audio shares $S_1, S_2 \ldots S_k$
Output: Secret Audio $A = \{A_1, A_2 \ldots A_m\}$
Description:
1. Reconstruct the polynomial $f(x)$ from shares $S_1, S_2 \ldots S_k$ using Lagrange interpolation in (2) in a finite field $GF(q')$

2. **for** $i = 1$ to length of share **do**
   Obtain a_0' coefficient at evaluation point $f(0)$ i.e. a_0' is the reconstructed amplitude secret at the ith time interval
   $A'(i) = a_0'$
3. **end for**
4. $A = (A' -$ absolute of the minimum value of A from Algorithm 1, step 2$)/10^d$;
5. **return** A;

3.3 Security Analysis

The proposed method is based on the SSS (k, n) threshold scheme which is proven to be information theoretically secure [8]. SSS has perfect secrecy when applied to independent input sequences, however our scheme preprocesses the audio signal before generating shares so it is imperative to examine the impact on information theoretically security.

Theorem 1. *Information theoretical security of SSS is preserved if the probability of revealing an audio secret sample a_0 shared under $GF(q)$ is the same as the probability of determining $a_0' = (a_0 \times 10^d) + \alpha$ shared under $GF(q')$ (where $\alpha = (\epsilon \times 10^d) + \gamma$ from Equation (4) and q' is a prime number greater than $(q \times 10^d) + \alpha)$*

Proof. For each plaintext of audio secret $a_0 \in A$ there is an equal probability that it can be any value from the set $0 \leq a_0 \leq q - 1$ of q values since SSS encryption is upper bounded by q. This probability is given by:

$$Pr(a_0)_{0 \leq a_0 \leq q-1} = \frac{1}{q} \tag{5}$$

Similarly, for each plaintext a_0' of the preprocessed audio secret A' where $a_0' = (a_0 \times 10^d) + \alpha$ there is also an equal probability of being any value from the set $0 \leq a_0' \leq q' - 1$ of q values with probability given as:

$$Pr(a_0')_{0 \leq a_0' \leq q'-1} = \frac{1}{q} \tag{6}$$

The probability of revealing the secret a_0 and a_0' in the above cases is the same $\frac{1}{q}$. Thus, our scheme preserves information theoretical security after preprocessing the original audio secret. An adversary in both cases will have to guess the secret with a probability of $\frac{1}{q}$.

3.4 Data Overhead

Our proposed scheme introduces some data overhead due to the preprocessing step. This data overhead is the number of bits used to represent the maximum preprocessed audio sample. Since the generation of shares under a finite field $GF(q')$ is upper bounded by q' (where q' is the first prime number greater than

$maximum[(a_0 + \epsilon) \times 10^d + \gamma])$ we can conclude that the data overhead is also upper bounded by the number of bits used to represent q'. If b is the number of bits to represent this value then:

$$b = log_2(q') \qquad (7)$$

Due to the dynamic range of audio signals, q' will always vary for different audio signals depending on the quantization level ($8bit, 16bit$ etc.) of the ADC converter used during quantization. From Equation (3), it can be seen that increasing d during preprocessing will yield minimal round-off error but higher data overhead so d should be chosen to maintain a balance between the two.

4 Experimental Results

Table 3 details the 6 audio files obtained from [9] that we use to test the proposed audio secret sharing method. In the (k, n) threshold scheme, we set $k = 2$ and $n = 3$ implying that 2 out of 3 created secret shares are required to reconstruct the secret audio.

Table 3. Data set

Test file (.wav)	length(secs)	Bits/sample	sampling frequency (Hz)
audio1	2	16	16000
audio2	43	16	8000
audio3	8	8	22050
audio4	14	8	44100
audio5	4	8	8000
audio6	2	32	8000

We implemented the audio secret sharing method using MATLAB14 on a 2.53GHz i5 CPU with 4GB RAM. Table 4 details the processing time for creating secret shares and reconstructing the original audio secret. The time information in the table suggests that the complexity of reconstructing the secret is relatively lower than creating secret shares. Since the proposed method is applied at an audio sample level, the processing time is directly proportional to the audio bit rate, which is associated with the sampling frequency and number of bits per sample.

Audio signals by nature have correlating adjacent samples and the use of random coefficients as a blinding factor in Equation (1) to generate shares eliminates this correlation. Thus, individual shares do not reveal information about the secret audio. Time domain plots of one of our test audio files (audio1) in Fig. 4 illustrates: 1) the difference between the audio secret and its noisy shares and 2) the similarity between the reconstructed and original secret audio. Fig. 3 shows the similarity scores between original secret audio, and 1st share, 2nd share,

Table 4. Average processing time to create shares and reconstruct the secret

Test file	length(secs)	Share creation (ms)	Secret reconstruction (ms)
audio1	2	152	7
audio2	43	1614	50
audio3	8	929	29
audio4	14	2770	80
audio5	4	150	12
audio6	2	83	5

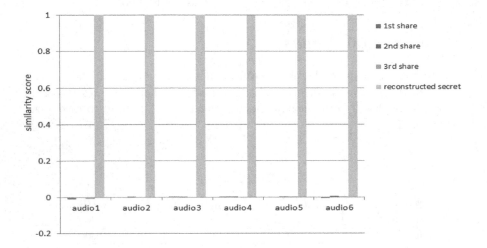

Fig. 3. Similarity Score

$3rd$ share and reconstructed secret audio. The similarities were computed using pearson's correlation method. Results suggest less than 1% correlation between the original secret audio and its shares.

It is also evident that the reconstructed secret audio is about 100% correlated with the original secret audio; suggesting minimal information losses due to rounding error in the preprocessing step.

We also performed a listening study to evaluate perceptual security which was conducted online [3]. User scores are summarized in Table 5. 20 subjects in the age range of $20-40$ years participated in the survey. The similarity score is captured in a 4 point scale where the value 3 is given when two audio files are exactly the same content wise and the value 0 is given when two audio files are not similar at all content wise. As expected, all the participants agreed that both the share and audio secret are completely dissimilar in terms of content. However, about 92% of average similarity score was achieved for content similarity between the

---

[3] https://az1.qualtrics.com/SE/?SID=SV_0AHmNAbzvekk
weN&Preview=Survey&BrandID=qtrial2014

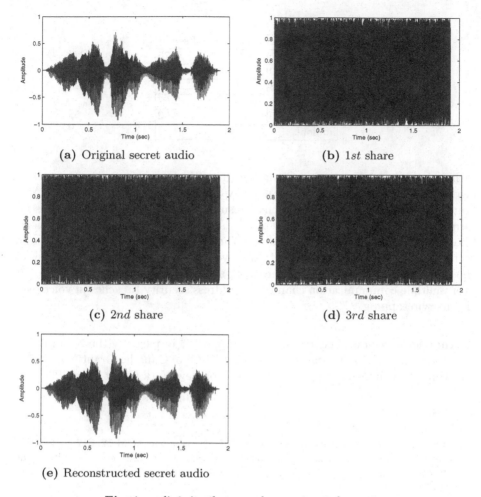

(a) Original secret audio

(b) 1st share

(c) 2nd share

(d) 3rd share

(e) Reconstructed secret audio

Fig. 4. audio1, its shares and reconstructed secret

original audio secret and the reconstructed audio secret which confirms that our proposed scheme is perceptually secure. However, as depicted in Fig. 3, using pearson's correlation analysis, we were able to establish about 100% similarity score between the original audio secret and the reconstructed audio secret. In the future we would like to investigate the disparity of human judgment (Table 5) vs machine evaluation (Fig. 3) of similarity.

5   Conclusion

In this paper we propose an audio secret sharing technique using the Shamir's secret sharing (SSS) scheme. Compared to existing techniques, the proposed

Table 5. User study

	share	reconstructed secret
audio1	0	2.75
audio2	0	2.67
audio3	0.08	2.75
audio4	0.08	2.67
audio5	0	2.92
audio6	0	2.83

technique is (k, n) threshold, information theoretically secure and computationally efficient decryption which does not rely on Human Auditory System (HAS). Our security analysis and experimental results also show that our scheme is information theoretically secure, perceptually secure and computationally efficient. Short time framing can be combined with the proposed scheme to share an audio secret of several hours long. In the future, we will experiment the capabilities of the proposed scheme for modeling network architecture to encrypt voice signals over multiple communication channels to address security breaches in voice over IP and wiretapping.

Acknowledgement. This research was supported in parts by the Natural Sciences and Engineering Research Council (NSERC) of Canada, Grant No. 371714 and University at Albany - State University of New York Grant No. 640075.

References

1. Shamir, A.: How to share a secret. Communications of the ACM 22, 612–613 (1979)
2. Lin, C.C., Laih, C.S., Yang, C.N.: New Audio Secret Sharing Schemes With Time Division Technique. Journal of Information Science and Engineering 19, 605–614 (2003)
3. Desmedt, Y.G., Hou, S., Quisquater, J.-J.: Audio and optical cryptography. In: Ohta, K., Pei, D. (eds.) ASIACRYPT 1998. LNCS, vol. 1514, pp. 392–404. Springer, Heidelberg (1998)
4. Nishimura, R., Fujita, N., Suzuki, Y.: Audio Secret Sharing for 1-Bit Audio. In: Khosla, R., Howlett, R.J., Jain, L.C. (eds.) KES 2005. LNCS (LNAI), vol. 3682, pp. 1152–1158. Springer, Heidelberg (2005)
5. Ehdaie, M., Eghlidos, T., Aref, M.R.: A novel secret sharing scheme from audio perspective. In: International Symposium on Telecommunications, pp. 13–18. IEEE, Tehran (2008)
6. Yoshida, K., Watanabe, Y.: Security of audio secret sharing scheme encrypting audio secrets. In: International Conference for Internet Technology and Secured Transactions, pp. 294–295. IEEE, London (2012)
7. Washio, S., Watanabe, Y.: Security of audio secret sharing scheme encrypting audio secrets with bounded shares. In: International Conference on Acoustics, Speech and Signal Processing (ICASSP 2014), pp. 7396–7400. IEEE, Italy (2014)
8. Stallings, W.: Cryptography and Network Security Principles and Practice, 5th edn. Prentice-Hall (2010)

9. SVV media audio database,
   http://download.wavetlan.com/SVV/Media/HTTP/http-wav.htm
10. Atrey Pradeep, K., Alharthi, S., Hossain, M.A., AlGhamdi, A., El-Saadik, A.:
    Collective control over sensitive video data using secret sharing. Multimedia Tools
    and Applications (2013), doi:10.1007/s11042-013-1644-0
11. Chan, K.F.P.: Secret Sharing in Audio Steganography. In: Information Security
    South Africa, ISSA. South Africa (2011)

Live Version Identification with Audio Scene Detection

Kazumasa Ishikura, Aiko Uemura, and Jiro Katto

Waseda University, 3-4-1 Okubo, Shinjuku-ku, Tokyo, 169-8555, Japan
{ishikura,uemura,katto}@katto.comm.waseda.ac.jp

Abstract. This paper presents a live version music identification system by modifying the conventional cover song identification system. The proposed system includes two stages: a live version identification phase and an audio scene-detection phase. We improve the accuracy of the system by weighting similarity scores in the live version identification phase and discriminating scenes by using RMS, pulse clarity and similarity scores. Results show that the proposed method performs better than the previous method. The final algorithm achieves 70% accuracy on average.

Keywords: live version, music identification, audio scene detection.

1 Introduction

Music repositories and databases are readily available for everyone, both in private collections and on the internet. It has become important to browse music collections by contents. Therefore, music information retrieval (MIR) is attracting widespread interest from all users of computers and the internet.

Music identification, a technique for MIR, matches an input query to information contained in a database and thereby finds appropriate music. Much research in recent years has specifically investigated cover song identification. Audio cover song identification has been benchmarked in the annual Music Information Retrieval Evaluation eXchange (MIREX) [1] tasks. The task of identifying cover songs presents numerous difficulties for content-based music retrieval because performances are often different from the original in various attributes including tempo, instrumentation or key. Live performance identification is also expected to be affected by several attributes such as the tempo, key, or noise.

This paper presents an audio identification system with scene detection from live performance. The method is superior to previous systems of cover song identification. We input the live performance music from a concert, including several songs, in telling contrast to cover-song identification systems, which examine single songs.

Our results are expected to contribute to annotation of titles and artist. They will also assist digital rights management for streaming music.

1.1 Previous Works and Systems

The Query by Humming (QBH) approach was used to incorporate audio into early MIR systems. Gradually the format has become available from polyphonic audio to

X. He et al. (Eds.): MMM 2015, Part I, LNCS 8935, pp. 408–417, 2015.

music. It is used for queries of a database. An earlier report [2] described the use of cosine distance between the most repeated melodic fragments of songs to calculate similarity. Tsai specifically investigated QBM and proposed a matching method by vocal melody extraction from music [3].

Alternatively, music similarity can be characterized by harmony rather than melody. Features that are less affected by sound quality, such as LPC and MFCC are often used for identification. Shazam [4] is a well known application for iOS and Android developed by Shazam Entertainment Ltd. An audio fingerprint is extracted from an input signal and is compared with the fingerprints in the database. If a match is found, then the title associated with the original is obtained from the database [5]. Some studies have specifically addressed cover songs resembling the original version in chord sequences [6–8]. Ellis showed a chroma vector that is synchronized to [6]. A chroma vector, a 12-dimensional vector that represents the intensity of each of the 12 semitone pitch classes of the chromatic scale irrespective of octaves, is often used for chord recognition. He solved the problem of the key and the time lag using cross-correlation and circulation cross-correlation for similarity calculation. In addition, Ravuri set the three standards for beats per minute (BMP) to remedy the failure of beat tracking [7]. He added dynamic programming (DP) matching to calculate similarity. The output was identified based on weights using Support Vector Machine (SVM). Serra et al. used a tonality descriptor based on chroma. They proposed a new chroma similarity measure and a dynamic programming local alignment algorithm [8].

Mahieux specifically explored the concept of high-speed processing and access of vast amounts of data. He applied principal component analysis to 2D Fourier Transform Magnitude and calculated the similarity by Euclidean distance [9].

In contrast to cover songs, only slight differences exist between an original song and a live version because the original song artists themselves perform the live version. Riley et al. proposed a system that was specialized for live versions. It can absorb the differences of effects and tempo [10]. They separated music into small pieces and extracted a chroma vector for each segment. Then vector quantization (VQ) is applied to the chroma vectors for comparison with prepared audio-words. Rafii et. al applied image processing approach to audio fingerprinting system that can deal with live version identification. The fingerprints were derived using constant-Q spectrogram and template matching was executed using the Hamming distance and the Hough Transform [11].

When a signal of the entire concert performance, including several songs is inputted, these works will identify one song that has the highest similarity score. Our method performs identification by parsing the audio signal into segments of several tens of seconds.

1.2   Organization of This paper

This paper is organized as follows. Section 2 presents our proposed system. Section 3 explains effects of our processing and presents discussion. Then section 4 introduces our implementation example. Section 5 concludes this report and describes future work.

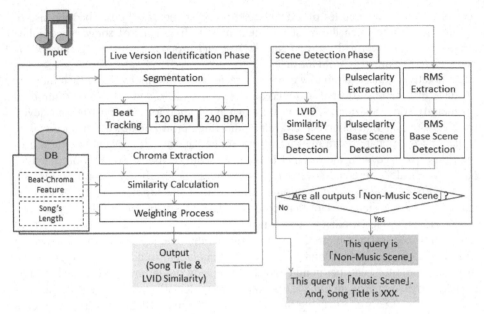

Fig. 1. Overview of our proposed system

2   Methods

This section presents our proposed system. Fig. 1 portrays an overview. The proposed system comprises two stages: live version identification and audio scene detection. In the live version identification phase, features extracted from the input query (segmentation) are compared with those in the database. Then the applicable music is identified through post-processing. These processes are a modified version of cover song identification presented in earlier reports [6–7]. In the audio scene detection phase, we discriminate the music scene using root mean square (RMS) and pulse clarity features in addition to similarity scores and the music title identified according to previous results. Finally, the output of non-music scene or original title is identified in each query.

2.1   Live version identification (LVID)

Fig. 2 portrays the flow of live version identification. First, three beat-synchronized chroma features are calculated from the input query. Then, each chroma vector is compared with those in the database and a similarity score is calculated. Subsequently, the score is weighted to judge the music resemblance with the previous version. The song candidate is excluded from the song list when the continuous putout is interrupted. Finally, an applicable music title is outputted.

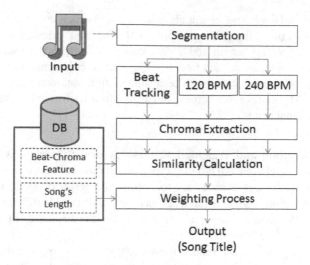

Fig. 2. Flow of live version music identification

Segmentation. Setting the appropriate segmentation length is important. If the query length is too short, then it is easy to find audio scene change. However, the length should include several chords to extract chord sequences. Therefore, we determine the query length using an experiment to resolve the tradeoff between the query length and the identification accuracy. We changed the query length and evaluated the music identification accuracy using Ellis's approach [6]. Fig. 3 shows the music identification accuracy for each query length.

For the sake of ensuring the accuracy, it is apparent that a length of 40 s or 80 s is better for identification because other application systems [4, 12] require about 20 s as the input length. Finally, we set 40 s as the query length.

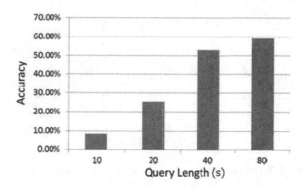

Fig. 3. Music identification accuracy for each query length

Chroma Vector Extraction and Beat Synchronization. Chroma features introduced by Ellis are based on spectrograms. The instantaneous frequency is used to enhance spectral resolution. Herein, we obtain chroma vectors, which are calculated using the Intelligent Sound Processing (ISP) tool box [13]. The DFT calculation is done using window frames of 93 ms length and 75% overlap. To obtain the final features, we adjust the frames averaged in each beat by reference to beat synchronization [6]. For this study, we used three features with beat information obtained using beat tracking, 120 and 240 BPM, because live version music is often performed at the same tempo as that of the original song.

Similarity Calculation and Weighted Process. A similarity score is calculated by combining cross-correlation and circular cross-correlation reference to [6–7]. Because the query length is 40 s, it is necessary to output as a result of identifying the same original song at various times. A tendency arises to fail the identification, especially the latter part in the song including a solo and ad lib. Performances often differ from the original music in the latter part.

To resolve this problem, we reduce the weight by estimating the timing at which the playing is switched to the next music using the signal length of the original music. We define the similarity score as shown below.

$$Similarity\ Score[n] = Ellis's\ Method[n] + Weight[n-1] \tag{1}$$

Therein, n is the query number, *Ellis's method*[n] is the original score, and *Weight*[$n-1$] is the changing value by 10 points in the case of the same song title with the previous query. Fig. 4 shows an example of weight change in a query series.

Subsequently we excluded song candidates from the song list when the continuous putout is interrupted because only rarely does the same music get played many times in a concert. This process reduces matching target songs and speeds up processing.

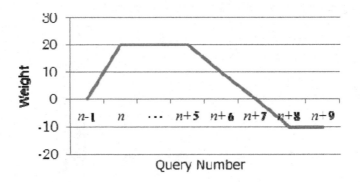

Fig. 4. Weight change based on the original song length (290 s)

2.2 Audio Scene Detection

Fig. 5 portrays the flow of audio scene detection. A non-music scene in a concert includes talking or silent intervals in the change of the performance. This study is intended

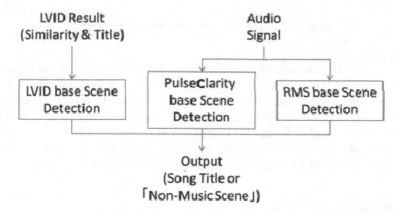

Fig. 5. Flow of audio scene detection

to detect those scenes. We simply specifically examine RMS and pulse clarity. Fig. 6 shows the values of RMS, pulse clarity, and LVID similarity in time series. It is readily apparent that the values differ between the music scene and non-music scene.

Actually, RMS is significantly different in terms of how music is recorded and the genre of music. Therefore, we normalize RMS to resolve this problem. Normalization is performed by dividing the maximum value of the RMS to the RMS value of currently playing from the beginning of the audio per second. The normalization is defined as shown below.

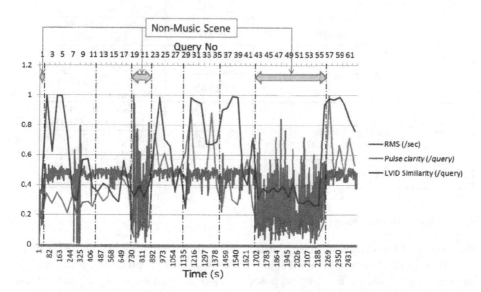

Fig. 6. Values of RMS, pulse clarity, and LVID similarity in a time series

$$Normalized\ RMS\,[n] = \frac{RMS\,[n]}{max\,(RMS\,[1],RMS\,[2],\cdots RMS\,[n])} \qquad (2)$$

where n is the query number.

Pulse clarity reflects how easily a listener can perceive the underlying rhythmic or metrical pulsation [14]. The value of pulse clarity is low in the non-music scene because the background music (BGM) has smaller amplitude than the music to be played or music that is not played. Additionally, the similarity score is low in the query including solo play, an ad lib, or a non-music scene.

The scene determination is executed using the threshold process. We determined the non-music scene that is obtained if three values are below the threshold.

3   Experimental Results

We manually annotate eight music concerts, which have total length of 403 min. The audio signals are encoded to mp3 and are downsampled to 22,050 Hz. For matching the original songs, we prepare 644 songs of 20 artists including five artists who appear in the concert. Table 1 presents dataset details.

Table 1. Dataset details

Audio number	Artist	Length (s)	Scene contents (s)
α1	Acid Black Cherry	2498	Music: 1774 (29.0%) Non-music: 724 (71.0%)
α2		2043	Music: 1815 (88.8%) Non-music: 228 (11.2%)
β	Avril Lavigne	4491	Music: 3958 (88.1%) Non-music: 533 (11.9%)
γ1	Matenrou Opera	2634	Music: 1976 (75.0%) Non-music: 658 (25.0%)
γ2		2940	Music: 2463 (83.8%) Non-music: 477 (16.2%)
δ1	Ringo Shiina	7139	Music: 6565 (92.0%) Non-music: 574 (8%)
δ2		1807	Music: 1642 (90.9%) Non-music: 165 (9.1%)
ε	Bump of Chicken	603	Music: 470 (77.9%) Non-music: 133 (22.1%)

3.1   Audio Scene Detection Results

We evaluated the effects of pulse clarity, LVID similarity and RMS on audio scene detection. For evaluation, we used four measures: accuracy, precision, recall and *F*-measure. Fig. 7 shows that our proposed method is superior to the conventionally used method. It is apparent that the method using three values yields the highest performance except recall. Because we have set higher threshold values, many music scenes are regarded as non-music scenes in the case of only one value.

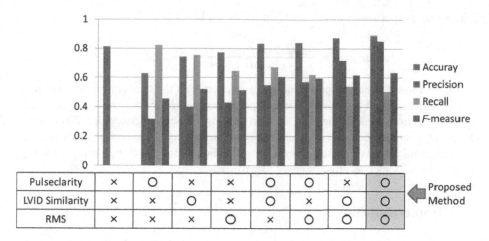

Fig. 7. Scene detection results obtained using pulse clarity, LVID similarity, and RMS

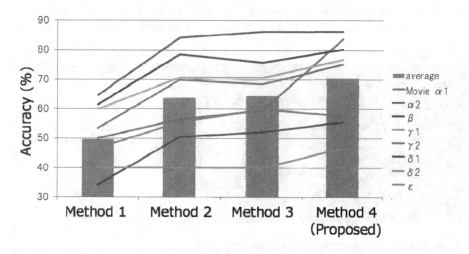

Fig. 8. Identification results obtained using method 1 (minor change of the previous method [6]), method 2 (combination of the method 1 and weighted process using constant value), method 3 (combination of the method 1 and weighted process using changing value), and method 4 (method 3 with scene detection).

3.2   Live version Identification Results

We evaluate the next three methods described in 2.

- Method 1: minor change of the previous method [6] which used the three features of estimated beat, 120 and 240 BPM.
- Method 2: combination of the method 1 and constant weight.
- Method 3: combination of the method 1 and changing weight.
- Method 4: Method 3 with scene detection.

The final output accuracy is calculated as shown below.

$$Accuracy = \frac{The\ number\ of\ correct\ outputs}{The\ number\ of\ queries} \times 100 \qquad (3)$$

Fig. 8 shows that our proposed method is superior to the conventionally used method. Results show that proposed method produces the best performance for live music identification because the weighted processing reduces unconcerned candidates. Moreover, scene detection avoids unnecessary processing. However, it takes time to calculate similarity. We should reduce song candidates for matching gradually using other techniques such as artist or genre identification.

4   Implementation

We implement the MATLAB application using the GUIDE tool. Fig. 9 shows our application for live version identification. This application displays the song title by extracting the audio signal streamed in Windows Media Player and other video playback software. The time lag is about 40 s close to the query length.

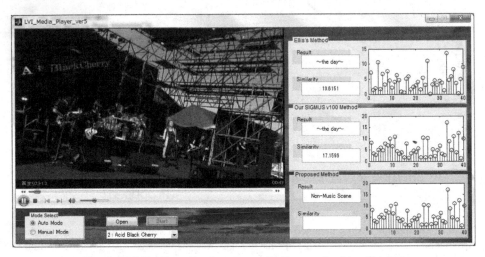

Fig. 9. MATLAB application system for live version identification

5   Conclusions

This study proposed a live version music identification system by modifying the conventional cover song identification system. We presented a system using a scene discrimination method that determines the music scene as well as the music title. Our results demonstrate that the proposed method performs better than the previous method and that the algorithm achieves 70% accuracy on average.

Future work shall realize real-time processing and shall gradually reduce song candidates for matching using other techniques such as artist or genre identification. We plan to use visual feature characteristics for music video identification.

References

1. Music Information Retrieval Evaluation eXchange, http://www.music-ir.org/mirex/wiki/MIREX_HOME
2. Marolt, M.: A mid-level melody-based representation for calculating audio similarity. In: Proceedings of International Conference on Music Information Retrieval, ISMIR (2006)
3. Tsai, W.H., Yu, H.M., Wang, H.M.: A query-by-example technique for retrieving cover versions of popular songs with similar melodies. In: Proceedings of International Conference on Music Information Retrieval, ISMIR (2005)
4. Shazam, http://www.shazam.com/
5. Haitsma, J., Kalker, T.: A Highly Robust Audio Fingerprinting System. In: Proceedings of International Conference on Music Information Retrieval, ISMIR (2002)
6. Ellis, D., Poliner, G.: Identifying Cover Songs with Chroma Features and Dynamic Programming Beat Tracking. In: Proceedings of ICASSP, pp. 1429–1432 (2007)
7. Ravuri, S.: Automatic Cover Song Detection: Moving from High Scores to General Classification. In: Proceedings of Music Information Retrieval Evaluation eXchange, MIREX (2009)
8. Serra, J., Gomez, E., Herrera, P., Serra, X.: Chroma Binary Similarity and Local Alignment Applied to Cover Song Identification. IEEE Transactions on Audio, Speech, and Language Processing 16(6), 1138–1151 (2008)
9. Mahieux, T.B., Ellis, D.: Large-Scale Cover Song Recognition Using the 2D Fourier Transform Magnitude. In: Proceedings of International Conference on Music Information Retrieval, ISMIR (2012)
10. Riley, M., Heinen, E., Ghosh, J.: A Text Retrieval Approach to Content-Based Audio Retrieval. In: Proceedings of International Conference on Music Information Retrieval (2008)
11. Rafii, Z., Coover, B., Han, J.: An Audio Fingerprinting System for Live Version Identification using Image Processing Techniques, In: Proceeding of ICASSP, pp.644-648 (2014)
12. SoundHound, http://www.soundhound.com/
13. Intelligent Sound Processing, http://kom.aau.dk/project/isound/
14. Lartillot, O., Eerola, T., Toiviainen, P., Fornari, J.: Multi-feature modeling of pulse clarity: Design, validation, and optimization. In: Proceedings of International Conference on Music Information Retrieval, ISMIR (2008)

Community Detection
Based on Links and Node Features in Social Networks

Fengli Zhang[1], Jun Li[2], Feng Li[1], Min Xu[3], Richard Xu[3], and Xiangjian He[3]

[1] Department of Automation, USTC, Hefei 230027,
Anhui, China
[2] School of Computer Science and Communication, AHPU,
WuHu 241000, Anhui, China
[3] School of Engineering and Communications, Faculty of Engineering and IT,
UTS, Sydney,Australia
zhangfli@mail.ustc.edu.cn, fli@ustc.edu.cn, beniciolee@126.com,
{Min.Xu,YiDa.Xu,Xiangjian.He}@uts.edu.au

Abstract. Community detection is a significant but challenging task in the field of social network analysis. Many effective methods have been proposed to solve this problem. However, most of them are mainly based on the topological structure or node attributes. In this paper, based on SPAEM [1], we propose a joint probabilistic model to detect community which combines node attributes and topological structure. In our model, we create a novel feature-based weighted network, within which each edge weight is represented by the node feature similarity between two nodes at the end of the edge. Then we fuse the original network and the created network with a parameter and employ expectation-maximization algorithm (EM) to identify a community. Experiments on a diverse set of data, collected from Facebook and Twitter, demonstrate that our algorithm has achieved promising results compared with other algorithms.

Keywords: Community Detection, Social Network, EM algorithm, Node Similarity.

1 Introduction

Recently, with the exploration of Internet, social networking is becoming an increasingly significant application because it enables users from different places to connect with each other. Strong community structure [2] is one fundamental property of social network. A very meaningful task of social network analysis is community detection, which aims to partition the users who have denser connectivity into one cluster. Community detection is a powerful tool to understand the internal structure of the network, that is, how users interact with each other. If we use community as a basic unit when doing research on the social networks, the network can be simplified and compressed effectively so that we can mine useful information from complex network with acceptable computation cost. Community detection also has many other applications such as friend suggestion, product recommendation and link inference.

X. He et al. (Eds.): MMM 2015, Part I, LNCS 8935, pp. 418–429, 2015.

A number of algorithms have been proposed for community detection, such as G-N algorithm [2], Spectral Clustering [3], Neman's Mixture Model [4] and MMSB [5]. Most of these algorithms only focus on topological structure. To learn more about the related algorithms, we can see the recent surveys [6 ,7]. However, in a real social network, there always exists link noise (incorrect links and missing links). The presence of link noise makes identifying community more difficult. For example, some nodes with no link or weak link but sharing fairly similar features may be grouped into distinct communities, which is unreasonable. Therefore, only considering the network links is not enough. In real life, people in one community not only have denser links but also more or less similarities among them. According to observation, we can take the node attributes into consideration to help alleviate the noise and strengthen the community signal.

In recent years, various algorithms have been proposed to combine the links and content for community detection. Zhu et al [8] introduce a method that jointly factorizes the content matrix and link matrix for a spectral clustering. Cohn and Hofmann [9] present a joint probabilistic model of document content and connectivity, an extension of PLSA [10, 11] and HITS [12, 13]. Erosheva et al [14] describe a mixed-membership model to analyze both the content of a document and its citation. Nallapati et al [15] present two different models called Pairwise-Link-LDA and Link-PLSA-LD. The former one combines LDA [16] and Mixed Membership Block Stochastic Model [5] and the other combines the LDA and PLSA models into a single graphical model. In [17], the objects such as photos and articles two users shared are regarded as edge content between them and then edge content is incorporated into the matrix factorization. In the article [18], the author presents CODICIL, a family of highly efficient graph simplification algorithms leveraging both content and graph topology to identify and retain important edges in a network. McAuley and Leskovec [19] try to automatically discover users' social circles fusing link and users' profile.

In this paper, we propose a joint probabilistic model of combining link and node features for community detection. In this work, we first build a SPAEM model only with the network links. Next, we create a new feature-based weighted network whose edge weight is the node feature similarity between two nodes. Then, we fuse the original network and the created network. If two nodes have a strong similarity, the original link between the two nodes will be strengthened, otherwise it will be weakened. How much the node features have impact on the original links can be determined by introducing a parameter. Finally, an expectation-maximization algorithm (EM) is employed for the optimization.

The rest of the paper is organized as follows. In section 2, we first review the SPAEM model. Then, how to create feature-based network has been discussed. Finally, we present the method of combining links and node features. In section 3, experimental results tested on different data sets are presented. Conclusions are drawn in section 4.

2   Our Method

In this section, we first introduce the SPAEM model and then create a new feature-based network. Next, we present a joint model combining link and node features.

In the following we assume that the network in the paper is undirected and unweighted. Let A denotes the adjacent matrix; $A_{ij}=1$ if there is a link between node i and j, otherwise $A_{ij}=0$.

2.1   SPAEM Model

SPAEM [1] model regards community detection as a probabilistic inference problem. It utilizes the idea of the probabilistic latent semantic analysis [2] which is a powerful algorithm in text mining. Compared with other algorithms [20, 21], SPAEM model possesses the mathematical simplicity and hence is easy to understand.

We assume that $N_{(i)}$ denotes the set of the neighbors of node i. Suppose that: there is c latent communities to be detected; every node has probability π_r to fall in group r; community r selects node i with probability $\beta_{r,i}$ with constraint $\sum_{i=1}^{n} \beta_{r,i} = 1$.

The edge e_{ij} is generated by the following finite mixture model where the community r is latent variable.

(1) Select a community r with the probability π_r

(2) The node i with probability $\beta_{r,i}$ to be selected by Community r.

(3) The node j with probability $\beta_{r,j}$ to be selected by Community r.

Assume that community r selects node i and node j independently, the probability of choosing the node pair {i, j} is

$$P(e_{ij} \mid \pi, \beta)=P(\{i, j\} \mid \pi, \beta) = \sum_{r=1}^{c} \pi_r \beta_{r,i} \beta_{r,j} \tag{1}$$

High value of $P(e_{ij} \mid \pi, \beta)$ is regarded as a reliable edge. If there is a link in the node i and node j, they should have a high likelihood of joining in the same community, in other words, the nodes in the same community with high value of β should be connected.

The logarithm probability of network A under parameters π, β can be modeled as

$$L= \ln P(A \mid \pi, \beta)$$

$$= \sum_{i=1}^{n} \sum_{j:j \in N(i)} \ln P(e_{ij} \mid \pi, \beta) \ .$$

$$= \sum_{i=1}^{n} \sum_{j:j \in N(i)} \ln \sum_{r=1}^{c} \pi_r \beta_{r,i} \beta_{r,j} \tag{2}$$

In order to optimize the value of parameters π, β, we maximize the logarithm probability by expectation-maximization (EM) algorithm [22].

E-step:

The posterior probability $P(g_{ij}=r \mid A, \pi, \beta)$, denoted by q_{ij}, then

$$
\begin{aligned}
q_{ij} &= P(g_{ij} = r \mid A, \pi, \beta) \\
&= \frac{P(g_{ij} = r, A \mid \pi, \beta)}{P(A \mid \pi, \beta)} \\
&= \frac{\pi_r \beta_{r,i} \beta_{r,j}}{\sum_{s=1}^{c} \pi_s \beta_{s,i} \beta_{s,j}}
\end{aligned}
\tag{3}
$$

M-step:

The expected logarithm probability of the network is

$$
\begin{aligned}
\vec{L} &= \sum_{i=1}^{n} \sum_{j: j \in N(i)} \sum_{r=1}^{c} q_{ij,r} \ln P(e_{ij}, g_{ij} = r \mid \pi, \beta) \\
&= \sum_{i=1}^{n} \sum_{j: j \in N(i)} \sum_{r=1}^{c} q_{ij,r} \ln \pi_r \beta_{r,i} \beta_{r,j}
\end{aligned}
\tag{4}
$$

By maximizing $\vec{L}$ we can get

$$
\pi_r = \frac{\sum_{i=1}^{n} \sum_{j: j \in N(i)} q_{ij,r}}{\sum_{i=1}^{n} \sum_{j: j \in N(i)} \sum_{s=1}^{c} q_{ij,s}}, \quad \beta_{r,i} = \frac{\sum_{j: j \in N(i)} q_{ij,r}}{\sum_{k=1}^{n} \sum_{j: j \in N(k)} q_{kj,r}}
\tag{5}
$$

2.2   Create Feature-Based Network

Assume that F_i denotes the set of the features of node i. The node feature similarity between node i and node j is defined by Jaccard coefficient. Next we create a link between node i and node j and take the value of the node feature similarity as the edge weight w_{ij} of node i and node j, that is

$$
W_{ij} = Jaccard(v_i, v_j) = \frac{\mid F_i \bigcap F_j \mid}{\mid F_i \bigcup F_j \mid}
\tag{6}
$$

If the value of feature similarity does not equal to 0, the two nodes form an edge with weight w_{ij}. Otherwise, there is no link between the nodes. Hence we get a weighted new network based on node features.

The new created network owns the same nodes with the original network. For any node i, it is impossible for community r to select the node with two possibilities at the same time. So in the new network, the probability that community r is selected can be still denoted by π_r and the probability that community r select node i is still $\beta_{r,i}$. The probability of choosing the node pair {i, j} is the same as Equation (1).

For the new network, we use $N'_{(i)}$ to denote the neighbors of node i. Because the network is weighted, we replace A_{ij} with W_{ij}. The expected logarithm probability of the weighted network can be rewritten as.

$$\vec{L} = \sum_{i=1}^{n} \sum_{j:j \in N'(i)} \sum_{r=1}^{c} W_{ij} q_{ij,r} \ln \pi_r \beta_{r,i} \beta_{r,j} \tag{7}$$

The weighted network can be used to detect community just relying on the node features.

2.3   Combining Link and Node Features

The information of the social network cannot be fully utilized if just applying each separately. Similar to the work in [9] which is an influential algorithm of combining content and connectivity in text mining, it is reasonable to merge the two networks into a joint probabilistic model, therefore we propose maximizing the following expected logarithm probability with a parameter α.

$$\vec{L} = \alpha \sum_{i=1}^{n} \sum_{j:j \in N(i)} \sum_{r=1}^{c} q_{ij,r} \ln \pi_r \beta_{r,i} \beta_{r,j} +$$
$$(1-\alpha) \sum_{i=1}^{n} \sum_{j:j \in N'(i)} \sum_{r=1}^{c} W_{ij} q_{ij,r} \ln \pi_r \beta_{r,i} \beta_{r,j} \tag{8}$$

In this model, the original links have limited effect when detecting communities. Even though the link between node i and node j is weak, if the two nodes have strong similarity, the link will be strengthened. Thus, they may form more reliable edge and have high probability of belonging to the same community.

The value of α depends on different applications, that is, the importance one assigns to predict links and node features. When detecting community, if we think the link is more important, the value of α can be set with a higher value.

Next what we do is to calculate $q_{ij}, \beta_{r,i}$ with EM algorithm. In E-step, we compute the posterior probability q_{ij}. In M-step, substitute q_{ij} into Equation (8) and optimize $\beta_{r,i}$

by maximizing $\tilde{L}$. The posterior probability $P(g_{ij}=r \mid A, \pi, \beta)$, denoted by q_{ij}, can still be computed by equation (3).

Taking the constraints into consideration $\sum_{r=1}^{n} \pi_r = 1, \sum_{i=1}^{n} \beta_{r,i} = 1$, the Lagrange function is

$$D = \alpha \sum_{i=1}^{n} \sum_{j: j \in N(i)} \sum_{r=1}^{c} q_{ij,r} \ln \pi_r \beta_{r,i} \beta_{r,j} +$$

$$(1-\alpha) \sum_{i=1}^{n} \sum_{j: j \in N1(i)} \sum_{r=1}^{c} W_{ij} q_{ij,r} \ln \pi_r \beta_{r,i} \beta_{r,j} + \tag{9}$$

$$\lambda (\sum_{r=1}^{c} \pi_r - 1) + \sum_{r=1}^{c} \gamma_r (\sum_{i=1}^{n} \beta_{r,i} - 1)$$

Where λ, γ_r are Lagrange multipliers. The derivatives of D are

$$\frac{\partial D}{\partial \pi_r} = \alpha \sum_{i=1}^{n} \sum_{j: j \in N(i)} \frac{q_{ij,r}}{\pi_r} + (1-\alpha) \sum_{i=1}^{n} \sum_{j: j \in N1(i)} \frac{W_{ij} q_{ij,r}}{\pi_r} + \lambda$$

$$\frac{\partial D}{\partial \beta_{r,i}} = \alpha \sum_{j: j \in N(i)} \frac{q_{ij,r}}{\beta_{r,i}} + (1-\alpha) \sum_{j: j \in N1(i)} \frac{W_{ij} q_{ij,r}}{\beta_{r,i}} + \gamma_r \tag{10}$$

Combine with the constraints $\sum_{r=1}^{n} \pi_r = 1, \sum_{i=1}^{n} \beta_{r,i} = 1$, and let the derivatives of Equation (10) equal to 0. We can get as follows,

$$\pi_r = \frac{\alpha \sum_{i=1}^{n} \sum_{j: j \in N(i)} q_{ij,r} + (1-\alpha) \sum_{i=1}^{n} \sum_{j: j \in N1(i)} W_{ij} q_{ij,r}}{\alpha \sum_{i=1}^{n} \sum_{j: j \in N(i)} \sum_{s=1}^{c} q_{ij,s} + (1-\alpha) \sum_{i=1}^{n} \sum_{j: j \in N1(i)} \sum_{s=1}^{c} W_{ij} q_{ij,s}} \tag{11}$$

$$\beta_{r,i} = \frac{\alpha \sum_{j: j \in N(i)} q_{ij,r} + (1-\alpha) \sum_{j: j \in N1(i)} W_{ij} q_{ij,r}}{\alpha \sum_{k=1}^{n} \sum_{j: j \in N(k)} q_{kj,r} + (1-\alpha) \sum_{k=1}^{n} \sum_{j: j \in N1(k)} W_{ij} q_{kj,r}} \tag{12}$$

By iterating Equations (3), (11) and (12) until convergence, we can obtain $q_{ij}, \beta_{r,i}$.

The probability community s selects of node i is $u_{s,i} = \pi_s \beta_{s,i}$. If community r meets the following condition, the node i belongs to community r.

$$r = \arg\max_s \{u_{s,i} = \pi_s \beta_{s,i}, s = 1, 2, ..., r\} \qquad (13)$$

The algorithm can also be used to detect overlapping community. For node i, $r = \arg\max_s \{u_{s,i} = \pi_s \beta_{s,i}, s = 1, 2, ..., r\}$, if there exists another community s such that $\dfrac{u_{s,i}}{u_{r,i}} > \theta (0 < \theta < 1)$, node i also belongs to community s.

3   Experiment

In this section, experiments on a small real data set are firstly carried on in order to intuitively demonstrate the difference between our method and the existing methods. Then, we experiment on the public data sets, i.e. Facebook[1] and Twitter[2], to observe the effect of parameter α and how our algorithm outperforms compared to other methods.

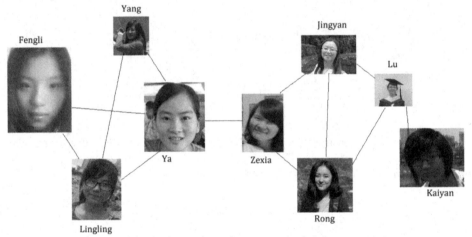

Fig. 1. the connections among the nine students

3.1   Experiment on Self-collected Data Set

In order to how node attributes and links affect the result and make the result of our algorithm more directed and visible, we apply the algorithm to a small real data set. The data set is collected by ourselves, which is about nine students in USTC: Lingling, Yang, Fengli, Ya, Zexia, Rong, Jingyan, Lu and Kaiyan. We investigate whether they have connection with each other when they just began their college life, as showed in Figure 1. The feature vectors of the nodes in Table 1 are their interests in music, dancing, reading, traveling and film. If one likes music, the value equals to 1. Otherwise,

---

[1] http://snap.stanford.edu/data/egonets-Facebook.html.
[2] http://snap.stanford.edu/data/egonets-Twitter.html.

the value is set to 0. We group them into different communities and verify that whether the obtained result is consistent with the communities they formed in later days in their college lives.

We test our algorithm on the above data set. In the experiments, α is set to 1.0, 0 and 0.7 respectively. The results are showed as Figure 2. α=1 means that community detection is only based on the network link. As shown in the left of Figure 2, the nodes in one community are linked more densely. The center shows the results when node attributes are the only consideration. "Kaiyan" is grouped into the "green" community because of their very strong similarity. Considering both structure and node features, we set α to 0.7. In this case, "Kaiyan" is grouped into two communities at the same time. There is no doubt that "Kaiyan" should belong to the "red" community because of their links. Besides, even though there is no link between "Kaiyan" and "Lingling et al, they have strong feature similarity. The node features strengthen the links between them. So it is reasonable for "Kaiyan" grouped into the "green" community. In fact, in the following years in USTC Kaiyan usually does some extracurricular activities with "green" group, and attends classes together with the other group. So Kaiyan connects with both the groups and should be assigned to the two communities simultaneously. Apparently, the result of α=0.7 agrees better with the reality.

Table 1. The feature vectors of the nine students

	Lingling	Yang	Fengli	Ya	Zexia	Rong	Jingyan	Lu	Kaiyan
music	1	0	0	0	0	0	0	0	1
dancing	0	0	1	0	0	0	0	0	0
reading	1	1	1	1	0	0	1	0	1
traveling	1	1	1	1	0	0	0	0	1
film	0	0	0	0	1	1	1	1	0

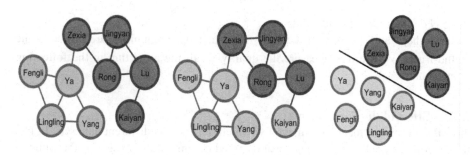

Fig. 2. Experiment on small real data. The value of α from left to right is 1.0, 0 and 0.7.

3.2   Experiments on Public Data Set

In this sub-section, we first introduce the criterion to evaluate the quality of the detected results. Then we compare our method with other methods on the date sets, i.e. Facebook and Twitter according to the evaluation criterion. Facebook data was collected from survey participants using this Face app[3]. Twitter data was crawled from public sources. Both the data set includes node features, ground-truth circles and networks.

3.2.1 Evaluation Criterion

To evaluate our algorithm, we maximize the consistency between the detected communities $\mathbb{C}=\{C_1...C_k\}$ and the ground-truth communities $\overline{\mathbb{C}}=\{\overline{C}_1...\overline{C}_k\}$.

The F-score of C on $\overline{C}$ is denoted as follows:

$$F(C,\overline{C}) = 2 \cdot \frac{precision(C,\overline{C}) \cdot recall(C,\overline{C})}{precision(C,\overline{C}) + recall(C,\overline{C})} \tag{14}$$

Precision and recall are defined as

$$precision(C,\overline{C}) = \frac{|C \cap \overline{C}|}{|C|}, recall(C,\overline{C}) = \frac{|C \cap \overline{C}|}{|\overline{C}|} \tag{15}$$

For each detected community C, we compute its F-score on $\overline{\mathbb{C}}$.

$$F(C,\overline{\mathbb{C}}) = \max_{\overline{C} \in \overline{\mathbb{C}}} F(C,\overline{C}) \tag{16}$$

Then the final F-score of $\mathbb{C}$ on $\overline{\mathbb{C}}$ is:

$$F(\mathbb{C},\overline{\mathbb{C}}) = \sum_{C \in \mathbb{C}} \frac{|C|}{|N|} F(C,\overline{\mathbb{C}}) \tag{17}$$

where N denotes the set of the nodes.

The higher value of $F(\mathbb{C},\overline{\mathbb{C}})$ denotes the detected communities are closer to the ground-truth. We use it to measure the quality of all the algorithm in the following experiments.

3.2.2 Effect of the Parameter α

The value of α is decided experimentally, which depends on different data sets. In this sub-section, we track how the quality of detected communities changes as the value of α varies from 0.1 to 0.9. Figure 3 shows the results of different α on Facebook and Twitter. For Facebook, the best quality is achieved when α=0.5, while for Twitter, when α=0.7, F-score is the highest. The weight value is determined by nodes and links' importance, which varies for different applications. Therefore, for different data sets, we can adjust the value of α to obtain the best result.

---

[3] https://www.facebook.com/apps/application.php?id=201704403232744.

3.2.3   Comparison with Other Methods

In this section, we compare our method with MMSB [5], SPAEM [1], K-means, and MaAuley and Leskovec's algorithm (MLA) [19]. MMSB and SPAEM only focus on network links; K-means is a classical algorithm that considers node features only; MLA is a new algorithm to discover social circles combining links and node features. We apply these algorithms on the data set of Facebook and Twitter. In the experiment, we set α of our method to 0.5 and 0.7 respectively for Facebook and Twitter.

From the encouraging results (Figure 4 and Figure 5), our algorithm outperforms the other four methods significantly.

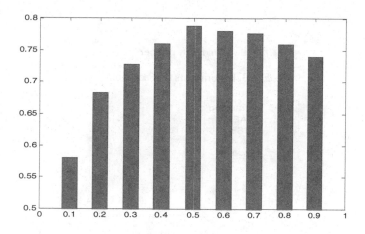

(a) Varying α on Facebook

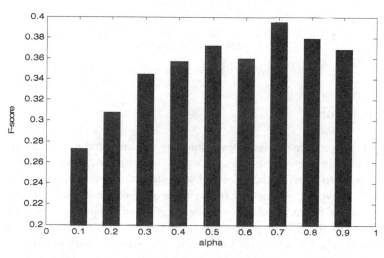

(b) Varying α on Twitter

Fig. 3 effects of varying α

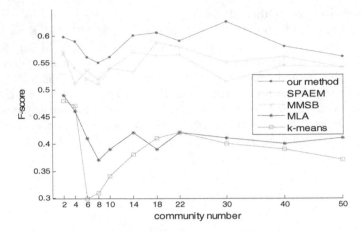

Fig. 3. Experiments on Facebook

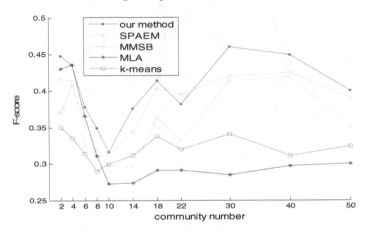

Fig. 4. Experiments on Twitter

4   Conclusion

In this paper, we propose an algorithm combining the links and node features. In the algorithm, we create a new feature-based network and fuse it with the original network with a parameter to alleviate the noise and strengthen the community signal. Experimental results show that our method outperforms state-of-the-art methods in clustering quality. For the future work, first, we plan to improve computing efficiency of our algorithm to adapt to the large scale networks. Then, we will try other algorithms to do optimization. The EM algorithm we adopt in our method is easy to fall into a local optimum. So we will utilize other algorithms to enhance the ability to search the global optimum.

References

1. Ren, W., et al.: Simple probabilistic algorithm for detecting community structure. Physical Review E 79(3), 036111 (2009)
2. Girvan, M., Newman, M.E.: Community structure in social and biological networks. Proceedings of the National Academy of Sciences 99(12), 7821–7826 (2002)
3. Pothen, A., Simon, H.D., Liou, K.-P.: Partitioning sparse matrices with eigenvectors of graphs. SIAM Journal on Matrix Analysis and Applications 11(3), 430–452 (1990)
4. Newman, M.E.J., Leicht, E.A.: Mixture models and exploratory analysis in networks. Proceedings of the National Academy of Science 104(23), 9564–9569 (2007)
5. Airoldi, E.M., et al.: Mixed membership stochastic blockmodels. Advances in Neural Information Processing Systems (2009)
6. Fortunato, S.: Community detection in graphs. Physics Reports 486(3), 75–174 (2010)
7. Xie, J., Kelley, S., Szymanski, B.K.: Overlapping community detection in networks: The state-of-the-art and comparative study. ACM Computing Surveys (CSUR) 45(4), 43 (2013)
8. Zhu, S., et al.: Combining content and link for classification using matrix factorization. In: Proceedings of the 30th Annual International ACM SIGIR Conference on Research and Development in Information retrieval. ACM (2007)
9. Hofmann, D.C.T.: The missing link-a probabilistic model of document content and hypertext connectivity. In: Proceedings of the 2000 Conference on Advances in Neural Information Processing Systems, The MIT Press (2001)
10. Deerwester, S.C., et al.: Indexing by latent semantic analysis. JASIS 41(6), 391–407 (1990)
11. Hofmann, T.: Probabilistic latent semantic indexing. In: Proceedings of the 22nd Annual International ACM SIGIR Conference on Research and Development in Information Retrieval. ACM (1999)
12. Cohn, D., Chang, H.: Learning to Probabilistically Identify Authoritative Documents. In: ICML (2000)
13. Kleinberg, J.M.: Authoritative sources in a hyperlinked environment. Journal of the ACM (JACM) 46(5), 604–632 (1999)
14. Erosheva, E., Fienberg, S., Lafferty, J.: Mixed-membership models of scientific publications. Proceedings of the National Academy of Sciences of the United States of America 101(Suppl. 1), 5220–5227 (2004)
15. Nallapati, R.M., et al.: Joint latent topic models for text and citations. In: Proceedings of the 14th ACM SIGKDD International Conference on Knowledge Discovery and Data Mining. ACM (2008)
16. Blei, D.M., Ng, A.Y., Jordan, M.I.: Latent dirichlet allocation. The Journal of Machine Learning Research 3, 993–1022 (2003)
17. Qi, G.-J., Aggarwal, C.C., Huang, T.: Community detection with edge content in social media networks. In: 2012 IEEE 28th International Conference on Data Engineering (ICDE). IEEE (2012)
18. Ruan, Y., Fuhry, D., Parthasarathy, S.: Efficient community detection in large networks using content and links. In: Proceedings of the 22nd International Conference on World Wide Web. International World Wide Web Conferences Steering Committee (2013)
19. Leskovec, J., Mcauley, J.J.: Learning to discover social circles in ego networks. Advances in Neural Information Processing Systems (2012)
20. Ramasco, J.J., Mungan, M.: Inversion method for content-based networks. Physical Review E 77(3), 036122 (2008)
21. Vazquez, A.: Population stratification using a statistical model on hypergraphs. Physical Review E 77(6), 066106 (2008)
22. Dempster, A.P., Laird, N.M., Rubin, D.B.: Maximum likelihood from incomplete data via the EM algorithm. Journal of the Royal Statistical Society. Series B (Methodological), 1–38 (1977)

Scaling and Cropping of Wavelet-Based Compressed Images in Hidden Domain

Kshitij Kansal[1,2], Manoranjan Mohanty[3], and Pradeep K. Atrey[4,2]

[1] International Institute of Information Technology, Hyderabad, India
[2] Department of Applied Computer Science, University of Winnipeg, MB, Canada
[3] Security Lab, SICS Swedish ICT AB, Lund, Sweden
[4] Department of Computer Science, University at Albany - State University of New York, Albany, NY, USA

Abstract. With the rapid advancement of cloud computing, the use of third-party cloud datacenters for storing and processing (e.g, scaling and cropping) personal and critical images is becoming more common. For storage and bandwidth efficiency, the images are almost always compressed. Although cloud-based imaging has many advantages, security and privacy remain major issues. One way to address these two issues is to use Shamir's (k, n) secret sharing-based secret image sharing schemes, which can distribute the secret image among n number of participants in such a way that no less than k (where $k \leq n$) participants can know the image content. Existing secret image sharing schemes do not allow processing of a compressed image in the hidden domain. In this paper, we propose a scheme that can scale and crop a CDF (Cohen Daubechies Feauveau) wavelet-based compressed image (such as JPEG2000) in the encrypted domain by smartly applying secret sharing on the wavelet coefficients. Results and analyses show that our scheme is highly secure and has acceptable computational and data overheads.

Keywords: Wavelet-Based Image Compression, Shamir's Secret Sharing, Hidden Domain Image Scaling/Cropping.

1 Introduction

The magnitude of digital images has boomed in recent years due to the creation of the digital imaging devices with increasing resolution. For example, in whole slide imaging, hospitals use histopathology images of a size of several gigabytes each [1]. With the emergence of cloud technologies, people are now storing these large images in third party cloud datacenters, which provide universal access, and are more flexible and economical alternative. For storage efficiency, the images are often compressed before being sent to the datacenters.

Security and privacy, however, are the major issues when an image is stored in third-party datacenters [2]. By accessing a datacenter, an adversary can know the image content (confidentiality issue), or tamper the image to provide misleading information (integrity issue). Since the stored image can be highly personal and important (such as medical images), the confidentiality issue can lead to privacy loss, and the integrity issue can be catastrophic (such as the detection of the wrong disease).

X. He et al. (Eds.): MMM 2015, Part I, LNCS 8935, pp. 430–441, 2015.

We can address the security and privacy issue in cloud-based imaging using secret image sharing to hide the image content from a datacenter [3]. Available image secret sharing schemes focus mainly on secret sharing uncompressed images, and can support image operations on share images [3] [4]. While these schemes are secure, they fail to address the fact that most of the images are compressed using JPEG, JPEG2000 etc.

Scaling and cropping a compressed shadow image (i.e., secret shared image) is challenging. We cannot use an additional non-homomorphic cryptosystem with secret sharing (as used by most secret image sharing schemes), since we want to preserve the additive and scalar multiplicative homomorphic property of secret sharing required in image scaling and cropping. In other words, secret sharing alone should be able to hide both the color and shape of the image by destroying the correlation among the image pixels. This requirement, however, conflicts with the condition that the correlation among pixels must be preserved to obtain compression.

In this paper, we address this challenge by proposing a secret image sharing scheme that can hide an image in such a way that both compression and image scaling/cropping are possible. We use Shamir's (k, n) secret sharing [5] to secret share an image, and CDF discrete wavelet transformation (DWT) based compression to compressing the image. The information in the low frequency LL wavelet band (we call it the *approximation band*) and all other high frequency wavelet bands (we call them *detailed bands*) are hidden by secret sharing the coefficient values and the coefficient positions respectively. Thus, scaling and cropping the approximation band can be done in the hidden domain. Cropping the detailed bands and the inverse DWT (IDWT) operations, however, is done in the plain-text domain. Our scheme can be used in cloud-based imaging, where a server can secret share the secret image; n cloud datacenters can store the shadow images and perform scaling and cropping of the secret shared approximation band coefficients; and the client can perform the reconstruction of the secret coefficient values and positions, the cropping of detailed bands, and the IDWT operations. Since the detailed bands are not required in the case of the scaling down operation, and the values of most of the coefficients of the detailed bands are zero, our scheme results in acceptable data overhead. Similarly, since secret reconstruction is the only additional operation, our scheme has low computation overhead at the client end.

The rest of the paper has been organized as follows. Section 2 presents prior works and highlights their issues. In Section 3, we highlight the challenges of integrating secret sharing with a compression scheme, such as with JPEG2000. Section 4 presents our scheme. Section 5 provides experimental results, and performs security and performance analysis. Section 6 concludes the work.

2   Background and Related Work

In this section, we will provide an overview of Shamir's secret sharing, list out some secret image sharing schemes, summarize wavelet-based compression, and discuss some existing approaches to integrate a cryptosystem with wavelet compression.

2.1   Shamir's Secret Sharing and Secret Image Sharing

Shamir's secret sharing [5] is a polynomial-based secret sharing scheme that hides a secret S by creating n shares of S in such a way that no information about S can be known from less than k (where $k < n$) shares. The shares are created by using the secret sharing polynomial

$$F(x) = S + \sum_{i=1}^{k-1} a_i x^i$$

in a finite field GF(p), where $p > S$ is a prime number, $a_i < p$ is a random number in GF(p), and x is the share number. The secret is then destroyed, and can only be reconstructed from at least k shares using the Lagrange interpolation formula

$$L(x) = \sum_{i=0}^{k-1} S_i t_i(x) \bmod p,$$

where S_i is the ith share and

$$t_i(x) = \prod_{j=0, j \neq i}^{k-1} \frac{x - x_j}{x_i - x_j}$$

is the Lagrange basis function. Being a perfectly secured (i.e., no information about the secret can be gathered from at most $k-1$ shares) and unconditional (no assumption about the computation capability of an adversary) cryptosystem, Shamir's secret sharing is highly secure. The data overhead, however, is a concern since n shares, where the size of a share is as large as the size of the secret, are created.

Secret image sharing based on Shamir's secret sharing is a thoroughly studied area. After Thien and Lin's seminal work [6], a number of researchers have proposed different secret image sharing schemes to hide both the color and shape of a secret image [7] [8] [9] [10] [3]. An exhaustive list of secret image sharing schemes can be found in the recent survey work by Tsai and Chen [4]. The majority of the existing works mainly focus on image storage, and use (k, k, n) multi-secret sharing to share an image. Since the use of (k, k, n) multi-secret sharing can disclose the shape of the secret image [3], an additional obfuscation technique, such as permutation [6], chaotic map [8], stenography [9], or matrix projection [7], is used along with secret sharing to hide the shape. The use of an additional scheme, however, can destroy the homomorphic property of secret sharing – making most of the existing secret image sharing schemes unsuitable for hidden domain processing.

Without using an extra obfuscation technique, Mohanty et al. [3] recently proposed a secret image sharing scheme that can perform scaling and cropping operations in the hidden domain. Although it is limited to bilinear scaling, the proposed scheme demonstrates that important image streaming operations can be performed in the hidden domain. However, their scheme is limited to uncompressed images, leaving the more realistic issue of scaling/cropping a compressed image in the hidden domain unaddressed. *In this paper, our objective is to design a secret image sharing scheme that can scale/crop a shadow image of a wavelet-based compressed image.*

2.2   Wavelet-Based Compression, and Its Integration with a Cryptosystem

Wavelet-based image compression schemes, such as JPEG2000 and EZW (Embedded Zerotrees of Wavelet), have become popular for image streaming since they provide multiresolution image representation. The compression pipeline consists of four main steps: wavelet transformation, quantization (for lossy compression), run-length encoding or zero trees, and and entropy coding. In the wavelet transform phase, DWT is typically used to represent a set of color values as high frequency coefficients and low frequency coefficients. Since the high frequency coefficients contain little information about the image, we can exclude these coefficients. Furthermore, due to spatial coherence, most of the wavelet coefficients are either zeros or near zeros. The near zero values can be replaced with zeros by thresholding. Thus, most of the wavelet coefficients can be efficiently packed using run-length encoding or zero trees. Further compression can be achieved by entropy encoding the run-length encoded or zero tree represented coefficients. There are a number of DWT techniques and entropy encoding schemes to choose form. The selection of a wavelet transform scheme and entropy encoding scheme is implementation dependent.

The use of a cryptosystem to hide wavelet-based compressed images is not new. For example, researchers have previously focused on the major aspects of encryption of JPEG2000-compressed image by trading off between security, image quality, and compression efficiency [11]. The main idea behind most of the previous works is to encrypt the significant information in a compressed images, such as low-frequency wavelet coefficients and important sub-bands, for improving compression efficiency [12] [13]. However, these works do not consider supporting processing in the hidden domain.

Recently, Zheng et al. proposed a method that uses the Paillier cryptosystem to hide DWT coefficients. Since the Paillier cryptosystem is homomorphic to addition and scalar multiplication, wavelet transformation and inverse transformation can be performed in the hidden domain. Thus, some image operations, such as scaling, can be performed in the hidden domain. Although Zheng et al.'s work is closely related to ours (our aim is to apply secret sharing to wavelet transformation), there are some differences between the two. First, Zheng et al.'s work focuses on hiding wavelet coefficients, but we concentrate on hiding coefficients and the spatial coherence among them (which is present in an image). Hiding the image coherence must be done in such a way that compression can be achieved. Second, Shamir's secret sharing, unlike the Paillier cryptosystem, provides perfect secrecy, and ensures data confidentiality and data integrity without using an extra data hiding technique. Third, besides image scaling, we aim to provide image cropping, to make the scheme more useful for image streaming.

3   Challenges in Integrating Image Hiding with Compression

Traditionally, the image compression principle conflicts with the principle of image hiding. While an image compression scheme utilizes the spatial coherence of an image, an image hiding scheme focuses on destroying the spatial coherence. Therefore, designing a scheme that will provide both compression and obfuscation is not easy. Furthermore, the aim of performing image operations in the hidden domain makes the problem more challenging, since the choice of selecting a cryptosystem is now restricted.

Although Shamir's secret sharing addresses the latter challenge, the former challenge persists. The problem here is now to secret share the quantized and encoded wavelet coefficients in a way that we are able to process the image in the hidden domain by minimally affecting the compression. For this task, there is a tradeoff between the compression and the level of encryption. We have tried various known techniques and their derivatives to address our problem. Both for sake of completeness and to pave the way for future research in this direction, we briefly summarize our findings below.

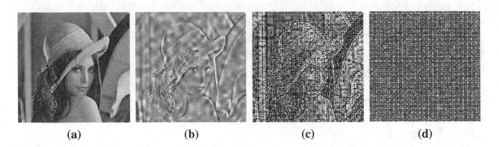

(a)   (b)   (c)   (d)

Fig. 1. Application of secret sharing to selected coefficients: (a) is the secret image, (b) is the shadow image when secret sharing is applied to an approximate region only, (c) is a shadow image obtained by applying secret sharing to nonzero coefficients, (d) is a shadow image obtained by applying secret sharing to all the coefficients.

A common way of applying encryption to the coefficients is to only hide the low frequency coefficients and keep the high frequency coefficients as they are. This scheme has been used for both DCT and DWT coefficients. Following this foot strip, we first applied secret sharing to low-frequency coefficients. As shown in Figure 1, the results, however, were unsatisfactory since the unshared coefficients represented the edges of the secret image. Although assuming all nonzero unshared coefficients as zeros can produce a noise-like image (since the exact positions of the zero coefficients are unknown), the image reconstructed from such a scheme is of bad quality (as shown in Figure 1). Furthermore, increasing the number of coefficients for secret sharing did not solve the problem, and rather increased the compression size (by randomizing more coefficients). Even secret sharing the nonzero coefficients also disclosed the shape of the image as most of the high frequency bands consist of unhidden zero values. A satisfactory result was found only by secret sharing all the wavelet coefficients. This naive scenario, however, did not provide any compression.

Note that an alternative scheme can be to probabilistically select a coefficient to secret share. Although this scheme can provide confidentiality by not secret sharing all the coefficients, it is not suitable for compression as it can break the correlation among neighbouring coefficients.

4   Proposed Scheme

In this section, we first describe how we integrate secret sharing with the compression pipeline, and then explain how the proposed scheme can be useful for cloud-based im-

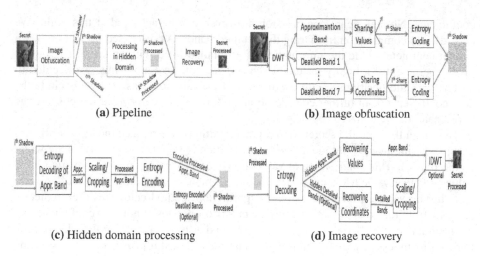

(a) Pipeline   (b) Image obfuscation

(c) Hidden domain processing   (d) Image recovery

Fig. 2.

age processing. Our scheme utilizes the fact that one can hide an image by hiding either the colors or the positions of the pixels. In our case, the act of hiding the pixel positions, however, is more difficult since we want to perform cropping operations in the hidden domain. To obtain compression, we utilize the property of wavelet transformation that the approximation wavelet band contains most of the information about the image, and the detailed wavelet bands contain multiresolution information. Figure 2a shows our scheme in three steps: image obfuscation, hidden domain processing, and image reconstruction. The image obfuscation step integrates secret sharing with the compression pipeline and creates n number of shadow images; the hidden domain processing step performs scaling/cropping operations on the shadow images, and the image recovery step recovers the secret scaled/cropped image from at least k scaled/cropped shadow images. In the following sections, we discuss these three steps in detail.

4.1   Image Obfuscation

As shown in Figure 2b, we integrate secret sharing after the quantization step but before the entropy encoding step of the compression algorithm. By applying wavelet transformation to an given image, we first get the approximation band and the detailed bands. Then, we apply quantization to the nonzero coefficients to obtain quantized coefficients, which will be input to the secret sharing process. Note that we do not perform run length encoding as it cannot compress a set of elements having random values (which is the case in the approximation band) or having unknown coordinates (which is the case in the detailed band).

To be able to achieve security and compression, we adopt different sharing approaches for the approximation band and the detailed bands. Since the approximation band contains a maximum of information, we secret share all the coefficients (i.e., nonzero quantized coefficients and zero coefficients) of this band. To break the correlation among the coefficients, we use different secret sharing polynomials for different coefficients. No significant compression, however, can be achieved at this point. We consider the compression in sharing the detailed bands. As discussed in Section 3, we cannot completely hide the information of a detailed band without secret sharing all of its coefficient – a highly unsuitable choice for compression. Thus, we decided to secret share the positions of the coefficients. To obtain compression, we exclude the zero coefficients from further consideration, and to obtain secrecy, we secret share the coordinates of the nonzero coefficients.

To efficiently store the secret shared pixel positions, we replace the xy-coordinates of a pixel with a single unique integer value (which is $Wy + x$, where W is the width of the image). This single integral value, however, is very large. To store such large values efficiently, we use a prediction method, which stores the difference of any two neighboring values. Mathematically, this prediction method calculates the prediction array A' from the array A that contains the pixel coordinates of nonzero coefficients, by the formula $A'[i] = A[i] - A[i + 1]$, for $i > 1$, and $A'[i] = A[i]$, otherwise. In addition to resulting in less storage bits, this method is also efficient for further encoding as any two elements of the prediction array can be closer to each other.

As the final step, we perform entropy encoding first on the secret shared coefficients of the approximation band, and then on the nonzero coefficients of the detailed band and on their secret shared coordinates. The entropy encoded j^{th} share of the coefficients of the approximation band, and the entropy encoded j^{th} share of the coordinates of the nonzero coefficients of the detailed bands, constitute the jth shadow image. Note that, for simplicity, our implementation uses the huffman entropy encoder. One can, however, choose any other entropy encoder.

4.2   Processing in Hidden Domain

This step performs certain basic image operations, such as scaling and cropping, on the shadow images (Figure 2c). Since the locations of the detailed bands are unknown, the processing is mainly done on the approximation band. Scaling/cropping of the detailed bands will be performed during the image recovery step, after finding the coordinates of their coefficients.

Depending on the scaling/cropping request, the compressed approximation band is first partially decompressed by performing entropy decoding on the compressed bits. After this step, the shared wavelet coefficients are known, and scaling/cropping operations can be performed on them. Note that we do not perform entropy decoding on the detailed bands since they will not be scaled/cropped at this stage.

In the case of a cropping request, mapping the region-of-interest (ROI) to the approximation band is first performed. Then, the coefficients of the ROI are selected, and the selected coefficients are entropy encoded. This entropy encoded ROI region and the entropy encoded detailed bands are sent to the image recovery step.

In the case of a scaling down request, DWT is performed on the approximation band in the hidden domain. The scaled down approximation band is then entropy encoded, and sent to the image recovery step.

In the case of a scaling up request, we do not process the shadow images. Rather, we send the approximation band and detailed bands to the image recovery step, since the required IDWT (i.e., inverse DWT) can be performed after finding the coordinates of the coefficients of the detailed bands.

4.3 Image Recovery

This step recovers the secret scaled/cropped image from k scaled/cropped shadow images. As shown in Figure 2d, we first decode the entropy encoded approximation band and entropy encoded detailed bands of each shadow image. Then, the position of the coefficients of the detailed bands and the values of the coefficients of the approximation band are reconstructed from their k shares using Lagrange interpolation. After these steps, a scaled down secret image can be available. However, in the case of cropping, we first crop the detailed bands, and then find the cropped image by using IDWT on the cropped approximation band (available from the hidden domain processing step) and the cropped detail bands. Similarly, a scaled up image can be found by first finding the secret image using IDWT on the approximation band and the detailed bands, and then performing higher level DWT on the secret image (assuming that the detail bands are zeros).

4.4 Secure Cloud-Based Image Scaling/Cropping

In this section, we discuss how our scheme can be useful for securing image scaling/cropping operations in cloud-based imaging. Our scheme can be used in a cloud-based framework that consists of a server (which holds the secret image, and will perform image obfuscation), n datacenters (which will perform scaling/cropping operations), and a client (who is authorized to recover the secret image from the shadow images). Our framework is based on the assumption that (i) the server and client are the trusted entities, (ii) the datacenters are connected among themselves and with the client via a high speed network, (iii) the inter-group communication among the datacenters is regulated, and (iv) an adversary cannot access k or more datacenters.

Table 1. Comparison between size of a secret image compressed by CDF wavelet-based compression and size of a shadow image compressed by our scheme

Name	Dimension	Size of uncompressed image	Size of compressed image (using CDF)	Size of compressed image (using our scheme)
Coin	206×244	147 KB	19.9 KB	19.86 KB
Lena_1	512×512	768 KB	122 KB	119 KB
Lena_2	204×204	122 KB	23.3 KB	24.1 KB
Pepper	384×512	576 KB	69.4 KB	61.3 KB

5 Results and Analysis

We evaluated our scheme by simulating the server, cloud datacenters, and a client in a PC powered by an Intel Core i3 2.20 GHz processor with 4 GB of RAM. The $(3, 5)$ Shamir's secret sharing was implemented in C, which was then linked to MATLAB by creating the mex file. The compression pipeline was implemented in MATLAB. The C and MATLAB were run on the Ubuntu 14.04 LTS platform. We used four different test images: two Lena images called Lena_1 and Lena_2, one coin image called Coin, and one pepper image called Pepper (Table 1).

Figure 3 shows the secret image, the compressed shadow images, and the reconstructed image. As shown in the figure, our scheme provides perceptual security, since nothing about the secret image can be known by looking at a shadow image. As demonstrated in Figure 4a and Figure 4b, our scheme can perform scaling and cropping operations in the hidden domain, without disclosing any critical information to a cloud datacenter.

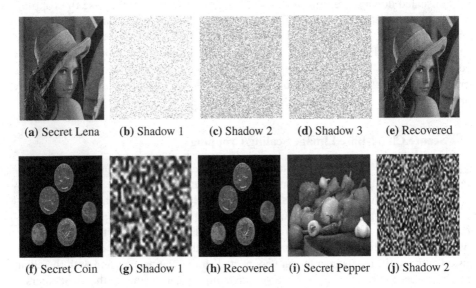

(a) Secret Lena   (b) Shadow 1   (c) Shadow 2   (d) Shadow 3   (e) Recovered

(f) Secret Coin   (g) Shadow 1   (h) Recovered   (i) Secret Pepper   (j) Shadow 2

Fig. 3. Secret sharing CDF wavelet-based compressed image. The shadow images are produced by applying IDWT on the shared approximation band (having hidden coefficient values) and shared detailed bands (having hidden coefficient coordinates).

5.1   Security Analysis

Confidentiality: By inheriting the perfect secrecy property of Shamir's secret sharing, our scheme guarantees high data confidentiality. The use of secret sharing ensures that an adversary having access to less than k shadow image can never know the value of a coefficient in the approximation band or the coordinate of a coefficient of a detailed band. Thus, even though the coordinates of the coefficients of the approximation band and the values of the coefficients of the detailed band are available, the image information cannot be known. The only way to know the image is to guess either the coefficient values of approximation band or the coefficient coordinates of the detailed bands.

From Section 3, we know that information about the image can be leaked if any of the band are known. Therefore, if p is the prime number used to share the coefficient

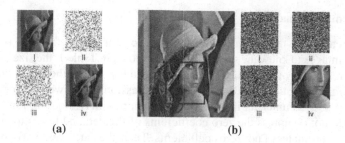

(a)   **(b)**

Fig. 4. Scaling and cropping of a compressed image in the hidden domain. Figure (a).(i) is the required scaled image, (a).(ii) and (a).(iii) are shadow scaled images, and (a).(iv) is the secret scaled image. Figure (b).(i), (b).(ii), and (b).(iii) are shadow cropped images, and (b).(iv) is the secret cropped image.

values of approximation band and q is the prime number used to share the coefficient coordinates of a detailed band, then the information of the image can be known with a probability $\mathrm{MIN}(\frac{1}{p^{W \times H}}, \frac{1}{q^N})$ respectively, where W is the width of image, H is the height of the image, and N is the number of nonzero coefficients of the detailed band having the minimum number of nonzero coefficients.

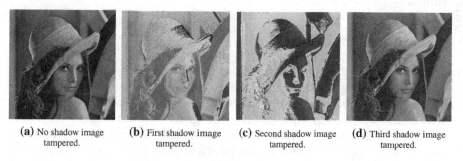

(a) No shadow image tampered.   **(b)** First shadow image tampered.   **(c)** Second shadow image tampered.   **(d)** Third shadow image tampered.

Fig. 5. Images recovered from tampered shadow images.

Integrity: By inheriting the property of (k, n) secret sharing, our scheme ensures the integrity of images. The $k < n$ condition provides $\binom{n}{k}$ different ways of recovering the secret image. Therefore, if any adversary changes the color values of at most $n - 1$ shadow images, then the recovered images from the tampered shadow images will differ from each other and from the secret image (as shown in Figure 5). As a result, by comparing at most $\binom{n-1}{k}$ recovered images, the client can detect tampering.

However, if all n shadow images are tampered by obeying the homomorphic property of secret sharing, then all the tampered recovered images will be alike. Thus, in this case, we cannot detect tampering. Similarly, if $n = k$, then tampering with even one share image is not detectable as there can be a maximum of one recovered image. Therefore, for applications requiring data integrity, we recommend $n > k$.

5.2   Performance Analysis

Data Overhead: The main data overhead of our scheme comes from the requirement of k shadow images to know the secret image. As shown in Table 1, the size of a shadow image compressed by our scheme is nearly equal to the size of the image compressed by non-secure, conventional wavelet-based compression, since we (i) do not use run length encoding to counter its inefficiency to compress neighbouring random values and (ii) we do not compress the zero coefficients of the detailed bands to compensate compressing coordinates of nonzero coefficients. Thus, the data overhead of our scheme is nearly $(k - 1)$ times more than the conventional scheme. Note that since our scheme is suitable for a hybrid streaming framework (scaling down by server, scaling-up by client, and cropping by both server and client), our scheme incurs more data overhead than the commonly used server-side scaling/cropping framework.

Computation Overhead: The computation overhead at the client end is also increased since the secret is reconstructed at the client-end. In addition, the client crops the detailed bands, and performs the IDWT operation for image cropping/scaling-up and DWT for image scaling up. For example, the client needs to work 8 ms, 590 ms, and 1930 ms more to scale down (by one level), crop (by 100×100), and scale-up (by one level) the Lena_2 image, respectively. We believe that this computation overhead is insignificant for a client device.

6   Conclusion and Future Work

The task of processing a compressed and encrypted image without affecting the compression efficiency and image quality is challenging. While compression of an image tries to exploit the spatial coherence of the image, encryption focuses on destroying the coherence. In this paper, we addressed this challenge for wavelet based compression by first separating the detail and approximation coefficient bands, and then secret sharing the coefficient values of the approximation band and coefficient coordinates of the detail bands. As a result, scaling down the image and cropping of the approximation band were possible in the hidden domain. Experiments and analysis showed that our scheme is highly secure, provides comparable compression, and results in acceptable overheads in a cloud-based streaming framework.

In future, the proposed work can be extended in a number of ways. For example, one approach can be to support hidden domain processing in other compression schemes, such as DCT-based compression scheme. Similarly, other image processing techniques, such as image segmentation and classification, can also be performed in hidden domain.

Acknowledgement. Majority of this work was performed when the first author, Kshitij Kansal, was a summer intern in the Department of Applied Computer Science at the University of Winnipeg, MB, Canada during May-June 2014. Third author Atrey's research contribution is supported in parts by the Natural Sciences and Engineering Research Council of Canada (Grant No. 371714) and the University at Albany - State University of New York (Grant No. 640075). Second author, Manoranjan Mohanty, an ERCIM Alain Bensoussan fellow at SICS Swedish ICT, would like to thank both ERCIM and SICS for supporting his part of work.

References

1. Sertel, O., Kong, J., Shimada, H., et al.: Computer-aided prognosis of neuroblastoma on whole-slide images: classification of stromal development. Pattern Recognization 42, 1093–1103 (2009)
2. Medical imaging in the cloud. Online Report (2012), http://www.corp.att.com/healthcare/docs/medical_imaging_cloud.pdf
3. Mohanty, M., Ooi, W.T., Atrey, P.K.: Scale me, crop me, know me not: Supporting scaling and cropping in secret image sharing. In: Proceedings of the 2013 IEEE International Conference on Multimedia & Expo, San Jose, USA (2013)
4. Tsai, M.-H., Chen, C.-C.: A study on secret image sharing. In: Proceedings of the 6th International Workshop on Image Media Quality and its Applications, Tokyo, Japan (2013)
5. Shamir, A.: How to share a secret. Communication of the ACM 22, 612–613 (1979)
6. Thien, C.-C., Lin, J.-C.: Secret image sharing. Computers & Graphics 26, 765–770 (2002)
7. Bai, L.: A reliable (k, n) image secret sharing scheme. In: Second IEEE International Symposium on Dependable, Autonomic and Secure Computing, Indianapolis, USA, pp. 31–36 (2006)
8. Li, L., El-Latif, A.A.A.: A novel secret image sharing scheme based on chaotic system. In: Proceedings of SPIE Fourth International Conference on Digital Image Processing, Kuala Lumpur, Malaysia (2012)
9. Ulutas, M., Ulutas, G., Nabiyev, V.V.: Medical image security and EPR hiding using Shamir's secret sharing scheme. Journal of Systems and Software 84, 341–353 (2011)
10. Alharthi, S.S., Atrey, P.K.: Further improvements on secret image sharing scheme. In: Proceedings of the 2nd ACM Workshop on Multimedia in Forensics, Security and Intelligence, Firenze, Italy, pp. 53–58 (2010)
11. Engel, D., Stütz, T., Uhl, A.: A survey on JPEG2000 encryption. Multimedia Systems 15(4), 243–270 (2009)
12. Martin, K., Lukac, R., Plataniotis, K.N.: Efficient encryption of wavelet-based coded color images. Pattern Recognition 38(7), 1111–1115 (2005)
13. Seo, Y.-H., Kim, D.-W., Yoo, J.-S., Dey, S., Agrawal, A.: Wavelet domain image encryption by subband selection and data bit selection. In: Proceedings of the World Wireless Congress, San Francisco, USA (2003)

MAP: Microblogging Assisted Profiling of TV Shows

Xiahong Lin[1], Zhi Wang[2], and Lifeng Sun[1]

[1] Tsinghua National Laboratory for Information Science and Technology
Department of Computer Science and Technology, Tsinghua University
[2] Graduate School at Shenzhen, Tsinghua University
{lin-xh12@mails., wangzhi@sz., sunlf@}tsinghua.edu.cn

Abstract. Online microblogging services that have been increasingly used by people to share and exchange information, have emerged as a promising way to profiling multimedia contents, in a sense to provide users a socialized abstraction and understanding of these contents. In this paper, we propose a microblogging profiling framework, to provide a social demonstration of TV shows. Challenges for this study lie in two folds: First, TV shows are generally *offline*, i.e., most of them are not originally from the Internet, and we need to create a *connection* between these TV shows with online microblogging services; Second, contents in a microblogging service are extremely noisy for video profiling, and we need to strategically retrieve the most related information for the TV show profiling. To address these challenges, we propose a *MAP*, a microblogging-assisted profiling framework, with contributions as follows: i) We propose a joint user and content retrieval scheme, which uses information about both actors and topics of a TV show to retrieve related microblogs; ii) We propose a social-aware profiling strategy, which profiles a video according to not only its content, but also the social relationship of its microblogging users and its propagation in the social network; iii) We present some interesting analysis, based on our framework to profile real-world TV shows.

Keywords: Online Microblogging, TV Show Profiling, Social Network, Data Visualization.

1 Introduction

Recent years have witnessed an increasing popularity of online microblogging services (e.g., Twitter and Weibo), and the rapid convergence between the online microblogging service and other multimedia services including online video streaming [13]. This new trend makes the online microblogging service a promising way to *profiling* multimedia contents, i.e., using the information from the microblogging systems to abstract, demonstrate and enrich multimedia contents [6]. In this paper, we study profiling one of the most important multimedia types — TV shows (videos) that are generally received by users on TVs, with online microblogging services.

Profiling TV shows is important to both content providers and users [10]. On one hand, an interesting profile of a TV show will allow a content provider to demonstrate the TV show more effectively, e.g., more users can be attracted to watch the TV show,

X. He et al. (Eds.): MMM 2015, Part I, LNCS 8935, pp. 442–453, 2015.

indicating more subscriptions from the content provider. On the other hand, an illustrating profile of a TV show also gives a user an efficient way to know about the TV show before actually watching it, improving the whole show viewing experience.

Today, it has become a norm rather than an exception, for users to share and receive information and use online social networks to search multimedia contents [1]. In this personal and social context, even TV show profiles have to be highly personalized and social-aware. Traditional video profiling is generally based on video content only [3], e.g., the profile is composed by the video title, actor information, etc., and has the following fundamental drawbacks: (1) For a content provider, the video profile is generated without any consideration about the *social effect* the profile may make, e.g., how the video is going to attract users in the online social network and how the profile will be correlated with other videos. (2) For users, the profile is *static* and composed in an one-size-fit-all manner, i.e., the same video profile is provided to all users without any user-specific information.

To address these challenges, we propose a TV show profiling framework called MAP by exploring the new design space of using online social network information. Our contributions in this paper are summarized as follows: (1) We design a joint user and content information retrieval scheme. We use information about both actors and topics in a TV show to retrieve microblogs from a microblogging system. (2) We propose a social-aware profiling scheme, which profiles a TV show according to not only its content, but also the social relationship of its microblogging users and its propagation in the social network. (3) We present some interesting results and analysis, using our framework to profile real-world TV shows.

The rest of this paper is organized as follows. We present the information retrieval strategy and the microblogging-assisted profiling framework in Sec. 2. Based on this framework, we present several analytical results in Sec. 3, using data from real-world TV show and microblogging systems. Finally, we conclude the paper in Sec. 4.

2 Profiling TV Shows with Microblogs

In this section, we present our profiling framework. First, to effectively utilize the noisy microblog information, we retrieve the most informative microblogs for a TV show, by jointly considering actors and topics of the show; Second, to efficiently provide a profile both informative to a TV show content provider and users, we create a profile for TV shows, from aspects of user, content, social relationship and propagation.

2.1 Background and Data Collection

We have collected data from a TV show service provider *BesTV* and an online microblogging service provider *Tencent Weibo* for our study. For simplicity, we refer to the TV show service as V, and the microblogging service as M. We have collected TV shows from the TV show company V. Our dataset contains $233,804,070$ valid user viewing records of 69 TV shows during June 1 through December 30, 2011. We have also collected the meta information of these TV shows, including the title, category, and the actors in the TV shows. In Table 1, we list the representative TV shows used in our study. We will use the indices (v1, v2, ..., v15) to denote these TV shows in this paper.

Table 1. Representative TV shows studied

ID	TV Show title	Category	Leading Actors	# microblogs	# microblog users
v1	*Aiqing Shuixingle*	Love/Idol	Stephy Qi/Tiffany Tang	571, 227	388, 109
v2	*Bubu Jingxin*	Love/Costume	Shishi Liu/Nicky Wu	2, 748, 223	1, 772, 973
v3	*Meiren Xinji*	Love/Costume	Ruby Lin/Mi Yang	169, 513	147, 812
v4	*Nanrenbang*	Idol/Modern	Di Yao/Lei Huang	1, 141, 362	844, 655
v5	*Qianshan Muxue*	Love/Idol	Yinger/Hawick Lau	271, 371	190, 288
v6	*Qingshi Huangfei*	Love/Costume	Ruby Lin	729, 010	544, 439
v7	*Xuebao*	War/Love	Zhang Wen	83, 835	67, 561
v8	*Wuxiekeji*	Love/Idol	Stephy Qi/Tiffany Tang	158, 534	116, 436
v9	*Aiqing Gongyu*	Comedy/Love	Loura/Jean Lee	1, 036, 144	775, 171
v10	*Aishang Zhameile*	Love/Idol	Cyndi Wong	78, 522	59, 080
v11	*Shachun*	Family/Tragedy	Helen Tao	56, 524	47, 582
v12	*Duanci*	War/Suspense	Lei Tong	14, 509	11, 410
v13	*Duanhounu*	War/Historical	Yunxiang Gao	5, 370	4, 260
v14	*Huangtuteng*	Love/Costume	Hans Zhang/Bing Bai	66, 881	48, 400
v15	*Luohun Shidai*	Love/Family	Zhang Wen/Di Yao	3, 544, 006	2, 789, 363

2.2 Informative Microblogs Retrieval

Microblogs from people are intrinsically noisy, as users have been using such online microblogging services to exchange all kinds of information. Besides, among these microblogs posted by users, only a little fraction of them are about TV shows, e.g., in popular microblogging systems including Twitter, only a very small fraction of microblogs are about TV shows[1]. It is a must to find the most informative contents when we use microblogs to profile TV shows. We propose to jointly use both actors and topics of a TV show to retrieve the most informative microblogs.

Microblogs from Actors. In our study, we focus on the professional TV shows, which are produced by large companies, instead of user-generated videos [7], as such professional videos (e.g., a movie or TV show) are the mainstream type of videos on a video service. For such a TV show v, we are able to find an actor list $\mathbf{A}_v$, in which actors of the show are included. According to our measurement results, there is over $1/3$ of the leading actors having a microblogging account on average. We collect microblogs posted by accounts in $\mathbf{A}_v$, if they have social accounts. For example, a TV show $v6$ in our study named "*Qingshi Huangfei*" has an actor named "Ruby Lin", who has a social account ID on the microblogging system M called "linxinru", and we have collected microblogs posted by this ID for profiling this TV show.

Microblogs from Topics. For a TV show v, we are also able to find a set of topics $\mathbf{O}_v$, including the title of the TV show, category, abstract, etc. In our study, such topics are used to search for microblogs, in a keyword-matching manner, i.e., if the content of a microblog contains any keyword in $\mathbf{O}_v$, this microblog will be used for profiling

---

[1] http://en.wikipedia.org/wiki/Twitter

TV show v. For example, the same TV show $v6$ "*Qingshi Huangfei*" has the leading roles "Mafuya", "Mengqiyou", "Liuliancheng", which are used as the topics to retrieve microblogs.

Using the above microblogs as "seed" microblogs, we further collected microblogs that are repost and comments to them. In summary, we have collected overall $23,963,680$ valid microblogs posted by $11,685,768$ users in M. In Table 1, we show the number of microblog users and microblogs for each of the representative TV shows. Each microblog record contains the follows ID, name, IP address of the publisher, time stamp when the microblog was posted, IDs of the root microblog if it is a reposted microblog, and contents of the microblog. Interested readers are referred to our previous studies on collecting the microblog traces [11,12].

2.3   MAP: Microblogging-Assisted Profiling of TV Shows

Based on microblogs, user profiles, and social connections collected using the retrieval scheme above, we profile TV shows from the following aspects. (1) *User aspect:* We present the demographics of microblogging users of a TV show, and their preferences; (2) *Content aspect:* We present the sentiment analyses of TV shows by the microblog content posted to the shows, and the "social network" of TV shows; (3) *Social relationship aspect:* We present the social topology of microblogs views of a TV show, as well as the social network and social influence of actors of TV shows, who have online microblogging accounts; (4) We present the propagation of microblogs users across different TV shows, i.e., how users' attention is shifting from a TV show to another.

Next, we present a case study using MAP, including the insights from the observations.

3   Insights Learnt from MAP

In this section, we present several results based on our profiling framework, from the aspects of user, content, social relationship and social propagation.

3.1   Demographics and Preference of Viewers

Demographics of Microblogging Viewers. Using the microblogging user information, we are able to demonstrate TV shows from a user aspect. Fig. 1(a) shows the age distribution of normal users. It forms a normal-like distribution, and we observe that the average age is 20, indicating the major of the users are quite young. Fig. 1(b) shows the top 15 province-level regions where the microblog users are located. Each bar in the figure is a top-10 participation index, defined as

$$PI_i = \frac{UN_i - UN_{10}}{UN_{10}} \tag{1}$$

where UN_i is the numbers of users in region i. From this demographics profiling, we are able to show the regional distribution of users. For example, we observe that for this particular TV show provider, Guangdong Province ranks the first, i.e., it has the most microblog users. In summary, we observe that the young users in top-tier cities in China are the major viewers of these TV shows.

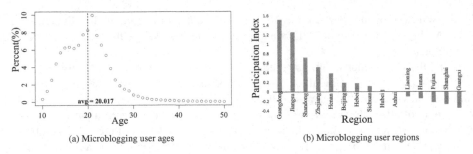

(a) Microblogging user ages

(b) Microblogging user regions

Fig. 1. Demographic of users

User Preference Clustering. A TV show provider gives 3 labels to each TV show to categorize its content. We use such content labels to infer users' preferences of video contents. According to the microblogs posted by a user, we are able to assign each user with a weight list of content labels. Using the weighted label lists of users, we are able to calculate a "distance" between any two users. Then, we cluster these users using a K-means clustering algorithm [5] using based on the label distance.

In Fig. 2(a), we plot the social connections between users that clustered into one group. Each circle represents a user, and a connection between two users indicates that one of them is following the other. We observe that in this figure users are closely connected to each other with a large clustering coefficient [9], when they are clustered into the same group by K-means according to their preference.

Using the same approach to cluster users, we further find that users' preference can be indicated from who they are socially connected in a microblogging system. In Fig. 2(b), we have plotted several VIP microblog users, i.e., *Tencent Video*, *Hunan TV*, and *Tencent Entertainment*. We observe that when users are clustered into the same group, they only follow the VIP users in that group, but have no connection with other users in the same group. This observation indicates that a TV show provider is able to identify users that may be interested in a TV show, by looking at the followers the VIP accounts.

3.2 Sentiment and Social Network of TV Shows

Online microblogging services also allow us to explore the sentiment and social network of TV shows.

Sentiment Analysis. Microblogs allow us to use the content of microblogs for sentiment analysis, i.e., we label each microblog "positive", "negative", or "non-sentiment", according to its content. After analyzing all the posts, we get the following observations: (1) Initial microblog posts on TV shows contain more emotional contents than reposts and comments, as illustrated in Fig. 3(a); (2) There are much more positive microblogs on TV shows than negative ones, as illustrated in Fig. 3(b).

Next, we study the sentiment distribution of microblogs in different TV shows. We find that most of them have a higher positive proportion except for two TV shows.

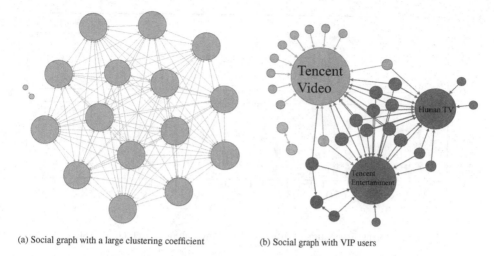

(a) Social graph with a large clustering coefficient

(b) Social graph with VIP users

Fig. 2. Social relationship among users sharing similar preference

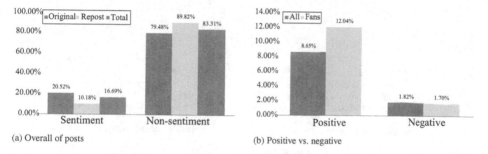

(a) Overall of posts

(b) Positive vs. negative

Fig. 3. Sentiment distribution of microblogs on TV shows

As shown in Fig. 4, each bar is the fraction of positive microblogs versus the TV show rank, and the curve is the view number of each TV show versus the show rank. We observe that a TV show that has too many positive microblogs or too many negative microblogs tend to have a small number of views. The reason may be that highly-viewed shows will attract more diverse discussions on the microblogging system.

"Social Network" Between TV Shows. Users can post microblogs about more than one TV shows, as a result, we are able to connect different TV shows by their common microblog users. Using this idea, we create a social network of TV shows, i.e., two shows are connected if they share a microblog user. In this graph, Fig. 5 shows the statistics of nodes (TV shows). Fig. 5(a) illustrates the degree distribution of TV shows. We observe that about 90% of the shows have a degree higher than 40, and the mean is 45.8, indicating that the graph is almost fully-connected. Fig. 5(b) illustrates the CDF of local clustering coefficients of the shows. We observe that the TV show graph has a high level clustering coefficient, indicating that these TV shows have many common users.

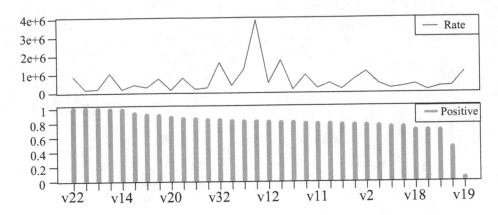

Fig. 4. Sentiment distribution of different TV shows, along with their TV rating

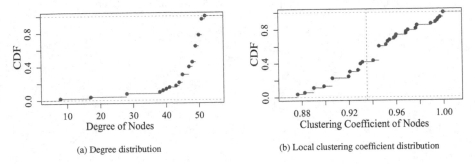

(a) Degree distribution (b) Local clustering coefficient distribution

Fig. 5. Statistics of Nodes

Next, we study the community structure of the social network of TV shows, as illustrated in Fig. 6. Since the average clustering coefficient of the network is 0.94, which is 2x of the same scale of a random network. The average diameter of the graph is 1.12, which is at the same scale of a random network. Further, we study its modularity using a heuristic approach [2]. We have modularity value of 0.05 (Using the community structure detecting algorithm proposed by Du et al. [4]) for the same social network of the TV shows. The very small value indicates that the social network of TV shows does not have a clear community structure [8].

We also explore the degree distribution of the network. Based on the topology, we have that the degree distribution of nodes in the TV show social network follows a power-law distribution $f(x) = -52x^{-0.5} + 58$ ($R^2 = 0.98$), indicating that the TV show social network can be regarded as a scale-free network[2].

---

[2] http://en.wikipedia.org/wiki/Scale_free_network

3.3   Social Networks of Viewers and Actors

An online microblogging service also allows us to explore the social relationship between the microblog users who have posted microblogs on the TV shows.

Social Topology Among Viewers. We first study the following relationship among microblog users. As shown in Fig. 7, most of the users (about 76%) are isolated (with no social connection to other users), and a small fraction (about 24%) of them have social connections.

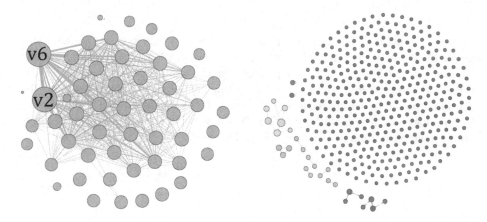

Fig. 6. Communities in the social network of TV shows

Fig. 7. Social graph among users of TV show v1

Social Networks among Actors. Today, online microblogging services are not only used by ordinary users, but also popularly used by celebrities and actors. We next study the social networks of actors of TV shows. Fig. 8(a) is a graph created according to microblogs posted by actors: there is an edge between two actors if one of them has posted a microblog on a TV show in which the other acts in. In this figure, we observe that some actors tend to help others to broadcast their shows, even they are not acting in those shows. In Fig. 8(b), we have plotted the following relationship between actors on the microblogging system. We observe that actors also share similar social topologies as ordinary users [1]. In Fig. 8(c), we plot an intersection of (a) and (b), i.e., only edges that appear in both graph (a) and (b) are plotted. We observe very few edges remaining in this figure, indicating that it is very common that actors post microblogs on shows of other actors they are not following or friending on a microblogging system.

Social Influence of Actors. Next, we study the social influence of actors to the shows they star in.

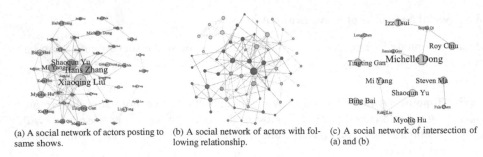

(a) A social network of actors posting to same shows.

(b) A social network of actors with following relationship.

(c) A social network of intersection of (a) and (b)

Fig. 8. Social networks among actors

Fig. 9(a) shows the CDF of the fraction of microblogs by actors over all microblogs about a TV show. We have the following observations: (1) Actors post a large fraction of microblogs in a TV show; (2) Though more than half actors post more than 5% TV-related microblogs, there is almost no actor posting more than 20%. Next, we study the microblogs posted by fans of the actors of a show. Fig. 9(b) plots the CDF of fractions of microblogs posted by fans of actors in a TV show. We observe that the fans of actors also input a large portion of microblogs for a TV show.

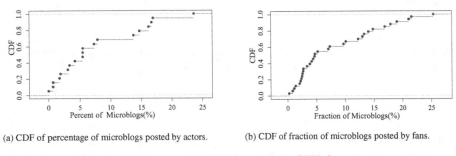

(a) CDF of percentage of microblogs posted by actors.

(b) CDF of fraction of microblogs posted by fans.

Fig. 9. Social influence from actors of TV shows

Finally, we study the heterogeneous influence of different actors. We calculate actors' average and variance influence on the shows they act in. The results are presented in Fig. 10, with their fans number. (1) We observe that for most of the actors, they tend to have a higher average influence if they a larger number of fans. (2) The influence variance is generally small.

3.4   Viewer Propagation across TV Shows

Viewers of TV shows generally enjoy different shows at different periods. Based on our profiling scheme, we are able to study the propagation of users across different shows.

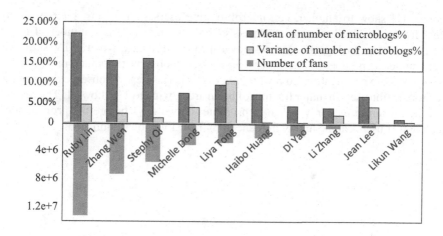

Fig. 10. Actors' social influence and number of fans

Propagation of Viewers Over Time. We study how users join the same show played at different times. In this experiment, shows are played in two different periods, denoted as the first round and second round. In Fig. 11, each bar plots the number of users in the first round, second round, and both rounds. We observe that though the number of users in both rounds accounts a small proportion. This observation indicates that new users tend to join a TV show if the show is played again.

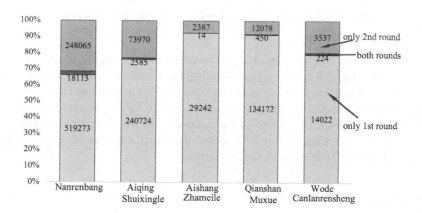

Fig. 11. Propagation of viewers over time

Propagation of Viewers Across Shows. Next, we study user propagation between TV shows. In Fig. 12, each node represents a TV show, and its diameter indicates the number of in-degrees. There is a directed edge between two shows if users "propagate" from one TV show to another, i.e., microblogging users start to post microblogs to another from one TV show. The thickness of an edge indicates the number of users being propagating. In Fig. 12(a), we study a propagation case when a TV show v2 has a plummeting number of microblogging users. We observe that much more social-network users are propagating from v2 to other shows. Besides, we observe that among these propagation users, a major fraction of them are moving to TV show v1. The reason is that the content provider V began to promote v1 two days before the propagation, indicating that users can be highly influenced by new TV shows published.

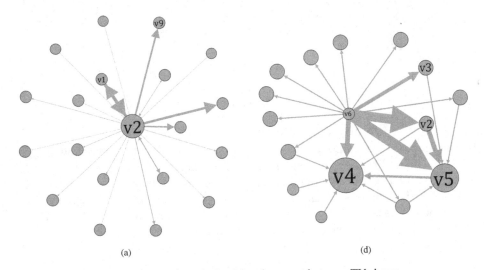

(a)   (d)

Fig. 12. Propagation of microblogging users between TV shows

In Fig. 12(b), we study another type of propagation. This figure illustrates a case when v6 is about to finish (delivering its last episode) and v4 and v5 are the newly released shows scheduled on the next day. We observe that a dominate fraction of users are propagating from v6 to v5. The reason is that v6 and v5 are in the same TV category. On the other hand, v4 is of a different TV category, and we observe that much fewer users are propagating from v6 to v4. This observation indicates that that users tend to propagate between TV shows with similar content.

4   Concluding Remarks

In this paper, we study the profiling of TV shows using online microblogging services, in a sense to provide users with a socialized profile of TV shows. It is challenging to profile TV shows with microblogs since TV shows are generally offline, and an effective information retrieval scheme is in demand to find the most useful microblogs for the profiling; meanwhile, information on a microblogging system is noisy for the profiling and an efficient profiling scheme is in demand. To address these two challenges, we propose a joint actor and topic information retrieval scheme for searching microblogs to profile a particular TV show. Based on that, we propose *MAP*, a microblog-assisted profiling framework, in which a TV show is profiled from the aspects of user, content, social relationship and social propagation. Using *MAP*, we profile real-world TV shows and present several analysis results, which are not only interesting to the ordinary viewers, but also important to content providers for better operation.

Acknowledgments. This research is supported in part by NSFC under Grant No. 61272231, 61472204, and 61402247, SZSTI under Grant No. JCYJ20140417115840259, and Beijing Key Laboratory of Networked Multimedia. We thank Tsinghua-Tencent Joint Lab and BesTV for providing the traces used our study.

References

1. Benevenuto, F., Rodrigues, T.: Characterizing user behavior in online social networks. In: Proceedings of the 9th ACM SIGCOMM, ACM (2009)
2. Blondel, V.D., Lambiotte, R.: Fast unfolding of communities in large networks. J. Stat. Mech. Theor. Exp. 2008(10), P10008 (2008)
3. Dimitrova, N., Zhang, H.: Applications of video-content analysis and retrieval. IEEE Multi-Media 9(3), 42–55 (2002)
4. Du, H., Feldman, M.W., Li, S., Jin, X.: An algorithm for detecting community structure of social networks based on prior knowledge and modularity. Complexity 12(3), 53–60 (2007)
5. Hartigan, J.A., Wong, M.A.: Algorithm as 136: A k-means clustering algorithm. J. R. Stat. Soc. 28(1), 100–108 (1979)
6. Holotescu, C., Grosseck, G.: M3-learning-exploring mobile multimedia microblogging learning. World Journal on Educational Technology 3(3), 168–176 (2011)
7. Milliken, M.C., Gibson, K.: User-generated video and the online public sphere: Will youtube facilitate digital freedom of expression in atlantic canada? American Communication Journal 10(3), 1–14 (2008)
8. Newman, M.E.: The structure and function of complex networks. SIAM Review 45(2), 167–256 (2003)
9. Soffer, S.N., Vázquez, A.: Network clustering coefficient without degree-correlation biases. Physical Review E 71(51), 057101_1–4 (2005)
10. Wakamiya, S.: Lee, R. Twitter-based tv audience behavior estimation for better tv ratings. DEIM Forum (2011)
11. Wang, Z., Sun, L., Chen, X., Zhu, W., Liu, J., Chen, M., Yang, S.: Propagation-based Social-aware Replication for Social Video Contents. In: ACM International Conference on Multimedia, Multimedia (2012)
12. Wang, Z., Sun, L., Zhu, W., Yang, S., Li, H., Wu, D.: Joint Social and Content Recommendation for User Generated Videos in Online Social Network. IEEE Transactions on Multimedia 15(3), 698–709 (2013)
13. Xu, T., Chen, Y.: Cuckoo: towards decentralized, socio-aware online microblogging services and data measurements. In: Proceedings of the 2nd ACM HotPlanet. ACM (2010)

Improved Rate-Distortion Optimization Algorithms for HEVC Lossless Coding

Fangdong Chen and Houqiang Li

CAS Key Laboratory of Technology in Geo-spatial Information Processing and
Application System
University of Science and Technology of China, Hefei, China
fangdong@mail.ustc.edu.cn,lihq@ustc.edu.cn

Abstract. To avoid distortion, the quantization is not implemented on residues for lossless mode in HEVC. As a result, the conventional lambda model in Rate-Distortion Optimization (RDO), where lambda is related to the quantization parameter (QP), is unreasonable for lossless coding. This paper first demonstrates the role that lambda value plays in the rough RDO of HEVC lossless coding, and a Simulated Annealing algorithm based approach is proposed to obtain the most appropriate lambda for each largest coding unit. Considering the computational complexity, the other simplified method using least square errors prediction is proposed to improve the rough RDO process. Experimental results reveal that on top of the lossless coding mode, the improved method offers the performance with a 1.0%, 1.3% and 1.1% bit-rate reduction on average for Random-access Main, Low-delay B Main and Low-delay P Main configurations, respectively, while brings negligible increases of computational complexity in the encoder.

Keywords: lossless coding, HEVC, lambda model, RDO.

1 Introduction

In many application areas, such as medical imaging and remote sensing, lossless video coding is playing a more prominent role. A set of standards is being constituted to accommodate these demands [1]. Among them, High Efficiency Video Coding (HEVC) [2] standard with lossless mode outperforms in both complexity and efficiency. HEVC is the newest video coding standard jointly developed by the Joint Collaboration Team - Video Coding (JCT-VC), which is a collaboration of ISO/IEC MPEG and ITU-T VCEG. For similar video quality, HEVC can provide more than 50% bit-rate reduction when compared with the prior standard H.264/AVC [3].

In the current design of HEVC, lossless coding is intended to be an extension of lossy coding, and can be signaled for the entire picture or at the individual coding unit (CU) level [4]. Different from lossy coding, no distortion occurs between original encoding pictures and reconstructed ones. For HEVC in lossy mode, it is mainly consisted of five processes, namely, prediction, transform,

X. He et al. (Eds.): MMM 2015, Part I, LNCS 8935, pp. 454–465, 2015.
© Springer International Publishing Switzerland 2015

quantization, entropy coding and loop filters [5]. For these five components, both transform and quantization are lossy, and loop filters are not necessary since no noise exists in reconstructed pictures. Hence, transform, quantization and loop filters are bypassed in HEVC when doing lossless coding. With this method, lossless coding of HEVC has little additional burden on implementation of the encoder and decoder.

To improve the efficiency, a more flexible coding structure is designed in HEVC with variable size of coding unit (CU), prediction unit (PU) and transform unit (TU). Besides, up to 35 prediction modes are defined for intra modes, and more complicated motion estimation is also undertaken to bring efficiency increases. To obtain the optimal modes from numerous candidates, the Rate-Distortion Optimization (RDO) [6] that selects coding mode with the smallest Rate-Distortion cost (RD cost) is employed for intra and inter prediction. The accurate RD cost calculation, however, involves dramatically increased complexity. To reduce computational complexity, several simplified RD cost calculation is achieved with acceptable decreases of coding efficiency. Some methods research on wiping off some non-essential candidate modes [7], and the others focus on accurately estimating the rate or distortion [8]. Nonetheless, all these simplified methods are developed for lossy coding, and hardly suitable for lossless coding. That's because the RDO must have a different meaning in lossless coding as there is no distortion. As a result, the compression ratio of HEVC lossless coding is still unsatisfactory.

In this paper, two novel schemes for rough RDO are proposed to further reduce bit-rate of the lossless mode. Firstly, an approach using simulated annealing (SA) algorithm [9] is promoted to help encoder achieve corrected mode with appropriate lambda value. Considering the computational complexity of SA algorithm, the further improved method is also proposed by predicting parameters from neighboring coding units.

The rest of this paper is organized as follows. Structure of RDO in HEVC lossy mode is briefly introduced in Section 2. Section 3 analyses the difference of RD-cost principle between lossy and lossless coding, and two schemes are proposed to improve the RDO in lossless mode. To evaluate the efficiency of the proposed methods, lots of experiments that anchor on the latest reference software are performed, and the results are presented in Section 4. Finally, Section 5 concludes the paper.

2   RDO in HEVC Lossy Mode

In the lossy mode of HEVC, minimization of distortion and minimization of bit-rate are two basic coding targets. However, these two targets are typically conflicted, on account of that less encoding bits usually inevitably come with more distortion. To achieve a tradeoff between them, RDO is utilized in the encoder side for the HEVC mode decision. Concretely, the mode with minimal RD-cost (J) in (1) will be designated as the final encoding mode. In (1), R is the bits to encode mandatory mode information, such as transformed residues, intra mode indexes, motion vector difference, etc., and D, which is often defined as the sum of square errors (SSE), stands for the distortion between the original pixels and reconstructed ones.

$$J = D + \lambda R. \tag{1}$$

As shown in (1), the optimization target is to find an optimal tradeoff between coding bits and reconstructed video quality. And the Lagrange multiplier λ is used to adjust the balance. λ in (1), whose value is $-\frac{\partial D}{\partial R}$, is recognized to express the amount of distortion that decreasing an encoding bit can generate. Since quantization produces the major distortion, the λ value can be roughly determined by the QP used for encoding, which has been shown in [6].

The flowchart of original RDO based mode decision is illustrated in Fig.1. As shown in Fig.1, to generate R and D in (1), the progresses of transform, quantization, entropy coding, inverse quantization, inverse transform and necessary loop filters have to be conducted, which must be of high complexity. Obviously, if all steps of mode decision employ the accurate calculations of RD cost, the complexity of the encoder should be far from acceptable. To deal with this problem, the other relatively rough RD cost (J_{rough}) is taken for most of non-critical decisions, whose expression is shown as

$$J_{rough} = D_{rough} + \lambda R_{pred}, \tag{2}$$

where D_{rough} is the difference between the original pixels and predicted ones, and R_{pred} stands for the bits to encode prediction information, excluding the residues. Obviously, both of them require little computation because no further processes of reconstruction and residual entropy coding need to be performed. Since these two principles have strengths and weaknesses, both of them are adopted in the current HEVC design to obtain the trade-off between coding efficiency and computational complexity. In the premier stage of mode decision, the rough principle is taken for vast modes to achieve several candidates that are of high probability to be the optimal encoding modes. After that, the best mode is determined among these candidates with the calculations of accurate RD-cost. Thanks to these two processes, HEVC outperforms in both complexity and efficiency.

3   Improved Rough RDO for Lossless Coding

3.1   Lambda Value of RDO in Lossless Mode

As mentioned above, no distortion happens in the duration of lossless encoding in HEVC. Hence, different from lossy coding, minimizing the encoding bit-rate is the only purpose of lossless coding. Furthermore, since the lossless mode is just applied on the range extension profile, which is not the mainstream technology of HEVC, we should not modify the lossy mode so much to make it suitable for lossless mode. In other words, the technologies developed for lossy mode need to be made use of when we are designing the lossless mode.

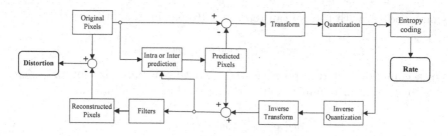

Fig. 1. Accurate RDO in HEVC

In the lossless mode, D in (1) equals zero and lambda value is positive, both of which result in that minimizing J is equivalent to minimizing R. Therefore, the accurate principle still works for lossless coding. However, with non-zero rough distortion, the rough RD cost will be influenced with the variation of lambda value, and the lambda value set by QP is no more appropriate for lossless mode. To be specific, D_{rough} in (2) is essentially the function of difference between the original pixels and predicted ones, we denote their relationship as follows,

$$D_{rough} = f(P_{org} - P_{pred}), \tag{3}$$

where P_{org} is defined as original pixels, and P_{pred} stands for predicted ones. Since original pixels consist of reconstructed residues, distortion and predicted pixels, (3) can be rewritten as

$$D_{rough} = f(P_{resi} + P_{dst} + P_{pred} - P_{pred}) = f(P_{resi} + P_{dst}) = f(P_{resi}), \tag{4}$$

where P_{resi} stands for reconstructed residues, and distortion P_{dst} equals zeros for lossless mode. It is shown in (4) that D_{rough} in lossless mode is only influenced by the residues, on the contrary, distortion is the main point for D_{rough} in lossy mode. It is the primary reason why lambda that decided by QP is suitable for lossy coding in rough RDO, but not for lossless coding. It is obvious that the higher the distortion is, the more residual encoding bits are. Therefore, we assume that the bits to encode residues in lossless mode $R_{residue}$ and D_{rough} have the following correlation,

$$R_{residue} = \alpha D_{rough}, \alpha > 0, \tag{5}$$

where α is defined as the slope. Although (5) may not be precisely accurate, it can generally reflect an authentic relationship and be accurate enough for rough decision when α is correctly set.

Apart from the bits to encode residues, bits of prediction information are the other part. For intra mode, intra mode index is the principal information, and motion vector difference and reference index are significant for inter prediction mode. Overall, encoding bits of lossless mode, which is denoted as R_{total}, consist of two portions as expressed in (6)

$$R_{total} = R_{residue} + R_{pred}. \tag{6}$$

Combined (6) with (5), we can represent (6) as (7) in another way.

$$R_{total} = \alpha D_{rough} + R_{pred}. \qquad (7)$$

On account of that α is positive, the objective of minimization of R_{total} is equivalent to that of minimization of R_{total}/α, that is, minimizing

$$J'_{rough} = D_{rough} + R_{pred}/\alpha. \qquad (8)$$

Comparing the representation between (2) and (8), we have the conclusion that if $\lambda = 1/\alpha$, minimization of RD cost in (2) corresponds to minimization of R_{total}, which is exactly the objective of lossless coding. Therefore, if the proportional relation of $R_{residue}$ and rough distortion D_{rough}, that is, the lambda value, is precisely acquired, the original rough RD-cost principle can be still adopted in lossless coding. To still keep compatibility and use rough principle in the lossless mode, the corresponding lambda needs to be modified, and two schemes for that will be introduced in the subsections followed.

3.2   Obtaining Optimal Lambda with Simulated Annealing Algorithm

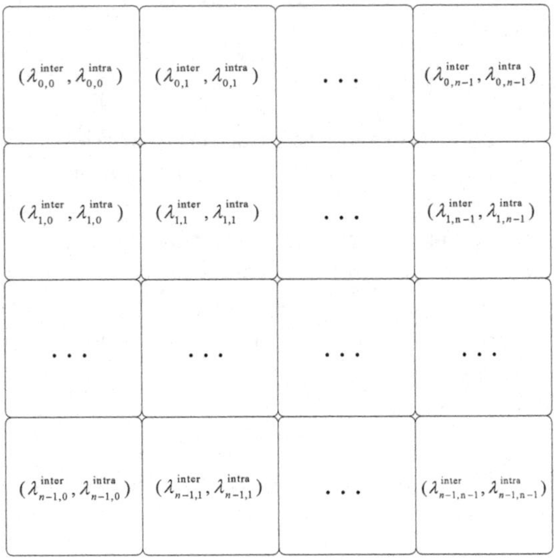

Fig. 2. The lambda of each LCU for intra and inter modes

Since residues in lossless coding are directly transmitted to the entropy coder, bits of coding prediction residues are just determined by the method of entropy coding. In the current design of HEVC, only one algorithm called context-based adaptive binary arithmetic coding (CABAC) is employed for residual entropy coding. Although many factors may influence the final coding bits, quantity of residues and the redundancy among residues are the two main points [8]. In (5), D_{rough} stands for the total quantity of residues, while the factor α indicates the correlation between neighboring residues. For CABAC entropy coding, higher redundancy among residues can help to bring less coding bits. Thus, α with smaller value indicates higher redundancy among residues. Since inter prediction further eliminates video redundancy than intra prediction, the value of α in (5) should be larger for the inter prediction, which results in that the value of lambda for the inter prediction becomes smaller. For this reason, we set a relatively larger value of lambda in inter prediction, while a smaller value is employed for lambda in intra prediction. Experiments based on different sequences are conducted, and the results show that the optimal lambda values for intra and inter prediction mode are around 5 and 1, respectively.

Although the fixed version of lambda value setting is simple, it should be inappropriate for some coding units, which leads to decreases in coding efficiency. Then, the improved method, which is similar to SA algorithm, is adopted to obtain optimal lambda value for each largest coding unit (LCU).

Simulated Annealing algorithm, which is derived from the physical process of cooling molten material down to the solid state, is an optimization process that has been widely used for various combinatorial problems. As shown in Fig. 2, SA algorithm is employed to acquire the optimal combination of lambda values for each LCU. At the beginning, the known solution above, i.e., 5 and 1 separately for intra and inter prediction mode, is configured as the initial solution for each LCU. Next, a constraint-based new solution is generated. If the new combination of lambda value for each LCU achieves less coding bits, it is accepted unconditionally as the next solution. However, if the new combination is worse, it is accepted with a certain probability rather than simply rejecting. To avoid a local optimum, the probability of acceptance of a worse solution is kept elevated at first, while lowered as the simulation progresses. When iteration times or rejecting times exceed some point, the algorithm is ended, and the current combination of lambda value is the optimal solution.

3.3   Estimating Model Parameters from Neighboring Coding Units

With SA algorithm, acquired lambda values are global optimum, and the coding efficiency achieved should be better than others. However, the complexity of the algorithm can be unacceptable with large iteration times. Considering that the feature of residues can be similar to that of neighboring CUs, another approach to derive relationship of D_{rough} and $R_{residue}$ from neighboring CUs are proposed followed.

Different from SA algorithm above, where kinds of lambda value setting can be tested, more complicated relationship of D_{rough} and $R_{residue}$ is required for accurate prediction. Hence, the correlation of rough distortion and residual coding bits in (5) should be rewritten as

$$R_{residue} = \alpha D_{rough} + \beta, \tag{9}$$

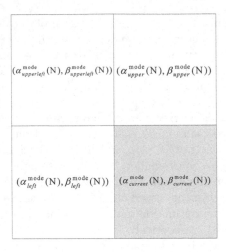

Fig. 3. Estimating model parameters from neighboring CUs

where β is the intercept. Then, the rough RD cost for lossless, which is distinct from (2), should be redefined as

$$J_{rough} = \alpha D_{rough} + \beta + R_{pred}, \tag{10}$$

As mentioned above, after rough mode decision, accurate RD cost should be calculated among several candidates. Consequently, when the encoding process of a certain coding unit is completed, several data of D_{rough}, R_{pred} and R_{total} can be obtained, where $R_{total} - R_{pred}$ is regarded as $R_{residue}$. With the data of $R_{total} - R_{pred}$ and D_{rough}, the model parameters α and β in (9) for current coding unit, which will be utilized to predict parameters of neighboring CUs, can be trained by linear regression algorithm based on least squares as described in [10].

As shown in Fig.3, when CUs on the left, upper and upper left are coded, that is, the model parameters α and β of these CUs have been acquired, the parameter α of current CU $\alpha_{current}^{mode}(N)$ should be set as follows,

$$\alpha_{current}^{mode}(N) = \begin{cases} min(L,U), & if\ C \geq max(U,L) \\ max(L,U), & if\ C \leq min(U,L) \\ median(L,U,C), & otherwise \end{cases} \tag{11}$$

In (11), L, U and C are defined as $\alpha_{left}^{mode}(N)$, $\alpha_{upper}^{mode}(N)$ and $\alpha_{upperleft}^{mode}(N)$, respectively, where mode is either intra or inter prediction mode, and the size of current CU is $N \times N$. When the parameter α has been designated, the parameter β in the same position will be set as that of the current CU. Noted that, for CUs with two available neighbors, α with a smaller value will be chosen for intra prediction, and rough RDO of inter prediction will choose α with a larger value. For CUs with single available neighbor, the neighbor's parameters will be the choice. Finally, for CUs without available neighbor, α of intra and inter prediction mode will be set as 0.2 and 1, respectively, while β is set as zero.

Table 1. Bit-rate variation (%) of RA-Main when compared to the HM13.0 lossless coding

Sequences	Method 1	Method 2
Class A	-0.9%	-0.9%
Class B	-0.8%	-0.6%
Class C	-0.4%	-0.4%
Class D	-1.2%	-1.1%
Class F	-2.5%	-2.1%
Average	-1.2%	-1.0%
Encoding Time Ratio	997%	103%

Table 2. Bit-rate variation (%) of LDB-Main when compared to the HM13.0 lossless coding

Sequences	Method 1	Method 2
Class B	-0.7%	-0.7%
Class C	-1.0%	-1.0%
Class D	-1.7%	-1.6%
Class E	-0.9%	-0.9%
Class F	-2.9%	-2.4%
Average	-1.4%	-1.3%
Encoding Time Ratio	1001%	105%

4   Experimental Results

The proposed algorithm has been implemented on the latest HEVC reference software HM13.0 [11], and the common test conditions in [12] are followed. HM13.0 in the lossless mode is chosen to be the anchor in the comparisons. Three coding configurations, namely, Random Access Main Profile (RA-Main) encoding, Low Delay B Main Profile (LB-Main) encoding, and Low Delay P Main Profile (LP-Main) encoding are followed and all the sequences specified in [12] are tested across the configurations. Moreover, both of two flags that separately signal whether lossless coding is applied for a particular picture and for the individual CU should be set as 1.

The experimental results in Table 1, 2 and 3 are measured in the form of rate variation $\triangle R$, which is defined by

$$\triangle R = \frac{R_{prosposed} - R_{anchor}}{R_{anchor}} \times 100\%, \tag{12}$$

where a negative number means bit-rate savings and a positive number means bit-rate increasing. The coding efficiency differences of two proposed methods are both summarized in these tables, for which Method 1 employs the SA algorithm, and parameters prediction from neighbors is utilized in Method 2. It reveals that both of two proposed methods work well under all the three coding structures and

Table 3. Bit-rate variation (%) of LDP-Main when compared to the HM13.0 lossless coding

Sequences	Method 1	Method 2
Class B	-0.6%	-0.6%
Class C	-0.9%	-0.8%
Class D	-1.2%	-1.2%
Class E	-0.6%	-0.5%
Class F	-2.6%	-2.2%
Average	-1.2%	-1.1%
Encoding Time Ratio	1015%	106%

on average provide a 1.0% to 1.4% additional bit-rate reduction over the anchor. Especially for Class F sequences, the bit-rate reduction for all coding structures reaches more than 2.0% on average. The reason is that the residual features of screen content sequences in Class F are extremely irregular, and an invariant lambda value for all frames is unreasonable. While our proposed methods can help to find the appropriate relation between residues and their related coding bits, which leads to performance improvements for lossless coding.

In addition, the results also reveal that the method using SA provides more satisfactory results than the method using prediction parameters from neighboring coding units. However, the latter scheme increases scarcely the encoding complexity, while the former one with SA algorithm brings substantial increases of computational complexity. It should be admitted that the proposed parameters prediction algorithm, in principle, should not work better than SA algorithm, when all possible lambda values are tested. Considering the complexity introduced by the SA algorithm and the coding efficiency, the proposed parameters prediction algorithm according to (9) can achieve most of the performance gain while exert little influence upon the encoding complexity. It is a good tradeoff between coding efficiency and encoding complexity.

In order to demonstrate that the performance of the proposed methods are always better than the anchor, bits per frame (BPF) of Kimono (Class B) and RaceHorses (Class D) sequences [12] for RA-Main, LDB-Main and LDP-Main configurations are illustrated in Fig. 4 (a), (b) and (c), respectively. As shown in the figures, the proposed algorithms always provide more satisfactory results than the anchor in all frames, and keep a stable bits reduction rate over them. In summary, the lossless coding process in HEVC can benefit from the proposed methods that achieve higher energy compaction and decorrelation, as well as no changes in syntax and semantics.

5   Conclusions

In this paper, an improved RD-cost principle is proposed for lossless coding in HEVC. Two schemes, i.e., a complicated version and a simplified one are both proposed. Based on the analysis of rough version of RD-cost principle, a

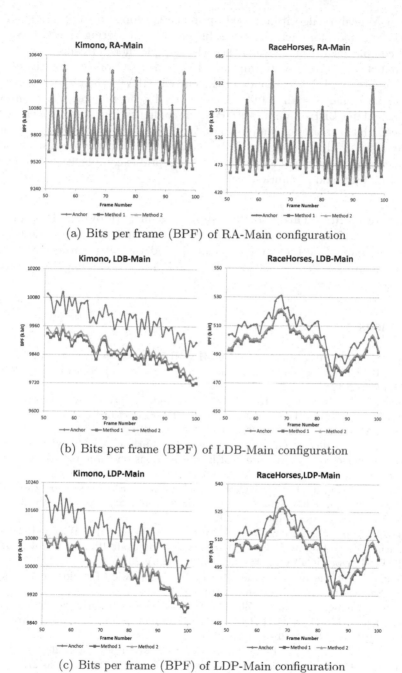

(a) Bits per frame (BPF) of RA-Main configuration

(b) Bits per frame (BPF) of LDB-Main configuration

(c) Bits per frame (BPF) of LDP-Main configuration

Fig. 4. Bits per frame (BPF) of Kimono (Class B) and RaceHorses (Class D) sequences (from the 51th frame to the 100th frame)

Simulated Annealing algorithm based approach is proposed to obtain the most appropriate lambda for each largest coding unit. Considering the high computational complexity of the scheme, the other simple method using least square errors prediction is proposed to improve the rough RDO process. In the experiments compared with lossless mode of the current HEVC, the improved method offers the performance with a 1.0%, 1.3% and 1.1% bit-rate reduction on average for RA-Main, LDB-Main and LDP-Main configurations, respectively, while brings negligible increases of computational complexity in the encoder. In conclusion, the lossless coding process in HEVC can benefit from the proposed method that achieves higher energy compaction and acceptable complexity, as well as no changes in syntax and semantics.

Acknowledgments. This work was supported in part by 973 Program under Contract 2013CB329004, the Fundamental Research Funds for the Central Universities under Contract WK2100060011, and the open project of Beijing Multimedia and Intelligent Software Key laboratory in Beijing University of Technology.

References

1. Zhou, M., Gao, W., Jiang, M., Yu, H.: HEVC lossless coding and improvements. IEEE Transactions on Circuits and Systems for Video Technology 22(12), 1839–1843 (2012)
2. Sullivan, G.J., Ohm, J.-R., Han, W.-J., Wiegand, T.: Overview of the High Efficiency Video Coding (HEVC) standard. IEEE Transactions on Circuits and Systems for Video Technology 12(22), 1649–1668 (2012)
3. Wiegand, T., Sullivan, G.J., Bjontegaard, G., Luthra, A.: Overview of the H. 264/AVC video coding standard. IEEE Transactions on Circuits and Systems for Video Technology 13(7), 560–576 (2003)
4. Chen, F., Zhang, J., Li, H.: Hybrid transform for HEVC-based lossless coding. In: IEEE International Symposium on Circuits and Systems, ISCAS 2014, pp. 550–553. IEEE Press, Melbourne (2014)
5. Bross, B., Han, W.J., Ohm, J.R., Sullivan, G.J., Wang, Y.K., Wiegand, T.: High Efficiency Video Coding (HEVC) text specification draft 10. JCTVC of ISO/IEC and ITU-T, Doc. JCTVC-L1003, Geneva (January 2013)
6. Sullivan, G.J., Wiegand, T.: Rate-distortion optimization for video compression. IEEE Signal Processing Magazine 15(6), 74–90 (1998)
7. Zhao, L., Zhang, L., Ma, S., Zhao, D.: Fast mode decision algorithm for intra prediction in HEVC. In: 2011 IEEE Visual Communications and Image Processing (VCIP), pp. 1–4. IEEE Press, Tainan (2011)
8. Hahm, J., Kyung, C.-M.: Efficient CABAC rate estimation for H.264/AVC mode decision. IEEE Trans. Circuits and Systems for Video Technology 20(2), 310–316 (2010)
9. Kirkpatrick, S., Gelatt, C., Vechhi, M.: Optimization by simulated annealing. Science 220(4598), 671–680 (1983)
10. Draper, N.R., Smith, H.: Applied regression analysis. Series in Probability and Mathematical Statistics. Wiley. Wiley (1981)

11. HM Reference Software Version 13.0,
    http://hevc.kw.bbc.co.uk/trac/browser/jctvc-hm/tags
12. Bossen, F.: Common test conditions and software reference configurations. JCTVC
    of ISO/IEC and ITU-T, Doc. JCTVC-L1100, Geneva (January 2013)

A Novel Error Concealment Algorithm for H.264/AVC

Jinlei Zhang and Houqiang Li

Chinese Academy of Sciences Key Laboratory of Technology in Geo-Spatial Information Processing and Application System, University of Science and Technology of China, Hefei 230027, China
jlzhang@mail.ustc.edu.cn, lihq@ustc.edu.cn

Abstract. To benefit network transmission, the bit stream of the whole frame coded by H.264/AVC is usually grouped into one packet. However, the packet loss during transmission will lead to the distortion of the reconstructed video and the error propagation. To deal with this problem, error concealment (EC) strategy is widely used to recover the lost frames and to weaken the effect of error propagation. In this paper, we propose a new EC method which uses both the forward and backward motion information to recover the motion information of the current lost frame. Besides, for the pixels whose motion information is quite different compared with its neighboring pixels, we propose to use the spatial correlation to fill up the pixels by minimizing the total variation (TV) norm. With the help of our proposed algorithm, we can obtain the motion vectors of the lost frames and improve the accuracy of motion vectors derived by the unidirection recovery method. Besides, the optimization strategy, minimizing the TV-norm for the pixels with quite different motion vectors, can help the decoder to recover the reconstructed frames more accurately. Experimental results show that the proposed algorithm can achieve both better objective performance and subjective performance compared to well-known schemes.

Keywords: Video Coding, Error Concealment, Whole Frame Recovery, H.264/AVC.

1 Introduction

With the rapid development of digital technology, video communications over networks have received increasing attention from the communication applications. During the transmission process in the network environment, various channel errors may lead to damage or loss of compressed video data packets. Besides, as we known, video coding uses the temporal prediction which may cause the errors propagate easily to succeeding frames. As a consequence, it is necessary to design appropriate error resilience and concealment tools for video encoders and decoders in order to counteract such losses and alleviate the error propagation.

Many related works [1,2,3,4,5] have been proposed. However, most of these methods assume that only a few blocks or slices in a video frame are missing,

X. He et al. (Eds.): MMM 2015, Part I, LNCS 8935, pp. 466–476, 2015.

and they are incompetent for the case when the data of a whole frame is lost. Actually, in video coding, the temporal correlation is so strong that the compression performance of inter prediction is very effective and a whole frame can be encapsulated into one packet. Several algorithms have also been proposed to deal with whole frame loss. The simplest whole frame loss error concealment (EC) method[6] used in H.264 is the frame copy (FC) method which directly copies the previous reconstructed frame to recover the lost frame. However, the FC method may lead to the frame un-continuity especially for the high motion videos. Another method called motion copy (MC)[6] duplicates the motion vectors (MVs) of the previous reconstructed frame and recover the missing frame through motion compensation. However, the MC method is based on an assumption that the collocated blocks in successive frames have similar motion activity. Actually, the assumption is not suitable for all the blocks in the frame especially for the high motion videos.

Some other algorithms [7,8,9,10,11,12] try to generate a better motion vector for the lost frame. The main ideas can be classified into three typical categories.

1. Taking full advantage of the previous reconstructed frames' MV information instead of only one previous frame. [7] and [8] use the concept of optical flow to generate the MVs of each pixel in the missing frame. And based on the theory of multiple hypotheses, [9] devises an adaptive integration scheme to make full use of each hypothesis' strength, which fully exploits the correlation between consecutive frames.

2. Making use of both the forward and backward frames' MV information. [10] first proposes to generate the MV of the missing frame based on pixel level bi-direction frames. And [11] proposes another algorithm to recover the MV of each pixel in the missing frames based on pixel level bi-direction frames' MV information.

3. Classifying the frame into different types, and use different filtering method to recover the missing frame. [12] classifies the blocks into three types (i.e. covered state, conflict state, and non-covered state), and designs different EC method for each type.

All of the algorithms above focus on recovering the MV information using the temporal correlation, which could only provide limited information for frame recovery. In this paper, not only the temporal MV correlation, we also make use of the spatial correlation. Firstly, we classify the blocks into three classes (i.e., new scene, uniform linear motion and non-uniform linear motion), and using different MV recovery methods and motion compensation methods to recover the pixels in the missing frame. Secondly, we pick the pixels with incorrect MVs based on the previous estimated MV of each pixel, and recover them by minimizing the total variation (TV) norm.

The rest of the paper is organized as follows. In Section 2, the main idea of the MV generation and the minimizing TV-norm algorithm are elaborated. In Section 3, the objective and subjective results of the proposed algorithm are presented. At last, we conclude this paper in Section 4.

2   Proposed Error Concealment Algorithm

The existing methods which make use of the bi-directional MV information or multiple previous frames are able to provide a better performance compared with the MC method. However, they also have the shortcoming. The MVs of the lost blocks are generated based on the assumption that all the blocks are uniform linear moving. Actually, it is not always satisfied. Some MVs are very likely to be estimated inaccurately, especially in large motion activity. This shortcoming will damage the accuracy of the MV for the block, which will result in the degraded performance. In order to overcome this problem, we propose a hybrid EC method based on bi-directional MV information and the spatial correlation, which uses not only the bi-directional MVs, but also the spatial correlation to get rid of the inaccurate MV pixels. Thus, the proposed algorithm is able to discard the pixels with inaccurate MVs and recover them by exploiting spatial correlation, which is helpful to protect the spatial smoothness of the frame.

2.1   Blocks Classification

As we known, the content of the sequences is always changed. Sometimes new scenes come into the content, and sometimes cover phenomenon happens. Thus, the uniform linear motion assumption is always incorrect for these blocks. In this paper, we first classify the blocks into three types, i.e. new scene, uniform linear motion and non-uniform linear motion. The sorting rules are as follows:

1. If the collocated block in the forward frame is intra predicted, then we define the missing block as the new scene for the forward frames, and the forward MV information is neglected and marked as the unavailable MV; otherwise, the forward MV is marked as available MV which will be used to estimate the MV of the missing block.
2. If the collocated block in the backward frame is intra predicted, then we define the missing block as the new scene for the backward frames, and the backward MV information is neglected and marked as the unavailable MV; otherwise, the backward MV is marked as available MV.
3. We first scale the MV information according to the reference index as follows,

$$MV_x = \frac{MV_x}{ref+1}; MV_y = \frac{MV_y}{ref+1} \qquad (1)$$

where ref means the reference index of the block. And then according to the scaled MV information of the two directions, if the difference of these two scaled MVs is greater than a threshold, then we define the block as non-uniform linear motion block; otherwise, the block is marked as uniform linear motion block.
4. For the non-uniform linear motion blocks, if the difference of the residuals in the two directions is larger than a threshold, then the corresponding MV information of the block with larger residuals is ignored and marked as the unavailable MV, while the other MV information is marked as available MV.

2.2   Bi-directional Motion Compensation

According to the classification of the blocks in Subsection 2.1, we recover the missing pixels using bi-directional motion compensation. As discussed in [10], the block-based motion vector estimation (MVE) method provides similar performance as pixel-based MVE method in little motion scenes. However, for the scenes with high motion activity, pixel-based MVE has an obvious better performance than block-based MVE. Thus, in this paper, we also use the pixel-level motion compensation method to recover the missing pixels.

On one hand, if the forward MV is available, and it is $f = (f_x, f_y)$, then the lost pixel $p_f(x, y)$ can be recovered as follows,

$$p_f(x, y) = p_r(x + f_x, y + f_y) \tag{2}$$

where $p_r(x, y)$ means the pixels in position (x, y) of the previous frame.

On the other hand, similar to the forward recovery method, if the backward MV is available, and it is $b = (b_x, b_y)$, then the lost pixel recovered by backward prediction $p_b(x, y)$ is as follows,

$$p_b(x, y) = p_r(x + b_x, y + b_y) \tag{3}$$

where $p_r(x, y)$ is the pixels in the last correctly reconstructed frame, and the (b_x, b_y) is the scaled MV corresponding to the reference frame.

At last, we combine the forward and backward pixel-based MVE methods. For the pixel in position (x, y) of the lost frame, its recovery value is estimated as follows,

$$p(x, y) = \omega(x, y) \times p_f(x, y) + (1 - \omega(x, y)) \times p_b(x, y) \tag{4}$$

where the pixel-based weight $\omega(x, y)$ is used to adjust the weights of the forward and backward methods. In this paper, we set $\omega(x, y)$ for all the pixels within the block according to the classification of the block as follows,

$$\omega(x, y) = \begin{cases} 1, & only\, f = (f_x, f_y)\, is\, available \\ 0, & only\, b = (b_x, b_y)\, is\, available \\ 0.5, & both\, f = (f_x, f_y)\, and\, b = (b_x, b_y)\, are\, available \end{cases} \tag{5}$$

2.3   Error Concealment Using Spatial Correlation

As the bi-directional MVE presented above, the pixels have the corresponding MV information. However, the MV of some pixels may be estimated inaccurately. In this subsection, we judge the accuracy of the estimated MV for each pixel as follows,

$$idx(x, y) = \begin{cases} 0, & if\, MV(x, y) - MV(x \pm 1, y \pm 1) > th \\ 1, & otherwise \end{cases} \tag{6}$$

where $MV(x \pm 1, y \pm 1)$ means the MV information of the pixels in the 8 neighboring area of current pixel. $idx(x, y) = 0$ means that the recovered pixel of (x, y) is inaccurate, and $idx(x, y) = 1$ means the recovered pixel is accurate.

According to the judgement method discussed above, there will be some holes in the missing frame since some of the recovered pixels are inaccurate. In other word, once the bi-directional MVE is completed, all elements in the missing frame are supposed to be trustworthy except for few remaining hole regions. In this subsection, we propose a method to recover the inaccurate pixels by making use of the spatial correlation. Let the current frame be denoted by $Y(x, y) \in R^{m \times n}$, and let Ω denote the set of trustworthy entries in $Y(x, y)$. Then the next task can be mathematically expressed as: given the set Ω, we shall recover $M \in R^{m \times n}$ from its incomplete observation $Y \in R^{m \times n}$. Since the temporal correlation has been exploited at subsection 2.2, we assume that the temporal smoothness has been satisfied. Thus, in this step, the spatial smoothness shall be enforced during the process of completion to reduce the unpleasant artifacts. Therefore, a 2-D TV regularized completion model is designed to achieve this objective.

We define the operator $O = [O_x^T, O_y^T]^T$, as the sub-operators of horizontal and vertical directions. The individual sub-operators are defined as follows,

$$
\begin{aligned}
O_x(M) &= vec[M(x + 1, y) - M(x, y)] \\
O_y(M) &= vec[M(x, y + 1) - M(x, y)]
\end{aligned}
\tag{7}
$$

where $vec[\cdot]$ stands for the vectorization operation. The definition of 2-D TV norm of a space volume M is then expressed as,

$$
||M||_{2-DTV} := \sum_i (|[O_x(M)]_i| + |[O_y(M)]_i|)
\tag{8}
$$

where $[\cdot]_i$ means the ith element from the 2-D volume. Since the definition in Eq.(8) has a similar formulation with the l_1 norm, we express $||M||_{2-DTV}$ in a l_1 norm-like manner as,

$$
||M||_{2-DTV} = ||O(M)||_1
\tag{9}
$$

From the definition of 2-D TV norm, we can see that it provides a natural piecewise smoothness in space domain for the missing frame, with edge preserving property. Then, we formulate a convex program as follows,

$$
\begin{aligned}
&\min_M ||M||_{2-DTV} \\
&s.t. \, P_\Omega(M - Y) = 0
\end{aligned}
\tag{10}
$$

$P_\Omega(M - Y)$ shall project into $R^{m \times n}$ and the projection is the $(i, j)^{th}$ entry of $M - Y$ if $(i, j) \in \Omega$, otherwise 0.

To solve Eq.(10), we introduce an auxiliary variable $Z \in R^{m \times n}$ and reformulate Eq.(10) into

$$
\begin{aligned}
&\min_{M,Z} ||M||_{2-DTV} \\
&Y = Z + M \\
&P_\Omega(Z) = 0
\end{aligned}
\tag{11}
$$

Algorithm 1. ADM algorithm to solve Eq.(11)

Input: Y, *tol*

1: Initialize M, Z, Λ, β and p.

2: **while** $\frac{||M+Z-Y||_F}{||Y||_F} \geq tol$ **do**

3:     Update M^{k+1}: solved the 2-D total variation regularized least square problem
     using Algorithm 2.

4:     Update Z^{k+1}:$Z^{k+1} = P_\Omega(\frac{1}{\beta}\Lambda^k - M^{k+1} + Y)$.

5:     Update Λ^{k+1}:$\Lambda^{k+1} = \Lambda^k - \beta(Y - Z^{k+1} - M^{k+1})$.

6:     Update β:$\beta = p \times \beta$

7: **end while**

Output: M

To solve Eq.(11), we apply the alternating direction method (ADM) [13], and we consider the augmented Lagrangian function of Eq.(11) as follows,

$$L_A(M, Z, \Lambda, \beta) = ||M||_{2-DTV} -$$
$$< \Lambda, Y - Z - M > + \frac{\beta}{2}||Y - Z - M||_F^2 \tag{12}$$

where $\Lambda \in R^{m \times n}$ is the Lagrange multiplier attached to the equality constraint $Y = Z + M$, β is a positive value either constant or varying with the optimization process, $< \cdot >$ denotes the matrix inner product and $|| \cdot ||_F$ denotes the Frobenius norm. Each iteration of the ADM optimizes the augmented Lagrangian function with respect to M and Z in a coordinate descent manner, and updates the Lagrange multiplier accordingly. Therefore, the original problem is now decomposed into several simpler subproblems. Assume that after the kth iteration, the parameters are (M^k, Z^k, Λ^k), then the ADM generates the $(k+1)^{th}$ iterate $(M^{k+1}, Z^{k+1}, \Lambda^{k+1})$ as

$$M^{k+1} = arg \min_M ||M||_{2-DTV} + \frac{\beta}{2}||M - Y + Z^k + \frac{1}{\beta}\Lambda^k||_F^2 \tag{13}$$

$$Z^{k+1} = arg \min_{Z:P_\Omega(Z)=0} \frac{\beta}{2}||Z - Y + M^{k+1} + \frac{1}{\beta}\Lambda^k||_F^2 \tag{14}$$

$$\Lambda^{k+1} = \Lambda^k - \beta(Y - Z^{k+1} - M^{k+1}) \tag{15}$$

The detailed ADM algorithm to solve the problem is shown as Algorithm 1.
   Each subproblem is solved as follows:

1. M-subproblem: The first subproblem to get the new iteration of M^{k+1} is a 2-D TV regularized least squares problem. A similar problem has been investigated in [14], in which a 3-D TV problem is formulated. Different to the problem in [14], in this paper, it is a 2-D TV problem, and we only need to compute the matrices on the spatial dimension. The detailed algorithm is shown as Algorithm 2.

Algorithm 2. The algorithm to update M^{k+1}

Input: $Q^k = Y - Z^k - \frac{1}{\beta}\Lambda^k$, tol_M

1: Initialize M^0, u^0, P^0, μ, q and $i = 0$.
2: Compute the matrices $\mathcal{F}[O_x]$, $\mathcal{F}[O_y]$ and $\mathcal{F}[I]$
3: **while** $\frac{||M_{i+1} - M_i||_F}{||M_i||_F} \geq tol_M$ **do**
4:       Update M_{i+1}: $M_{i+1} = \mathcal{F}^{-1}[\frac{\mathcal{F}[\beta Q^k + \mu O^T u - O^T P]}{\beta|\mathcal{F}[I]|^2 + \mu(|\mathcal{F}[O_x]|^2 + |\mathcal{F}[O_y]|^2)}]$.
5:       Update $u_{i+1} = (u_x, u_y)$:$u_x = max(|O_x M_{i+1} + \frac{1}{\mu}P_x| - \frac{1}{\mu}, 0) \cdot sign(O_x M_{i+1} + \frac{1}{\mu}P_x)$.
         u_y is derived similar to u_x.
6:       Update the Lagrange multiplier P_{i+1}:$P_{i+1} = P_i - \mu(u_{i+1} - OM_{i+1})$.
7:       Update μ:$\mu = q \cdot \mu$.
8:       $i = i + 1$.
9: **end while**
10: $M^{k+1} = M_i$
Output: M^{k+1}

---

2. Z-subproblem: The second subproblem is to update Z^{k+1}, it is obtained as,

$$Z_{i,j}^{k+1} = \begin{cases} \frac{1}{\beta}\Lambda^k - M^{k+1} + Y, & if (i,j) \notin \Omega \\ 0, & otherwise \end{cases} \tag{16}$$

3   Experimental Results

To prove the effectiveness of the proposed algorithm, we first compare the algorithm with FC and MC methods. The proposed algorithm is implemented on the H.264/AVC reference software JM18.5. Five test sequences $foreman(30Hz$, 300 frames), $bus(15Hz$, 75 frames), $mobile(30Hz$, 300 frames), $basketballdrive(50Hz$, 500 frames) and $soccer(30Hz$, 300 frames) are chosen to evaluate the performance of the proposed algorithm. All of the test sequences are in QCIF size. The period of intra frame reset is 30 and the number of reference frames is 5. A constant QP of 28 is maintained for all the frames.

In this simulation, we test two cases: 1) one frame is dropped in every 5 pictures; 2) one frame is dropped in every 20 pictures. And the dropped frames are concealed with EC algorithms. The peak signal-to-noise ratio(PSNR) is chosen as the objective measurement, which is computed using the original video sequence as the reference. The PSNR values of each method for the first 50 frames of bus and $foreman$ sequences are plotted in Fig. 1. It is obvious that the proposed algorithm outperforms the existing method under different dropping cases. Besides, the average PSNR performances over all the erroneous frames are shown in Table 1 and Table2, from which we can see that the proposed algorithm yields higher PSNR performances than the other methods. For the first case, the proposed algorithm achieves 4.662dB and 1.336dB in average performance gain compared with FC and MC, and for the second case, it achieves about 4.334dB

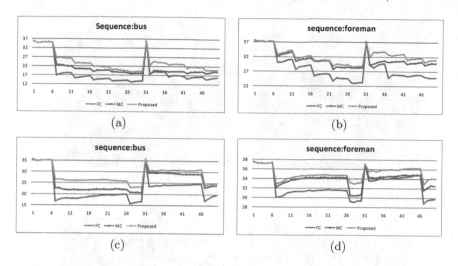

Fig. 1. PSNR results of different EC algorithms: the top two figures correspond to one frame dropped every 5 frames, and the bottom two figures correspond to one frame dropped every 20 frames

Table 1. Compared the proposed algorithm with FC and MC, one frame is dropped every 5 frames

Sequences	FC	MC	Proposed	Proposed-FC	Proposed-MC
foreman	23.06	25.50	27.09	4.03	1.59
bus	15.63	18.99	21.02	5.39	2.03
mobile	20.68	25.96	26.60	5.92	0.64
basketballdrive	19.75	23.04	24.88	3.29	1.84
soccer	17.44	19.70	20.28	2.26	0.58
average	19.312	22.638	23.974	4.662	1.336

and 1.466dB in average. For some sequences with new scene coming into the content, such as *bus*, the performance gain achieves to 5.39dB and 6.05dB compared with FC under two test cases.

For subjective evaluation, one original frame and three recovered frames reconstructed by FC, MC and the proposed algorithm are demonstrated in Fig.2. The 6th frame of *bus* and the 239th frame of *foreman* are chosen. From Fig.2, we can observe that the proposed algorithm has an obvious subjective quality improvement compared with FC and MC.

Since the proposed algorithm can be easily transplanted to other versions of the codec and achieve an equally good performance, we transplant the codec into JM10.2 to compare with the result reported in [11]. All the test conditions are the same as that used in [11]. The number of reference frames is 1, QP is 22, the sequence size is QCIF, intra period is 15, and one frame is dropped in

Fig. 2. Subjective quality results of different EC algorithms: the top four figures are sequence *foreman*, and the bottom four figures are sequence *bus*

Table 2. Compared the proposed algorithm with FC and MC, one frame is dropped every 20 frames

Sequences	FC	MC	Proposed	Proposed-FC	Proposed-MC
foreman	27.75	31.80	32.51	4.76	0.71
bus	19.07	22.86	25.12	6.05	2.26
mobile	27.00	29.26	30.37	3.37	1.11
basketballdrive	23.59	26.44	29.00	5.41	2.56
soccer	22.15	23.54	24.23	2.08	0.69
average	23.912	26.78	28.246	4.334	1.466

Table 3. Compared the proposed algorithm with the method proposed in [11]

Sequences	Error Free	FC	MC	[11]	Proposed	Proposed-[11]
mobile	38.80	26.69	30.68	31.82	32.71	0.89
bus	39.08	18.97	25.84	26.43	27.15	0.72

every 15 frames. The results are shown as Table 3, from which we can see that the proposed algorithm also outperforms the method proposed in [11].

4 Conclusion

In this paper, we propose a whole frame loss error concealment algorithm to recover the lost frames. Although the follow-up frames in temporal layer can not be decoded correctly, the motion information and the residuals can be decoded, Thus, we propose to use the bi-directional motion information to recover the MVs of the missing frames. Besides, we use the MV and residuals information to classify the blocks into three types, and we design different methods to deal with each type. Therefore, the proposed algorithm can achieve better MV information for the missing frames. Besides, by exploiting the spatial correlation, we propose to improve the recovered frames by minimizing the TV-norm of some incorrect pixels, which enhances the spatial smoothness of the recovered frames and provides a better subjective visual quality. Experimental results show that our proposed whole frame loss error concealment algorithm can achieve significant PSNR improvement compared with previous works and the subjective quality also outperforms previous works.

Acknowledgments. This work was supported in part by 973 Program under Contract 2013CB329004, the Fundamental Research Funds for the Central Universities under Contract WK2100060011, and the open project of Beijing Multimedia and Intelligent Software Key laboratory in Beijing University of Technology.

References

1. Kumwilaisak, W., Kuo, C.J.: Spatial error concealment with sequence-aligned texture modeling and adaptive directional recovery. Journal of Visual Communication and Image Representation 22(2), 164–177 (2011)
2. Zhang, R.F., Zhou, Y.H., Huang, X.D.: Content-adaptive spatial error concealment for video communication. IEEE Transactions on Consumer Electronics 50(1), 335–341 (2004)
3. Zhai, G., Cai, J., Lin, W., Yang, X., Zhang, W.: Image Error-concealment via Block-based Bilateral Filtering. In: Proceeding of IEEE International ICME, pp. 621–624 (2008)
4. Zhang, Y.B., Xiang, X.G., Ma, S.W., Zhao, D.B., Gao, W.: Aoto Regressive Model and Weighted Least Squares based Packet Video Error Concealment. In: Proceeding of IEEE DCC, pp. 455–464 (2010)
5. Zhou, J., Yan, B., Gharavi, H.: Efficient Motion Vector Interpolation for Error Concealment of H.264/AVC. IEEE Transactions on Broadcasting 57(1), 75–80 (2011)
6. Wu, Z., Boyce, J.M.: An Error Concealment Scheme for Entire Frame Losses Based on H.264. In: Proceeding IEEE ISCAS (2006)
7. Belfiore, S., Grangetto, M., Magli, E., Olmo, G.: Concealment of Whole-Frame Losses for Wireless Low Bit-Rate Video Based on Multiframe Optical Flow Estimation. IEEE Transactions on Multimedia 7(2), 316–329 (2005)
8. Baccichet, P., Bagni, D., Chimienti, A., Pezzoni, L., Rovati, F.S.: Frame Concealment for H.264/AVC Decoders. IEEE Transactions on Consumer Electronics 51(1), 227–233 (2005)
9. Zhao, C., Ma, S.W., Zhang, J., Gao, W.: A Highly Effective Error Concealment Method for Whole Frame Loss. In: Preceeding of IEEE ISCAS, pp. 2135–2138 (2013)
10. Chen, Y., Yu, K., Li, J., Li, S.: An Error Concealment Algorithm for Entire Frame Loss in Video Transmission. In: Preceeding of IEEE PCS (2004)
11. Yan, B., Gharavi, H.: A Hybrid Frame Concealment Algorithm for H.264/AVC. IEEE Transactions on Image Processing 19(1), 98–107 (2010)
12. Wu, Z.Y., Boyce, J.M.: An Error Concealment Scheme for Entire Frame Losses Based on H.264/AVC. In: Preceeding of IEEE ISCAS, pp. 4463–4466 (2006)
13. Bertsekas, D.: Nonlinear Programming. Athena Scientific (1999)
14. Chan, S.H., Khoshabeh, R., Gibson, K.B., Gill, P.E., Nguyen, T.Q.: An Augmented Lagrangian Method for Total Variation Video Restoration. IEEE Transactions on Image Processing 20(11), 3097–3111 (2011)

Edge Direction-Based Fast Coding Unit Partition for HEVC Screen Content Coding

Mengmeng Zhang[1] and Yangxiao Ou[2]

[1] North China University of Technology, Beijing, P.R. China
zmm@ncut.edu.cn
[2] North China University of Technology, Beijing, P.R. China
nvniaobi@sina.com

Abstract. High efficiency Video Coding is a new video coding standard that presents numerous advantages over previous video coding standards. However, Rate distortion optimization (RDO) complexity is extremely high for screen content coding, which cannot adjust to the real-time performance. This paper proposed a fast and efficient algorithm based on edge direction to partition coding units (CUs) based on their relationship with edge direction. Sobel operator is used to determine the edge direction from the total image before intra prediction. The key point of this algorithm is to determine the relationship between the edge direction and CUs. Experimental results show that the proposed edge direction-based CU partition algorithm provides a decrease in the screen content coding processing time up to 39%, with a little increase in bit-rate(0.7% on average) and a negligible reduction in the PSNR value.

Keywords: HEVC, edge direction, Sobel, coding unit, screen content coding.

1 Introduction

High-efficiency Video Coding (HEVC) is a novel international video coding standard developed by the Joint Video Team from ISO/IEC MPEG and ITU-T VCEG. HEVC adopts numerous new tools, including a highly flexible data structure representation, which includes the coding unit (CU), prediction unit (PU), and transform unit (TU). RDO is applied, during the partitioning of largest CUs. HEVC supports four intra modes in intra coding, including intra 64x64, intra 32×32, intra 16×16, intra 8×8. Some new tools are also adopted in screen content coding, Intra Block Copy (IntraBC) reuses the notion of Sliding Window Lempel-Ziv coding, which represents a block of pixels by its length and a displacement pointing to a previous occurrence in spatially reconstructed region, to search frequently occurred patterns in the current frame. The complexity of mode decision and the intra predictor generation comprise 57% of intra coding, which serves as a processing bottleneck. In the Serial Real-time Communications System, for example the video conference, the high complexity of RDO is the biggest barrier.

Fast decision algorithms have been recently proposed to reduce computation complexity. A smooth CU usually selects a lager intra coding block, whereas a complex

X. He et al. (Eds.): MMM 2015, Part I, LNCS 8935, pp. 477–486, 2015.

CU usually chooses a smaller coding block because the block size of intra coding is dependent on CU smoothness. Designs [1] facilitated an improvement in the intra block size decision by using the number of edge points. This algorithm reduced computation complexity with a slight Peak Signal to Noise Ratio (PSNR) decrease in H.264 and a significant increase in the coding stream when applied to the screen content coding in HEVC. Designs [2] proposed a fast intra prediction mode based on edge direction, but efficiency is insufficient in terms of the mode decision. Design [3] proposed an edge mode algorithm in intra prediction and achieves sufficient efficiency, but this algorithm is done in HM7.0 which is not the best reference   software for screen content coding. Design [4] proposed a content-based method with efficient inter prediction mode decision. In design [5], the author proposed an algorithm which can decide the prediction mode rapidly and the algorithm have a decrease of 50% in complexity of HEVC, but it has an big increase in bit-rate(about 2.5%). Paper [7] presented a design for a self-adaptive algorithm based on edge information and MAD which achieves sufficient efficiency and reduces intra prediction complexity. Design [8] provides a fast encoding techniques for skipping of IntraBC search which result in about 21%-24% encoding time reduction for intra coding.

This paper proposes an CU partition algorithm which is based on edge direction. This partition mode can provide structural support for some transform, for example, contourlet transform, bandelet transform and so on. Edge direction of the current CU is extracted by using the Sobel mask which has low complexity. In screen content coding, CUs usually have more complex but more regular edges than natural sequence and this is why edge information is more effective in intra prediction. Thus, the edge direction aids in selecting which CU requires partitioning. We introduce the concept of feature value to intuitively illustrate the relationship between edge direction and the partition mode. We achieve a range of 31% to 64% reduction in complexity when using the proposed algorithm compared with the use of HM12.1+RExt5.1 with a PSNR drops of only 0.03dB on average. Compared to the result in paper [9] which results in about   27.6% encoding time reduction and 10.9% BD-rate increasing in screen content coding, the proposed algorithm is more suitable for screen content coding.

The remainder of this paper is organized as follows. Section II presents the proposed algorithm. Section III discussed the experimental results and a simple conclusion is provided in Section IV.

2   Proposed Algorithm

2.1   Overview

Figure 1 illustrates that CU partition size selection problem can be reduced to a set of the most probable partitions, relative to the texture of the video frames. A large CU with high edge complexity must be subdivided to decrease individual complexity, whereas small partitions with low complexity can be combined to form large partitions. We note that highly detailed regions (blocks B and C having high edge complexity) are encoded using small CU sizes, whereas less detailed regions are encoded using large CU sizes (block A).

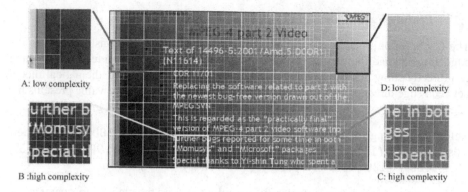

A: low complexity

D: low complexity

B :high complexity

C: high complexity

Fig. 1. CU borders in the first frame of sc_WordEditing

The flowchart of our proposed scheme is shown in Figure 2. As previously mentioned, the key point of the proposed algorithm is to determine the relationship between the selected CUs and their edge direction. Thus, a mathematical model must be built between the edge direction and the partition mode of CUs. In next section, we introduce a feature value to present the edge complexity of CUs.

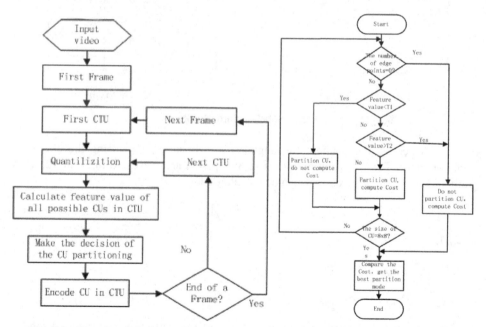

Fig. 2.(a) Flowchart of proposed algorithm   **Fig. 2.(b)** Flowchart of mode decision

Fig. 2. Flowchart of the proposed algorithm

2.2   Feature Value Calculation

The key point of proposed algorithm is to determine out the relationship between the edge direction and the partition modes of CUs. We need to obtain a feature value that can directly show this correlation because we cannot establish a mathematical model between edge direction and the partition mode of CUs. The feature value presents the edge complexity of CU. The method to calculate the feature value is given follows.

Before calculating the edge information, we split the range $(0°, 180°)$ into 12 intervals:. $(0°, 15°)$ $(15°, 30°)$ $(30°, 45°)$ $(45°, 60°)$ $(60°, 75°)$ $(75°, 90°)$ $(90°, 105°)$ $(105°, 120°)$ $(120°, 135°)$ $(0°, 150°)$ $(150°, 165°)$ $(165°, 180°)$.

To reduce computation complexity, we adopted 6 as the stepper to quantize the 256 pixel values, which will not result in too much details losing. We use Sobel operator to calculate the edge information of all the pixels for every CU.

-1	0	1
-2	0	2
-1	0	1

(a)Horizontal Mask

-1	-2	-1
0	0	0
1	2	1

(b)Vertical Mask

Fig. 3. Sobel operator masks

First, we calculate the gradient Rectangular Coordinate System using the classical Sobel operator. Two convolution masks of 3×3 pixels are presented in Figure 2. The gradient vector $D_{i,j}$ of pixel $p_{i,j}$ can be calculated as;

$$D_{i,j} = \{D_{x_{i,j}}, D_{y_{i,j}}\} \tag{1}$$

And;

$$D_{x_{i,j}} = p_{i+1,j-1} + 2 \times p_{i+1,j} + p_{i+1,j+1} - p_{i-1,j-1} - 2 \times p_{i-1,j} - p_{i-1,j+1} \tag{2}$$

$$D_{y_{i,j}} = p_{i-1,j-1} + 2 \times p_{i,j-1} + p_{i+1,j-1} - p_{i-1,j+1} - 2 \times p_{i,j+1} - p_{i+1,j+1} \tag{3}$$

Where $D_{x_{i,j}}$ and $D_{y_{i,j}}$ represent the differences in the vertical and horizontal directions respectively. Therefore, the amplitude of the gradient vector can be roughly estimated by $G_{i,j}$, which is defined as

$$G_{i,j} = \sqrt{D_{x_{i,j}}^2 + D_{y_{i,j}}^2} \tag{4}$$

We then choose an appropriate threshold T. If $G_{i,j} > T$, then pixel $p_{i,j}$ is considered as an edge point. Using the information shown above, we can get the gradient direction θ by using Formula (5).

$$\theta = arccos(\frac{D_{x_{i,j}}}{G_{i,j}})$$ (5)

When we get the gradient direction θ of all the pixels in a CU, we do statistics on the number of edge points whose gradient direction fall within the 12 intervals shown above separately. The number of edge points in the 12 intervals are recorded as I_1 $I_2 I_3 \dots I_{12}$. Then we choose the largest from $I_i (i = 1,2,3 \dots\dots 12)$ and call it I. We then take ΔI as the feature value which represent the edge complexity of this CU. We calculate ΔI by using Formula (6).

$$\Delta I = I / \sum_{i=1}^{12} I_i$$ (6)

The sum of I_i $(i =1,2,3\dots\dots.12)$ is not equal to zero; otherwise, we consider that this CU has no edge points. At this point, we can obtain the feature value that presents the edge complexity of CUs. We then have to decide on which CU requires partitioning and which does not.

2.3   Threshold Decision and Judgment of Proposed Algorithm

We collected sufficient CUs which are partitioned by HEVC from different screen sequences under different Quantization Step (QP 22, 27, 32, 37), different frame rate (20, 30 ,60). These CUs are partitioned by RDO to obtain the threshold of the proposed algorithm. The CUs include 8×8, 16×16, 32×32 and 64×64. The sequences include six different screen content sequences which represent most of the sequences of screen content coding. With the work in the previous section, we can easily build a gradient line chart based on the correlation between the feature value and CUs shown in Figure 4.

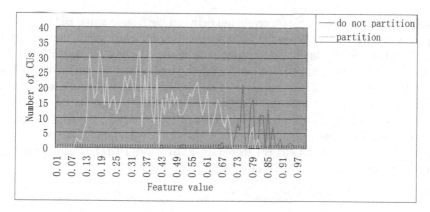

Fig. 4. (a) CUs of 64×64 size

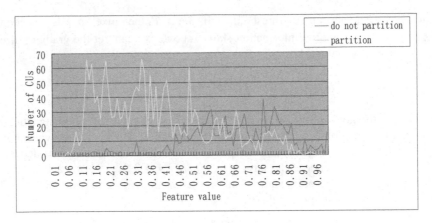

Fig. 4. (b) CUs of 32×32 size

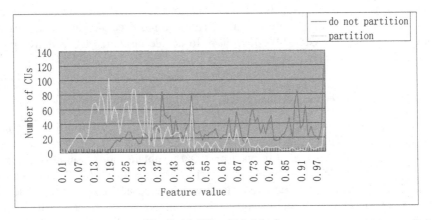

Fig. 4. (c) CUs of 16×16 size

Fig. 4. Relationship between CUs and feature value

Figure 4 (a), (b), (c) show a clear correlation between the feature value and CUs which are partitioned by HEVC. When the feature value is sufficiently small, the CU will be partitioned. Conversely, when feature value is large, the CU will not be partitioned. In Figure 4, we can see when the feature value is not large enough or sufficient small, we cannot decide this CU will be partitioned or not. In this case, we should adopt RDO as the principle to decide whether this CU will be partitioned or not. Based on this relationship, we can obtain the threshold of the CUs of 16×16, 32×32 and 64×64 separately, through which we can determine whether CUs require partitioning. We compute the integral for Figure 4 (a), (b) and (c) to obtain $F(\Delta I)$ and $D(\Delta I)$ which present the ratio of partitioned CUs and the CUs will not be partitioned.

$$F(\Delta I) = \sum_{i=0.00}^{\Delta I} q_i / (\sum_{i=0.00}^{\Delta I} n_i + \sum_{i=0.00}^{\Delta I} q_i) \qquad (7)$$

$$D(\Delta I) = \sum_{i=\Delta I}^{1} n_i / (\sum_{i=\Delta I}^{1} n_i + \sum_{i=\Delta I}^{1} q_i) \qquad (8)$$

As introduced in the previous section, ΔI is the feature value. n_i presents the number of CUs with the same feature value that will not be partitioned. q_i presents the number of CUs with the same feature value that will be partitioned (i=0, 0.01, 0.02.......1). Through our research, we derived these principles for selecting the threshold:

- When $F(\Delta I)$ is close to 1, then the threshold T_1 is equal to ΔI.
- When $D(\Delta I)$ is close to 1, then the threshold T_2 is equal to ΔI.

Based on these rules shown above, we can easily obtain thresholds T_1 and T_2 based on CU sizes. The best values of T_1 and T_2 are as follows.

- When the size of CU is 64×64, T_1 is equal to 0.7, and T_2 is equal to 0.97.
- When the size of CU is 32×32, T_1 is equal to 0.4, and T_2 is equal to 0.94.
- When the size of CU is 16×16, T_1 is equal to 0.17, and T_2 is equal to 1.

By now we get the feature value and threshold, so we can describe this algorithm in a CTU which is the basic partition unit in video frames. The proposed algorithm is as follow.

1. If no edge points are found in this CU, we will compute for the cost and decide that this CU will not be partitioned, then jump to step 6. Otherwise jump to step 2.
2. If the feature value (shown in section 2.2) of this CU is smaller than T_1 which is modified with the size of CU changing, the CU is partitioned without cost calculation, then jump to step 5. If not, jump to step 3.
3. If the feature value of this CU is greater than T_2 which is modified with the size of CU changing, we compute for the cost and do not partition the CU, then jump to step 6. If not, jump to step 4.
4. We partition this CU and compute for the cost and jump to step 5.
5. When this decision is completed and the decision is to partition this CU, we determine whether the size of   this CU is 8×8. If not, we will skip to step 1. Otherwise we jump to step 6.
6. If the size of current CU is 8×8 or the decision is not to partition this CU, we choose the best partition mode based on the cost of the different partitioning structures.

3   Experimental Results

Up to all the frames of each screen content sequence are coded to test the performance of the proposed algorithm, and the test condition is "encoder_intra_main_rext". QP values are set to 22, 27, 32, and 37. A computer with a 2.8 GHz core was used in this experiment. We used HM12.1+RExt5.1 for the comparison to determine the performance of the proposed algorithm.

$$T = \frac{T_{proposed} - T_{HM12.1+RExt5.1}}{T_{HM12.1+RExt5.1}} \times 100\% \tag{9}$$

$$\Delta Y_{PSNR} = Y_{PSNR_{proposed}} - Y_{PSNR_{HM12.1+RExt5.1}} \tag{10}$$

$$\Delta Bitrate = \frac{Bitrate_{proposed} - Bitrate_{HM12.1+RExt5.1}}{Bitrate_{HM12.1+RExt5.1}} \tag{11}$$

In the equation (9), (10) and (11), $T_{proposed}$, $Y_{PSNR_{proposed}}$ and $Bitrate_{proposed}$ represent the encoding time, PSNR and the bitrate of the proposed algorithm. $T_{HM12.1+RExt5.1}$, $Y_{PSNR_{HM12.1+RExt5.1}}$ and $Bitrate_{HM12.1+RExt5.1}$ represent the encoding time, PSNR and the bitrate of HM12.1+RExt5.1. BD-PSNR/Rate measures are used to evaluate the changes in coding efficiency. The percentage difference in bit-rate ($\Delta Bitrate$) and the luminance PSNR difference ($\Delta PSNR$) are also used to compare the proposed algorithm with HM 12.1+RExt5.1. Table 1 and Figure 5 shows the results comparison of the proposed algorithm with the design [1] which is optimized by us to adapt to the screen content coding, the proposed algorithm achieves an encoding time reduction of 39% compared with the anchor(the default algorithm of HM12.1+RExt5.1) with only a 1.1% loss of BD-rate on average. In Table 1, we can see the reduction time T

Table 1. Results of the proposed algorithm

Sequence	Proposed algorithm					Algorithm in paper [1]				
	QP	ΔBitrate (%)	T (%)	ΔY-PSNR (dB)	Y-BD-rate (%)	QP	ΔBitrate (%)	T (%)	ΔY-PSNR (dB)	Y-BD-rate (%)
sc_map_1280 x720_60_8bit _444	22	0.6	-31	-0.03	0.6	22	0.9	22.1	-0.05	1.0
	27	0.5	-30.4	-0.01		27	0.7	22.0	-0.04	
	32	0.5	-30.1	0		32	0.6	23.7	-0.01	
	37	1.1	-28.7	0.02		37	1.0	23.9	0	
sc_web_brow sing_1280x72 0_30_8bit_44 4	22	1.1	-37.1	0.02	1.2	22	4.7	29.1	-0.16	5.1
	27	1.2	-41.3	0.07		27	3.9	29.4	-0.14	
	32	1.1	-38.3	-0.01		32	2.8	29.5	-0.18	
	37	1.3	-37.4	-0.06		37	2.5	30.6	-0.25	
sc_wordEditi ng_1280x720 _60_8bit_444	22	0.6	-32.2	-0.07	1.3	22	3.0	26.5	-0.26	4.9
	27	0.5	-28.5	-0.07		27	2.9	27.3	-0.22	
	32	0.8	-32	-0.08		32	3.0	27.6	-0.19	
	37	0.8	-31.5	-0.1		37	2.7	27.7	-0.29	
sc_SlideShow _1280x720_2 0_8bit_444	22	1.2	-60.6	-0.08	2	22	10.2	57.5	-0.08	11.1
	27	1	-62.8	-0.07		27	9.9	58.3	-0.19	
	32	1.5	-64.7	-0.02		32	8.4	59.1	-0.12	
	37	2.7	-63.1	0		37	7.8	61.0	-0.15	
sc_programm ing_1280x72 0_60_8bit_44 4	22	0.9	-34.3	-0.09	1.3	22	2.6	29.3	-0.18	3.2
	27	0.9	-31.9	0.01		27	2.3	28.5	-0.07	
	32	1	-33.7	-0.04		32	2.0	27.3	-0.13	
	37	1.1	-34	-0.07		37	1.7	27.5	-0.14	
sc_ppt_doc_x ls_1920x1080 _20_8bit_444	22	0.4	-31.6	-0.04	0.5	22	2.2	19.6	-0.20	2.7
	27	0.3	-27.3	0.02		27	1.9	18.7	-0.25	
	32	0.5	-28.1	-0.01		32	1.7	19.4	-0.10	
	37	0.5	-24.5	-0.08		37	1.4	22.2	-0.18	
Average		0.7	-39	-0.03	1.1		2.2	32	-0.15	4.7

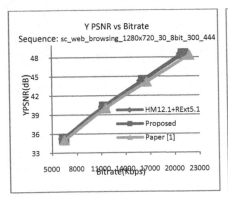
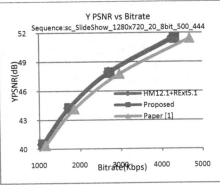

Fig. 5. Y PSNR vs Bitrate

of sequence "sc_ppt_doc_xls_1920x1080_20_8bit_444" is less than the others, because the edge direction of this sequence is more complex and it has more complex edge direction than the other sequences.

4 Conclusions

In this paper, we presented a new algorithm for partitioning CTU in the screen sequence by using the correlation between the edge direction and CUs. A strong correlation between edge direction and the partition mode of CUs was determined through analysis. Such analysis results in a simple decision that can be used for CU partitioning. The proposed algorithm aims to reduce computation complexity significantly with an acceptable loss of BD rate. Results show that the proposed algorithm significantly reduced coding time with an acceptable decrease in quality, which indicates that RDO computation complexity was significantly reduced. Thus, the proposed algorithm meets the requirements of screen content encoding. Next we will focus on transform and quantization, which can provides a significant bit-rate saving.

Acknowledgements. This work is supported by the Natural National Science Foundation of China (No.61370111, No. 61103113) and Beijing Nova Programme (Z14111000180000).

References

1. Chang, C.-Y., Chien, C.-A., Chang, H.-C., Chen, J.-W., Guo, J.-I.: A two level mode decision algorithm for H.264 high profile intra encoding. In: 2012 IEEE International Sysmposium on Circuits and Systems (ISCAS) (May 2012)
2. Ye, T., Zhang, D., Dai, F., Zhang, Y.: Fast mode decision algorithm for intra prediction in HEVC. ACM (August 2013)
3. Hu, S., Cohen, R.A., Vetro, A., Kuo, C.-C.J.: Screen Content Coding For HEVC Using Edge Modes. In: 2013 IEEE International Conference on Acoustics, Speech and Signal Processing (ICASSP) (May 2013)
4. Lin, Y.K., Chang, T.S.: Fast block type decision algorithm for intra prediction in H.264 FRext. In: Proc. ICIP, vol. 1, pp. 585–588 (2005)
5. Na, S., Lee, W., Yoo, K.: Edge-Based Fast Mode Decision Algorithm for Intra Prediction in HEVC. In: 2014 International Conference on Cpnsumer Electronics (ICCE) (January 2014)
6. Yuan, Y., Sun, X.: Edge Information Based Effective Intra Mode Decision Algorithm. In: 2012 IEEE International Conference on Signal Processing, Communication and Computing (ICSPCC) (2012)
7. Bai, C., Yuan, C.: Fast coding tree unit decision for HEVC intra coding. In: 2013 IEEE ICCE-China Workshop, ICCE-China (April 2013)
8. Kwon, D.-K., Budagavi, M.: Fast intra block copy (IntraBC) search for HEVC screen content coding. In: 2014 IEEE International Symposium on Circuits and Systems (ISCAS (June 2014)
9. Jiangyong, S., Xiling, L., Jun, Z.: An Edge-based Approach for Video Text Extraction. In: International Conference on Computer Technology and Development, ICCTD 2009 (2009)

10. Wang, X., Zhai, C.: An Improved Fast Intra-prediction Algorithm in H.264/AVC Based on Edge Feature. In: 2008 International Conference on Computer Science and Software Engineering (2008)
11. Yan, S., Hong, L., He, W., Wang, Q.: Group-Based Fast Mode Decision Algorithm for Intra Prediction in HEVC. In: 2012 Eighth International Conference on Signal Image Technology and Internet Based Systems (SITIS) (2012)
12. Dung, T., Sole, J., Yin, P., Gomila, G., Nguyen, T.Q.: Selective Data Pruning-Based Compression Using High-Order Edge-Directed Interpolation, doi: 10.1109/Tip.2009.2035845 (2010)
13. Leng, J., Sun, L.: Content Based Hierarchical Fast Coding Unit Decision Algorithm For HEVC. IEEE Computer Society (May 2011)
14. Bai, H., Zhu, C., Zhao, Y.: Optimized Multiple Description Lattice Vector Quantization for Wavelet Image Coding. IEEE Transactions on Curcyuts and Systems for Video Technology (2007)

Signal-Aware Parametric Quality Model for Audio and Speech over IP Networks

SongBo Xie, Yuhong Yang, Ruimin Hu, Yanye Wang, Hongjiang Yu,
ShaoLong Dong, Li Gao, and Cheng Yang

National Engineering Research Center for Multimedia Software, Computer School,
Wuhan University, Wuhan, 430072, China
Research Institute of Wuhan University in Shenzhen, China
SaberXie@whu.edu.cn, ahka_yang@yeah.net, hrm1964@163.com,
944708668@qq.com, 1739593607@qq.com, 3659239@qq.com, gaoli_02@126.com,
yangcheng41506@126.com

Abstract. This paper proposes a new signal-aware parametric quality assessment model for audio and speech over IP. The perceptual importance of the reproducing packets and relevant neighborhoods are included in the model. The model is developed from a built audio and speech quality assessment framework using the Artificial Neural Networks(ANN). The overall quality is evaluated by combining the signal-aware parameters with network parameters using a large set of audio and speech samples. It is shown that the signal-aware approach gains higher correlation with the improved PEAQ outputs compared with other parametric method such as E-model.

Keywords: speech/audio over IP, quality measurement, AVS-P10, signal-aware, PEAQ.

1 Introduction

With the rapid growth of the internet, audio services are becoming more and more prevalent, and the streaming audio over mobile and IP services has gained increasing usage to provide high quality audio feeds over the Internet [1]. Audio and speech codecs, such as 3GPP AMR-WB+(WB+) [11] and AVS-P10 in China[2], are developed to meet the demand of high quality and low bandwidth requirement. There is also an increasing demand to quantify the quality of audio over Mobile and IP network in order to ensure better audio and speech performance perceived by end users.

The evaluation of audio and speech service can be performed using either objective or subjective methodologies[3]. The subjective methodology quantifies users' satisfaction using a panel of human subjects. It is well-recognized that subjective tests are time consuming, cumbersome, and expensive. So, the objective methodologies which aim to quantify the perceived quality using auto-executable quality assessment algorithm are an effective solution. The objective quality assessment can be divided into intrusive (full-reference, double-ended)

X. He et al. (Eds.): MMM 2015, Part I, LNCS 8935, pp. 487–497, 2015.

algorithms, which need original and degraded sequences both, and non-intrusive (non-reference, single-ended) ones, which only need degraded sequences or network parameters to estimate the perceived quality. In the context of audio over mobile and IP, non-intrusive approaches are more suitable because their ability to manage and monitor live audio services.

A well-known non-intrusive model is called E-Model defined by ITU-T G.107 [4],which computes the MOS-CQE using a full dedicated network parameters, such as packet loss, delay, codec schema and so on. To make this calculation simpler, RG Cole in 2001 suggests a curve fitting expression to both delay and equipment impairment factor based on the subjective measurements[5]. And L.Sun proposed an objective methodology that evaluates speech quality using PESQ rather than subjective measurements in deriving regression models in 2004[6], this is more efficient than tradition training methods. Most speech quality assessment tools assume that all missing packets have an equal effect on final perceived quality, which however, is not coincident with empirically recongnition. So a new set of quality assessment models that is sensitive of loss patterns have been proposed to evaluate the perceptual speech quality considering importance of missing packets[7][8].These approaches refered are all aimed at speech quality assessment only.

To estimate audio quality, A new model was proposed to measure MOS of audio over IP networks based on E-model and WB-PESQ/PEAQ[9]. Then in 2013, Z. Li presented a methodology to rank music quality over IP using machine learning techniques[10]. But they both estimate audio quality without considering the perceputal improtance of missing packets.

In these paper, we propose a new signal-aware parametric approach for measuring both audio and speech quality over mobile and IP. The main contributions of this paper can be concluded as follows:

1. This paper proposes an ANN technique, which accounts for both the packet loss patterns, signal features of dropped fragments and packet loss concealment (PLC) at the receiver to evaluate audio and speech quality with improved PEAQ.

2.We argue the different coding modes in hybrid codecs, like AMR-WB+ and AVS-P10, and why they contribute to different levels of degraded quality while losing particular combination of them. We quantify the degradation level of a frame with its neighborhood frames by segmental weighted SNR (weightedSNR).

3. We conduct verification tests and also compare with E-model to show the effectiveness of the proposed approach.

The rest of the paper is organized as follows, Section 2 outlines our proposed framework to assess audio/speech quality based on the perceived importance of reproducing packet. And Section 3 presents how to identify the perceived importance of a particular packet by using mode information retrieved from the received coding bitstreams, and how to conduct ANN trainning on AVS-P10 codec. Section 4 verifies the ANN model and compares it with well-known E-model method. Section 5 concludes our work and explores future directions.

2   Proposed Signal-Aware Approach

Our approach predicts the audio quality over IP using Artificial neural network(ANN) technique based on the perceptual importance of reproducing lost packets and relevant neighborhood. The main procedure, similar to [13][14], is to have the specific degraded sequences evaluated using intrusive assessment approaches, then use the measured MOS to train a ANN model with the relation between the impaired parameters and the measured perceived quality. The procedure is presented in Fig. 1.

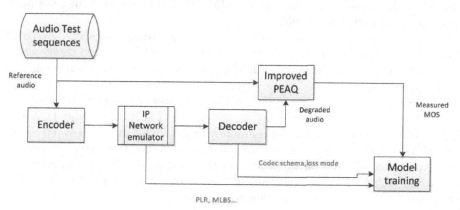

Fig. 1. Overview of signal-aware approach

Firstly, this procedure need to identify impairment parameters in the given environment. We classified all the impairments into two sets of relevant parameters, one is coding impairments (signal related) and the other is transmission impairments (network related). In all transmission impairments, we focus on the packet loss rate(PLR) and Mean Loss Burst Size(MLBS), because these two parameters can mainly impact the transmission quality. Other impairments, like jitter which can be translated into extra PLR by jitter buffer technique, and delay which has no serious influence on transmission quality degradation while it is lower than 400ms, will not be considered in this paper. MLBS, here, is to represent the burstiness of loss. While the transmission impairments are all directly related to network services, the coding impairments are measured through the analysis of received packets. As shown in Fig. 1, this impaireness contain the type of codec used and the loss modes combination to specify the perceived importance of loss packets.

A series of degraded sequences are generated by simulating impairments after choosing representative parameter values. Then QoE of the specific degraded sequences are objectively measured by using the improved PEAQ[15]. It should be noticed that this improved PEAQ is developed to suit lower bandwidth audio/speech sequences and network impairments, which is more appropriate to use it here other than original PEAQ. What we get now is a large set of impairment parameters and its corresponding QoE value. Finally, some of the data are used to train ANN model and remainning ones are used for validation.

2.1   Loss mode classification

Codec Mode for Audio/Speech. Generally, hybrid codecs, like AMR-WB+, AVS-P10, use mixed ACELP/TCX algorithm to deal with audio and speech signals. The codec modes in AMR-WB+ can be divided into four categories, ACELP256, TCX256, TCX512 and TCX1024 [11]. Each of them refer to different coding modes and frame length, which also present their corresponding signal features. Generally speaking, ACELP256 codec is more accurate and efficient for speech signals while TCX more likely be applied to audio signals. TCX256 is more suitable for noisy transient audio while TCX512 and TCX1024 are more likely to fit with stationary audio.

In these modes, the frame length can refer the steady state of the signal. Short window coding modes, like ACELP256 and TCX 256, last 20ms which only occupy one quarter of a full super-frame, means it is more changeable and unsteady. Serious voice degradation must be considered if this nonstationary signal are lost. What may make the degradation worse is that the packet loss occur while nonstationary frames switch frequently. The TCX512 and TCX1024 frames refer to more steady signal characteristics, which can be recovered better through the PLC methods. These features will be applied to cluster the perceptual importance of reproducing frames. In this paper, we make use of AVS-P10 and define ACELP256,TVC256,TVC512,TVC1024 [2] as mode 0,1,2,3 respectively and * represent all the corresponding four modes.

Mode Clustering. In [12], Hu Y introduced serveral objective measures and compared their performance. From all this objective measures, we choose the weighted segSNR as our overall quality measurement for its accuracy and convenience.

The classification procedure is shown in Fig. 2. Firstly, by discarding particular packets and recording modes information, a series of degraded sequences can be made with its lost packet information recorded. To avoid the discrepancy brought by silent frame, we decide whether to use the degraded sequences or not by analyzing mean energy of the special lost packets. Then the weightedsegSNR of degraded sequences is calculated and they referred to the level of impairments, that is, if the weighted segSNR is smaller, it means the impairment is obvious and this particular packet can be seen as more important. Packet modes is clustered to different levels in coincidence with the value of weighted segSNR. We record not only the current mode but also its neighborhood modes while classifing because of the correlation between consecutive packets used in PLC technique.

2.2   Codec Related Parameters

The codec related parameters are necessary in the training process. Corresponding bitrate and PLC techniques should be specified for the specific codec.

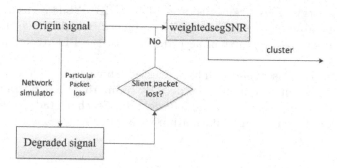

Fig. 2. Precedure of classification

3 Mode Classifications And ANN Model Building

The mode classifications use twelve different audio/speech sequences as reference signal to produce various levels of degraded sequences. These sequences selected from MPEG test sequences are shown in Table 1. To simplify the algorithm, we choose AVS-P10 as our codec and each frame, occupied 20ms, as a single packet during transmission.

Table 1. Test sequence from MPEG

Items	Sequences	Sequence type
Item 1	es01	female voice
Item 2	es02	male voice
Item 3	es03	female voice
Item 4	sc01	instrument
Item 5	sc02	instrument
Item 6	sc03	instrument
Item 7	si01	instrument
Item 8	si02	instrument
Item 9	si03	instrument
Item 10	sm01	instrument
Item 11	sm02	instrument
Item 12	sm03	instrument

3.1 Mode Classification

In order to ensure the importance of each mode, we produce the degraded sequences by discarding every consecutive frames with the same loss length. The loss length here defined as 1, 3 or 5, representing how many packets lost in one situation. It means if es01 have 532 frames, we will produce about 532*3=1596 degraded es01 sequences. Notice that we discriminate silent packets with the energy threshold 15dB, which means, these degraded sequences will be used only if the mean energy of lost frames is large than 15dB. While recording the loss mode corresponding with every degraded sequence, we also take its previous three good frames into account in accordance with the PLC technique applied in AVS-P10.

After having origin sequences all be processed, about 6,000 different sequences were produced with its loss mode recorded. By comparing these degraded and its corresponding origin sequences, weighted segSNR of each degraded sequence was calculated. And the clustering information with different loss mode is display in Fig. 3.

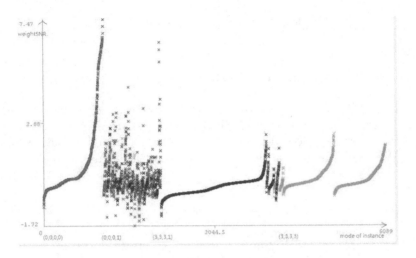

Fig. 3. Clustering results

From the Fig. 3, it can be informed that mode 0(ACELP256) is indeed more sensitive than mode 3, which support the hypothesis given in section 2.1. And our classification clustered by modes is illustrated in Table 2.

Table 2. Classification result

Class	Previous three mode	Current mode	Mean weightSNR
Class 1	3,3,3	*	0.05
	3,3,1	1	
Class 2	*,*,0	0	0.90
Class 3	else		0.34

3.2 Signal-Aware Model Trainning

In order to train the Neural Networks, we conduct a set of objective tests with PEAQ for different audio/speech sequences under different conditions. As refered in the beginning of this section, we used AVS-P10 (packetized per frame, referring to 20ms packets) with two different coding bitrates(12kbps,24kbps) as our codec parameters.

The test sequences we used is listed in Tab. 1. As shown in Fig. 1, the main input network parameters are packet loss rate(PLR) and their distribution(MLBS) for Neural Network trainning. And mode parameters ,according to Tab. 2,depend on the mode combination where loss occured. Firstly for PLR, we considered it from 1% up to 30% with the step of 1%. And the MLBS range from 1 to 7. The loss mode was spread to three different classes. Some combinations may be impossible, while MLBS can not be set to high if the PLR is low with the limitation of the sequence duration. And some mode classes may not exist in the particular sequence for that speech sequence almost consist of mode 0. So only valid combination would be taken into account.

As the same way to handle the silence packets above, we will not consider the sequence quality if the losses occur in the silent period because it hardly impact on the perceived quality. Therefore, we only cosider those packets which energy were higher than 15dB.

For each combination of PLR, MLBS and loss mode, we generated five different loss patterns using all twelve origin sequences respectively. Then these loss patterns will be measured by improved PEAQ and recorded automatically.

To simplifiy the architecture of Neural Network, we decided to generate different NN for each coding bitrate, that means trainning two NN corresponding to two different bitrates. Then we took the PLR, MLBS and loss modes as the input parameters for the NN and corresponding MOS received from PEAQ as the target parameters. To verify the effectiveness of loss mode parameters, we also train the NN by taking the PLR and MLBS as input parameters only.

4   Verification and Comparison

4.1   Verification

We use the multilayer Perceptron feed-forward neural network to build our quality assessment model with three hidden layers. While constructing the NN, we decomposed the data sets into two parts: 80% of the data for training, and remainning 20% for validation. Here, Weka software[16] was used for training and measuring the correlation automatically. For simplification, we only show the NN result while sequences coding at 24 kbps. The result is depicted in Fig. 4. From Table 3, the proposed signal-aware trained NN model shows higher correlation with the measured PEAQ scores. That means the prediction of audio/speech sequences is more accurate and effective by using our model based on the objective improved PEAQ. Also the correlation without loss mode parameters is lower than the signal-aware approach by 9%, which represent the necessity and efficiency to take the mode information into quality evaluation.

4.2   Comparision to E-model

In this section, we compare our proposed model with the well-known parametric model:E-model. However E-model proposed initially aim to evaluate speech

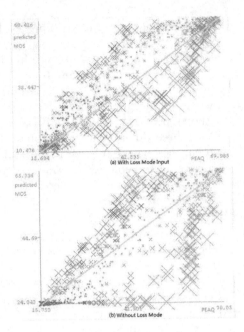

Fig. 4. Test result

Table 3. Performance of signal-aware model

Model	Correlation
ANN model without mode parameters	78.9%
Mode-aware parametric model	87.2%

quality only, the model we used has been extended in [9] to audio quality measuring. The parameters of the E-model are given as follows:

$$R = 93.2 - I_d - I_{e-eff} - A; \qquad (1)$$

$$I_{e-eff} = I_e + (95 - I_e)\frac{P_{loss}}{\frac{P_{loss}}{BurstR} + Bpl}; \qquad (2)$$

$$I_d = 0; A = 0; \qquad (3)$$

$$MOS_c = \begin{cases} 1 & if\ R \leq 0 \\ 4.5 & if\ R > 100 \\ 1 + 0.05R + R(R - 60)(100 - R) * 7 * 10^{-6} & if\ 0 < R \leq 100 \end{cases} \qquad (4)$$

In the equation (2), BurstR parameters means the burstiness of loss, which can be calculate by two-state Markovian model[17]. The Bpl and I_e represent device impairement, related to what specific codec used. The equation (4) was used to translate calculated R factor into objective MOS.

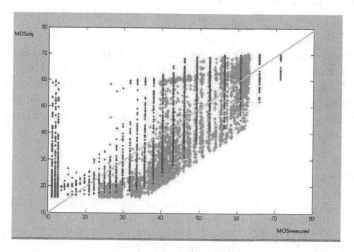

Fig. 5. Performance of different prediction model(green spots is our proposed ANN dataset while red ones represent E-model)

The Fig. 5 shows the correlation difference between the two different models. We noticed that E-model may misjudge some sequences when there is no serious distortion occured, because E-model only considers packet loss and loss bursti- ness parameters without evaluating the perceptual importance of loss packets. The loss may happens exactly where the packet is silent or not so perceptual im- portant to end users. It can be observed that the signal-aware ANN model have a good correlation with the objective PEAQ scores compared with E-model.

5   Conclusions and Future Works

In this paper, we present a signal-aware parametric QoE evaluating approach for both audio and speech over mobile and IP. It can be applied to monitor transmission quality and adjust related parameters to get the high perceptual quality for end-users. Input parameters to measure distortion are packet loss rate, mean loss burst size and the class of loss modes, which can be easliy retrieved from either network or coding bitstream themselves. Experiment shows its high correlation with the objective PEAQ score. Futhermore, we compared it with other nonintrusive method E-model and the constructed ANN model without loss mode parameters. Results well support our hypothesis, that type of loss packets play an important role in the perceived quality and our proposed model is more effective than E-model.

The proposed evaluating approach only consider transmission impairments and coding impairments, which might not reach the full identity with subjective opinion. Future works will focus on expanding the trainning database in realistic transmission and take other impairments, like delay into cosideration. In the mean time, by applying this approach to monitor audio/speech quality in real time transmission, we want to find some effective approaches to adjust the coding bitrate or jitter buffer technique to preserve good Quality of Experience.

Acknowledgement. This work is supported by National Nature Science Foundation of China (No.61102127, 61231015, 61201340, 61201169, 61471271, 61401319), Guangdong-Hongkong Key Domain Breakthrough Project of China (No. 2012A090200007), The major Science and Technology Innovation Plan of Hubei Province (No. 2013AAA020) and Science and Technology Foundation of Guizhou Province (No. LKS[2011]1).

References

1. Jonsson, L., Coinchon, M.: Streaming audio contributions over IP (PDF). EBU Technical Review (2008) (retrieved December 27, 2010)
2. AVS_M2421, AVS P10: Mobile audio and speech codec Committe Draft v1.0," AVS beijing Ad-hoc Meeting, Beijing, China (July 2008)
3. Jelassi, S., Rubino, G., Melvin, H., Youssef, H., Pujolle, G.: Quality of experience of VoIP service: A survey of assessment approaches and open issues. IEEE Communications Surveys & Tutorials 14(2), 491–513 (2012)
4. Rec, I.G.: 107-The E Model, a computational model for use in transmission planning. International Telecommunication Union 8, 20–21 (2003)
5. Cole, R.G., Rosenbluth, J.H.: Voice over IP performance monitoring. ACM SIGCOMM Computer Communication Review 31(2), 9–24 (2001)
6. Sun, L., Ifeachor, E.: New models for perceived voice quality prediction and their applications in playout buffer optimization for VoIP networks. In: 2004 IEEE International Conference on Communications, vol. 3, pp. 1478–1483. IEEE (2004)
7. Jelassi, S., Youssef, H., Hoene, C., et al.: Voicing-aware parametric speech quality models over VoIP networks. In: Global Information Infrastructure Symposium, GIIS 2009, pp. 1–8. IEEE (2009)
8. Hoene, C.: Internet telephony over wireless links. PhD, Technical University of Berlin, TKN (2005)
9. Graubner, M., Mogre, P.S., Steinmetz, R., et al.: A new QoE model and evaluation method for broadcast audio contribution over IP. In: Proceedings of the 20th International Workshop on Network and Operating Systems Support for Digital Audio and Video, pp. 57–62. ACM (2010)
10. Li, Z., Wang, J.C., Cai, J., et al.: Non-reference audio quality assessment for online live music recordings. In: Proceedings of the 21st ACM International Conference on Multimedia, pp. 63–72. ACM (2013)
11. 3GPP TS 26.290: Extended Adaptive Multi-Rate-Wideband (AMR-WB+) codecs(Release 10) (April 2011)
12. Hu, Y., Loizou, P.C.: Evaluation of objective quality measures for speech enhancement. IEEE Transactions on Audio, Speech, and Language Processing 16(1), 229–238 (2008)
13. Cherif, W., et al.: A_PSQA: Efficient real-time video streaming QoE tool in a future media internet context. In: 2011 IEEE International Conference on Multimedia and Expo (ICME). IEEE (2011)

14. Cherif, W., et al.: A_PSQA: PESQ-like non-intrusive tool for QoE prediction in VoIP services. In: 2012 IEEE International Conference on Communications (ICC). IEEE (2012)
15. Yuhong, Y., et al.: A new mobile audio quality assessment using Jitter Distortion Measure approach. In: 2013 Fifth International Workshop on Quality of Multimedia Experience (QoMEX). IEEE (2013)
16. Hall, M., et al.: The WEKA data mining software: an update. ACM SIGKDD Explorations Newsletter 11(1), 10–18 (2009)
17. Yajnik, M., et al.: Measurement and modelling of the temporal dependence in packet loss. In: Proceedings of the Eighteenth Annual Joint Conference of the IEEE Computer and Communications Societies, INFOCOM 1999, vol. 1. IEEE (1999)

Patch-Based Disparity Remapping
for Stereoscopic Images

Dawei Lu, Huadong Ma, Liang Liu, and Huiyuan Fu

Beijing Key Lab of Intelligent Telecomm. Software and Multimedia,
Beijing University of Posts and Telecomm., Beijing 100876, China
ludawei19@gmail.com,mhd@bupt.edu.cn,
liangliu82@gmail.com,fhy@bupt.edu.cn

Abstract. Post-production and processing for stereoscopic 3D are attracting a lot attention in recent years. In particular, the acquired disparity in most situations requires further manipulation to adjust different view conditions. This paper proposes a novel method to address the issue of disparity remapping of stereoscopic images. We present a nonlinear disparity mapping model to adjust the depth range of the whole image as well as the special visual important regions. To implement this model, our method compute saliency maps for the stereoscopic images. Then we extend the PatchMatch algorithm to search for the proper patches in both the left and the right images by visual combined constraints, and use them to iteratively refine the images to meet the target depth range. Our method is capable of minimizing the distortion of the images and ensuring the correct stereo consistency after disparity remapping. The experimental results demonstrate that the proposed approach can adjust the depth range to improve the stereoscopic effects while preserving the naturalness of the scene.

Keywords: Stereoscopic 3D, disparity remapping, visual importance, image patch.

1 Introduction

With a substantial success in the 3D digital cinema industry, more and more stereoscopic 3D images and movies are being produced. In general the stereoscopic content is captured by placing two cameras side by side for the same scene; the natural and comfortable viewing experience is determined by proper settings for the geometry of the scene, the location of viewer and the size of display device. However, the complex and combined conditions are difficult to handle in the production step so that many undesirable results are created. Then, stereoscopic 3D post-production becomes a popular research issue recently [1].

Disparity/depth remapping is one of the most important and difficult issues in the post-production process. A captured stereo content requires further manipulations because of cameras misalignment, incorrectly prediction for the perceived depth, and different display devices adjustment. While various view conditions

X. He et al. (Eds.): MMM 2015, Part I, LNCS 8935, pp. 498–509, 2015.

require different depth ranges for the proper view experience, disparity remapping may have several emphases including global disparity range adjustment, object depth mapping , depth manipulation between some objects, etc.

There are several methods proposed to address this issue. Based on image warping method, some existing works [2] [3] [4] [5] [6] made mesh for image pairs and warp them in a continuous manner. This kind of methods can prevent results from creating new occlusions and is always robust, but it can not completely avoid distortion in some regions, especially in object edges. For other works [7] [9] [12], they addressed this issue by using the estimated disparity maps or user interaction. These methods can minimize distortion , but the computational cost is very high or the application field is limited.

In this paper, we propose a novel disparity remapping method for adjusting the depth range of stereoscopic 3D images as well as selectively adapting depth of the local important regions. As our approach of disparity manipulation can handle different image regions, we detect and extract the visual importance regions for two views by saliency detection and image segmentation at first. Then, with the unconnected saliency regions we propose a nonlinear disparity remapping model to adjust the disparity values for different depth ranges occupied by different image regions. The disparity ranges including saliency regions are remapped according to saliency cues; the created depth overlapping are handled by additional constraints; the depth range in other regions are adjusted by typical linear depth remapping method. In order to implement our disparity remapping model, we extend the PatchMatch algorithm [10] to present a patch-based image editing method. By utilizing this method, the local image regions in two views are rearranged with stereo consistency, and the missing regions are completed from other regions in the same view and the corresponding regions in the other view.

The main contributions of this paper are summarized as follows:

– We propose a patch-based method to remap the disparity range of stereoscopic images according to the view conditions. The method can produce stereo consistent results and minimize the shape distortion after the remapping procedure;
– We remap the unconnected saliency regions and the background respectively and propose a check algorithm to avoid depth overlapping for the results;
– We consider the tradeoff between user interactions and computation constraint complexity. This make our method implement efficiently and preserve stereo quality.

2   Related Work

Stereoscopic 3D content post-production for displaying on different devices is attracting many research interests in recent years. Especially, depth/disparity adapting is an important and challenging research field which makes more and more works focus on it.

The common approach to address this issue is inspired by 2D image warping technique. Lang et al. [2] proposed a nonlinear disparity remapping approach which utilizes the saliency maps extracted from the scene and sparse feature correspondences between two images. This method introduces four mapping operators to address the depth adjusting problems of stereoscopic images/videos as a warping issue so that it can avoid some complicated steps in previous works. However, they did not consider the relationship of perceive depth among objects in the stereo images. Moreover, like many other warping methods, they can not avoid the distortion of the shape with sparse features. Yan et al. [4] introduced a linear depth mapping method for stereoscopic videos. This approach creates a linear depth mapping model to adjust the depth range of stereoscopic videos, and it preserves the 3D structure as well as ensures that the video changes smoothly. Besides the deficiency of image warping technique, the limitation of this method is that it can not handle some special regions which have inappropriate disparity values.

Another kind of disparity mapping methods requires estimated disparity maps. A depth manipulation method by using disparity histogram analysis was presented in [12]. This approach treats disparity remapping as a layer-based issue through analyzing the disparity maps and using nonlinear disparity adjustment functions. The local depth range of stereo images can be well handled in this method while the original depth order can not be destroyed, but for some images with complex scene the depth layers are classified difficultly and the depth change in videos is not addressed accurately. Yan et al. in [7] proposed stereo image editing method based on shift-map method. This method can preserve 3D features and minimize shape distortion which limits the warping technique. However, the approach edits the images in pixel level so that it is difficult to apply to the videos by high computational cost.

In summary, several previous works are presented for disparity remapping of stereoscopic 3D. Although these works applied various constraints to ensure the stereo quality, some works did not consider content preservation and image coherence, and the others reguired many information from the users or only handled the overall depth range. In contrast, we aim at addressing this issue for selectively local image regions as well as the whole image, and we focus on efficiently remapping the disparity with proper user interactions and simple constraints.

3   Our Disparity Remapping Approach

In this section we propose our disparity remapping approach. Given the input image pair (I_L, I_R) in which I_L and I_R are the left image and right image respectively, we compute their disparity maps by the approach suggested in [11]. First, we detect and extract the saliency regions in two views for the local disparity manipulation. Then we propose a nonlinear disparity remapping model for stereoscopic images. This model can change the overall depth range as well as the disparity ranges of different local regions. Finally, we propose a patch-based image editing method based on [10] to achieve the depth changing. Moreover,

Fig. 1. Demonstration of the saliency extraction: (a)original left image of a stereo pair, (b)saliency map using histogram based contrast algorithm, (c)saliency map using region based contrast algorithm, (d) user strokes on the original image, (e)extracted result for one saliency region, (f)final extracted saliency result. Here in (d), the green strokes and the yellow strokes indicates two unconnected saliency regions.

this editing method could check stereo consistency by the remapped disparity so that the inconsistent and incorrect values from the input disparity maps can be corrected in the results. After all the steps we can get the output stereo image pair(I'_L, I'_R).

3.1 Image Saliency Regions Detection

While stereoscopic 3D images are displayed, the viewers always pay attention to the saliency regions. Thus we need detect and extract the saliency regions for the stereo image pair. First we estimate two saliency maps for each image using the approach in [13]. Then we extract the saliency regions using GrabCut method in [14]. Although Cheng et al. [13] show that their results are better than many previous works and also use GrabCut to extract saliency regions, they can only create proper results when the images have one saliency object or connected saliency region.

In this paper we extract the unconnected saliency regions respectively by the two saliency maps and user interactions. Here we combine two saliency maps as image segmentation cues and use it to iteratively run GrabCut algorithm. For one saliency region, at each iteration we make strokes on the current segmentation result to facilitate next iteration. After iteratively refining the segmentation results, the saliency region can be extracted. Then we address the other saliency regions with the same steps and combine them to obtain the final saliency result of a stereo image pair. Figure 1 shows the components of the procedure of the saliency extraction.

3.2   Disparity Remapping Model

In this paper, we remap the original disparity range of stereoscopic images to a new disparity range for preferable stereo perception. From a given stereo image pair (I_L, I_R) and the computed disparity maps, we have the disparity value $d(x, y)$ as the distance from a pixel (x, y) in one image to the corresponding pixel in the other image. Specifically, we define $d_L(x, y)$ as the disparity from the left image to the right image, and $d_R(x, y)$ is from the right image to the left image. Then the disparity range between two images can be denoted as $\Phi = [d_{min}, d_{max}]$, the new target disparity range can be denoted as $\Phi' = [d'_{min}, d'_{max}]$, and our disparity remapping function can be denoted $\eta : \Phi \to \Phi'$.

For globally linear disparity scaling model, the scaling factor which maps the orginal disparity range to a new disparity range are defined as:

$$\lambda = \frac{d'_{max} - d'_{min}}{d_{max} - d_{min}} \tag{1}$$

With the scaling factor, the remapping function η adapts a disparity value $d \in \Phi$ to a target range Φ':

$$\eta_l(d) = \lambda(d - d_{min}) + d'_{min} \tag{2}$$

The linear disparity remapping functions can scale the overall depth range of stereo images, but for many created image pairs only the local regions need this operation. Thus we use nonlinear disparity remapping function to address this issue which is similar to the method in [2]. We define the locally nolinear disparity remapping functions $\eta_n(d)$ as follows:

$$\eta_n(d) = \begin{cases} \eta_1(d), & d \in \Phi_1 \\ \cdots & \cdots \\ \eta_k(d), & d \in \Phi_k \\ \eta_z(d), & d \in \Phi_z \end{cases} \tag{3}$$

where $\Phi_1, ..., \Phi_k$ is a set of different disparity ranges that are occupied by different saliency regions, and Φ_z corresponds to the regions which belongs to background areas(or perspective). For each disparity range including saliency regions, the scaling function is defined as:

$$\eta_i(d) = d + \omega_i \cdot \upsilon_i(d), \quad i = 1, ..., k, \ d \in \Phi_i, \tag{4}$$

where $\upsilon(d)$ is the average disparity magnitude of the saliency regions in the disparity range Φ_i, and ω is the weight that controls the amount of disparity change. For background areas, the mapping procedure can be handled by Eq. 2.

In this way we can selectively adjust the disparity of local regions in stereo images, but it may create depth overlapping or inversion. Thus we add a overlapping check to the nonlinear disparity remapping functions, we assume that $\{\Phi_1, ..., \Phi_k, \Phi_z\}$ has a ascending order and show the overlapping check procedure in Algorithm 1.

Algorithm 1. Check algorithm for depth overapping

---
1. **for** $i = 1$ to $k - 1$ **do**
2.   **if** $max(\eta_i(d)) > min(\eta_{i+1}(d))$ **then**
3.   Change ω_{i+1} in Eq.4 until $max(\eta_i(d)) \leqslant min(\eta_{i+1}(d))$
4.   **end if**
5. **end for**
6. **if** $max(\eta_k(d)) > min(\eta_0(d))$ **then**
7.   Use Eq.2 to change $\eta_k(d)$ until $max(\eta_k(d)) \leqslant min(\eta_0(d))$
8. **end if**

3.3 Patch-Based Image Editing

Now our aim is to apply the remapping functions presented above to the input image pair, so we propose a patch-based method to extend the PatchMatch algorithm [10] to edit the images in image patch level. The editing steps include image reshuffling and image completion. In traditional 2D image processing, image reshuffling(also known as image content rearrangement) is to move the local regions to new positions, and image completion is to fill the missing regions from the rest regions of the image. Barnes et al. [10] achieves this procedure in a single image as a global optimization problem, they measure the image completeness and coherence by seeking to minimize the following objective functions [15] [18]:

$$D_{comp}(S, T) = \sum_{s \in S} \min_{t \in T} D(s, t), \qquad (5)$$

$$D_{coh}(S, T) = \sum_{t \in T} \min_{s \in S} D(t, s), \qquad (6)$$

where S is the source region in image reshuffling(the rest regions of the image in image completion), T is the target region in image reshuffing(the hole in image completion), $s \in S$ and $t \in T$ are patches in the source and target regions respectively, and $D(s, t)$ measures the patch distance between s and t. For image reshuffling, the image completeness and coherence are both considered to ensure that the target region contains as much information from the source region as possible, and any artifacts are penalized for the similarity between the input and the output. But for image completion, only the image coherence should be considered.

While the input becomes the stereoscopic images, the extra view image and disparity between two views should be considered. In the image reshuffling step, the source region(saliency region) is horizontally moving to the target region that satisfies the remapped disparity. So with the new depth information we define the new disparity of the target regions in both two images by the following function:

$$d(T_L, T_R) = \eta(d(S_L, S_R)), \qquad (7)$$

where S_L and S_R are the corresponding source regions with the original disparity d in left and right images respectively, while T_L and T_R are the corresponding target regions with the remapped disparity $\eta(d)$.

Then we have the total similarity measure between source region and target region:

$$D_{total}(S,T) = \gamma D_{comp}(S,T) + (1-\gamma)D_{coh}(S,T). \tag{8}$$

As the new positions to which the original regions move are defined by stereo correspondence in two views, we handle the measure problem in the left and the right views respectively. To arrange the source regions to the target regions correctly and completely, we correct the offsets to the new positions at each iteration, and we use $\gamma = 0.75$ in the whole implementation procedure.

In the image completion step, for one of the stereoscopic images the missing region is filled from the rest of the image and the other view. First, as the viewpoint difference between two stereo images, part of the missing region in one view may be visible in the other. Then with the remapped disparity information we can mutually fill these parts for two images similar to [17]. Let (Ω_L, Ω_R) denote the missing regions in the left and right views, and let (Ψ_L, Ψ_R) denote the rest of the two images. So if pixel $(x,y) \in \Psi_L$ and $(x - \eta(d_L(x,y)), y) \in \Omega_R$, the pixel in the right view can be filled by the equation:

$$I_R'(x - \eta(d_L(x,y)), y) = I_L(x,y), \tag{9}$$

where d_L is the disparity from the left view to the right view. Similarly, if the pixel $(x,y) \in \Psi_R$ and $(x + \eta(d_R(x,y)), y) \in \Omega_L$, we have:

$$I_L'(x + \eta(d_R(x,y)), y) = I_R(x,y), \tag{10}$$

Then, for the other missing regions we denote them as (Ω_L', Ω_R') in two views, we address this issue by using a patch-based completion method similar to [16]. Here, let (Ψ_L, Ψ_R) denote the source regions, and let (Ω_L', Ω_R') denote the target regions and $F(t)$ denote the mapping function between I_L' and I_R' with the remapped disparity. In particular, this function includes the mapping from patch $t \in \Omega_L'$ centered at (x,y) to the corresponding patch centered at $(x - d_L(x,y), y)$ and the patch $t \in \Omega_R'$ centered at (x,y) to the corresponding patch centered at $(x + d_R(x,y), y)$.

Thus we extend Eq.6 to stereoscopic media to define the stereo completion coherence as minimization of the flollowing objective function:

$$D_{coh-stereo}(\Psi_L, \Psi_R, \Omega_L', \Omega_R') =$$
$$\sum_{t \in \Omega_L' \cup \Omega_R'} \min_{s \in \Psi_L, \Psi_R} D(t,s) + \sum_{t \in \Omega_L' \cup \Omega_R'} D(t, F(t)). \tag{11}$$

where the first term is similar to Eq.6 and ensures the image coherence when the target regions are filling in each view, the additional term ensures stereo consistency in the filled regions with the remapped disparity.

We also extend the PatchMatch algorithm [10] to search for the patches which can fill the target regions by propagating the candidate patches from neighbors in the source regions and the other view. The patches we search for in the propagation step have the following constraints:

- The patches which match the current patch and the neighbors of the patches which match the neighbors of the current patch;
- The patches which are in the same image regions with the current patch by classification of the remapped disparity ranges are considered first, then to the regions of neighbor disparity ranges;
- The corresponding patch and its neighbors in the other view by $F(t)$.

Then for the color computation of the missing regions, we apply the blending manner presented in [16] to achieve it.

4   Experimental Results and Discussion

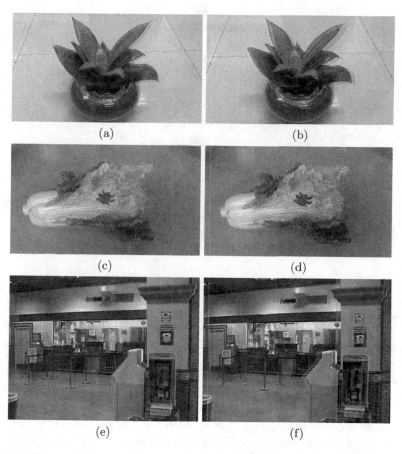

(a)   (b)

(c)   (d)

(e)   (f)

Fig. 2. Demonstration of our disparity reampping method with comparing between the original stereo images and the remapped results: (a), (c) and (e) are the input stereo images; (b), (d) and (f) are the corresponding output images.

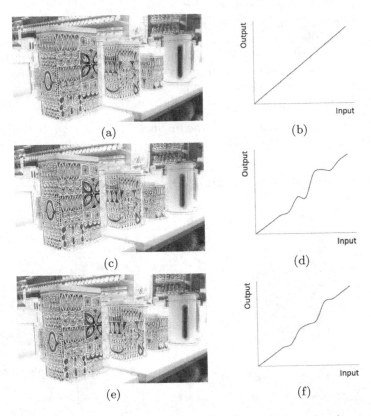

Fig. 3. Examples of handling the depth overlapping and inversion in our disparity reampping method: (a) is the input stereo image; (c) is the disparity remapped image before depth overlapping processing; (e)is the result after depth overlapping processing; (b), (d)and (f) are the disparity mapping functions for the corresponding images

In this section, we evaluate the performance of the proposed method by comparing the results with various situations. In our implementation, the weight ω for controlling the change of the disparity in different depth ranges is set to $[-4.5, 7.5]$, and by this way we can adapt the stereoscopic images to different resolutions on various display devices. The results in this section are presented as red-cyan anaglyph images.

Figure 2 shows our disparity remapping results. Fig.2(b) increases the disparity range to 200 % by Fig.2(a) and Fig.2(d) reduces the disparity range to 50 % by Fig.2(c), and we can see that only the disparity range of the foreground object in each group is changed. But we also observe that in Fig.2(f) the disparity range of foreground regions is increased and that of background regions is decreased, by this way the whole scene is nearer to the viewers and the stereo visual experience can be improved.

Figure 3 demonstrates the results for treating the depth overlapping and inversion in depth manipulation procedure. Given the input stereo image in Fig.3(a), we increase the depth range for it and get the result in Fig.3(c). Due to the limitation

Fig. 4. Comparison between our disparity remapping results and the results of method in [2]: (a) and (b) are original stereoscopic images; (c) and (d) are the results of [2]; (e) and (f) are our results. Here the both two methods increase the original disparity ranges to 200 %.

similar to many other nonlinear disparity mapping methods, the depth inversion may be created after the mapping procedure, in Fig.3(c) we can observe that the second and third nearest boxes from the viewer have the overlapped depth ranges comparing to the kettle, and we observe the disparity remapping function including depth overlapping in Fig.3(d). The corrected result by check and modification of the depth overlapping is shown in Fig.3(e) and the corrected disparity remapping function is shown in Fig.3(f).

We also compare our method with nonlinear disparity mapping method in [2]. Figure 4 shows the comparison results. Given the two original stereoscopic images in Fig.4(a) and Fig.4(b), we apply the method in [2] to increase the depth ranges and show the results in Fig.4(c) and Fig.4(d). From Fig.4(c) we can boserve that the result in most of saliency regions is acceptable, but the far right eave of the pavilion is obviously distorted compared to original image and the buildings in the background areas of right image region are also deformed.

On the contrary, our results in Fig.4(e) do not have these problems. Similarity, From Fig.4(d) we observe the distortion and deformation in the roller coaster in the far distance. Our results in Fig.4(f) can also avoid creating the distortion.

5   Conclusion

In this paper we propose a disparity remapping method for stereoscopic images, which can adjust the disparity ranges of the local image regions separately as well as the whole depth range of the image pair. We use saliency maps and GrabCut approach to detect and extract the visual important regions and present a nonlinear disparity remapping model to address the disparity adjustment in various regions respectively. In order to implement the model we propose a patch-based image editing technique based on PatchMatch algorithm, which allow us to accurately and efficiently rearrange the image regions and complete the image for meeting a desired disparity range.

Our method considers various stereo display devices, view environments and image resolutions, so it flexibly handles the change of disparity for stereoscopic images so that the outputs can appropriately adapt to different application scenarios. In the processing procedure, our method preserves image coherence and 3D structures and ensures the stereo consistency in the images with the remapped disparity. Moreover, our approach can minimize the shape distortion and avoid creating new artifacts. The experimental results illustrate the effectiveness and accuracy of the presented method. As a future work, we will extend the method to stereoscopic video depth manipulation and add visual fatigue prediction to improve stereo quality.

Acknowledgments. The research reported in this paper is supported by The National Key Technology R&D Program of China under Grant No. 2012BAH39F07; jiangsu Provincial Natural Science Foundation of China under Grant No. BK2011170; the National Special Foundation for Internet of Things; the Funds for Creative Research Groups of China under Grant No.61121001; the Cosponsored Project of Beijing Committee of Education.

References

1. Smolic, A., Kauff, P., Knorr, S., Hornung, A., Kunter, M., Muller, M., Lang, M.: Three-dimensional video postproduction and processing. Proceedings of the IEEE 99(4), 607–625 (2011)
2. Lang, M., Hornung, A., Wang, O., Poulakos, S., Smolic, A., Gross, M.: Nonlinear disparity mapping for stereoscopic 3D. ACM Transactions on Graphics 29(4), 75 (2010)
3. Chang, C.H., Liang, C.K., Chuang, Y.Y.: Content-aware display adaptation and interactive editing for stereoscopic images. IEEE Transactions on Multimedia 13(4), 589–601 (2011)

4. Yan, T., Lau, R.W., Xu, Y., Huang, L.: Depth mapping for stereoscopic videos. International Journal of Computer Vision 102(1-3), 293–307 (2013)
5. Kim, D., Choi, S., Sohn, K.: Depth adjustment for stereoscopic images and subjective preference evaluation. Journal of Electronic Imaging 20(3), 033011 (2011)
6. Sohn, H., Jung, Y.J., Lee, S.I., Speranza, F., Ro, Y.M.: Disparity remapping to ameliorate visual comfort of stereoscopic video. In: Proceedings of SPIE 8648. Stereoscopic Displays and Applications XXIV (2013)
7. Yan, T., He, S., Lau, R.W., Xu, Y.: Consistent stereo image editing. In: ACM International Conference on Multimedia, pp. 677–680 (2013)
8. Pritch, Y., Kav-Venaki, E., Peleg, S.: Shift-map image editing. In: The 12th IEEE International Conference on Computer Vision, pp. 151–158 (2009)
9. Tasli, E.H., Alatan, A.A.: User assisted disparity remapping for stereo images. Signal Processing: Image Communication 28(10), 1374–1389 (2013)
10. Barnes, C., Shechtman, E., Finkelstein, A., Goldman, D.: PatchMatch: A randomized correspondence algorithm for structural image editing. ACM Transactions on Graphics 28(3), 24 (2009)
11. Yang, Q.X.: Hardware-Efficient Bilateral Filtering for Stereo Matching. IEEE Transactions on Pattern Analysis and Machine Intelligence 36(5), 1026–1032 (2014)
12. Lee, S., Kim, Y., Lee, J., Kim, K., Lee, K., Noh, J.: Depth manipulation using disparity histogram analysis for stereoscopic 3D. The Visual Computer 30(4), 455–465 (2014)
13. Cheng, M.M., Mitra, N.J., Huang, X., Torr, H.S., Hu, S.M.: Global contrast based salient region detection. IEEE Transactions on Pattern Analysis and Machine Intelligence (to appear, 2014)
14. Rother, C., Kolmogorov, V., Blake, A.: "Grabcut" - Interactive foreground extraction using iterated graph cuts. ACM Transactions on Graphics 23(3), 309–314 (2004)
15. Simakov, D., Caspi, Y., Shechtman, E., Irani, M.: Summarizing visual data using bidirectional similarity. In: IEEE Conference on Computer Vision and Pattern Recognition, pp. 1–8 (2008)
16. Morse, B., Howard, J., Cohen, S., Price, B.: Patchmatch-based content completion of stereo image pairs. In: The Second International Conference on 3D Imaging, Modeling, Processing, Visualization and Transmission, pp. 555–562 (2012)
17. Wang, L., Jin, H., Yang, R., Gong, M.: Stereoscopic inpainting: Joint color and depth completion from stereo images. In: IEEE Conference on Computer Vision and Pattern Recognition, pp. 1–8 (2008)
18. Wexler, Y., Shechtman, E., Irani, M.: Space-time completion of video. IEEE Transactions on Pattern Analysis and Machine Intelligence 29(3), 463–476 (2007)

3D Depth Perception from Single Monocular Images

Hang Xu, Kan Li, FuYu Lv, and JianMeng Pei

Beijing institute of technology, Beijing, China
{xuhang,Likan}@bit.edu.cn
lvfuyu91@sina.com, bitpeijianmeng@gmail.com

Abstract. Depth perception from single monocular images is a challenging problem in computer vision. Since the single image is lack of features of context, we only find all the cues from the local image. This paper presents a novel method for 3D depth perception from a single monocular image containing the ground to estimate the absolute depthmaps more accurately. Different from previous methods, in our method, we first generates the ground plane depth coordinate system from a single monocular image by image-forming principle, and then locates the objects in image with the coordinate system using the geometric characteristics. At last, we provide an method to estimate the accurate depthmaps. The experiments show that our method outperforms the state-of-the-art single-image depth perception methods both in relative depth perception and absolute depth perception.

1 Introduction

An increasing attention in computer vision field is paid to RGB-D image, which represents RGB images with depth information [6,7]. The depth information accommodates images more suitable for various computer vision tasks [8]. Before RGB-D images came into wide spread use, there are large numbers of images without depth information on the Internet and on some open datasets. We try to estimate the depth information from single monocular images in order to solve more computer vision tasks. But estimating the depth information from single monocular images is still a challenging task. As we know, humans appear to be extremely good at judging depth from single monocular images because of their common sense and life experience. However, it is impossible for a computer vision system to learn all these kinds of prior knowledge. Besides, for the single images, extracting the features can rely on nothing but the image itself. This paper focuses on the images containing the ground, since the ground plane will be a superior reference frame of the depth, and most scenes in images include the ground plane.

In the last decade, some methods have been proposed to estimate depths from monocular images, but almost all the methods based on Markov Random Field (MRF) are sensitive to multicolored objects in an image, and need many learning parameters, which are difficult to be learned. Some methods using geometric

X. He et al. (Eds.): MMM 2015, Part I, LNCS 8935, pp. 510–521, 2015.
© Springer International Publishing Switzerland 2015

characteristics only estimate the relative depth of objects or scene in an image, but can't obtain their absolute depth.

In order to resolve the problems above, we propose a novel approach based on the image-forming principle and consider the geometric characteristics. Our method need less parameters, while it can obtain the absolute depth information and reduce the effects of multicolored objects in images. Our method is based on the assumption that no objects are hung in the air and one object's depth can be determined by the points it touches the ground plane or other objects. We also assume that the optical axis of the image-forming device is always parallel to the ground plane.

2   Related Work

Depth estimating from single monocular images is the object of current attention in the literature. One of the popular methods is based on MRF proposed by Saxena's group. Saxena presented an improvement MRF algorithm to capture depths and relationships between depths at different points in images [19,20]. Schulte and his partners used an image set collected by a 3D distance scanner to train MRF and predicted depth [21]. Sun and Ng improved the MRF method and fixed the parameters to make it more suitable for 3D reconstruction [22]. Saxena et.al. showed that the MRF method can do well on images captured by robotic camera. They used the method to achieve robotic grasping [18]. Betra's group used a new learning model to develop the MRF method to estimate depthmaps [1]. In addition, Kratz and his partners used Expectation-maximization algorithm based on MRF to calculate the depth-energy for defogging the images, while the algorithm can also get the depth information [13]. However, MRF-based methods ignore the geometry of the object which are easily affected by the color changing of multicolored objects, resulting in a large depth error. Another disadvantage of MRF-based methods is difficult to confirm their parameters in learning phase.

Methods using the geometric characteristics of objects only can obtain the relative depth from monocular images. Hoiem and Efros made 3D reconstruction from monocular images, but they focused on generating 3D graphical images rather than calculating the absolute depth [12]. Hebert's group took a method to construct the surface layout of objects, which helps predict the depth relationship of objects [10]. Hedau et.al. developed an algorithm to recover the free space of indoor scenes using the relative depth information by exploiting the box like geometric structure in images [5]. However, almost all approaches using geometric characteristics cannot get the accurate depthmaps.

Other main depth perception methods from monocular images also have their limitations. Lin et.al. provided a method based on Shape From Defocus (SFD) [14,16] to estimate depth from single defocused image [15], but it is unable to distinguish the difference between blurred textures and defocused sharp textures. Methods based on T-junction cues can only obtain the relative depth of objects, not the absolute depth [2]. In addition, Nicolas et.al. presented an unsupervised method to approximate basic scene geometry properties [17]. Their method can obtain the absolute depth of objects, but it has a high requirement for imaging device. Cherian's work is very similar to ours, but it is not accurate enough especially on vertical regions [3].

In this paper, we present a novel 3D depth perception method from single monocular images. Our contributions are: 1. Derive the relationship of points between the 3D real world and the 2D image by image-forming principle and propose a method to generate the depth coordinate system on images. 2. Propose the image depth perception method to obtain the accurate depthmaps of images. 3. Present a measure of relative depth perception performance. 4. Verify the performance of our method in both relative depth accuracy and absolute depth accuracy.

3   Ground Plane Depth Coordinate System

As we know, the relationship of the 3D real world and 2D digital image can be described by a coordinate system. In our approach, we first analyze the image-forming principle and then propose a method to generate the ground plane depth coordinate system on images.

3.1   Image-Forming Principle

In Fig. 1, point O is the position of image-forming device lens. Now we use depth coordinate system $X_C Y_C Z_C$ to describe the real world space. In this coordinate system, a point $P(X, Y, Z)$ in real world is transformed to a 2D point $p(x, y)$ on the real imaging plane. In image-forming principle, device can receive light from outside and form a upside down image on the real imaging plane behind lens with a distance F, which is called focal distance. In order to help understanding, we add a virtual imaging plane in front of lens with a distance F. An erected image will be formed on the virtual imaging plane. Obviously, it is impossible to calculate the precise depth value between the two different coordinate systems, but we can get their relationship which is helpful to our work.

In the condition of Fig. 1, we can easily prove that $\triangle OO'p$ is similar to $\triangle OO''P$ and $\triangle O'pq$ is similar to $\triangle O''PQ$. Then we get the relationship in Formula (1).

$$\frac{OO'}{OO''} = \frac{O'p}{O''P}, \frac{O'p}{O''p} = \frac{pq}{PQ} = \frac{O'q}{O''Q}$$

$$\frac{F}{Z} = \frac{y}{Y} \tag{1}$$

Formula (1) shows the relationship between the point 2D coordinates and its 3D coordinates, which helps us figure out how the depth changing in the real world react on the image.

We assume that there are two points $P_0(X, Y, Z_0)$ and $P_1(X, Y, Z_1)$ in 3D real world coordinate system, and the distance between the two points on axis Z_C is

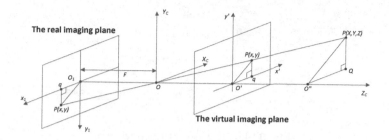

Fig. 1. Optical image-forming principle

d. In Fig. 2, point P_0 is mapped to $p_0(x_0, y_0)$ on the imaging plane, and point P_1 is mapped to $p_1(x_1, y_1)$. Based on Formula (1), we can obtain the Formula (2) and the relationship between the two points' 2D coordinates and their depth in real world as Formula (3).

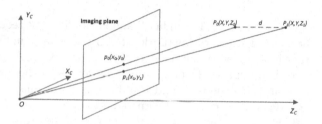

Fig. 2. The relationship between 3D and 2D coordinates

$$\frac{F}{Z_0} = \frac{y_0}{Y}, \frac{F}{Z_1} = \frac{y_1}{Y} \tag{2}$$

$$\frac{y_1}{y_0} = \frac{Z_0}{Z_1} \tag{3}$$

3.2 Depth Coordinate System Generating Algorithm

In this subsection, we will propose a method to generate the depth coordinate system. In the Formula (3), the relationship between the two points' 2D coordinates and their depth in real world is provided. From this relationship, we can get an inference that all points on the ground plane in the image have the similar relationship. So we can use the relationship to generate the depth coordinate system of the ground plane in the image.

We suppose there is a set P, which contains n points in 3D real world. These points always have the same coordinates on axis X_C and Y_C, and every two neighboring points $P_i(X, Y, Z_i)$ and $P_{i+1}(X, Y, Z_{i+1})$ have the same distance d (actually, these points belong to the same line, and the line is on the ground

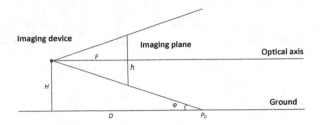

Fig. 3. The features of the device

plane and parallel to the optical axis of the device). We can represent the depth coordinate system on the image using a set C.

$$C = \{y_i | P_i \in P, i = 0, 1, \cdots, n\} \tag{4}$$

From the previous conclusion in Formula (3) and Formula (4), it seems not difficult to propose the method of generating the depth coordinate system. But another problem is here that how to determine the depth of the first point $P_0(P_0 \in P)$.

In Fig. 3, the vertical distance from the imaging device to the ground is H. The height of the imaging plane is h, and the distance from imaging device to the imaging plane is focal distance F. The angle between visual range and optic axis is represented as φ. We assume that the depth of the bottom most row of the image on the ground plane from the imaging device is D. So we can obtain the expression of D as Formula (5).

$$D = \frac{H}{\tan \varphi} = \frac{2FH}{h} \tag{5}$$

From Formula (5), D is only affected by the focal distance F, the altitude H of the device and the altitude h of the imaging plane. When processing robotic vision images or images with the device parameters, we can easily get these statuses to calculate the value of D. We can also use some experienced values to determine the value of D if there is no information about device status. The value of $D \in [5, 10](meters)$ can be used in most images on the Internet or in the open datasets. Then we provide a recurrence formula in general case as follows.

$$y_{n+1} = (\frac{D + nd}{D + (n + 1)d})y_n \tag{6}$$

Then we give the 3D coordinate generating algorithm as follows.

Algorithm 1. Depth coordinate system generating algorithm.	
Input:	The feature parameters of device h, F; The state parameters of device H; The unit depth d and the point set P.
Output:	The coordinate set C.
Step 1:	Calculate the horizontal distance D by Formula (5).
Step 2:	Set $C = \emptyset$, $i = 0$, $y_0 = -\frac{ROW}{2}$, put y_0 into set C. (ROW is the total rows of the processed image.)
Step 3:	Use y_i to calculate the coordinate of the next point $P_{i+1} \in P$ with further unit depth, y_{i+1}, by Formula (6), then put y_{i+1} into set C
Step 4:	If we have the next point $P_{i+2} \in P$, set $i = i + 1$ and go step 3; otherwise, go step 5.
Step 5:	Return the coordinate set C.

By Algorithm 1, we obtain the coordinate set C, which contains all 2D coordinates of 3D points on ground plane with different depth. We can get different accurate-scale coordinates by adjusting the unit depth d. In Fig. 4, we set the unit depth $d = 2(m)$ of the coordinate, and set the middle one as $d = 1(m)$, the right one as $d = 0.5(m)$. The performance shows that the smaller the d is, the more accurate the coordinate is. But setting the value of d too small is unnecessary and inefficient. Based on a large number of experiments, we find that $d = 1(m)$ can obtain a good performance and keep a high efficiency.

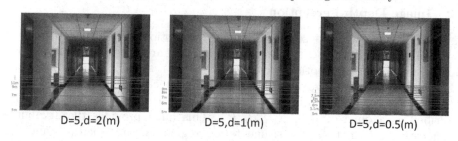

D=5,d=2(m)   D=5,d=1(m)   D=5,d=0.5(m)

Fig. 4. Coordinates with different unit depth

4   Image Depth Perception Algorithm

In order to determine the depth of every point in an image, we should figure out the different areas in an image firstly. We use a texture segmentation algorithm based on [9,4,11], which is using the geometric characteristics. Then we propose a depth perception algorithm to obtain the depthmap of the image based on segmentation preprocessing result.

4.1   Texture Segmentation

Since the depth coordinate system is used for mapping the ground plane to different depth, the images should be preprocessed to obtain the ground area firstly. Based on the geometric characteristics, the objects in images are categorized precisely [9], where the sky, the ground and vertical in images are divided from each other. In Fig. 5, image A is the original image, image B is the result of categorized. In image B, the blue part is the sky area, the green one is the the ground area, and the red one is vertical area. The arrows "↑", "←" or "→" on the objects show the depth changing direction. The marks "○" or "×" on the objects represent that those objects are solid or hollow respectively.

After the sky, ground and vertical areas have been separated, we divide the vertical area more precisely [4,11] to get the location of the objects (see Fig. 5, image B and C).

Fig. 5. Texture segmentation process. A: original image. B: the image is categorized into sky, vertical and ground. C: the vertical is segmented more precisely.

4.2   Image Depth Perception

In this subsection, we propose an image depth perception method, which includes the ground plane depth perception and the vertical area depth perception. Obviously, we can set the depth of sky infinite.

After the preprocessing, we obtain a image in which every objects and areas are separated from each other. Any parts of the segmentation result can be described as a quad of $R_i(C_i, M_i, B_i, P_i)$. This quad provides the attributes and location information of a segmentation region. In this quad, C_i and M_i both represent the category of the region. C_i shows the region belonging to sky, ground plane or vertical, and M_i shows which kind of mark is on the region. B_i is the bottom line number of the region. P_i is the 2D coordinate of the top-left point of the region.

We suppose a set R that includes all the regions of the segmented image. The set R can be described as $R = \{R_i | i = 0, 1, \cdots, n\}$, where n is the total number of regions in a image.

By the assumption that no objects are hung in the air and one object's depth can be determined by the points it touches the ground plane, we can easily estimate the depth information of most regions. But there exist some regions covered by other regions or not touching the ground plane directly, so it is difficult to estimate their depth information by their bottom line. Since the work of [12] shows that neighboring regions can provide important cues to infer the depth information of one region. Inspired by [12], we propose a method to estimate the region's depth information by its neighboring regions based on their touching length. Every neighbors having been determined their depth should make a contribution to deciding the depth of this region. We use the following equation to determine the depth of regions which are not touching the ground directly:

$$dep = \sum_{j=1}^{m} r_j dep_j \qquad (7)$$

In the Formula (7), m is total number of neighboring regions having been determined the depth, r_j is the rate of two regions' touching length. The value of r_j is given by $r_j = \frac{TL_j}{\sum_{j=1}^{m} TL_j}$. TL_j in the equation means the touch length of two regions. Meanwhile, it must satisfy that $\sum_{j=1}^{m} r_j = 1$.

Now we give the image depth perception algorithm as follows.

Algorithm 2. Image depth perception Algorithm.	
Input:	The coordinate set C; The region set R.
	($R = \{R_i(C_i, M_i, B_i, P_i)|i = 1, 2, \cdots, n\}$)
Output:	The depthmap of the image.
Step 1:	Set a set $RF = \emptyset$; Initialize the depthmap empty.
Step 2:	Traverse R, find all $R_i \in R$, where $C_i = SKY$.
	Set depth of each R_i the max value, remove R_i from R
	and put it into RF.
Step 3:	Traverse R, find all $R_i \in R$, where $C_i = GROUND$.
	Set depth of each R_i the corresponding value based on
	the coordinate set C, remove R_i from R and put it into RF.
Step 4:	Scan the other elements in R and find $R_i \in R$ which touches
	the ground plane and has minimum B_i.
	If it succeeds, go step 5; otherwise, go step 6.
Step 5:	According to the M_i and the touching points of the R_i,
	set depth of the R_i, remove R_i from R and put it into RF.
	Go step 4.
Step 6:	Scan the other elements in R and find $R_i \in R$ which
	has minimum B_i. If it succeeds, go step 7; otherwise, go step 8.
Step 7:	Calculate the depth of the R_i base on Formula (7),
	set depth of the R_i, remove R_i from R and put it into RF.
	Go step 6.
Step 8:	Return the depthmap.

5 Experiments

5.1 Dataset

In this paper, we use an image set of 350 images/depthmaps provided by Stanford (http://make3d.cs.cornell.edu/data.html) to finish the experiments. The data is collected by a 3D distance scanner, and comprised of a large set of

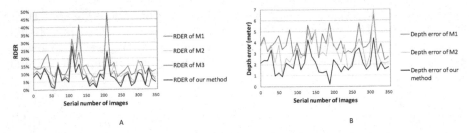

Fig. 6. Experimental results. A: The $RDER$ of different methods. B: The depth error of different methods.

original images and their corresponding ground-truth depthmaps. Each image is of size 1704×2272 and each depth map is of size 55×305.

In the experiments on depth perception, we provide some other researchers' method as contrast. The MRF method [1] is a state-of-the-art method in this filed, which is called M1. Cherian's work [3] is very similar to ours, which is called M2. Hoiem's method [12] has a good performance, which is called M3. But it cannot obtain the depthmaps, and we only use it for relative depth experiment.

5.2   Relative Depth Perception

Since many applications of computer vision only need relative depth of images to finish their tasks, the error rate of relative depth is also a valuable indicator of the depth perception. We firstly make the experiment on relative depth perception.

We propose the Relative Depth Error Rate ($RDER$), a measure of the relative depth perception performance. On the dataset, we already have the ground-truth depthmaps. Based on the ground-truth, we sort every pixel in the ascending order, and treat the original image's sequencing result as the standard depth order. By using depth perception method, we obtain the image's depth information and sort its pixels to get a new depth order. Then the inverted sequence number between the new depth order and the standard depth order shows the relative depth errors in the depth estimating result. As we know, every order has a maximal inverted sequence number, which shows the maximal value of relative depth errors in the image.

We define the $RDER$ as $RDER = \frac{ISN}{MISN}$, where the numerator is inverted sequence number (ISN) of the estimating result and the denominator is the maximal inverted sequence number ($MISN$) of the standard order.

Fig. 6 A shows that the $RDER$ of all images on the dataset, which is obtained by our method, M1, M2 and M3. The figure demonstrates that our method outperforms M1 and M2 on the relative depth perception, and it is about the same between our method and M3. Since our method and M3 take the object's geometric characteristics into consideration, which make sure the relative depth order between objects rarely change.

As the result, the average $RDER$ of our method is 8.173%, and the average $RDER$ of M1, M2 and M3 are 15.359%, 11.872% and 9.367% respectively.

5.3   Absolute Depth Perception

In this subsection we evaluate the absolute depth perception performance of our approach on the image dataset. We use our method, M1 and M2 to estimate the depthmap of each image. Then we compute the average difference of the ground-truth depthmap with the result depthmaps, and the difference shows the absolute depth estimating error. Apparently, the smaller depth error is, the more accurate the depth estimating method is.

Fig. 6 B shows that the depth error on all the images obtained by our method, M1 and M2. We observe in the figure that our method shows better performance on absolute depth perception than other methods. In our method, the geometric characteristics based texture segmentation make sure the atomicity of every objects, and avoid the influence of the color changing in an object perfectly. The results reflect this effort clearly.

The experiment results show that the average depth error of our method is 2.1359 meters, while the average depth error of M1 and M2 is 3.7843 meters and 3.1970. Our method has succeeded in improving the accuracy of the depth information perception.

5.4   Experimental Results

Fig. 7 shows a number of test results of experiments. The column A of each row is the original image, the column B is the relative depth obtained by M2, the column C is the absolute depth estimated by M1, and the column D is the absolute depth result of our method. Comparing with M2's approach, we can see that our approach outperforms in many detail areas, such as top of the tree and the forest behind the building. M2's approach is good at obtain the relative depth, but it cannot estimate the accurate depthmap. Comparing with the M1, our approach outperforms both in relative depth perception and absolute depth perception obviously. Since the M1 is sensitive with color changing, many depth estimating errors happen as shown in the figure.

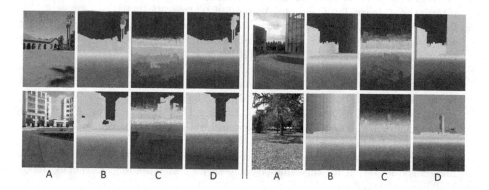

Fig. 7. A: original image. B: relative depth obtained by M2. C: absolute depth obtained by M1. D: absolute depth obtained by our method.

6　Conclusions

In this paper, we firstly analyze the image-forming principle and propose a method to generating the depth coordinate system in image. Then we use the geometric characteristics to avoid the influence of color changing in an object and propose a novel image depth perception method to obtain the accurate depthmaps of images. In experiment, a measure is provided to judge the relative depth perception performance. The experiments show that our method outperforms the-state-of-art methods both in relative depth perception and absolute depth perception.

In the future, we will consider the priori information of objects to help us compute the depth of the images especially in vertical area. Considering the angel changing of device is another field we can accommodate in the future. In addition, we will also use more diverse datasets to testify our method's robustness.

Acknowledgments. The Research was supported in part by Natural Science Foundation of China(No.60903071), Specialized Research Fund for the Doctoral Program of Higher Education of China, and Training Program of the Major Project of BIT.

References

1. Batra, D., Saxena, A.: Learning the right model: Efficient max-margin learning in laplacian crfs. In: 2012 IEEE Conference on Computer Vision and Pattern Recognition (CVPR), pp. 2136–2143 (2012)
2. Calderero, F., Caselles, V.: Recovering relative depth from low-level features without explicit t-junction detection and interpretation. International Journal of Computer Vision 104(1), 38–68 (2013)
3. Cherian, A., Morellas, V., Papanikolopoulos, N.: Accurate 3d ground plane estimation from a single image. In: 2009 IEEE International Conference on Robotics and Automation (ICRA), pp. 2243–2249. IEEE (2009)
4. Felzenszwalb, P.F., Huttenlocher, D.P.: Efficient graph-based image segmentation. International Journal of Computer Vision 59(2), 167–181 (2004)
5. Hedau, V., Hoiem, D., Forsyth, D.: Recovering free space of indoor scenes from a single image. In: 2012 IEEE Conference on Computer Vision and Pattern Recognition (CVPR), pp. 2807–2814. IEEE (2012)
6. Henry, P., Krainin, M., Herbst, E., Ren, X., Fox, D.: Rgb-d mapping: Using kinect-style depth cameras for dense 3d modeling of indoor environments. The International Journal of Robotics Research 31(5), 647–663 (2012)
7. Henry, P., Krainin, M., Herbst, E., Ren, X., Fox, D.: Rgb-d mapping: Using depth cameras for dense 3d modeling of indoor environments. In: Experimental Robotics, pp. 477–491. Springer (2014)
8. Hoiem, D., Adviser-Efros, A.A., Adviser-Hebert, M.: Seeing the world behind the image: spatial layout for three-dimensional scene understanding. Carnegie Mellon University (2007)
9. Hoiem, D., Efros, A.A., Hebert, M.: Geometric context from a single image. In: 2005 10th IEEE International Conference on Computer Vision, pp. 654–661. IEEE (2005)

10. Hoiem, D., Efros, A.A., Hebert, M.: Recovering surface layout from an image. International Journal of Computer Vision 75(1), 151–172 (2007)
11. Hoiem, D., Efros, A.A., Hebert, M.: Closing the loop in scene interpretation. In: 2008 IEEE Conference on Computer Vision and Pattern Recognition (CVPR), pp. 1–8. IEEE (2008)
12. Hoiem, D., Efros, A.A., Hebert, M.: Recovering occlusion boundaries from an image. International Journal of Computer Vision 91(3), 328–346 (2011)
13. Kratz, L., Nishino, K.: Factorizing scene albedo and depth from a single foggy image. In: 2009 12th IEEE International Conference on Computer Vision, pp. 1701–1708. IEEE (2009)
14. Levin, A., Fergus, R., Durand, F., Freeman, W.T.: Image and depth from a conventional camera with a coded aperture. ACM Transactions on Graphics (TOG) 26(3), 70 (2007)
15. Lin, J., Ji, X., Xu, W., Dai, Q.: Absolute depth estimation from a single defocused image. IEEE Transactions on Image Processing: a publication of the IEEE Signal Processing Society 22(11), 4545 (2013)
16. Namboodiri, V.P., Chaudhuri, S.: Recovery of relative depth from a single observation using an uncalibrated (real-aperture) camera. In: 2008 IEEE Conference on Computer Vision and Pattern Recognition (CVPR), pp. 1–6. IEEE (2008)
17. Nicolas, H.: Depth analysis for surveillance videos in the h. 264 compressed domain. In: 2012 Proceedings of the 20th European Signal Processing Conference (EUSIPCO), pp. 146–149. IEEE (2012)
18. Saxena, A.: Monocular depth perception and robotic grasping of novel objects. Ph.D. thesis, Citeseer (2009)
19. Saxena, A., Chung, S.H., Ng, A.Y.: Learning depth from single monocular images. In: Neural Information Processing Systems Conference (NIPS), vol. 18, pp. 1–8 (2005)
20. Saxena, A., Chung, S.H., Ng, A.Y.: 3-d depth reconstruction from a single still image. International Journal of Computer Vision 76(1), 53–69 (2008)
21. Saxena, A., Schulte, J., Ng, A.Y.: Depth estimation using monocular and stereo cues. In: International Joint Conference on Artificial Intelligence (IJCAI), pp. 2197–2203. Morgan Kaufmann Publishers Inc. (2007)
22. Saxena, A., Sun, M., Ng, A.Y.: Make3d: Depth perception from a single still image. In: AAAI Conference on Artificial Intelligence (AAAI), pp. 1571–1576 (2008)

Muscular Movement Model Based Automatic 3D Facial Expression Recognition

Qingkai Zhen[1], Di Huang[1,*], Yunhong Wang[1], and Liming Chen[2]

[1] IRIP Lab, School of Computer Science and Engineering, Beihang Univ.,
Beijing 100191, China
[2] LIRIS UMR5205, Ecole Centrale de Lyon, Lyon 69134, France
dhuang@buaa.edu.cn

Abstract. Facial expression is the most important channel for human nonverbal communication. This paper presents a novel and effective approach to automatic 3D Facial Expression Recognition, FER based on the Muscular Movement Model (MMM). In contrast to most of existing methods, MMM deals with such an issue in the viewpoint of anatomy. It first automatically segments the input face by localizing the corresponding points around each muscular region of the reference face using Iterative Closest Normal Pattern (ICNP). A set of shape features of multiple differential quantities, including coordinates, normals and shape index values, are then extracted to describe the geometry deformation of each segmented region. Therefore, MMM tends to combine both the advantages of the model based techniques as well as the feature based ones. Meanwhile, we analyze the importance of these muscular areas, and a score level fusion strategy which optimizes the weights of the muscular areas by using a Genetic Algorithm (GA) is proposed in the learning step. The muscular areas with their optimal weights are finally combined to predict the expression label. The experiments are carried out on the BU-3DFE database, and the results clearly demonstrate the effectiveness of the proposed method.

Keywords: 3D Facial Expression Recognition, Muscular Movement Model.

1 Introduction

The facial expression is one of the most important manners for communication in the daily life of human beings. During the last several decades, machine based automatic Facial Expression Recognition (FER) has received extensive attention in many communities, e.g. computer vision, pattern recognition, psychology, Human Computer Interaction (HCI), etc., and it has also involved in a variety of useful applications, such as facial animation, autism diagnosis, tiredness detection, and food tasting questionnaire.

Existing approaches in 3D FER can be roughly categorized into two classes, i.e. model-based and feature-based [12]. Model-based methods make use of a generic model, based on which common expressions can be indicated by measuring the parameters of shape deformation achieved in the fitting phase. They usually need to find facial surface correspondence by means of registration. For example, Mpiperis et al. proposed

---

* Corresponding author.

X. He et al. (Eds.): MMM 2015, Part I, LNCS 8935, pp. 522–533, 2015.

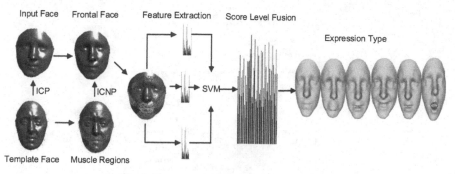

Fig. 1. Framework of the proposed approach

a bilinear model to simultaneously encode identities and expressions of the subjects [8]. In [14], Yin et al. presented a model based approach for 3D expression analysis, tracking and classification by using primitive surface features. Model-based methods are theoretically attractive, however, they tend to incur vital errors in convergence because of the utilization of insufficiently discriminative features as well as the change in face mesh topology. While there are more approaches belonging to feature-based that concentrate on the extraction of expression sensitive geometry features from input 3D face scans. For instance, Berretti [1] et al. calculated SIFT based features in the positions of given 3D facial landmarks, then selected the most relevant feature subset, and finally classified them with SVM. Soyel and Demirel [10] adopted six kinds of distances between 11 manually labeled landmarks as features, and then used the neural network for classification of seven kinds of expressions (six basic ones plus neutral). Feature based approaches are straightforward and their accuracies are directly related to the discriminative power of adopted facial features. Unfortunately, most of these techniques generally require a large number of pre-defined fiducial landmarks, which is not practical in the real world since it is quite difficult to localize enough landmarks accurately and automatically.

This paper presents a novel and efficient approach to automatic 3D FER. In contrast to the existing ones in literature, our method closely follows the discovery in anatomy that expression is the consequence of combined facial muscular movement. The Facial Action Coding System is expected to improve the ability of computers to understand facial expressions. Nevertheless, due to the steep demand in accurate landmarks and dedicated dataset, it is hard to bridge the gap between the recognition of facial expressions and that of action units in the 3D domain. While in this study, the Muscular Movement Model (MMM) is proposed to recognize expressions by fusing the deformation quantization in a set of geometry features of different regions, each of which contains an independent muscle. Therefore, MMM tends to combine the advantages of model-based methods as well as feature-based ones. Fig. 1 illustrates the overview of the proposed method. It first localizes landmarks by aligning corresponding points between the input and reference faces using Iterative Closest Normal Pattern (ICNP) [7]. Based on these landmarks, segmentation on input face can thus be automatically achieved. To comprehensively measure the deformation of each muscular region, the coordinate, normal and shape index of each vertex within the area are jointly employed to compose a powerful

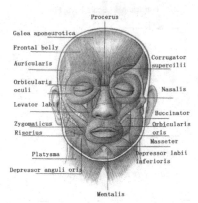

Fig. 2. Facial muscle anatomy model

feature set, since they reflect the shape characteristics in different levels, i.e. original data as well as its 1st and 2nd order deviations. The Genetic Algorithm (GA) based credits fusion is adopted at the score level to optimize the weights of different muscular regions and SVM is finally used to predict the expression. The classification accuracy achieved on BU-3DFE clearly highlights the effectiveness of the proposed method.

2   Muscular Movement Model (MMM)

The anatomy model of human facial muscles is shown in Fig. 2. Most of them are expression related, and the expression of human beings is essentially the combined result of the stretch and contraction by different facial muscles [12].

Inspired by such a fact, in this study, we propose the Muscular Movement Model (MMM) to represent expressions of 3D facial surfaces, measure their similarities with the template in different feature spaces, and finally predict their labels. In this section, we present the key steps of MMM, including facial muscle region segmentation, shape feature extraction, weighted sum fusion, and expression prediction in the subsequent.

2.1   Facial Muscle Region Segmentation

In this subsection, we introduce the algorithm of facial muscle segmentation. It is quite straightforward to segment these regions using landmarks. However, on the one hand, it consumes huge amount of manpower in manually labeling a great number of samples, which is not realistic in practical applications. On the other hand, automatically and accurately localizing so many fiducial points still remains a big challenge in the community. We hence design a model-based method to achieve automatic segmentation by finding the correspondence between the reference and input faces.

Segmentation of Reference Face. Specifically, we first randomly select a 3D face model of a neutral expression (with the scan ID F0001_NE00WH_F3D) and regard it as the template. We then manually annotate 69 landmarks around the eyes, nose, mouth, and face contour (see Fig. 3 (a) for an illustration). Next, the points along the geodesic path [4] between every two adjacent landmarks around each muscle are linked

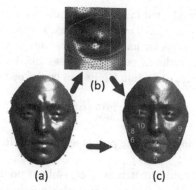

Fig. 3. (a) The selected 69 landmarks on the reference model; (b) The region of the left orbicularis oculi muscle; and (c) Segmented facial muscle regions on the reference face

to segment this muscular region into an independent closed region. Fig. 3 (c) displays the segmentation result on the 3D face template. After that, to efficiently find all the vertices within the region, we propose a seed fill algorithm which iteratively determines the points connected to a given node on the segmented face mesh. A seed vertex within this region is required for initialization. (the red point in Fig. 3 (b)). In each of the following iterations, the points adjacent to the seed point (denoted with the green points in Fig. 3 (b)) are located and regarded as new seeds in the next round. The procedure continues until it reaches the boundary of the region.

In our case, the initialized seed is manually marked since we only need to deal with one 3D model, i.e. the reference face.

Segmentation of Test Face. Recall that before recognizing the expression of an input face, we automatically segment it as on the reference face. In order to avoid localizing so many landmarks, we achieve the segmentation by finding the correspondence between the reference and input faces. To fulfill such a task, we make use of the Iterative Closest Normal Point (ICNP) method, originally proposed for 3D face recognition [7], since it proves that when the geometry of a face surface changes under expressions, it is able to effectively and efficiently find the same corresponding points that it usually finds from its neutral face.

Let $\{r_i\}_{i=1}^N$ represent the point set of a given segmented muscle region of the reference face and $\{p_i\}_{i=1}^M$ represent the point set of the entire input face, where r_i and p_i indicate 3D vectors containing the coordinates of the points. While $\{\beta_i\}_{i=1}^N$ and $\{\pi_i\}_{i=1}^M$ are used to denote the sets of unit normal vectors of the muscle region of the reference face and the whole input face, respectively. where N and M are the numbers of the vertices of the two surfaces.

Each point r_i of a given segmented muscle region of the reference face can find its corresponding point through the following steps. For the first, the point with the minimum value of Euclidean distance is selected. Let $\{p_i^c\}_{i=1}^N$ denote the closest points, and it is a subset of $\{p_i\}_{i=1}^M$ such that,

$$p_i^c = p_k, k = arg \min_{1 \leq j \leq M} \|r_i - p_j\|_2 \qquad (1)$$

The KD-tree algorithm is conducted to accelerate the search process. We search within a window of pre-defined size (10x10 *mm* in our case) centered at the closest point to find the point whose normal vector has the smallest angle with the normal vector at the reference point. These produced points are thus named as Closest Normal Points (CNPs). Formally, the process of finding CNPs can be defined as:

$$p_i^{cn} = p_l, l = \arg \min_{k \in \eta} \left(arc \cos \left(\beta \cdot \pi_k \right) \right) \tag{2}$$

where η denotes the set of indices of the points located within the pre-defined window centered at the point p_i^c.

The CNPs cannot be directly treated as the resultant points because they are not smoothly corresponding to the reference point. A smoothing process is then implemented to solve this problem. Let $\{d_i\}_{i=1}^{N_{reg}}$ denote the distance vectors, generated by subtracting the coordinates of the reference points from the ones of their closest normal points.

$$d_i = p_i^{cn} - r_i \tag{3}$$

To smooth these distance vectors, a 2D Wiener filter (with a window of 20×20) is applied to each of the three components of the distance vector. Let $g(d_i)$ represent the set of filtered distance vectors. Then, the smoothed closest normal points $p_i^{cn'}$ can be approximately achieved by

$$p_i^{cn'} = r_i + g(d_i) \tag{4}$$

Note that the approximated points may not lie exactly on the input face. To deal with such a problem, the KD-Tree algorithm is exploited again to search the closest points of $p_i^{cn'}$ on the input facial surface as:

$$p_i^{cn''} = p_k, k = arg \min_{1 \leq j \leq M} \|p_i^{cn'} - p_j\|_2 \tag{5}$$

where $p_i^{cn''}$ denotes the set of smoothed closest normal points on the input surface.

The smoothed CNPs of each muscle region on the reference model make up of the corresponding region on the input face, thereby segmenting it into 11 pre-defined regions. Each muscle region on the input face is then rotated and translated to reduce the distance between the CNPs and the reference points, and a more accurate correspondence can thus be achieved. The operation is repeated until the distance reaches the convergence, i.e. no more significant rotation is applied to the input face. The rotation matrix and translation vector are produced by Singular Value Decomposition (SVD) [7]. Fig. 4 demonstrates the process of facial muscle segmentation. Fig. 5 shows the example of segmentation on 3D face scans of the same subject of different expressions.

2.2   Feature Extraction

To comprehensively measure the deformation of the segmented muscular regions on the input face with the corresponding ones on the reference face, the coordinate, normal,

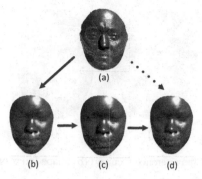

Fig. 4. (a) The reference face model; (b) The closest points measured by Euclidean distance; (c) The closest points measured by angles of normal vectors; and (d) The CNPs after 2D wiener filter.

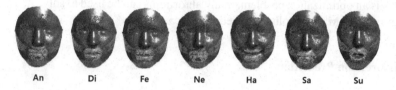

An   Di   Fe   Ne   Ha   Sa   Su

Fig. 5. The automatic segmentation result on the 3D facial surfaces belonging to the same subject under different expressions (An: Anger; Di: Disgust; Fe: Fear; Ha: Happy; Sa: Sad; Su: Suprise)

and shape index values of each vertex are jointly used to compose a powerful feature set to capture shape characteristics.

Shape Index (SI) at point p can be calculated using maximum(k_1) and minimum(k_2) local curvatures:

$$SI(p) = \frac{1}{2} - \frac{1}{\pi} tan^{-1} \frac{k_1(p) + k_2(p)}{k_1(p) - k_2(p)} \tag{6}$$

where $k_1(p)$ and $k_2(p)$ are principal curvatures at a point p of a surface S [9]. $k_1(p)$ and $k_2(p)$, are calculated as:

$$k_1(p) = H(p) + \sqrt{H(p)^2 - K(p)}; \tag{7}$$

$$k_2(p) = H(p) - \sqrt{H(p)^2 - K(p)}; \tag{8}$$

Figure 6 shows the two kinds of geometry features of the reference face.

2.3 Optimized Weighted Score Sum Fusion

Considering that different facial muscles contribute non-equally to expressions, we investigate in the subsequent the importance of regions to each facial expression, aiming to further improve the performance in FER by a weighting scheme. The weighted sum rule is generally defined as:

$$S = \sum_{i=1}^{N} \omega_i * S_i \tag{9}$$

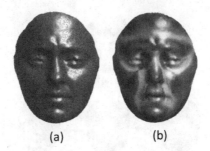

(a)   (b)

Fig. 6. (a) Normal of the reference face and (b) Shape Index of the reference face

where ω_i is the weight, S_i is a similarity score, and N is the number of regions. A bigger weight indicates a higher importance, and a smaller one indicates a lower importance. Since (9) is an optimization problem, many algorithms can be used to solve it, The basic GA algorithm in Matlab toolbox is used in our case.

2.4   Expression Prediction

Given a muscle region represented as follows,

$$P = \{p_i\}_{i=1}^{M} \tag{10}$$

where $p_i = (xi, yi, zi)$ denotes the 3D coordinates of each vertex; M denotes the number of vertices (corresponding to the ones of the reference face) within the region. Each normal component of the p_i can be represented as:

$$N = \{n_i\}_{i=1}^{M} \tag{11}$$

where $n_i = (n_{x_i}, n_{y_i}, n_{z_i})$ and $-1 \leqslant n_{x_i}, n_{y_i}, n_{z_i} \leqslant 1$. Its SI value can be represented as:

$$I = \{I_i\}_{i=1}^{M} \tag{12}$$

where $0 \leqslant I_i(P) \leqslant 1$. P, N, and I thus represent the feature vector of the coordinates, normals and shape index values, respectively. For a test face, the feature vectors of P, N, and I are extracted from each muscle region, and then fed into an SVM classifier separately to generate its similarity S_i. In each training stage, the weight ω_i of corresponding muscle region is learned, and the final S is computed according to (9). We then predict expression label of the test face to the one with the maximum value in probability as

$$Ex = \max_{1 \leq j \leq 6} Prob_j \tag{13}$$

where $Prob_j$ represents the probability value of the six basic expressions at the score level. Fig.7 illustrates the process of weight optimization and label prediction in each independent experiment.

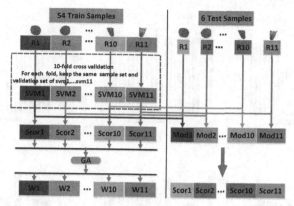

Fig. 7. Process of the weight optimization and label prediction. R_i, $Score_i$ and W_i represent the muscle region, similarity score , and optimal weight, respectively.

3 Experimental Results

To validate the effectiveness of the proposed method in 3D FER, we carry out the experiments on the BU-3DFE dataset [13]. To make the proposed method more practical, especially to deal with the models that are not in the frontal upright position as the ones in BU-3DFE, we regard all test 3D face models in arbitrary poses. We launch the automatic landmarking technique [11] to first localize three main fiducial points, i.e. two inner corners of the eyes and nose tip, based on which ICP is then exploited for coarse alignment to roughly correct poses.

Table 1. Matrix on BU-3DFE using Setup (I)

%	AN	DI	FE	HA	SA	SU
AN	**79.5**	8.6	1.4	1.6	7.5	1.4
DI	1.5	**85.7**	2.6	4.5	2.8	3.2
FE	3	2.8	**63.3**	21.9	3.8	5.3
HA	0	1.2	3.6	**94.6**	0.2	0.4
SA	15.8	0.7	2.4	0.4	**79.2**	1.6
SU	0	1.4	1.8	0.5	0.2	**96.1**
Average:			83.2			

In literature, there are two main protocols to evaluate methods to perform six prototypical expression classification on 3D-BUFE: (I) choosing fixed 60 subjects [2] and (II) randomly selecting 60 subjects in each round of the experiment [1].

We carry out the experiments using both protocols for fair comparison with the state of the art. Each time, to the 60 subjects involved in the experiments, only the two models with higher expression intensities, i.e. level 3 andlevel 4, are employed. They are randomly split to 54 vs. 6 as in the training and test phase. This experiment is repeated 1,000 times in both protocols, and the average classification performance is reported.

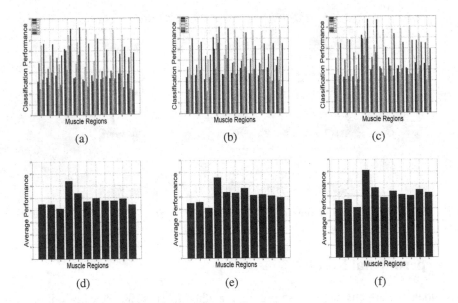

Fig. 8. The top row is the performance of each facial muscle to different types of expression by using different features (a) coordinate features (b) normal features (c) shape index features. The bottom row is the average performance of each facial muscle by using each features (d) coordinate features (e) normal features (f) shape index features.

First, we evaluate the individual importance of each facial muscle area to different expressions using those features (i.e. coordinate, normal and SI), and average results are given in Fig. 8 and 9.

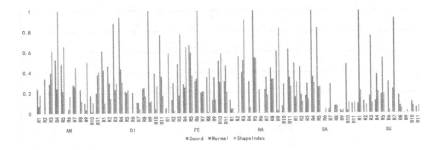

Fig. 9. Importance analysis (measured by weight value) of each facial muscle region to predict different expressions using individual geometry features, the standard deviation is less than 0.05

The upper row of Fig. 8 depicts the results of each facial region based on the feature of (a) coordinate, (b) normal, (c) shape index, while its bottom row displays the average performance of each facial region based on the same features. From this figure, we can see that these facial muscle regions provide different impacts to the six typical expressions, and the region of orbicularis oris (the muscle surrounding mouth)

Table 2. Performance of our method using different components on BU-3DFE

Method	Accuracy
Region ICP	66.4%
Region ICNP (Coordinate)	74.1%
Whole Face ICNP (Coordinate)	70.2%
Region ICNP (Normal)	75.3 %
Region ICNP (Shape Index)	80.2 %
Region ICNP (Coord & Normal & Shape Index)	**83.2%**

Table 3. Performance comparison with the state of the art on BU-3DFE

Method	Orig.	Setup I	Setup II
Berretti et al. [1]			77.5%
Gong et al. [2]		76.2%	
Wang et al. [13]	83.6% (20 T)	61.8%	
Soyel et al. [10]	91.3% (10 T)	67.5%	
Li et al. [6]			82.0%
Tang et al. [3]	95.1% (10 T)	74.5%	
Pierre et al. [5]			75.8%
Our Result	96.4% (10 T)	84.5%	**83.2%**

influences expression most, corresponding to the fact that some approaches in 3D face recognition eliminate the mouth region to deal with the problem caused by expression variations. Figure 9 illustrates the weights (after normalization) of muscular regions based on individual features for different expression recognition. We can see that even though some region achieves better classification performance, it contributes less in distinguishing between different expressions when combined with others, indicated by the smaller values in its weight. For example, in Fig. 8 (a), we can observe that the region of zygomaticus (Region 7) achieves the first place in performance to recognize fear based on the coordinate feature. However, it is not so important indicated by its weight (Fig. 9), and the region contributes most is depressor anguli oris (Region 5).

Secondly, from Table 2, we can see that the region based analysis is more effective than the holistic one demonstrated by the comparison of ICNP using coordinate features. The result achieved by region ICNP using normal or shape index features is better than the one of region ICNP using coordinate features, indicating that embedding more discriminative features (normal or curvature information) to the model-based analysis leads to an improved performance. The result of combining the coordinate, normal and shape index features is the best, which shows that those features provide complementary cues. Meanwhile, these facts also demonstrate that the proposed method possesses the advantages of the feature based and model based methods in 3D FER.

Finally, we compare our approach with the state of the art in Table 3. [13], [10], [3] are three early studies on the exact same topic. All these works used Setup I in their experiments, but they only repeated a few times (10-20), where performance stability cannot be guaranteed. When we employ such a protocol, the average of the FER accuracies reaches 96.4%. In fact, Gong et al. [2] reproduced the approaches of [13], [10], and [3] using the same experimental protocol but repeating 1,000 times, and the results are largely degraded. We can see that our result is better than [2] [13] [10] [3] [6] using the protocol of Setup (I) repeating 1,000 times, and better than [1] using the protocol of Setup (II) repeating 1,000 times. It should be noted that [6] and [1] depend on a large number of manual landmarks, while the proposed method is totally automatic.

4   Conclusions

In this paper, we propose a novel, fully automatic approach for 3D FER, which is based on the Muscular Movement Model. It automatically segments the muscle regions on the input face by aligning the corresponding ones on the reference face, and does not need manual landmarks. The coordinate, normal and shape index are jointly used to compose a powerful feature set to capture shape characteristics, which improves the distinctiveness in deformation of facial surface regions. GA is used for seeking optimal weights of muscle regions, and a score level fusion scheme is adopted to recognize expressions. The accuracy achieved on the BU-3DFE dataset clearly demonstrates the effectiveness of the proposed method.

Acknowledgment. This work was supported in part by the National Natural Science Foundation of China (NSFC) under Grant 61202237 and 61273263; the Specialized Research Fund for the Doctoral Program of Higher Education (No. 20121102120016); the research program of State Key Laboratory of Software Development Environment (SKLSDE-2013ZX-31); the joint project by the LIA 2MCSI lab between the group of Ecoles Centrales and Beihang University; and the Fundamental Research Funds for the Central Universities.

References

1. Berretti, S., Bimbo, A.D., Pala, P., Amor, B.B., Daoudi, M.: A set of selected sift features for 3D facial expression recognition. In: ICPR (2010)
2. Gong, B., Wang, Y., Liu, J., Tang, X.: Automatic facial expression recognition on a single 3D face by exploring shape deformation. ACM MM (2009)
3. Hao, T., Huang, T.S.: 3D facial expression recognition based on automatically selected features. In: CVPRW (2008)
4. Klassen, E., Srivastava, A.: Geodesics between 3D closed curves using path-straightening. In: Leonardis, A., Bischof, H., Pinz, A. (eds.) ECCV 2006, Part I. LNCS, vol. 3951, pp. 95–106. Springer, Heidelberg (2006)
5. Lemaire, P., Amor, B.B., Ardabilian, M., Chen, L., Daoudi, M.: Fully automatic 3D facial expression recognition using a region-based approach. J-HGBU (2011)
6. Li, H., Jean-Marie, M., Chen, L.: 3D facial expression recognition based on histograms of surface differential quantities. ACIVS (2011)
7. Mohammadzade, H., Hatzinakos, D.: Iterative closest normal point for 3D face recognition. IEEE TPAMI (2013)
8. Mpiperis, I., Malassiotis, S., Strintzis, M.G.: Bilinear models for 3D face and facial expression recognition. IEEE TIFS (2008)

9. Nair, P., Cavallaro, A.: 3D face detection landmark localization and registration using a point distribution model. IEEE TMM (2009)
10. Soyel, H., Demirel, H.: Facial expression recognition using 3D facial feature distances. ICIAR (2007)
11. Szeptycki, P., Ardabilian, M., Chen, L.: A coarse-to-fine curvature analysis-based rotation invariant 3D face landmarking. In: BTAS (2009)
12. Fang, T., Zhao, X., Ocegueda, O., Shah, S.K., Kakadiaris, I.A.: 3D facial expression recognition: A perspective on promises and challenges. In: FG (2011)
13. Wang, J., Yin, L., Wei, X., Sun, Y.: 3D facial expression recognition based on primitive surface feature distribution. In: CVPR (2006)
14. Yin, L., Wei, X., Longo, P., Bhuvanesh, A.: Analyzing facial expressions using intensity-variant 3D data for human computer interaction. In: ICPR (2006)

Azimuthal Perceptual Resolution Model Based Adaptive 3D Spatial Parameter Coding

Li Gao, Ruimin Hu, Yuhong Yang, Xiaocheng Wang,
Weiping Tu, and Tingzhao Wu

National Engineering Research Center for Multimedia Software, Computer School,
Wuhan University, Wuhan, 430072, China
Research Institute of Wuhan University in Shenzhen, China
gllynnie@gmail.com, hurm1964@163.com,
ahka_yang@yeah.net,clowang@163.com,echo_tuwp@163.com,785860285@qq.com

Abstract. The spatial perceptual feature of human ears plays an important role on the quantization of spatial parameter. Human ears have the most sensitive feature for sound in frontal direction, less for the rear area, and least for the lateral sides. Traditional quantization method of spatial parameter for frontal channel pair in stereo spatial audio coding is not suitable for that of channel pairs in different directions such as lateral or rear directions in multichannel audio coding. An azimuthal perceptual resolution model based adaptive spatial parameter coding for 3D audio multichannel signals is proposed in this paper. Based on the omnibearing and non-uniform azimuthal perceptual resolution model of human ears for sound sources in different directions, quantization values of spatial parameters can be estimated adaptively according to the location and configuration of channel pairs in arbitrary directions. The density features of quantization steps for spatial parameters of channel pairs in arbitrary directions are corresponding to the non-uniform azimuthal perceptual resolution of human ears. So that the quantization noise can be effectively reduced under the directional perceptual threshold with the improved reproduced spatial sound quality.

Keywords: 3D audio, multichannel audio coding, spatial parameter, azimuth, perceptual resolution.

1 Introduction

Parametric coding methods encode two or more channel signals into a downmix accompanied with spatial parameters, which can provide preferable compression ratio with considerable sound quality in 3D multichannel audio coding. The coding of the downmix signal can follow the traditional audio codecs. As for the coding of spatial parameters such as interchannel level difference (ICLD), interchannel time difference (ICTD) and interchannel coherence (ICC), which represent the spatial location information of virtual sound source generated by channels, traditional spatial hearing features of human ears were introduced to remove redundancy of spatial parameters in current Spatial Audio Coding (SAC) methods [1,2].

X. He et al. (Eds.): MMM 2015, Part I, LNCS 8935, pp. 534–545, 2015.
© Springer International Publishing Switzerland 2015

The locating capability of human ears for the sound at horizontal plane mainly depends on binaural cues: Interaural Time Difference (ITD) and Interaural Level Difference (ILD), which are caused by the difference of the sound propagated to two ears [3]. Although the binaural cues are the key factors for human ears to locate the direction of sound source, there is still limitation in perceptual sensitivity of binaural cues. Only when the change of binaural cues exceed the minimum threshold value human ears can perceive the change, which called Just Notice Difference (JND) of binaural cues. The JND values of binaural cues depend on the perceptual sensitivity of human ears. As the approximate expressions of binaural cues ILD and ITD without considering the transmitting procedure from sound source to human ears, the spatial parameters ICLD and ICTD are often quantized according to the JND features of binaural cues, based on which Binaural Cue Coding (BCC) was proposed accompanied with the basic framework of SAC. While the perceptual sensitivity of binaural cues is nonuniform (for example, different ILD value has different JND), but the quantization step sizes of spatial parameters are uniform in BCC, so the reproduced quality of spatial sound is still unsatisfactory.

Human ears have nonuniform sensitivities for sounds in different directions: the most sensitive fot the direction of right ahead, less the rear area, the least in the lateral directions. This sensitivity difference features of spatial cues are added in Parametric Stereo (PS) coding to give nonuniform quantization for spatial parameters [4]. In 2005 MPEG Surround extended spatial audio coding from stereo signals to multichannel signals [5], however, the same quantization table of ICLD from stereo signals is continued to be used for all pairs of channel signals in different directions

Han-gil Moon [6] extracted virtual source location information (VSLI) as spatial parameter instead of ICLD in multichannel coding. But the quantization step of VSLI was uniform 3° for all horizontal directions, which was not accord with the nonuniform perceptual directional resolution of human ears. Yang-Won Jung [7] also used horizontal angular resolution of 3° to guide the quantization of ICLD and set nonuniform quantization tables of spatial parameter for different pairs of channel signals. However the density characteristics of quantization step of all channel pairs are still uniform with that of the frontal channel pairs.

Hu [8] firstly tested and analyzed the JND of binaural cues in full spectrum band to guide the quantization of spatial parameter, and the bit rate of spatial parameter is reduced by 15% maintaining the nearly considerable stereo sound quality. Chen [9] extracted the angle information of virtual sound as spatial parameter and squeezed the angle information in the horizontal range from 360° to 60° for quantization. Although different quantization were used for different channel pairs in different directions, the quantization steps of spatial parameter in all channel pairs were uniform.

MPEG Surround was accepted as a part of MPEG 3D audio multichannel standard in 2013, however in which the quantization method of spatial parameters for different channel pairs in different directions still followed the quantization method of current spatial audio coding for stereo systems.

As for the stereo audio systems with two main speakers located in front, when ICLD is around 0, the virtual sound source generated by two channels is about to locate quite in front of the head with ILD is around 0 too. The JND of ILD of the sound quite in front of the head is the smallest, which means little change of ILD will be perceived. So according to the JND feature of ILD, ICLD values around 0 should be most finely quantized as current SAC methods do.

But for the pair of channels located at lateral area (such as channel pair L and Ls in 5.1 system), when ICLD of two channels is around 0, the virtual sound locates center between two speakers and around the extension of linking line between two ears. Under this situation although ICLD of two speakers is around 0, ILD will be much more bigger than 0. That means for the lateral pair of channels, the ICLD values around 0 will correspond to ILD values much bigger than 0. This lateral situation is just the opposite to the frontal situation. So the traditional quantization method of spatial parameter based on the auditory perceptual feature for frontal channel pair in stereo spatial audio coding is not suitable for that of channel pairs in different directions such as lateral or rear channel pairs in multichannel audio coding.

Based on an azimuthal perceptual resolution model, an adaptive spatial parameter coding method for 3D audio multichannel signals is proposed in this paper. An omnibearing perceptual model based on Minimum Audible Angle (MAA) is used to guide the quantization of spatial parameters for channel pairs in different directions. The quantization method is consistent with the perceptual feature of human ears for sound source in different directions, so the quantization noise is effectively reduced under the directional azimuthal perceptual threshold to improve the reproduced spatial sound quality.

2   Related Work of Quantization of Spatial Parameter

Given a pair of speakers at arbitrary locations, spatial parameters are extracted from the two speakers signals. ICLD is one of the most important spatial parameters to describe the location of the virtual sound source generated by two speakers. Due to that human ears have different perceptual sensitivities for sound sources in different directions, this sensitivity difference feature is often used to guide the quantization of spatial parameters. As current spatial audio codecs do, the sensitivity feature of binaural cues such as ILD is often used to guide the quantization of spatial parameters such as ICLD.

Human ears have the most sensitive directional perceptual feature for the sound source located at quite in front of the head, where the ILD corresponds to 0. The small change of ILD will attract attention of human ears. Along with the location of the sound source change from front to lateral, the absolute value of ILD will increase as well as the directional sensitivity of human ears will decrease.

For the stereo audio systems with two main speakers located in front, when ICLD is about 0 the virtual sound source will locate quite in front of the head, and the ILD value is near 0 too. If ICLD increases then ILD will increase too, as well as the sound evoked in direction from front to lateral area. Under the circumstances of stereo audio systems, ICLD has the same variation tendency with ILD for the location of the sound source change from front to lateral. So the perceptual sensitivity feature of ILD is used in spatial audio codecs to guide the quantization of ICLD to make sure that the ICLD values near 0 are more finely quantized.

In MPEG Surround multichannel audio codec, spatial parameters are extracted from pairs of channel signals. The quantization of ICLD used in MPEG Surround for different pairs of channel signals from different directions follows the same quantization method of ICLD used in stereo codec. The ICLD of frontal channel pair has the same variation tendency with ILD. However the ICLDs from different channel pairs located in different directions do not always have the same variation tendency with ILD, especially for the ICLD of lateral pairs, which even have the opposite results with the frontal pairs. Spatial location distortion of virtual sound will be caused by improper uniform quantization of spatial parameters for all channel pairs.

The noise-to-mask ratio (NMR) of azimuth is to evaluate the spatial location distortion introduced by current spatial parameter quantization method. For current subband k the NMR is computed as

$$NMR(k, i, j) = \left| \frac{\varphi(k, i, j) - \varphi_q(k, i, j)}{MAA(\varphi(k, i, j))} \right|, \tag{1}$$

in which $\varphi(k, i, j)$ and $\varphi_q(k, i, j)$ is the original and quantized azimuth of sound source respectively, $MAA(\varphi(k, i, j))$ is the minimum audible angle of $\varphi(k, i, j)$ and regarded as mask threshold of azimuth, i and j is the index number of two channels that form the virtual sound source.

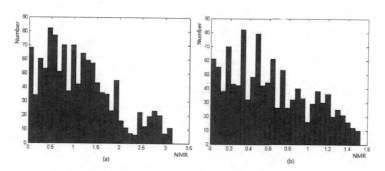

Fig. 1. NMR of azimuth for (a) frontal and (b)lateral pair of speakers

Fig. 1 further plots the azimuthal NMR of m_ot_x_9_org.wav (3GPP test sequence) by current spatial parameter quantization method for frontal and lateral channel pairs respectively. It is observed that with current spatial parameter quantization method, a large number of NMRs of azimuth are bigger than 1 for both frontal and lateral pair of speakers. It clearly reflects that the current quantization method of spatial parameter brings severe quality degradation of directional information of reproduced sound source for it is not accord with the spatial hearing feature of human ears. Differentiated quantization of spatial parameters should be provided for channel pairs in different directions, otherwise directional perceptual distortion of the virtual sound source will be caused.

3   Proposed Adaptive Spatial Perceptual Coding

3.1   Structure of Adaptive Spatial Perceptual Coding

Fig. 2 illustrates the structure of proposed adaptive spatial perceptual quantization and coding of spatial parameter in 3D audio multichannels. Based on the spatial hearing mechanism of human ears for sound in different directions, the omnibearing perceptual characteristic of azimuth is modelled. Then the quantization table of azimuth for different channel pairs in different directions can be adaptively derived from the perceptual model of azimuth according to arbitrary speakers configurations. Through the mapping from azimuth to spatial parameter, the quantization table of spatial parameter can be attained. Meanwhile the density characteristics of quantization steps of spatial parameter are accord with the spatial directional perceptual characteristics of human ears.

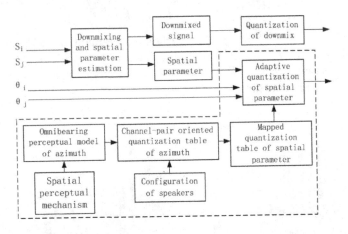

Fig. 2. Structure of proposed adaptive spatial perceptual coding of spatial parameter

3.2   Omnibearing Perceptual Model Based on MAA

Researches on JND of binaural cues indicate that: when ILD is 0 the JND value is the smallest; The bigger the ILD is, the bigger the JND is. It agrees with the fact that the most perceptual sensitive direction for human ears is in front of head. The perceptual sensitivity feature of ILD is important for the quantization of ICLD in stereo audio coding, however, it is not suitable for ICLD computed from channel pairs located outside the frontal area.

Besides ILD, there are still some other parameters concerned with the directional perception of human ears, such as the azimuthal perception of sound. In 1958, Mills [10] put forward a classical method to measure the human auditory localization sensitivity and defined the human auditory sensitivity as Minimum Audible Angle (MAA). MAA is the minimum spatial interval angle that one can distinguish two sound sources with different locations. The results showed that the MAA was about 1° in horizontal plane. Recently the feature of MAA has already been used in audio coding such as [11], in which MAA is used as a threshold to determine the truncation order of HOA signals and control the truncation error within MAA.

Referring to current available MAA results, we organized a large amount of subjective tests of MAA for different azimuths at different elevations. A total of 32 directional test points non-uniformly located from right ahead to rear directions for each elevation are choosed in our experiment and a MAA database of 3D sound in different directions with different elevations is established as our previous fundamental work [12]. As shown in shown in Fig. 3, it has the same variation trend of MAA along with different azimuths as the existing MAA data and can be used as a supplement of current available MAA data.

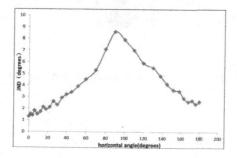

Fig. 3. JND of horizontal azimuth

With the horizontal MAA (similar to the JND of azimuth) shown in Fig. 3, the process of get the omnibearing quantization table of horizontal azimuth is described in Algorithm 1. The non-uniform density distribution of quantization steps of azimuth in 360° horizontal plane is demonstrated in Fig. 4. From which it can be deduced that along with the azimuth changes from 0° to 90° the quantization step of azimuth should be gradually increased, and the smallest and largest quantization step of azimuth correspond to the directions around 0° and 90° respectively.

Algorithm 1 QofAzimuth(azimuth x,minimum audible angle $f(x)$, quantization value q, quantization boundary b)

```
1:    i = 0;              \\ Quantization index
2:    x = 0;              \\ Azimuth from 0° to 180°
3:    q[i] = x;           \\ Quantization value of azimuth
4:    b[i] = x + f(x);    \\ Boundary of quantization value
```

```
5:    Do
6:        If b[i] > x + 1 - f(x + 1)
              \\ Search for boundary of each quantization value
7:        Then x + +;
8:        Else
9:        i + +;
10:       x + +;
11:       q[i] = x;          \\ Get quantization value
12:       b[i] = x + f(x);   \\ Get quantization boundary
13:       End if
14:   While(X < 180°)
```

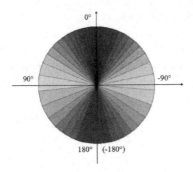

Fig. 4. Non-uniform density distribution of quantization steps of azimuth in 360° horizontal plane

3.3   Virtual Source Localisation Estimation

According to the tangent law [13], given azimuths of two speakers as in Fig. 5, the azimuth of the virtual sound source caused by the two speakers can be calculated by tangent law. The tangent law is described as :

$$\frac{g_L - g_R}{g_L + g_R} = \frac{\tan \varphi}{\tan \theta}, \tag{2}$$

$$g_L^2 + g_R^2 = g_S^2, \tag{3}$$

in which θ is half the angle between the two speakers and φ is the azimuth of the virtual sound source reproduced by the two speakers, g_S is to ensure the total energy. The gain of each speaker signal is

$$g_L = g_S \cdot \frac{\tan \theta + \tan \varphi}{\sqrt{2 \cdot \tan^2 \theta + 2 \cdot \tan^2 \varphi}}, \tag{4}$$

$$g_R = g_S \cdot \frac{\tan \theta - \tan \varphi}{\sqrt{2 \cdot \tan^2 \theta + 2 \cdot \tan^2 \varphi}}. \tag{5}$$

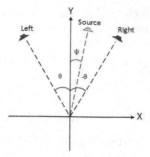

Fig. 5. Virtual source localisation estimation with tangent law

Then the azimuth of virtual sound source is estimated as:

$$\varphi = \tan^{-1}\left[\frac{g_L - g_R}{g_L + g_R}\tan\theta\right].\tag{6}$$

It can be concluded from (6) that given the azimuths and the gains of the two speakers the azimuth of virtual sound source can be estimated.

3.4 Azimuth Transformation Between Coordinates

Given two arbitrary speakers in multichannel audio system, the azimuth of virtual sound source can be estimated with (6) based the coordinate system as $X'Y'$ in Fig.6. This azimuth can be transformed to the main coordinate system of XY.

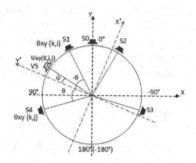

Fig. 6. Azimuth transformation between coordinates

For the current subband k, the azimuth of speaker S_i, S_j in the main coordinate XY is $\theta_{xy}(k,i)$ and $\theta_{xy}(k,j)$ respectively. the azimuth of speaker S_i, S_j in the new coordinate $X'Y'$ is

$$\theta(k,i) = -\frac{1}{2}(\theta_{xy}(k,j) - \theta_{xy}(k,i)),\tag{7}$$

$$\theta(k,j) = \frac{1}{2}(\theta_{xy}(k,j) - \theta_{xy}(k,i))\tag{8}$$

respectively. Given the azimuth of the virtual source S_{ij} in coordinate XY is $\varphi_{xy}(k, i, j)$, the azimuth of the virtual source S_{ij} in coordinate $X'Y'$ is

$$\varphi(k, i, j) = \frac{1}{2}(\theta_{xy}(k, j) + \theta_{xy}(k, i)) - \varphi_{xy}(k, i, j). \tag{9}$$

3.5   Quantization of ICLD Based on MAA Perceptual Model

ICLD is caculated from two speakers signals, which can be located in arbitrary directions.

$$ICLD = 10 \log_{10}(\frac{g_L^2}{g_R^2}) \tag{10}$$

With transformation of (6) and

$$\frac{g_L}{g_R} = \frac{\tan \frac{1}{2}(\theta(k, j) - \theta(k, i)) + \tan \varphi(k, i, j)}{\tan \frac{1}{2}(\theta(k, j) - \theta(k, i)) - \tan \varphi(k, i, j)}, \tag{11}$$

the correlation between ICLD and the estimated azimuth φ of virtual sound source is calculated by

$$ICLD = 20 \log_{10}(\frac{\tan \frac{1}{2}(\theta(k, j) - \theta(k, i)) + \tan \varphi(k, i, j)}{\tan \frac{1}{2}(\theta(k, j) - \theta(k, i)) - \tan \varphi(k, i, j)}), \tag{12}$$

or

$$\varphi(k, i, j) = \frac{10^{\frac{ICLD}{20}} - 1}{10^{\frac{ICLD}{20}} + 1} \cdot \tan \frac{1}{2}(\theta(k, j) - \theta(k, i)). \tag{13}$$

With (9) φ in coordinate $X'Y'$ and φ_{xy} in coordinate XY can be transformed to each other. Algorithm 1 gives quantization values and corresponding boundaries of azimuths in the main coordinate. After these quantization values of azimuths are transformed to the azimuths in the new coordinate they can be transformed to ICLD by (12). Then we can get the quantization values of ICLD adaptively according to two speakers located in arbitrary directions.

4   Experimental Results

To verify the performance of the proposed method, a series of subjective listening tests and objective evaluations have been carried out. Reference codec is MPEG Surround at bitrate of 48kbps, and comparison codec is the modified reference codec with the proposed quantization of spatial parameter instead. There are 6 test items selected from 3GPP standard test sequences including speech and music signals (m_ot_x_9.wav, s_cl_mt_1.wav, s_cl_mt_2.wav, s_no_2t_1.wav, s_no_ft_4.wav, sbm_sm_x_4.wav), which were allocated to speaker pairs located in different directions (mainly frontal and lateral speaker pairs considered as in 5.1 audio system).

According to algorithm 1, 25 quantization points of azimuth range from $0°$ (front) to $180°$ (rear) are obtained, 8 for $0°$ - $30°$, 8 for $30°$ - $110°$, and 9 for $110°$ - $180°$. Due to the bilateral symmetry, there are total 48 quantization points of

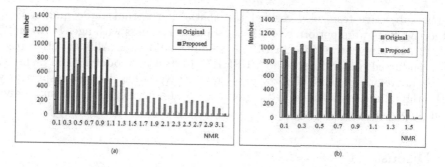

Fig. 7. Average NMR for (a) front and (b) lateral channel pair

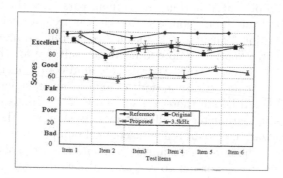

Fig. 8. MUSHRA evaluation results with 95% confidence interval and the mean

azimuth in 360° horizontal plane in the proposed quantization methods, which can be transformed to 48 corresponding ICLD values for channel pairs in all directions. But for the uniform ICLD quantization tables for all channel pairs (in four directions) in MPEG Surround, the total quantization number would come to 60 in 360° horizontal plane. Compared with current quantization method of ICLD, the coding bit-rate for ICLD quantization will be obviously reduced especially for lateral channel pairs and the proposed non-uniform quantization method can efficiently remove the perceptual redundancy in spatial parameters (although only ICLD is considered here, ICTD will be tested further).

The objective quality was measured with the NMR of spatial parameter as shown in Fig.7. NMR from equation (1) can be used to evaluate if the quantization noise of directional information is within the perceptual threshold. The less the NMR is, the less the quantization distortion of spatial parameter is, and the better reproduced spatial quality can be obtained. It is observed from Fig.7 that with the original improper quantization method of spatial parameter, a large number of NMRs that exceed 1 indicate that the quantization noise will severely decrease the reproduced spatial sound quality. However with the MAAs of all directions as a guide to the quantization step size of spatial parameter, the proposed method effectively constrains the quantization noise within the perceptual threshold of directional information.

The subjective tests also verified the spatial perceptual audio quality. The subjective tests were performed by 6 experienced listeners with ages between 22 and 32 who actively working in the domain of audio area. All subjective tests were conducted conforming to the MUSHRA [14] test methodology and included an original reference, a 3.5 kHz low-pass anchor, two signals decoded respectively with original codec and modified codec with proposed method. The subjective result is presented in Fig. 8.

5   Discussion

The spatial hearing perceptual feature of human ears plays an important role on the quantization of spatial parameter. Human ears have the most sensitive feature for sound in frontal direction, less for the rear area, and least for the lateral sides. The traditional quantization method of spatial parameter for frontal channel pairs in stereo spatial audio coding is not suitable for that of the lateral channel pairs. However current MPEG Surround coding method continues to apply the similar quantization method for multichannel signals as in stereo spatial audio coding. Although there are some different quantization tables by researchers for spatial parameter of different pairs of channels, the density characteristics of quantization step for different channel pairs are not consistent with perceptual feature of human ears. Improper quantization of spatial parameter would bring perceptual distortion to the spatial sound quality especially for multichannel signals with 3D speakers configurations. An azimuthal perceptual resolution model based adaptive spatial parameter coding method for 3D audio multichannel signals is proposed in this paper. An omnibearing perceptual model based on MAA is used to guide the quantization of spatial parameters for channel pairs in different directions. The quantization method is consistent with the perceptual feature of human ears for sound source in different directions, so the quantization noise can be effectively reduced under the directional perceptual threshold with the improved reproduced spatial sound quality.

Acknowledgments. The research was supported by NNSF of China (No. 61231015, 61102127, 61201340, 61201169,61471271)), Guangdong-Hongkong Key Domain Breakthrough Project of China (No. 2012A090200007), and STIP of Hubei Province (No. 2013AAA020).

References

1. Baumgarte, F., Faller, C.: Binaural cue coding-part I: psychoacoustic fundamentals and design principles. IEEE Trans. Speech Audio Process. 11(6), 509–519 (2003)
2. Faller, C., Baumgarte, F.: Binaural cue coding-part II: schemes and applications. IEEE Trans. Speech Audio Process. 11(6), 520–531 (2003)
3. Blauert, J.: Spatial Hearing: The Psychophysics of Human Sound Localization, sect. 4, ch. 2, pp. 137–177. The MIT Press, USA (1983)

4. Breebaart, J., van de Par, S., Kohlrausch, A., Schuijers, E.: Parametric Coding of Stereo Audio. EURASIP Journal on ASP (9), 1305–1322 (2005)
5. Hilpert, J., Disch, S.: The mpeg surround audio coding standard [standards in a nutshell]. IEEE Signal Processing Magazine 26(1), 148–152 (2009)
6. Moon, H., et al.: A multi-channel audio compression method with virtual source location information for MPEG-4 SAC. IEEE Transactions on Consumer Electronics 51(4), 1253–1259 (2005)
7. Jung, Y.-W., Oh, H.-O., Kim, H.J., et al.: New CLD Quantization Method for Spatial Audio Coding. In: Proceeding of 120th Audio Engineering Society Convention, vol. 6734 (May 2006)
8. Chen, S., Xiong, N., et al.: Spatial parameters for audio coding: MDCT domain analysis and synthesis. Multimedia Tools and Applications 48(2), 225–246 (2010)
9. Cheng, B.: Spatial squeezing techniques for low bit-rate multichannel audio coding (2011)
10. Mills, A.W.: On the minimum audible angle. J. Acoust. Soc. Am. 30(4), 237 (1958)
11. Daniel, A., McAdams, S., Nicol, R.: Multichannel Audio Coding Based on Minimum Audible Angles. In: Audio Engineering Society Conference: 40th International Conference: Spatial Audio: Sense the Sound of Space. Audio Engineering Society (2010)
12. Heng, W., Cong, Z., Ruimin, H., Weiping, T., Xiaochen, W.: The Perceptual Characteristics of 3D Orientation. In: Gurrin, C., Hopfgartner, F., Hurst, W., Johansen, H., Lee, H., O'Connor, N. (eds.) MMM 2014, Part II. LNCS, vol. 8326, pp. 353–360. Springer, Heidelberg (2014)
13. Pulkki, V.: Spatial sound generation and perception by amplitude panning techniques (2001)
14. ITU-T, Method for the subjective assessment of intermediate sound quality (MUSHRA) (2001)

Flat3D: Browsing Stereo Images on a Conventional Screen

Wenjing Geng, Ran Ju, Xiangyang Xu, Tongwei Ren, and Gangshan Wu

State Key Laboratory for Novel Software Technology
Nanjing University, China
jenngeng@gmail.com, {juran,xiangyang.xu}@smail.nju.edu.cn,
{rentw,gswu}@nju.edu.cn

Abstract. Expensive and cumbersome 3D equipment currently limits the popularization of emerging stereo media on the Internet. Particularly for stereo images, as a major kind of stereo media widespread on the Internet, there is not yet a good solution to show stereoscopy in conventional displays. By investigating the principles of human visual system (HVS), this paper proposes a method, called Flat3D for animating stereoscopy only through a conventional screen (2D) based on the motion parallax. The way for exhibition is dynamically transforming consistent views from left to right and then playing back reversely. The relative motion impresses viewers with strong depth perception. We investigate some factors which affect viewing experience in Flat3D and find that a reasonable fixation point and structure-preserved view transition contribute the most. Based on the above findings, we develop an adaptive fixation acquisition approach combining color and depth cues, as well as employing a probability-based view synthesis to generate the view sequences. Experiments which compared the above factors in and out of consideration show that our approach is a more convenient, effective and automatic alternative for browsing stereo images in common flat screens.

Keywords: Stereo images, fixation, view synthesis, transfer sequences.

1 Introduction

The amount of stereo images is soaring up on the Internet due to the popularization of 3D acquisition devices, for example Fujifilm 3D camera. Nevertheless, viewers cannot perceive 3D without the help of 3D equipment such as 3D glasses and 3DTV. Undoubtedly, there is a seemingly wide gap between viewers and stereo images because of extra demands for 3D equipment. In other words, stereo photographs captured by people or downloaded from the Internet, would degrade to side-by-side or red-cyan pictures in the eyes, as shown in Fig. 1 (a)-(b). Most people even do not understand what the meaning of stereo media is without glasses. This limitation is also another main reason that blocks the further widespread of 3D media. Therefore, a convenient and effective method to exhibit stereo images widespread on the Internet turns out to be an essential and

X. He et al. (Eds.): MMM 2015, Part I, LNCS 8935, pp. 546–558, 2015.

(a) side-by-side   (b) red-cyan   (c) wiggle stereoscopy

Fig. 1. Display of stereo images

significant problem. Besides, how to build a more pleasing 3D effects to cater for feelings of human vision through flat screens is another challenge.

To this problem, there are mainly two solutions separately from hardware [20] and software perspective. The former usually makes changes in the structure of screens based on optical principles, including lenticular lens, parallax barrier [14] and directional backlight [18]. Although these techniques have been applied in many devices, serving for advertising media and many other applications, it usually costs a lot, which makes it not applicable for the visualization of daily stereo images. Besides, there still exist some defects on resolution, visual angle and distance. The latter always utilizes computer graphics or visual rules to alter the content of images to create the third dimension, depth, such as wiggle stereoscopy[1], adding auxiliary lines[2] and perspective. Actually, these methods can indeed create stereoscopy, but there are something should not be ignored. First, the stereoscopic effect is usually achieved by human interference which is impossible to apply to the numerous stereo images on the Internet, let alone offers good and cosy user experience between left and right views based on uncertain fixation (Fig. 1 (c)). The 3D effect of this method inevitably flickers a lot. In a word, relying on photo editing and lacking complete automatic operating mechanisms make it impossible to apply to arbitrary stereo images. Second, some 3D effects generated by adding auxiliary or perspective cannot directly apply to stereo images at present and the generated results are obvious artifact.

In this paper, we propose a novel display approach named Flat3D for browsing stereo images on a conventional screen based on two findings of improving user experience for depth perception. First, considering the different influence upon region of interest (ROI) of human visual system (HVS), convergent point is calculated based on color and depth information, which is liable to find the optimal and pleasing focus point just as human do. Second, a probability-based rendering for view synthesis [2] is used to synthesize spatial-temporal consistent views for its insensitiveness to depth inaccuracy. Then the structure-preserved views are aligned based on the fixation. The format of demonstration is GIF or video sequences following visual persistence. By adaptively fixing human focus, the transfer sequences keep cyclical view transition from left to right and then from right to left. Compared with the existing methods, our approach is automatic, cheap, and efficient, and thus competent in exhibiting stereo images on a conventional screen

---

[1] http://en.wikipedia.org/wiki/Wiggle_stereoscopy
[2] http://www.mymodernmet.com/profiles/blogs/3d-gifs

such as PC screens and mobile screens, with good user experience. In summary, the major contributions include:

- An automatic and systematic approach for browsing stereo images is proposed by utilizing principles of motion parallax, fixation and visual persistence to improve user experience, which provides a cheap, convenient and comfortable alternative for demonstration of stereo images in daily life;
- Color and depth cues are combined to predict the fixation and structure-preserved views are synthesized by referring to the convergence of human eyes, which brings preferable 3D effects when browsing stereo images through common screens.

The rest of this paper is organized as follows. The background is briefly introduced in Sect. 2. Then a brief review of the related work is introduced in Sect. 3. The approach is detailedly described in Sect. 4 and evaluated by a few experiments shown in Sect. 5. Finally, we give a conclusion in Sect. 6.

2 Background

Main factors for HVS to produce stereoscopic impression are binocular parallax, motion parallax, accommodation and convergence [3]. The left and right eyes capture different content of the same scene. Due to accommodation and convergence, one image is composed in the brain. Existing 3D displays usually utilize horizontal parallax to create 3D mostly relying on expensive equipment. While we follow the motion parallax depending on relative movement, synthesize the intermediate views and align all the views based on the fixation marked as F_c in the Fig. 2. For stereo images are usually captured by binocular camera, we have got the first factor of binocular parallax from left and right views.

Motion parallax is a psychological phenomenon that observers view objects that are closer to them moving faster than objects that are further away from them. As to our framework, we utilize motion parallax to produce stereoscopy. We need to find the convergent point closer to HVS, or human would feel uncomfortable if the stationary point is not attractive. In other words, we need

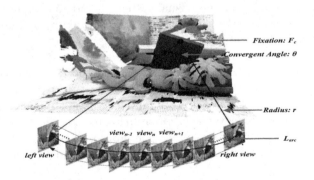

Fig. 2. The mechanism of the proposed approach

to automatically detect region of interest (ROI) first and calculate its centroid as fixation point then. For centroid is uniform distribution of quality, it is the optimal point to be fixation. We then transform this problem to saliency map calculation, because saliency determines attentional selection in psychology [11]. Furthermore, the transfer sequences are composed according to visual persistence that human can preserve images within the range of $0.1s$ to $0.4s$. In order to build spatial-temporal consistent stereoscopy in the display, the duration of each view should not be set too long. Eq. (1) shows that it would take more time to transfer the sequences from different views with increasing number of view frames. And the velocity of the transfer sequences is inverse proportion to number of views.

$$v = \frac{L_{arc}}{T_{total}} = \frac{\theta \pi r}{Num_{view} \times t_{per}}, \tag{1}$$

where L_{arc} is the arc path that views are transforming by and T_{total} is the total time cost of transfer sequences from left to right view. θ is the angle of view and r is the radius labeled in Fig. 2. Num_{view} denotes the number of view frames and t_{per} shows the time persistence of each frame.

Based on the above methodology of psychology, physiology and HVS, we propose an automatic display approach for browsing 3D in a flat screen which caters to the requirement for comfortable 3D effects.

3   Related Work

We briefly review the similar research on wiggle stereoscopy and also the relevant techniques on saliency detection and multi-view synthesis.

Wiggle Stereoscopy. Wiggle stereoscopy describes technique giving an illusion of 3D by showing two images in rapid alternation. With the spring up of stereo images, it has gradually become a useful method when amateurs want to create stereoscopy. Therefore, there are several step-by-step tutorials distributed in the Internet. Most of these procedures rely on image editing software with human interference, for example Photoshop, which means it is impossible for thousands of stereo images to generate stereoscopic wiggle sequences. Besides, the common practice is composing dynamic switching between left and right views, for example StereoPhoto Maker, which would cause flicker to the result images. Because it lacks comparatively complete theoretical instructions, the generation GIF results always cause a little dizzy when watching it, which limits the widely use of this technique. Different from wiggle stereoscopy, the proposed approach automatically generates the results by exploring the principles of HVS, eliminating strong flicker by adaptively calculating fixation and consistent transforming.

Saliency. Saliency is a physiological and psychological phenomenon of visual attention, and has been used as a fundamental in many vision and multimedia tasks[16]. Itti et al. [4] used color, intensity and orientation to compute the saliency map. Judd et al. [7] learnt a saliency model on a thousand images. Cheng et al. [1] introduced a global contrast based salient region detection by automatic estimating salient object regions. We choose the absorbing markov chain method to detect saliency proposed by Jiang et al. [5] for its efficiency and robustness to multi-focus images. Furthermore, it has been proved that depth perception has a strong impact on visual attention [15]. But most existing work always take a 2D color image as input which lacks depth cues. Lang et al. [9] modeled saliency as the conditional probability given depth and depth range. But the limitation is that global depth structure information is missing. Therefore, we utilize a depth saliency based on anisotropic center-surround difference proposed by Ju et al. [6] for its superiority to the state-of-the-arts.

Multi-View Synthesis. Multi-view synthesis is an essential step in many 3D display technologies. Min et al. [13] synthesized a virtual view by adapting a reverse warping instead of a forward warping. Mahajan et al. [12] interpolated a virtual view based on the idea that the given pixel to be synthesized in the virtual views traces out the path in reference images. These methods are both discrete formulation which would bring the problem of hole region. Lang et al. [10] presented a method free from hole filling, but the geometric distortion is till inevitable. Another challenge is depth inaccuracy on view interpolation. Kunita et al. [8] introduced the layered probability maps for dealing with depth ambiguities. Although the similar concept is used in the rendering process, including many geometric prior makes it not suitable for our framework. A probability-based rendering (PBR) method is proposed by Ham et al. [2] who addressed view synthesis as an image fusion. This method gives a good solution to depth inaccuracy and generates consistent images from different views.

4   Approach

Given an input of stereo images, we aim to automatically produce transfer sequences animating its stereoscopy in common displays. We formulate the problem by simulating HVS and utilizing motion parallax. The framework of our approach is shown in Fig. 3, which is made up of three main modules. The detailed procedures of the proposed approach are presented in Algorithm 1.

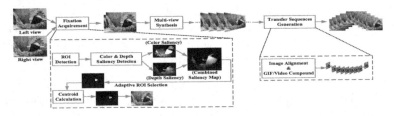

Fig. 3. The overview of the proposed approach

Algorithm 1. Outline of Flat3D

Input : A rectified stereo images
Output: Dynamic transfer sequences from different views
begin

   1. Calculate visual fixation to acquire the fixed point as reference. (Sect. 4.1)
     (a) Detect ROI based on saliency detection calculated by Eq. (2)-(5).
     (b) Calculate visual fixation, centroid of ROI according to Eq. (6).
   2. Synthesize multi-view images from left to right angel. A probability-based
     rendering method is adopted based on Eq. (7). (Sect. 4.2)
   3. Generate transfer sequences. (Sect. 4.3)
     (a) Align different views from left to right based on the fixation point
      according to the transformation matrix defined in Eq. (8).
     (b) Compound transfer sequences based on visual persistence.

end

4.1 Fixation Acquirement

Visual fixation is the maintaining of visual gaze on a single location and plays an important role in human stereo vision. Saliency detection, or gaze prediction, computationally detects the fixation. Different from existing methods working on single color images, we believe many factors having effects on fixation selection, such as color contrast, spatial structure and depth. Hence, a combination of color and depth cues calculated from the stereo image pairs is adopted. This module consists of ROI detection and centroid calculation.

ROI Detection. In our approach, ROI is detected by calculating saliency map, for saliency is the most informative and interesting region in a scene. Stereo images contain more information compared to monocular images. Not only color and structure, but also depth can contribute to saliency detection. Therefore, we present a method of color saliency plus depth saliency to acquire a compound saliency map and then get ROI. Considering the appearance divergence and spatial distribution of salient objects and background, we employ the saliency detection via absorbing markov chain for color saliency [5]. The saliency map calculated by [5] with factors of color and spatial distribution is $Sal_{c,s}$, and the equation is as follows. Where i indexes the transient nodes on graph and y_w denotes the normalized weighted absorbed time vector.

$$Sal_{c,s}(i) = y_w(i), i = 1, 2, \cdots, t. \tag{2}$$

For the other powerful cue, depth, we calculate the saliency based on anisotropic center-surround difference [6], denoted by Sal_d as shown in Eq. (3).

$$Sal_d(i) = \bar{D}_{acsd}(i) = \sum_{k=1}^{n} \sum_{t=1}^{8} D_{acsd}^t(p) / \sum_{i=1}^{m} D_{acsd}(i), \tag{3}$$

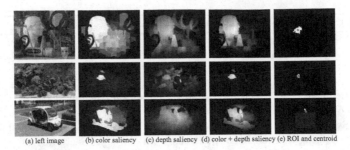

(a) left image   (b) color saliency   (c) depth saliency (d) color + depth saliency (e) ROI and centroid

Fig. 4. Results of saliency detection and centroid calculation

where $\bar{D}_{acsd}(i)$ dedicates the normalized ACSD value of superpixel i and $D_{acsd}(i)$ is the ACSD value of superpixel i. $D^t_{acsd}(p)$ represents the ACSD value of pixel p in the tth direction. The number of pixels in the superpixel is n and the number of superpixel is m.

Then the candidate saliency map Sal is calculated as Eq. (4) and normalized to $[0, 1]$. Then a threshold function is used to determine the interest region, shown in Eq. (5). $I_{bw}(\cdot)$ is the binary image labeled the connected region, τ is an adaptive threshold calculated using OTSU. Results of saliency maps for each processing step are listed in Fig. 4 (b)-(d).

$$Sal = Sal_{c,s} \times Sal_d. \tag{4}$$

$$I_{bw}(x) = \begin{cases} 1 & Sal(x) > \tau \\ 0 & \text{otherwise} \end{cases}. \tag{5}$$

Centroid Calculation. We choose centroid to represent visual fixation point, which is the point representing mass balance in physics. Let $c(x, y)$ denotes centroid of a region, we calculate its coordinate as Eq. (6), where $I(x, y)$ is the intensity of the binary image.

$$\begin{cases} x_c = \sum_{(x,y)\in T} xI(x,y) / \sum_{(x,y)\in T} I(x,y) \\ y_c = \sum_{(x,y)\in T} yI(x,y) / \sum_{(x,y)\in T} I(x,y) \end{cases}. \tag{6}$$

Besides, connected regions obeying the rule defined in Eq. (6) may surpass one. We firstly sort the connected regions and choose the largest one as the ROI and compute the centroid of it. Some examples of binary images marked with red centroid are shown in Fig. 4 (e).

4.2 Multi-view Synthesis

The aim of our framework is generating consistent view transition in a specific time according to visual persistence. In fact, fusion based image morphing (FBIM) and motion based image morphing (MBIM) could also create motion parallax which seems to be unnecessary for multi-view synthesis. However, image wiggling only between left and right views would be too flickering. We resort to an image fusion method which is performed in a probabilistic way [2] for its insensitivity to depth inaccuracy to preserve change of view structure.

Let $I_l(m)$ and $I_r(m)$ represent left and right images separately. Their corresponding steady state matching probability defined in [2] is $P_l(m,d)$ and $P_r(m,d)$. Assuming that the baseline between left and right images is normalized to 1, then the location of virtual view $I_v(m)$ is denoted by α, where $0 < \alpha < 1$. In order to minimize the inaccuracy brought by depth map, the synthetic process is transformed to image fusion. First candidate re-sampled color images on the virtual view $I_l^v(m)$ from left images and virtual view $I_r^v(m)$ from right images are calculated. Then view synthesis turns to image fusion, shown in Eq. (7).

$$\begin{aligned} I_v(m) &= \alpha I_l^v(m) + (1-\alpha)I_r^v(m) \\ &= \sum_d (\alpha I_l^v(m,d)P_l(m,d) + (1-\alpha)I_r^v(m,d)P_r(m,d)), \end{aligned} \quad (7)$$

where $I_l^v(m,d)$ and $I_r^v(m,d)$ are image intensity along the different disparity hypothesis d. The detailed description of this method is described in [2]. By utilizing PBR method, we calculate the intermediate views and some of them are shown in Fig. 5. The red straight line is a fixed marker used for showing the consistent view change from right to left.

Fig. 5. Results of synthesized views

4.3   Transfer Sequences Generation

The centroid of salient region is chosen as the stationary point in which the disparity should be zero. The reason is that if the most attractive point has a frequent change, human would feel uneasy and difficult to perceive 3D. Admittedly, multi-view based image change without convergence(MBICWC) could also bring 3D effect, it is simply image switching and easily degrades stereoscopy for not obeying rules of stereo vision. In order to make the disparity of fixation to be zero, each view should be aligned based on the transformation matrix:

$$T = \begin{bmatrix} 1 & 0 & 0 \\ 0 & 1 & 0 \\ disp & 0 & 1 \end{bmatrix}. \quad (8)$$

$$disp = D_{target}(centroid), \quad (9)$$

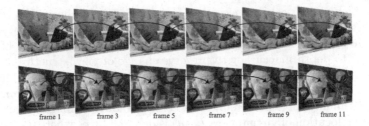

frame 1   frame 3   frame 5   frame 7   frame 9   frame 11

Fig. 6. Examples of image alignment based on fixation

where *disp* is disparity of centroid from target image to reference image derived from depth map D. The direction is negative in our method because right images are used as reference images. We trace the trajectory of fixation listed in Fig. 6, align them and make their disparity to be zero.

The final processing is generating dynamic sequences of all the aligned views. Either GIF or video format is feasible based on the actual requirement. However, on the basis of visual persistence about $0.1s$ to $0.4s$, the frame rate should obey this rule which has been discussed in Sect. 2.

5   Experiments

5.1   Datasets and Experimental Settings

Our approach is implemented using Matlab on a desktop PC with an Intel i7_4770 CPU and 16GB memory. It is noted that the generated results of our approach contain continuous transfer sequences from left to right view and then from right to left. Hence the 3D effect could be dynamically displayed in any flat screen as GIF or video format[3].

To show the efficiency and robustness, we evaluate our approach on Middlebury Stereo Datasets [17] and OBSIR datasets [19]. The details of the datasets are listed in Table 1, where GT is the ground truth. We use 0.1s for each frame and 20 frames are used to create smooth view transition but only 9 views need to be interpolated. The frame rate of the generated transfer sequences is 10fps.

Table 1. Description of Middlebury and OBSIR dataset

Dataset	Num.	Num. of test	GT. of depth map	GT. of multi-view images
Middlebury	39	39	Y	Y
OBSIR	10513	1382	N	N

5.2   Results and Discussion

We make a thorough comparison by considering the influence of fixation, multi-view synthesis and a user study is conducted to evaluate the 3D effect.

---

[3] http://mcg.nju.edu.cn/publication/MMM15-GengWJ.html

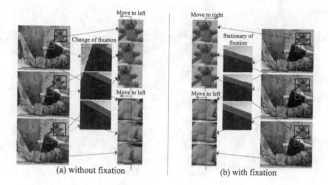

Fig. 7. Comparison with view change without image alignment based on fixation

Automatic Fixation. Fig. 7 gives a concrete comparison between MBICWC and Flat3D. We extract the same contents from three views, marked by black, red and blue windows. Red stars mark the visual fixation in each frame. It can be seen that all the pixels are moving from right to left in Fig. 7 (a) because of no fixation. On the contrary, in Fig. 7 (b) objects behind the fixation move from left to right and objects in front of the fixation move from right to left. Above all, it is the fixation that causes the relative motion before and after the fixation in 3D space. Objects behind the fixation, e.g. pink bear, move to the right while the objects in front of fixation, e.g. green legs move to the left.

View Interpolation. Not every image movement can create stereoscopy in flat screens. The proposed method is not a plain change among images. It is an imitation of human stereo vision which makes flat screens to be retina. In order to show the efficiency and reasonability, we also compare our method with FBIM and MBIM. The first line of Fig. 8 shows several frames generated by FBIM from six views separately. The middle frames become obscure due to the image fusion of linear interpolation, which looks like playing slides and hardly generates 3D effects. The second line is created by MBIM. Although frames are more distinct than the above line, there are inevitable holes caused by depth inaccuracy. The third line is the same frames built by the proposed approach, which has consistent trajectory to ensure smooth transition on views and one stationary fixation to create 3D effects.

User Study. To further evaluate the efficiency, we conducted a 3D effects user study aiming at whether the transfer sequences automatically produced by our method are preferred by users to those of sequences generated by other methods. We invite 36 subjects (20 males and 16 females with ages from 16 to 47) with normal vision and different educational background participated in the experiment. The dynamic sequences produced by 4 different approaches are compared, including FBIM, MBIM, MBICWC and the proposed approach. Each type of

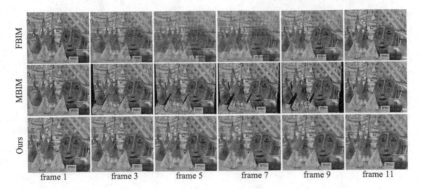

Fig. 8. Comparison results

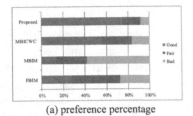

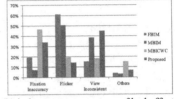

(a) preference percentage   (b) inducement percentage of bad effects

Fig. 9. The comparison of the four evaluated approaches on user ratings

dynamic sequences is numbered as 1, 2, 3, 4 with GIF format for its convenience. In the user study, the feelings for 3D effects are divided into 3 quality levels: Good, Fair, Bad, according to the observers' own perspective. After above process, the ratings from the users on each level are accumulated and the proportions of each quality level by the different approaches are listed in Fig. 9 (a). The Good rates are separately 75.13%, 55.56%, 13.89% and 27.78%. Our approach received most preference for its good performance in 3D effects. Besides, we offer 4 main options of affecting stereoscopy for subjects to choose if bad 3D effects are labeled. The options are fixation inaccuracy, flicker, view inconsistent and something others. The statistics are the average percentage based on each subject's option within bad levels, shown in Fig. 9 (b). 3D effects of MBICWC are close to ours because it is also a smooth view transition, but no fixation degrades its 3D effect. MBIM received the most Bad for annoying holes during view change which easily brings strong flicker. And FBIM is accurately a process of image fusion with inevitable flicker.

6   Conclusion

This paper proposes an automatic and systematic approach for browsing stereo images in flat screens by generating transfer sequences. Based on findings of human stereo vision, we believe our method is a good solution to exhibit stereo images stereoscopic in daily screens, e.g., mobile screens. The experiments and user evaluations demonstrate the reasonability and effectiveness of our method.

Acknowledgement. This work is supported by the National Science Foundation of China (No.61321491, 61202320), Research Project of Excellent State Key Laboratory (No.61223003), Natural Science Foundation of Jiangsu Province (No.BK2012304), and National Special Fund (No.2011ZX05035-004-004HZ). It was also partially supported by Collaborative Innovation Center of Novel Software Technology and Industrialization.

References

1. Cheng, M., Mitra, N., Huang, X., Torr, P., Hu, S.: Global contrast based salient region detection. PAMI (2014)
2. Ham, B., Min, D., Oh, C., Do, M., Sohn, K.: Probability-based rendering for view synthesis. TIP (2014)
3. Howard, I.P.: Binocular vision and stereopsis. Oxford University Press (1995)
4. Itti, L., Koch, C., Niebur, E.: A model of saliency-based visual attention for rapid scene analysis. PAMI (1998)
5. Jiang, B., Zhang, L., Lu, H., Yang, C., Yang, M.H.: Saliency detection via absorbing markov chain. In: ICCV. IEEE (2013)
6. Ju, R., Ge, L., Geng, W., Ren, T., Wu, G.: Depth saliency based on anisotropic center-surround difference. In: ICIP. IEEE (2014)
7. Judd, T., Ehinger, K., Durand, F., Torralba, A.: Learning to predict where humans look. In: CVPR. IEEE (2009)
8. Kunita, Y., Ueno, M., Tanaka, K.: Layered probability maps: basic framework and prototype system. In: VRST. ACM (2006)
9. Lang, C., Nguyen, T.V., Katti, H., Yadati, K., Kankanhalli, M., Yan, S.: Depth matters: Influence of depth cues on visual saliency. In: Fitzgibbon, A., Lazebnik, S., Perona, P., Sato, Y., Schmid, C. (eds.) ECCV 2012, Part II. LNCS, vol. 7573, pp. 101–115. Springer, Heidelberg (2012)
10. Lang, M., Hornung, A., Wang, O., Poulakos, S., Smolic, A., Gross, M.: Nonlinear disparity mapping for stereoscopic 3D. TOG (2010)
11. Ma, L., Xu, K., Wong, T., Jiang, B., Hu, S.: Change blindness images. TVCG (2013)
12. Mahajan, D., Huang, F.C., Matusik, W., Ramamoorthi, R., Belhumeur, P.: Moving gradients: a path-based method for plausible image interpolation. TOG (2009)
13. Min, D., Kim, D., Yun, S., Sohn, K.: 2D/3D freeview video generation for 3dtv system. SPIC (2009)
14. Neil, A.: Autostereoscopic 3d displays. Computer (2005)
15. Niu, Y., Geng, Y., Li, X., Liu, F.: Leveraging stereopsis for saliency analysis. In: CVPR. IEEE (2012)
16. Ren, T., Ju, R., Liu, Y., Wu, G.: How important is location in saliency detection. In: ICIMCS. ACM (2014)
17. Scharstein, D., Szeliski, R.: High-accuracy stereo depth maps using structured light. In: CVPR. IEEE (2003)
18. Wetzstein, G., Lanman, D., Hirsch, M., Raskar, R.: Tensor displays: compressive light field synthesis using multilayer displays with directional backlighting. TOG (2012)

19. Xu, X., Geng, W., Ju, R., Yang, Y., Ren, T., Wu, G.: Obsir: Object-based stereo image retrieval. In: ICME. IEEE (2014)
20. Zhang, Y., Ji, Q., Zhang, W.: Multi-view autostereoscopic 3D display. In: OPEE. IEEE (2010)

Online 3D Shape Segmentation by Blended Learning

Feiqian Zhang, Zhengxing Sun*, Mofei Song, and Xufeng Lang

State Key Laboratory for Novel Software Technology, Nanjing University, P.R. China
szx@nju.edu.cn

Abstract. This paper presents a novel online 3D shape segmentation framework, which blend two learning methods together: unsupervised clustering based method, and supervised progressive learning method. The features of this method lie in four aspects. Firstly, we use weighted online learning to train a segmentation model to achieve the blended learning framework. Secondly, we perform co-segmentation based on unsupervised clustering to analyze the shape set, and initialize this segmentation model. Thirdly, based on this segmentation model, users can segment new shapes by using supervised progressive learning method. And this segmentation model can also be incrementally updated by weighted online learning during the progressive segmentation. Finally, the segmentation of shapes in the initial set can be corrected based on the updated segmentation model. Experimental results demonstrate the effectiveness of our approach.

Keywords: 3D shape, Co-segmentation, Online learning.

1 Introduction

With the fast development of 3D scanning techniques, 3D shapes are nowadays intensively used as an emerging type of digital multimedia. Techniques for efficient and robust analysis and processing of 3D shapes have also become an active area of research in computer graphics [1]. Segmentation of 3D shapes into meaningful parts is a fundamental problem in shape analysis and processing [2]. A large number of segmentation methods have been proposed, and most of them pay attention to segmenting an individual shape based on geometric features such as convexity and curvature [2]. Although a variety of geometric features have been investigated, no single feature or collection of features is known to produce high-quality results for all classes of shapes [2], and consistently segmenting a set of shapes remains challenging [3]. Therefore, more and more researchers have been focusing on co-segmentation which simultaneously segments a set of shapes of the same family into consistent parts has been attracting more and more attention [4][5][6].

The existing co-segmentation methods obtain some common composition information by clustering [4][6][7][8][9][10][11] or optimization [5][12] on the whole

---

* Corresponding author.

X. He et al. (Eds.): MMM 2015, Part I, LNCS 8935, pp. 559–570, 2015.

shape set, and meanwhile use it to achieve the consistent segmentation. Unfortunately, it is difficult for users to convey their intentions during these methods. Although in some semi-supervised methods [7][13][14], users can segment the shapes according to their intentions, they could hardly change the number and categories of the consistent parts easily after co-segmentation results are presented. Moreover, they are confined in the stationary shapes set, and are lack of some updating strategies for the variation of the shape set. They need to be performed over the whole shape set, when there are only a small amount of new shapes added, which is very inefficient.

Recently, in our previous work, we have proposed a progressive 3D shapes segmentation method [15]. It allows users to label and segment the 3D shapes progressively by using online learning. And as the increasing of the shapes, the segmentation can become more accurate according to users' intention. However, at the beginning of the stages, there are no segmentation suggestions to users. Thus, they have to do extensive work for the first few shapes to accomplish the segmentation.

Accordingly, in this paper, we propose a novel online 3D shape segmentation framework, which blend two learning methods together: unsupervised clustering based method, and supervised progressive learning method. The features of this method lie in four aspects. Firstly, we use weighted online learning to train a segmentation model to achieve the blended learning framework. Secondly, we perform co-segmentation based on unsupervised clustering to analyze the shape set, and initialize this segmentation model. Thirdly, based on this segmentation model, users can segment new shapes by using supervised progressive learning method. And this segmentation model can also be incrementally updated by weighted online learning during the progressive segmentation. Finally, the segmentation of shapes in the initial set can be corrected based on the updated segmentation model.

Our approach has three advantages. Firstly, the initial segmentation of 3D shapes can be obtained during the analysis by unsupervised clustering based method, which reduces the interactions in the subsequence segmentation. This initial segmentation can be corrected based on the updated segmentation model, so it would not limit the results of subsequence progressive segmentation. Secondly, users' intention can be learned in the supervised progressive learning method. They can direct the segmentation through the whole pipeline with their interactions. Thirdly, we can blend the above two learning methods by using weighted online learning. So the 3D shapes segmentation can be accumulated and updated effectively.

We evaluate the presented approach on several 3D shapes, and the experimental results demonstrate the above advantages of our method.

2   Overview

In our method, unlabeled shapes can be segmented based on a segmentation model. This segmentation model can be online trained, when new shapes are progressively segmented with users' interactions. Therefore, the segmentation can be accumulated and updated effectively.

Feature Descriptors. Our method is performed based on the feature descriptors of the shapes. We compute some shape descriptors for each facet. According to the study on feature selections in the supervised approach [16], we choose five shape descriptors including Gaussian curvature (GC) [17], shape diameter function (SDF) [18], average geodesic distance (AGD) [19], shape contexts (SC) [20], and the geodesic distance to the base of the shape (GB) [4]. Then we can get the feature vector $x_i \in X$ for each facet i in the shapes.

Online Segmentation. Figure 1 illustrates the pipeline of our online segmentation method. At the very beginning, we perform initial co-segmentation based on unsupervised clustering on the unlabeled shapes in the initial set to get the initial labels of these shapes. According to these initial labels, a segmentation model can be initialized through weighted online learning. Next, based on this segmentation model, users can segment new shapes interactively through progressive segmentation, which is a supervised progressive learning method. And this segmentation model can also be incrementally updated during the progressive segmentation by using weighted online learning. Then, the segmentation of shapes in the initial set can be corrected based on this updated segmentation model.

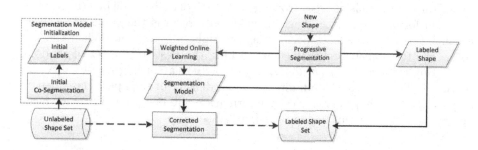

Fig. 1. The pipeline of the proposed method

3   Algorithm

In this section, we will describe the whole pipeline in details. First of all, we will introduce our Weighted Online Learning in section 3.1. Then, the following three sections present the remaining steps.

3.1   Weighted Online Learning

This step is the core of our online framework, which blends unsupervised clustering based method with supervised progressive learning method together. In this step, based on the labeled facets, an Online Random Forest (ORF) [21] is trained as the segmentation model. The labels of these facets are composed of three kinds of labels: initial labels, interaction labels and submitted labels. Initial labels are the labels obtained from initial co-segmentation. These labels are uncertain. They may be incorrect or do not match the users' segmentation requirements. While interaction labels and submitted labels are both confident. They are from progressive segmentation. Interaction labels are the labels which users have imputted to correct the segmentation results. And submitted labels are the labels of the shapes when their segmentation results are accepted by users. These three kinds of labeled facets are attached with different weights w_t: w_t^C, w_t^I and w_t^S respectively, to indicate the confidence of the labels. These weights will be introduced in the following sections.

Next, we train an Online Random Forest to get the probabilities that facets belong to all the part classes. For each labeled facet t (x_t, w_t, y_t), the Online Random Forest will be updated one time, where x_t is the feature vector of facet t, $y_t \in C$ is the label of this facet, and C is the set of labels ($K = |C|$). To achieve the online learning, the algorithm should combine the online bagging and online decision trees with random feature selection. For the bagging part, the arrival of sequential data is modeled by a Poisson distribution. Therefore, the trees are retrained k times on each facet t (x_t, w_t, y_t) in the marked set, where k is a random number generated by Poisson(λ) and usually λ is set to a constant (typically $\lambda = 0.1$ in our approach). In this way, the node gathers the statistics online.

For the online decision trees growing part, each decision node in a tree contains a test in form of $g(x) > \theta$, where $g(x)$ usually returns the value of a selected feature, θ is a threshold to decide the left/right propagation. When the node splits, the test can be determined by picking the best from a set of randomly generating tests $S = \{(g_1(x), \theta_1), \ldots, (g_N(x), \theta_N)\}$ according to a quality measurement. The node also maintains the label density of each part class, denoted by $p_t = [p_{t,1}, \ldots, p_{t,K}]$. This density information will be updated by each facet t (x_t, w_t, y_t) falling in this node in the marked set:

$$p_{t,k} = \begin{cases} p_{t-1,k} + w_t & \text{if } k = y_t \\ p_{t-1,k} & \text{if } k \neq y_t \end{cases}, \tag{1}$$

and the total weighted number n_R of this node will be updated as $n_R = n_R + w_t$.

In online mode, the statistics are gathered gradually, therefore, the decision when to perform the splitting depends on:

1. if there has been enough facets in a node to have a robust statistics: $n_R > \alpha$, where α is the minimum number of facets a node has to observe before splitting. (We set $\alpha = 200$ in our approach.)

2. if the splits are good enough for the classification purpose according to the quality measurement. In our approach, we ensure that the facets in a node should not come from the same part class.

If the above conditions are satisfied, we choose the test s^* with highest quality measurement gain as the main decision test of node R, and create two corresponding children nodes R_{ls^*} and R_{rs^*}.

After retraining the Online Random Forests t times with labeled facet t, the estimated probability for k-th part class of each other facet i can calculated by

$$g_t(k|x_i) = \frac{1}{T} \sum_{j=1}^{T} \frac{p_{t,j}(k|x_i)}{n_R}, \qquad (2)$$

where $p_{t,j}(k|x_i)$ is the estimated density of k-th part class in the leaf n_R of the j-th tree where x_i falls, and T is the number of trees. We set $T = 10$ in our approach.

3.2   Segmentation Model Initialization

In this step, we perform an initial co-segmentation on the unlabeled shapes in the set to get initial labels of these shapes. The initial co-segmentation method is an unsupervised clustering based method. We can use any co-segmentation algorithms in this step. Specifically, we use algorithm in Zhang et al. [11] to co-segment the unlabeled shapes. This algorithm is performed directly on facet, and does not depend on any per-object segmentation algorithms.

Considering that the initial labels obtained by initial co-segmentation are uncertain. They may be incorrect or do not match the users' segmentation requirements. We set a small weight for each facet to avoid the initial labeling result to heavily affect the segmentation. This weight can be determined dynamically. According to the principle of Online Random Forest, to get more accurate training result, the number of each part class should keep balance. So we define a total initial co-segmentation influence parameter W^C, and define the weight w_i^C of each facet i as:

$$w_i^C = \frac{W^C}{n_{y_i}^C}, \qquad (3)$$

where y_i is the initial label of facet i obtained in initial co-segmentation, and n_y^C is the number of facets whose initial labels are y after initial co-segmentation.

Then, based on the feature vectors x_i , weight w_i^C and initial labels y_i of these initial labeled facets, we perform the weighted online learning to initialize the segmentation model.

3.3   Progressive Segmentation

In this step, a progressive learning based segmentation method [15] is performed. Users can submit a new shape to the shape set. This shape can be segmented progressively with users' interactions based on the segmentation model. And this segmentation model can also be incrementally updated during the progressive segmentation.

Firstly, each new shape can be segmented based on the segmentation model to obtain the initial result. Next, users can simply click on the false-labeled regions of the corresponding parts, and input the corrected labels of these parts. Facets in these regions are used to train the segmentation model, and the inputted labels are the interaction labels. We define a total interaction influence parameter W^I, and the weight w_i^I of each facet i can be defined as:

$$w_i^I = \begin{cases} \frac{W^I}{n_{y_i}^I} & \text{if } n_{y_i}^C \neq 0 \\ \frac{W^C}{n_{y_i}^I} & \text{if } n_{y_i}^C = 0 \end{cases} \tag{4}$$

where y_i is the interaction label of facet i during correcting, and n_y^I is the number of facets in the corrected regions whose interaction labels are y. If users add a new part class during correcting, i.e. $n_{y_i}^C = 0$, the total influence of this part class should be proportionate with the other classes. In this condition, we set weight $w_i^I = \frac{W^C}{n_{y_i}^I}$, and update the initial number of the new part class as $n_{y_i}^I$.

Then, based on the feature vectors x_i, weights w_i^I and interaction labels y_i of these facets, we perform the weighted online learning (section 3.1) to retrain the segmentation model. The segmentation model is further used in the corrected segmentation (section 3.4) to get the refining result of the shape. Users can continue to correct the refining result until it is accepted by them, and the final labeled shape can be got.

Finally, to learn a more accurate segmentation model, it is retrained according to the final labeled shape. The label of each facet is the submitted label. We define a total submitted influence parameter W^S, and the weight w_i^S of each facet i can be defined as:

$$w_i^S = \frac{W^S}{n_{y_i}^S}, \tag{5}$$

where y_i is the submitted label of facet i in the labeled shape, and n_y^S is the number of facets in the labeled shape whose submitted labels are y.

The segmentation model thus can be used to segment other new shapes. As more shapes are segmented occasionally, the segmentation model can be updated continuously during the progressive segmentation process.

3.4   Corrected Segmentation

To make our method insensitive to the initial co-segmentation, we perform this step to correct the segmentation of shapes in the initial set based on the updated segmentation model. Similarly to [4][8], for each shape in the set, we construct the dual graph $G = (V, E)$, with vertices $v \in V$ to represent facets of the shape, and edges $e \in E \subseteq V \times V$ link the adjacent facets. This step can obtain the label of each facet in the shape by minimizing the following energy

$$\mathcal{E}(l) = \sum_{u \in V} \mathcal{E}_D(u, l_u) + \sum_{\{u,v\} \in E} \mathcal{E}_S(u, v, l_u, l_v), \tag{6}$$

where l_u and l_v are the labels assigned to facets u and v, respectively. The data term $\mathcal{E}_D(u, l_u)$ describes the penalty of assigning a part class label l_u to a facet

u, and the smoothness term $\mathcal{E}_S$ depicts the penalty for assigning different labels to two adjacent faces.

The data term is given by $\mathcal{E}_D(u, l_u) = -\gamma \log(g_t(l_u|x_u) + \epsilon)$, where $g_t(l_u|x_u)$ is the probability that facet u is labeled as l_u-th part class under the current segmentation model, $\gamma > 0$ (We set $\gamma = 0.1$ in our approach.) is a constant that regulates the influence of the data term in the total energy, and ϵ ($\epsilon = 1e - 6$) is a small threshold to avoid zero value in the logarithm function.

The smoothness term is defined as

$$\mathcal{E}_S(u, v, l_u, l_v) = \begin{cases} 0 & \text{if } l_u = l_v \\ -\log(1 - \min(\theta_{uv}/\pi, 1) + \epsilon)l_{uv} & \text{otherwise} \end{cases} \quad (7)$$

to penalize boundaries between facets with the exterior dihedral angle θ_{uv}. l_{uv}. l_{uv} indicates the length of the edge between adjacent facets u and v. ϵ is still the threshold to avoid zero value.

Finally, we employ graph cuts optimization [22] to minimize the energy $\mathcal{E}$ and finish the segmentation.

4 Results

In this section, we describe the experimental results and demonstrate the performance of our approach.

Data Set. We evaluate our online segmentation method on 6 sets of shapes from several typical object categories, where 4 sets (Human, FourLeg, Ant and Chair) come from Princeton Segmentation Benchmark (PSB) [2], and the rest (Candelabra and Vases) are from the Shape Co-segmentation Benchmark (COSEG) [7]. Shapes from these object categories include organic shapes and man-made shapes. And we also collect a large set: Large Chairs from COSEG to examine the performance on efficiency of our method. The ground-truth of the shapes from COSEG [7] is provided by the authors. And for shapes from PSB, we use the labels created by [16].

Online Segmentation. Figure 2 shows an example of our online segmentation process. We set the parameters $W^C = 2000$, $W^I = 4000$ and $W^S = 8000$ respectively. Figure 2(a) is an initial co-segmentation result for some shapes from Candelabra set. The number of part classes in the initial co-segmentation is 3, and notice that this result is not good. The handle and fire part have not been segmented, and the remaining parts have been labeled incorrectly.

Firstly, users can choose some shapes in the set to correct. The shape marked with red box in Figure 2(a) is the one users have selected. Figures 2(d)-2(e) show the progressive segmentation of this shape. Users have interacted to correct the segmentation of the shape twice as shown in Figures 2(d) and 2(f). They click on the corresponding regions, and the click positions are shown using colored points (marked with colored cycles). The interaction labels are indicated with different colors. During this process, a new part is added (the handle part marked with cyan),

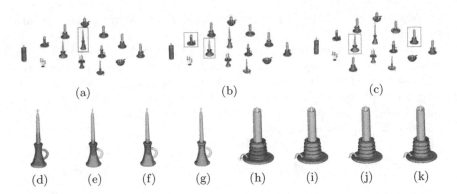

Fig. 2. An example of the online segmentation in our method. (a) Initial co-segmentation result. (b) Refined segmentation after correcting one shape. (c) Refined segmentation after adding a new shape. (d)-(g) Progressive segmentation for the shape selected from the set. (h)-(k) Progressive segmentation for the shape new added.

and the refined segmentation results of these two interaction processes are shown in Figures 2(e) and 2(g). Segmentation in Figure 2(g) is accepted by users, so we obtain the final labeled shape as shown in Figure 2(g). As shown in Figure 2(b), after the above shape is re-segmented in the progressive segmentation, the segmentation of the whole set is refined based on the updated segmentation model. We can notice that the fire parts of many shapes can be successfully segmented, and the handle parts of certain shapes, such as the ones marked with green boxes, can also be segmented.

Next, we add a new shape into the set, which is shown in Figure 2(h) with the initial segmentation result. This initial segmentation of this new shape can be got based on the current segmentation model. Similarly, users can correct this result in progressive segmentation (shown in Figures 2(h)-2(k)). And after the shape is added to the shape set, the segmentation of the whole set can be updated. As shown in Figure 2(c), some shapes can be well-segmented. Representative shapes are marked with green boxes in Figure 2(c).

This example shows that users can change the number and categories of parts through the segmentation. The segmentation of the shapes in the initial set can become better in vision during the progressive segmentation of some shapes. And it would not limit the results of subsequence progressive segmentation.

Accuracy and Time. To evaluate our method, we use the classification accuracy criterion presented in [4][16], which measures the percentage of the shape's surface area which is correctly labeled.

We randomly select 10 shapes in each category as the initial shape set, and perform initial co-segmentation on them. The initial co-segmentation results are usually bad, which may have not the enough number of the part classes. And we update the set by randomly adding the remaining shapes one by one. For each added shape, we use their ground-truth labeling to retrain the segmentation model to represent users' correction, and average the accuracies over the shapes

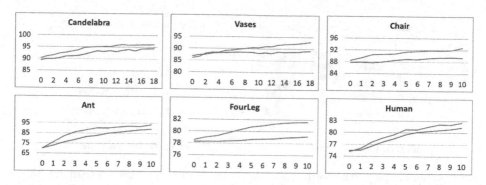

Fig. 3. Segmentation accuracies after adding new shapes for weighted method (blue curve) and unweighted one (red curve). X-axis indicates the number of shapes increases and y-axis indicates the accuracies.

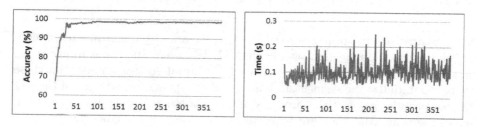

Fig. 4. Segmentation accuracies and updating time for large set: Large Chairs

in the set to represent the accuracy of the set. We repeat this experiment for 5 times, and use the average accuracy to evaluate our method. Figure 3 presents the accuracies of the 6 small sets, where the blue curve shows the result of our method, and the horizontal axis indicate the number of the added shapes.

We also compare our method with the unweighted one, in which the facet weights are always set to 1, no matter what kind the labels are. The accuracy of the unweighted method is shown in the red curve in Figure 3. We can see that, the segmentation results can become more precise with new shapes added. And weighted method can get higher accuracies than the unweighted one.

Then we perform this experiment on Large Chairs set, which have 400 shapes. Figure 4 illustrates the accuracies and updating time for this set. Experiments were performed on an Intel(R) Core(TM) 3.10GHz CPU with 8GB RAM.

Overall, the accuracies are increased with new shapes added. The updating time is lower than 1 min throughout the procedure, and it does not rapidly grow as the number of added shapes increases. It demonstrates that our updating strategy is effective and efficient.

Comparison to the State-of-the-art. To examine the segmentation quality of our method, we compare our approach to Kalogerakis et al. [16] and Hu et al. [6]. We use the segmentation accuracy after submitting 19 shapes as the accuracy of our method. For method in [16], we use the accuracy when the training set size

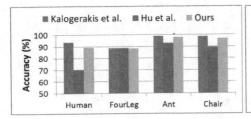
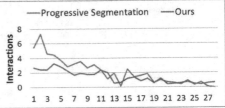

Fig. 5. Comparisons   **Fig. 6.** Interactions

is 19 as its accuracy. And for method in [6], the accuracy of the co-segmentation is used here directly. Figure 5 shows the comparisons of shapes from PSB. The average accuracies of [16], [6] and ours are 94.9%, 85.4% and 93.0% respectively. It is seen that our method get a close performance to the offline supervised method [16], and outperforms the unsupervised method [6].

User Study. We perform a user study, where we ask 15 participants to segment the shapes from Candelabra set (28 shapes). We select 10 shapes from the set as the initial set in our method, and perform initial co-segmentation on these shapes to initialize the segmentation model. Participants can add one of the rest 18 shapes and segment them progressively, or choose one shape from the initial set to segment. And they are requested to segment each shapes according to their intention until they are satisfied with the segmentation. Figure 6 shows the number of interactions for each progressively segmented shape (curve "Ours"). We can see that the number of interactions needed reduces as the accumulation of the segmented shapes.

We also compare our method with the progressive segmentation method in [15]. The number of interactions in [15] is shown in curve "Progressive Segmentation" in Figure 6. The total number of the interactions during the whole experiment in this method [15] and ours are 55 and 39 respectively. Since we perform the clustering based co-segmentation during the initialization stage as the potential segmentation. Users can segment 3D shapes with the assistance of this potential segmentation. Therefore, they need fewer interactions at the initial stage compared to [15]. In addition, compared with method in Wang et al. [7], they need 26 link constraints (one link constraint connects two interaction points on the shapes, so is roughly

(a)   (b)   (c)

Fig. 7. Segmentation results according to users' intention. (a) 5 parts: lip (blue), neck (yellow), body (red), handle (green) and base (cyan). (b) 4 parts: top (blue), body (red), handle (green) and base (cyan). (c) Ground-truth segmentation.

equal to two interactions in our system) to get the accuracy of about 98%. So, our method is comparable with theirs in the aspect of interactive complexity.

Users can also control the segmentation according to their own requirements. Figures 7(a) and 7(b) show two segmentation results of some shapes in Vases category. As the reference, ground-truth segmentation is shown in Figure 7(c).

5 Conclusion

We present a novel online 3D shape segmentation framework in this paper, which has three advantages. Firstly, the initial segmentation of 3D shapes can be obtained during the analysis by unsupervised clustering based method, which reduces the interactions in the subsequence segmentation. Secondly, users' intention can be learned in the supervised progressive learning method. Thirdly, we can blend the above two learning methods by using weighted online learning. So the 3D shapes segmentation can be accumulated and updated effectively. Experiments demonstrate the above advantages.

However, there are still some other problems, such as noisy interactions, to be addressed in the online 3D shape segmentation framework. If there are conflicts between users' interactions, or there are even false labels inputted by users, the segmentation model can be trained incorrectly. We should explore some strategy to handle this problem in future.

Acknowledgments. This work is supported by the National Natural Science Foundation of China No. 61272219, 61100110, 61321491; the National High Technology Research and Development Program of China No. 2007AA01Z334; the Key Projects Innovation Fund of State Key Laboratory No. ZZKT2013A12; the Program for New Century Excellent Talents in University of China No. NCET-04-04605; the Graduate Training Innovative Projects Foundation of Jiangsu Province No. CXLX13 050; the Science and Technology Program of Jiangsu Province No. BE2010072, BE2011058, BY2012190.

References

1. Schröder, P., Sweldens, W.: Digital geometry processing. In: Frontiers of Engineering: Reports on Leading-Edge Engineering From the 2000 NAE Symposium on Frontiers in Engineering, p. 41. National Academies Press (2001)
2. Chen, X., Golovinskiy, A., Funkhouser, T.: A benchmark for 3d mesh segmentation. ACM Trans. Graph. 28(3), 73:1–73:12 (2009)
3. Golovinskiy, A., Funkhouser, T.: Consistent segmentation of 3d models. Computers and Graphics 33(3), 262–269 (2009)
4. Sidi, O., van Kaick, O., Kleiman, Y., Zhang, H., Cohen-Or, D.: Unsupervised co-segmentation of a set of shapes via descriptor-space spectral clustering. ACM Trans. Graph. 30(6), 126:1–126:10 (2011)
5. Huang, Q., Koltun, V., Guibas, L.: Joint shape segmentation with linear programming. ACM Trans. Graph. 30 (6), 125:1–125:12. (2011)

6. Hu, R., Fan, L., Liu, L.: Co-segmentation of 3d shapes via subspace clustering. Computer Graphics Forum 31(5), 1703–1713 (2012)
7. Wang, Y., Asafi, S., van Kaick, O., Zhang, H., Cohen-Or, D., Chen, B.: Active co-analysis of a set of shapes. ACM Trans. Graph. 31(6), 165 (2012)
8. Meng, M., Xia, J., Luo, J., He, Y.: Unsupervised co-segmentation for 3d shapes using iterative multi-label optimization. Comput. Aided Des. 45(2), 312–320 (2013)
9. Luo, P., Wu, Z., Xia, C., Feng, L., Ma, T.: Co-segmentation of 3d shapes via multi-view spectral clustering. The Visual Computer 29(6-8), 587–597 (2013)
10. Wu, Z., Wang, Y., Shou, R., Chen, B., Liu, X.: Unsupervised co-segmentation of 3d shapes via affinity aggregation spectral clustering. Computers and Graphics 37(6), 628–637 (2013)
11. Zhang, F.Q., Sun, Z.X., Song, M.F., Lang, X.F., Yan, H.: 3d shapes co-segmentation by combining fuzzy c-means with random walks. In: Proceedings of 13th International Conference on Computer-Aided Design and Computer Graphics, pp. 16–23. IEEE Press, Hong Kong (2013)
12. Kim, V.G., Li, W., Mitra, N.J., Chaudhuri, S., DiVerdi, S., Funkhouser, T.: Learning part-based templates from large collections of 3d shapes. ACM Trans. Graph. 32(4), 70:1–70:12 (2013)
13. Lv, J., Chen, X., Huang, J., Bao, H.: Semi-supervised mesh segmentation and labeling. Computer Graphics Forum 31(7), 2241–2248 (2012)
14. Wu, Z., Shou, R., Wang, Y., Liu, X.: Interactive shape co-segmentation via label propagation. Computers and Graphics 38, 248–254 (2014)
15. Zhang, F.Q., Sun, Z.X., Song, M.F., Lang, X.F.: Progressive 3D shape segmentation using online learning. Computer-Aided Design 58, 2–12 (2015)
16. Kalogerakis, E., Hertzmann, A., Singh, K.: Learning 3d mesh segmentation and labeling. ACM Trans. Graph. 29(4), 102:1–102:12 (2010)
17. Gal, R., Cohen-Or, D.: Salient geometric features for partial shape matching and similarity. ACM Trans. Graph. 25(1), 130–150 (2006)
18. Shapira, L., Shalom, S., Shamir, A., Cohen-Or, D., Zhang, H.: Contextual part analogies in 3d objects. Int. J. Comput. Vision. 89(2-3), 309–326 (2010)
19. Hilaga, M., Shinagawa, Y., Kohmura, T., Kunii, T.L.: Topology matching for fully automatic similarity estimation of 3d shapes. In: Proceedings of the 28th Annual Conference on Computer Graphics and Interactive Techniques, SIGGRAPH 2001, pp. 203–212. ACM Press, New York (2001)
20. Belongie, S., Malik, J., Puzicha, J.: Shape matching and object recognition using shape contexts. IEEE Transactions on Pattern Analysis and Machine Intelligence 24(4), 509–522 (2002)
21. Saffari, A., Leistner, C., Santner, J., Godec, M., Bischof, H.: On-line random forests. In: 2009 IEEE 12th International Conference on Computer VisionWorkshops (ICCV Workshops), pp. 1393–1400. IEEE Press, Kyoto (2009)
22. Boykov, Y., Veksler, O., Zabih, R.: Fast approximate energy minimization via graph cuts. IEEE Transactions on Pattern Analysis and Machine Intelligence 23(11), 1222–1239 (2001)

Factorizing Time-Aware Multi-way Tensors for Enhancing Semantic Wearable Sensing

Peng Wang[1]*, Alan F. Smeaton[2], and Cathal Gurrin[2]

[1] National Laboratory for Information Science and Technology
Department of Computer Science and Technology, Tsinghua University
pengwangnudt@sina.com
[2] Insight Centre for Data Analytics
Dublin City University, Glasnevin, Dublin 9, Ireland
alan.smeaton@dcu.ie, cathal.gurrin@computing.dcu.ie

Abstract. Automatic concept detection is a crucial aspect of automatically indexing unstructured multimedia archives. However, the current prevalence of one-per-class detectors neglect inherent concept relationships and operate in isolation. This is insufficient when analyzing content gathered from wearable visual sensing, in which concepts occur with high diversity and with correlation depending on context. This paper presents a method to enhance concept detection results by constructing and factorizing a multi-way concept detection tensor in a time-aware manner. We derived a weighted non-negative tensor factorization algorithm and applied it to model concepts' temporal occurrence patterns and show how it boosts overall detection performance. The potential of our method is demonstrated on lifelog datasets with varying levels of original concept detection accuracies.

Keywords: visual lifelogging, concept detection, NTF, concept semantics, wearable sensing.

1 Introduction

With the maturity of lightweight sensors and computing devices, and more recently the emergence of unobtrusive wearable visual sensing devices like Google Glass or Microsoft's SenseCam, the creation of large volumes of personal, first-person visual media archives for quantified-self applications has become feasible. Visual lifelogging is the term used to describe one class of personal sensing and digital recording of all our everyday behaviour which employs wearable cameras to capture image or video of everyday activities [1].

To manage what is in effect a new form of multimedia, the lifelog, state-of-the-art techniques suggest that we use statistical mapping from low-level visual

---

* This research work was part-funded by the National Natural Science Foundation of China under Grant No. 61272231, 61472204, Beijing Key Laboratory of Networked Multimedia and Science Foundation Ireland under grant SFI/12/RC/2289.

X. He et al. (Eds.): MMM 2015, Part I, LNCS 8935, pp. 571–582, 2015.

features to semantic concepts which are more appropriate to users' understanding of their lifelogs. According to the TRECVid benchmark, acceptable results in mapping low level features to semantic concepts have been achieved already, particularly for concepts for which there exists enough annotated training data [2]. However, unlike most other kinds of multimedia content, a wide range of semantic concepts will usually appear in visual lifelogs because of the wide variety of activities that people usually engage in and which are subsequently logged and recorded. In addition, due to the wearers' movements while capturing a visual lifelog, images captured within the same event or activity may have significant perceptual differences as, for example, users will turn around and face a window while still being in the same room. This poses many challenges for the organisation of wearable visual lifelogs which is essential if lifelogs are to be used to good effect.

In addition to visual media, a rich pool of information can be collected in wearable sensing by individuals to record their own activities and this can be used to build applications that enhance their quality of life in many ways including productivity, health monitoring and wellness, safety and security, social interactions, leisure and more. However, the raw lifelog data has comparatively little metadata and so performing content-based operations on the lifelog is problematic, especially as the archives become larger. Accurately structuring a lifelog into events [3] is considered crucial in managing visual logs for various applications, and the identification of events and event boundaries [4] is normally the first step in processing lifelogs. However, this alone doesn't offer a complete solution because we need to know what the contents of events actually are and how they relate to each other. Therefore, the focus in research has shifted towards mining deeper meanings from visual lifelogs and lifelog events i.e. determining the semantics reflected in lifelogs.

Concepts express the semantics of media in a useful way and are usually automatically detected by providing a meaningful link between low-level features like colours and textures, and high-level semantics. In [5], the semantic indexing method has shown potential for relating low-level visual features to high-level semantic concepts (such as indoors, outdoors, people, buildings, etc.) for visual lifelogs using supervised machine learning techniques. This is then applied in [6] to learn lifestyle traits from lifelogs collected by different users, based on the automatically detected everyday concepts. The accuracy of a concept detector/classifier is an important factor in the provision of satisfactory solutions to indexing visual media and it is also widely accepted that detection accuracy can be improved if concept correlation can be utilised. The utilization of correlation in multi-concept detection falls into two main categories: multi-label training and detection refinement/adjusting. A typical multi-label training method is presented in [11], in which concept correlations are modeled in the classification model using Gibbs random fields. Since all concepts are learned from one integrated model, the direct shortcoming is the lack of flexibility, which means the learning stage needs to be repeated when concept lexicon is changed. Because detection scores obtained by specific binary detectors allow independent and possibly specialized classification techniques to be leveraged for each concept [14], detection refinement using post processing attracts much research interest based on utilising concept correlations inferred from preconstructed knowledge [12,15] or annotation sets [16,17,18]. These methods highly depend on external knowledge such as WordNet or the training data. When concepts do not exist in the lexicon ontology or extra annotation sets are insufficient for correlation learning (limited size of corpus or sparse annotations), these methods can not adapt to these situations and obtain equally good results. In [19], a semantic enhancement method is proposed for lifelogging based on weighted none-negative matrix factorization (WNMF), but the temporal semantics can not been applied in this model.

Fig. 1. A variety of wearable visual lifelog devices through the ages including SenseCam (bottom right)

In this paper, we propose an enhancement to concept detection by using inherent inter-concept correlations. Based on the assumption that the scores from the initial detectors are reasonably usable for some concepts similar as in [16], our method exempts from using any extra annotation sets and includes concept detection results as the only input. To evaluate the effectiveness of our approach to enhancing concept detection, we employed SenseCam (shown in Figure 1) as a wearable device to log details of users' lives. SenseCam has a lightweight passive camera with several built-in sensors which captures the view of the wearer with its fisheye lens. By default, images are taken at the rate of about one every 50 seconds while the on-board sensors can help to trigger the capture of pictures when sudden changes are detected in the environment of the wearer.

2   Overview of Problem and Solution

We define the research problem as follows: given particular streams of everyday activities divided into discrete events with consecutive images each of which has some concepts detected, the task is to use each concepts' contextual semantics, embedded in the detection results, to improve the overall detection performance. We assume a lexicon of concepts L. Let $\{E_1, E_2, ..., E_n\}$ be the set of event streams in the dataset. Event E_i is represented by successive images $I^{(i)} = \{Im_1^{(i)}, Im_2^{(i)}, ..., Im_k^{(i)}\}$. Each image $Im_j^{(i)}$ might have several concepts detected. We assume the concepts appearing in image $Im_j^{(i)}$ are represented as a confidence vector $C_j^{(i)} = \{c_{j1}^{(i)}, c_{j2}^{(i)} ... c_{jM}^{(i)}\}$ for M concepts. The whole set of SenseCam images can be denoted as $I = \{I^{(1)}, I^{(2)}, ..., I^{(n)}\}$ which has dimension $\sum_{i=1}^{n} k_i$, where k_i is the number of images in each event E_i.

Concatenating confidence vectors from all SenseCam images represents detection results as a 2-dimensional matrix, however this loses information from event segmentation and the features of different events are not captured separately. To utilise the temporal features reflected in different events, a tensor is employed to formalize the above problem given its merit in representing the structure of multidimensional data more naturally than matrices. The algorithm for enhancing concept detection proposed in this paper requires a nonnegative tensor factorization (NTF) approach to capture latent feature structure. By introducing a new dimension, NTF can preserve and model temporal characteristics of each event and avoid significant information loss.

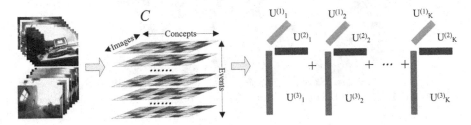

Fig. 2. NTF-based concept detection enhancement framework

The procedure for concept tensor construction and factorization is illustrated in Figure 2. As shown, our approach treats the concept detection results in a natural way which has the advantage of preserving local temporal constraints using a series of two-dimensional slices. Each slice is a segmented part of an event and is represented by a confidence matrix. In Figure 2, we use different colors of slices to show that they are the segments from different events. Meanwhile, the confidences of concept existences in each slice are represented by various gray levels. The slices are then stacked one below another to construct a three-dimensional tensor which preserves the two-dimensional characters of each segment while keeping temporal features along the event dimension and avoids significant loss of contextual information.

Assume each slice is a segment of N SenseCam images, each of which is represented by a vector of M concept detection confidences (i.e. concept vectors). The constructed concept detection tensor C has the dimensionality of $N \times M \times L$ for events with L slices in total. The task now is to modify the $N \times M \times L$ dimensional tensor C in order to keep consistency with the underlying contextual pattern of concepts. The factorization of weighted non-negative tensor C and the concept detection enhancement based on this WNTF method is now described.

3   Time-Aware Concept Detection Enhancement

3.1   Weighted Non-Negative Tensor Factorization (WNTF)

As we can see from Section 2, the confidence tensor C has a dimensionality of $N \times M \times L$ which consists of N neighborhood SenseCam images, M semantic concepts and L time intervals. The task of WNTF is to find the latent features to represent the three components of confidence tensor C. The tensor can then be approximated by the Tucker Decomposition (TD) [8] as

$$C \approx G \times_1 U^{(1)} \times_2 U^{(2)} \times_3 U^{(3)}$$

where $G \in \mathcal{R}^{R \times S \times T}$, $U^{(1)} \in \mathcal{R}^{N \times R}$, $U^{(2)} \in \mathcal{R}^{M \times S}$ and $U^{(3)} \in \mathcal{R}^{L \times T}$. The operator $\times_i (i = 1, 2, 3)$ denotes the tensor-matrix multiplication operators with the subscript i specifying which dimension of the tensor is multiplied with the given matrix. In Tucker Decomposition, the high-order tensor is factorized into a core tensor G and a factor matrix $U^{(i)}$ along each mode i [9]. In the TD model, each element in C is approximated by

$$\hat{C}_{ijk} = \sum_{r=1}^{R} \sum_{s=1}^{S} \sum_{t=1}^{T} G_{rst} U_{ir}^{(1)} U_{js}^{(2)} U_{kt}^{(3)}$$

As a particular case of the general Tucker Decomposition, the Canonical Decomposition (CD) [10] is derived from the TD model by constraining that each factor matrix has the same number of columns, i.e., the length of latent features has a fixed value of K. By setting G as a diagonal tensor

$$G_{ijk} = \begin{cases} 1, & \text{if } i = j = k \\ 0, & \text{else} \end{cases}$$

the CD model simplifies the approximation of tensor C as a sum of 3-fold outer-products with rank-K decomposition $\hat{C} = \sum_{f=1}^{K} U_{\cdot f}^{(1)} \otimes U_{\cdot f}^{(2)} \otimes U_{\cdot f}^{(3)}$, which means that each element $\hat{C}_{ijk} = \sum_{f=1}^{K} U_{if}^{(1)} U_{jf}^{(2)} U_{kf}^{(3)}$.

The CD approximation factorization defined above can be solved by optimizing the cost function defined to qualify the quality of the approximation. Different forms of cost function and corresponding optimization can be applied to this problem. Euclidian distance can be used to define the cost function, which has the form of $F = \frac{1}{2} \|C - \hat{C}\|_F^2$. However, in factorizing the confidence tensor, the weighted measure is more suitable since detection performance differs due to the characteristics of concepts and quality of the training set. To distinguish the contribution of different concept detectors to the cost function, the weighted cost function is employed as

$$F = \frac{1}{2} \|C - \hat{C}\|_W^2 = \frac{1}{2} \|\sqrt{W} \circ (C - \hat{C})\|_F^2$$

$$= \frac{1}{2} \sum_{ijk} W_{ijk} (C_{ijk} - \sum_{f=1}^{K} U_{if}^{(1)} U_{jf}^{(2)} U_{kf}^{(3)})^2$$

$$\text{s.t. } U^{(1)}, U^{(2)}, U^{(3)} \geq 0 \tag{1}$$

where $\circ$ denotes element-wise multiplication, $W = (W_{ijk})_{N \times M \times L}$ denotes the weight tensor and $\| \cdot \|_F^2$ denotes the Frobenius norm, i.e., the sum of squares of all entries in the tensor. The nonnegative constraints guarantees each component described by $U^{(1)}$, $U^{(2)}$, $U^{(3)}$ are additively combined.

A gradient descent method can be applied for optimizing this problem, implemented by updating each matrix $U^{(t)}$ in the opposite direction to the gradient at each iteration through

$$U^{(t)} \leftarrow U^{(t)} - \alpha_{U^{(t)}} \circ \partial F / \partial U^{(t)}, t = 1, 2, 3 \qquad (2)$$

To solve the partial differential $\partial F / \partial U^{(t)}$, we can rewrite Equation (1) as

$$F = \frac{1}{2} < C - \hat{C}, C - \hat{C} >_W = \frac{1}{2} < C - \sum_{f=1}^{K} \otimes_{t=1}^{3} U_{\cdot f}^{(t)}, C - \sum_{f=1}^{K} \otimes_{t=1}^{3} U_{\cdot f}^{(t)} >_W$$

where $< X, Y >$ denotes the inner product of two 3-way tensors [20] which is defined as $< X, Y >= \sum_{ijk} x_{ijk} y_{ijk}$. Hence we conduct the derivative

$$dF = \frac{1}{2} d < C - \hat{C}, C - \hat{C} >_W = < W \circ (C - \hat{C}), -d(\sum_{f=1}^{K} \otimes_{t=1}^{3} U_{\cdot f}^{(t)}) > \qquad (3)$$

Without losing generality, we focus on the update of the fth column in $U^{(1)}$ in the following derivation procedure and the update rule for other columns and matrices can be obtained in a similar manner. By taking the differential with respect to $U_{\cdot f}^{(1)}$, we can obtain the derivative of Equation (3) as

$$dF(U_f^{(1)}) = < W \circ (C - \hat{C}), -d(U_{\cdot f}^{(1)}) \otimes U_{\cdot f}^{(2)} \otimes U_{\cdot f}^{(3)} >$$
$$= < W \circ \hat{C}, d(U_{\cdot f}^{(1)}) \otimes U_{\cdot f}^{(2)} \otimes U_{\cdot f}^{(3)} >$$
$$- < W \circ C, d(U_{\cdot f}^{(1)}) \otimes U_{\cdot f}^{(2)} \otimes U_{\cdot f}^{(3)} >$$

Hence the differential with respect to an element $U_{if}^{(1)}$ can be represented as

$$\partial F / \partial U_{if}^{(1)} = < W \circ \hat{C}, e_i \otimes U_{\cdot f}^{(2)} \otimes U_{\cdot f}^{(3)} > - < W \circ C, e_i \otimes U_{\cdot f}^{(2)} \otimes U_{\cdot f}^{(3)} >$$
$$= \sum_{jk} (W \circ \hat{C})_{ijk} U_{jf}^{(2)} U_{kf}^{(3)} - \sum_{jk} (W \circ C)_{ijk} U_{jf}^{(2)} U_{kf}^{(3)}$$

where e_i is the ith column of the identity matrix and has the same dimension as $U_{\cdot f}^{(1)}$. By employing $\alpha_{U^{(1)}}$ as the form $\alpha_{U_{if}^{(1)}} = U_{if}^{(1)} / \sum_{jk} (W \circ \hat{C})_{ijk} U_{jf}^{(2)} U_{kf}^{(3)}$, where $/$ denotes element-wise division, and substituting into Equation (2), we obtain the multiplicative updating rule [21] as

$$U_{if}^{(1)} \leftarrow U_{if}^{(1)} \frac{\sum_{jk} (W \circ C)_{ijk} U_{jf}^{(2)} U_{kf}^{(3)}}{\sum_{jk} (W \circ \hat{C})_{ijk} U_{jf}^{(2)} U_{kf}^{(3)}}$$

The updating of $U^{(2)}$ and $U^{(3)}$ can be achieved in a similar manner. Note that it is not hard to prove that under such updating rules, the cost function in Equation (1) is non-increasing in each optimization step.

3.2 WNTF-Based Concept Detection Enhancement

To obtain a reconstruction of the underlying semantic structure that we can mine for co-occurrences and so enhance raw concept detection performance, the weights must be set in terms of concept accuracy. Because each confidence value C_{ijk} in tensor C denotes the probability of concept C_j occurring in the image, estimating the existence of C_j is more likely to be correct when C_{ijk} is high enough. Under this premise [16], we used the concept detection enhancement as in Algorithm 1:

Algorithm 1. WNTF-based detection enhancement

Input:
$C = (C_{ijk})_{N \times M \times L}$: original confidence tensor, *threshold*
Output:
$C_{new} \in \Re_{N \times M \times L}$: adjusted confidence tensor for C
Data:
$W \in \Re_{N \times M \times L}$: weight tensor
$U^{(1)} \in \Re_{N \times K}, U^{(2)} \in \Re_{M \times K}, U^{(3)} \in \Re_{L \times K}$

1 **begin**
2   Normalize C at each concept slice:
     $C(:, j, :) = normalize(C(:, j, :)), 1 \le j \le M$;
3   Initialized $U^{(1)}, U^{(2)}, U^{(3)}$ randomly with small numbers;
4   **for** *each C_{ijk} in C* **do**
5        $C'_{ijk} = C_{ijk}, W_{ijk} = 1$ if $C_{ijk} \ge threshold$;
       $C'_{ijk} = 0, W_{ijk} = w, w \in (0, 1)$; Otherwise;
6   **repeat**
7        $U_{if}^{(1)} \leftarrow U_{if}^{(1)} \sum_{jk} (W \circ C')_{ijk} U_{jf}^{(2)} U_{kf}^{(3)} / \sum_{jk} (W \circ \hat{C}')_{ijk} U_{jf}^{(2)} U_{kf}^{(3)}$
8        $U_{jf}^{(2)} \leftarrow U_{jf}^{(2)} \sum_{ik} (W \circ C')_{ijk} U_{if}^{(1)} U_{kf}^{(3)} / \sum_{ik} (W \circ \hat{C}')_{ijk} U_{if}^{(1)} U_{kf}^{(3)}$
9        $U_{kf}^{(3)} \leftarrow U_{kf}^{(3)} \sum_{ij} (W \circ C')_{ijk} U_{if}^{(1)} U_{jf}^{(2)} / \sum_{ij} (W \circ \hat{C}')_{ijk} U_{if}^{(1)} U_{jf}^{(2)}$
10   **until** *Converges*;
11   **for** *each $C'_{ijk} \in [C']_0$* **do**
12        $C'_{ijk} = \sum_{f=1}^{K} U_{if}^{(1)} U_{jf}^{(2)} U_{kf}^{(3)}$
13 Return $C_{new} = [average(C'_{ijk}, C_{ijk})]_{N \times M \times L}$;

Firstly, each concept-oriented slice $C(:, j, :)$ of tensor C is normalized at $Max - Min$ scale [14] for each specific concept j, which is indeed a lateral slice in the tensor visualized by Figure 2. This is then followed by constructing a new sparse tensor C' by thresholding C, whose element is

$$C'_{ijk} = \begin{cases} C_{ijk}, & \text{if } C_{ijk} \ge threshold; \\ 0, & \text{otherwise.} \end{cases}$$

The rationale for this is to retain elements with high confidence as "seeds" and use the contextual information modeled by non-negative tensor factorization to predict other concepts in correlation with these seed concepts. A sparse confidence tensor C' is achieved and we denote the non-zero element set in C' as $[C']_+$. Meanwhile, the set $[C']_0$ can be used to denote zero elements in C' which need to be estimated from $[C']_+$. C' is then factorized using the updating algorithm described in Section 3.1. This involves the iterative optimization of the cost function defined in Equation (1). In the optimization step, we configure the settings of weights as $W_{ijk} = 1$ if $C'_{ijk} \in [C']_+$, otherwise $W_{ijk} \in (0,1)$. In this step, the component matrices of $U^{(1)}$, $U^{(2)}$ and $U^{(3)}$ are returned as an estimate of the contextual structure of C'.

Finally, the approximation of elements in $[C']_0$ can be calculated using the refactorized features as $\sum_{f=1}^{K} U_{if}^{(1)} U_{jf}^{(2)} U_{kf}^{(3)}$, in which each component is the latent factor learned from the sparse tensor C'. The new confidence values for elements in $[C']_0$ form an estimate of concept detection to adjust the original detection result by averaging the original confidence and the new estimated value.

4   Results and Discussion

4.1   Experimental Setup and Dataset

To assess the performance of our algorithm, we used a set of 85 everyday concepts and a dataset including event samples of 23 activity types collected from 4 SenseCam wearers consisting of 12,248 SenseCam images [13]. Concept detectors with different accuracy levels were simulated and the metrics of AP and MAP were calculated for concepts based on a manual groundtruth. Different concept detection accuracies were provided in the dataset by varying the mean of the positive class μ_1 in the range [0.5...10]. The details of simulation are described in [13], following on from the work by Aly in [7]. For each setting of parameters, we executed 20 repeated runs to avoid random performance and the averaged concept AP and MAP were both calculated. The accuracy of the detection of original concepts is simulated with various accuracy levels and the MAPs are shown in Table 1 (first row) with the increased values of simulation parameter μ_1. The rationale for this is to test the performance of our algorithm at different concept detection accuracies. WNTF-based enhancement is carried out as described in Section 3 with concept detection confidence as the only input.

4.2   Detection Enhancement Analysis

Since averaging MAP over different detection accuracies is meaningless, pairwise comparison is depicted in Table 1 at different μ_1 values where detection enhancement ($K = 50$, $threshold = 0.3$) is applied. As shown by the improvement in Table 1, our algorithm can self-learn the contextual semantics of concepts and enhance the overall detection performance for various original detection accuracy levels. The highest overall improvement of 10.59% is achieved at $\mu_1 = 2.0$ when the original detection performance is neither too low nor too high. The improvement is shown to be significant and robust at various original detection accuracy levels in Table 1.

The less significant performance of our enhancement approach at $\mu_1 = 5.0$ makes sense as the initial detection accuracy is good enough. In this case, there is no space to improve detection accuracy, which is also the case when $\mu_1 = 4.0$

Table 1. Improved concept detections for various original accuracies

Value of μ_1	0.5	1.0	1.5	2.0	2.5	3.0	4.0	5.0
Original MAP	0.0946	0.1570	0.2645	0.4124	0.5797	0.7313	0.9251	0.9891
Adjusted MAP	0.0959	0.1640	0.2874	0.4560	0.6242	0.7744	0.9410	0.9912
Improvement	1.40%	4.48%	8.65%	10.59%	7.69%	5.89%	1.72%	0.21%

at which the original MAP has already reached 0.9. However, our approach can still enhance detection results with an improvement of 1.72% at $\mu_1 = 4.0$. On the other hand, when the original detection accuracy is too low, as shown in Table 1 at $\mu_1 = 0.5$, low accuracy detected elements can be selected and treated as "seed" candidates in our algorithm. Though this is an extreme, which is impractical in real world applications, our approach still works well with the average improvement of 1.40% achieved.

In many lifelog application scenarios, concept detection confidences need to be binarized to decide the existence or absence of concepts, instead of using the raw concept detection confidence values. Figures 3 and 4 illustrate the F-score, Recall-Precision improvement at different binarization levels $threshold_{bin}$, after applying our enhancement algorithm, taking the two concepts 'inside bus' and 'building' as instances. To consider the role of different filtering values of $threshold$ in Section 3.2, we assign $threshold = 0.5$ and $threshold = 0.8$ in Figure 3 and Figure 4 respectively. As shown by these two figures, the curves for two concepts are both enhanced. Since we use $threshold = 0.5$ in implementing the WNTF-based method in Figure 3, a large proportion of the adjusted concept detection confidences are below this threshold value. Hence the enhancements are significant for the parts of curves when the binarization $threshold_{bin} <$ 0.5. In this case, if we choose the binarizing threshold at higher values such as $threshold_{bin} \geq 0.5$, the use of WNTF will affect the result less significantly because most of the adjusted confidences are less than $threshold_{bin}$ and the corresponding concepts are still decided not to be present in the SenseCam images. Meanwhile, if we choose a higher value of $threshold = 0.8$, a larger range of enhancement for Recall and Precision can be achieved as shown by the curves in Figure 4.

Our algorithm has the advantage of enhancing a large number of concepts as demonstrated in Figure 5. In Figure 5, the performances of WNTF-based $(K = 50)$ and WNMF-based [19] methods are compared across all 85 concept APs using the same $threshold$. The detection of around 60 concepts are improved by our algorithm at $\mu_1 = 1.5$. In [19], the advantage of WNMF-based method has been demonstrated against ontological method for lifelogging concept enhancement. However, by utilising the temporal features, the WNTF-based method is more effective and the overall improvement is significant across all 85 concepts.

4.3   Impact of Parameters

The impact of parameters on enhancement performance is demonstrated in Figures 6 and 7, in which improvement is depicted at two different concept

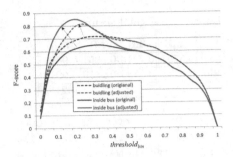

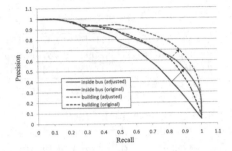

Fig. 3. F-score enhanced at $threshold = 0.5$ (for 'inside bus' and 'building')

Fig. 4. Recall-precision curve enhanced at $threshold = 0.8$ (for 'inside bus' and 'building')

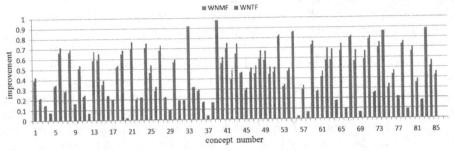

Fig. 5. Improvement comparison over all concepts

detection accuracies, determined by $\mu_1 = 1.5$ and $\mu_1 = 2.5$ respectively. In Figure 6, results for all settings of $K \in [10, ..., 80]$ and $threshold \in [0.1, ..., 0.9]$ are shown. All cases in Figure 6 are achieved by executing the algorithm in 20 runs and the averaged MAP improvement across all 85 concepts are obtained. We notice the robustness of the WNTF-based enhancement algorithm through the improvements achieved over different configurations of K and $threshold$.

As shown in Figure 6, detection performance is improved in most cases when the value of $threshold$ is not very high. The reason is because when $threshold$ is chosen as too high, there will be fewer correct concept detection results chosen, hence the potential for overall performance improvement is lessened. As shown in Figure 6, the best overall performances are achieved when $threshold = 0.3$ for $\mu_1 = 1.5$. In Figure 7, for which the original MAP is better as shown in Table 1, more correctly detected concepts can be used when higher $threshold$ is chosen to give better estimates on the others. That is why $threshold = 0.5$ achieves the best performance for Figure 7. The choice of "noisy" concepts can also degrade the improvement, as depicted when $threshold$ is small, say, $threshold = 0.1$ in both figures. In these cases, erroneous detection results are likely to be chosen to $[C']_+$ in the thresholding procedure which contaminates performance.

The impact of selected latent features is shown in Figure 7 in which the improvement in detection for different $threshold$ values are depicted across different K values. With the increase in K, performance improves gradually and converges

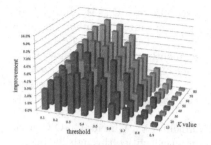

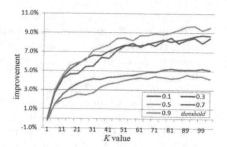

Fig. 6. *MAP* improvement with various parameter configures ($\mu_1 = 1.5$)

Fig. 7. Impact of feature number K ($\mu_1 = 2.5$)

at stable values. For poorly-chosen *threshold* values such as 0.1 and 0.9, the performance converges earlier, which reduces the potential for improvement if we increase the number of features. This implies that higher dimensionality is necessary to characterize the semantic features of concepts when more correct concept detection results are selected as "seeds" in the enhancement. For all settings of *threshold*, the performance keeps increasing and usually achieves satisfactory enhancement when about 50 latent features are selected.

5   Conclusions and Future Work

We present an algorithm to improve performance of semantic concept detection for wearable visual sensing. Based on non-negative tensor factorization, the algorithm models concept appearance patterns through partial concept detection results, which have better accuracy. For this purpose, we derived a weighted factorization method for updating latent features representing the structure of a multi-way confidence tensor. Based on this weighted nonnegative tensor factorization, local temporal constraints in each event segment are retained and reflected for the time-aware enhancement which uses the concept co-occurrence and re-occurrence patterns. The confidences of less accurate concept detections are then estimated and adjusted to enhance performance of overall concept detection. This method has been evaluated in experiments on datasets with various original detection accuracies. Since the factorization of time-aware WNTF also models the temporal structure of events, the application of this approach to event structuring and detection is a promising suggestion for future work.

References

1. Gurrin, C., Smeaton, A.F., Doherty, A.: LifeLogging: personal big data. Foundations and Trends in Information Retrieval 8(1), 1–127 (2014)
2. Smeaton, A., Over, P., Kraaij, W.: High level feature detection from video in TRECVid: a 5-year retrospective of achievements. In: Divakaran, A. (ed.) Multimedia Content Analysis, Theory and Applications, pp. 151–174. Springer (2008)

3. Doherty, A.R., Pauly-Takacs, K., Caprani, N., Gurrin, C., Moulin, C.J.A., O'Connor, N.E., Smeaton, A.F.: Experiences of aiding autobiographical memory using the SenseCam. Human-Computer Interaction 27(1-2), 151–174 (2012)
4. Doherty, A.R., Smeaton, A.F.: Automatically segmenting lifelog data into events. In: WIAMIS 2008, pp. 20–23. IEEE Computer Society, Washington, DC (2008)
5. Byrne, D., Doherty, A.R., Snoek, C.G.M., Jones, G.J.F., Smeaton, A.F.: Everyday concept detection in visual lifelogs: validation, relationships and trends. Multimedia Tools Appl. 49(1), 119–144 (2010)
6. Doherty, A.R., Caprani, N., O'Conaire, C., Kalnikaite, V., Gurrin, C., O'Connor, N.E., Smeaton, A.F.: Passively recognising human activities through lifelogging. Computers in Human Behavior 27(5), 1948–1958 (2011)
7. Aly, R., Hiemstra, D., de Jong, F., Apers, P.: Simulating the future of concept-based video retrieval under improved detector performance. Multimedia Tools and Applications 60(1), 1–29 (2011)
8. Tamara, G., Kolda, B.W.: Bader: Tensor decompositions and applications. SIAM Review 51(3), 455–500 (2009)
9. Tamara, G.: Kolda: Multilinear operators for higher-order decompositions. Tech. Report SAND2006-2081 (2006)
10. Rendle, S., Schmidt-Thieme, L.: pairwise interaction tensor factorization for personalized tag recommendation. In: WSDM 2010, pp. 81–90 (2010)
11. Qi, G.J., Hua, X.S., Rui, Y., Tang, J., Mei, T., Zhang, H.J.: Correlative multi-label video annotation. In: ACM MM 2007, pp. 17–26 (2007)
12. Wu, Y., Tseng, B., Smith, J.: Ontology-based multi-classification learning for video concept detection. In: ICME 2004, vol. 2, pp. 1003–1006 (2004)
13. Wang, P., Smeaton, A.F.: Using visual lifelogs to automatically characterise everyday activities. Information Sciences 230, 147–161 (2013)
14. Smith, J.R., Naphade, M., Natsev, A.: Multimedia semantic indexing using model vectors. In: ICME 2003, vol. 2, pp. 445–448 (2003)
15. Jin, Y., Khan, L., Wang, L., Awad, M.: Image annotations by combining multiple evidence & WordNet. In: ACM MM 2005, pp. 706–715 (2005)
16. Kennedy, L.S., Chang, S.F.: A reranking approach for context-based concept fusion in video indexing and retrieval. In: CIVR 2007, pp. 333–340 (2007)
17. Wang, C.H., Jing, F., Zhang, L., Zhang, H.J.: Image annotation refinement using random walk with restarts. In: ACM MM 2006, pp. 647–650 (2006)
18. Wang, C.H., Jing, F., Zhang, L., Zhang, H.J.: Content-based image annotation refinement. In: CVPR 2007, pp. 1–8 (2007)
19. Wang, P., Smeaton, A.F., Zhang, Y.C., et al.: Enhancing the detection of concepts for visual lifelogs using contexts instead of ontologies. In: ICMEW, pp. 1–6 (2014)
20. Shashua, A., Hazan, T.: Non-negative tensor factorization with applications to statistics and computer vision. In: ICML 2005, pp. 792–799 (2005)
21. Lee, D.D., Seung, H.S.: Learning the parts of objects by nonnegative matrix factorization. Nature 401, 788–791 (1999)

Author Index